DATE DUE

VOLUME FIVE HUNDRED AND TEN

Methods in
ENZYMOLOGY
Cellulases

METHODS IN ENZYMOLOGY

Editors-in-Chief

JOHN N. ABELSON AND MELVIN I. SIMON

*Division of Biology
California Institute of Technology
Pasadena, California*

Founding Editors

SIDNEY P. COLOWICK AND NATHAN O. KAPLAN

VOLUME FIVE HUNDRED AND TEN

Methods in
ENZYMOLOGY

Cellulases

EDITED BY

HARRY J. GILBERT
Institute for Cell and Molecular Biosciences
Medical School
Newcastle University
Framlington Place
United Kingdom

ELSEVIER

AMSTERDAM • BOSTON • HEIDELBERG • LONDON
NEW YORK • OXFORD • PARIS • SAN DIEGO
SAN FRANCISCO • SINGAPORE • SYDNEY • TOKYO
Academic Press is an imprint of Elsevier

Academic Press is an imprint of Elsevier
525 B Street, Suite 1900, San Diego, CA 92101-4495, USA
225 Wyman Street, Waltham, MA 02451, USA
32 Jamestown Road, London NW1 7BY, UK

First edition 2012

Copyright © 2012, Elsevier Inc. All Rights Reserved.

No part of this publication may be reproduced, stored in a retrieval system or transmitted in any form or by any means electronic, mechanical, photocopying, recording or otherwise without the prior written permission of the publisher

Permissions may be sought directly from Elsevier's Science & Technology Rights Department in Oxford, UK: phone (+44) (0) 1865 843830; fax (+44) (0) 1865 853333; email: permissions@elsevier.com. Alternatively you can submit your request online by visiting the Elsevier web site at http://elsevier.com/locate/permissions, and selecting *Obtaining permission to use Elsevier material*

Notice
No responsibility is assumed by the publisher for any injury and/or damage to persons or property as a matter of products liability, negligence or otherwise, or from any use or operation of any methods, products, instructions or ideas contained in the material herein. Because of rapid advances in the medical sciences, in particular, independent verification of diagnoses and drug dosages should be made

For information on all Academic Press publications
visit our website at elsevierdirect.com

ISBN: 978-0-12-415931-0
ISSN: 0076-6879

Printed and bound in United States of America
12 13 14   10 9 8 7 6 5 4 3 2 1

**Working together to grow
libraries in developing countries**

www.elsevier.com | www.bookaid.org | www.sabre.org

ELSEVIER   BOOK AID International   Sabre Foundation

Contents

Contributors	xiii
Preface	xxi
Volumes in Series	xxiii

1. Measurement of *endo*-1,4-β-Glucanase — 1
Barry V. McCleary, Vincent McKie, and Anna Draga

1.	Introduction	2
2.	Purity of EG Enzymes Included in the Study	2
3.	Polysaccharide and Oligosaccharide Substrates	3
4.	Viscometric Assay of EG Activity	4
5.	Reducing-Sugar Assays for EG Activity	6
6.	Soluble Chromogenic Substrates	10
7.	Insoluble Chromogenic Substrates	14
8.	Summary	15
	References	16

2. Biomass Conversion Determined via Fluorescent Cellulose Decay Assay — 19
Bente Wischmann, Marianne Toft, Marco Malten, and K. C. McFarland

1.	Introduction	20
2.	Correlation Between Cellulose Decay Monitored by HPLC and by Fluorescence	21
3.	Assay Development and Modeling of Response Using Statistics	24
4.	Final Assay Procedure	32
	Acknowledgment	36
	References	36

3. The Analysis of Saccharification in Biomass Using an Automated High-Throughput Method — 37
Caragh Whitehead, Leonardo D. Gomez, and Simon J. McQueen-Mason

1.	Introduction	38
2.	Automatic Saccharification Platform	39

3.	Enzyme Characterization	44
4.	Concluding Remarks	49
	Acknowledgments	49
	References	49

4. **Studies of Enzymatic Cleavage of Cellulose Using Polysaccharide Analysis by Carbohydrate gel Electrophoresis (PACE)**   51

Ondrej Kosik, Jennifer R. Bromley, Marta Busse-Wicher, Zhinong Zhang, and Paul Dupree

1.	Introduction	52
2.	The Principle of PACE	54
3.	Discussion: Other Possible Applications	64
4.	Summary	65
	Acknowledgments	65
	References	65

5. **Measuring Processivity**   69

Svein J. Horn, Morten Sørlie, Kjell M. Vårum, Priit Väljamäe, and Vincent G. H. Eijsink

1.	Introduction	70
2.	Classical Methods for Measuring Processivity	71
3.	Learning from Family 18 Chitinases	76
4.	Methods Based on Labeling—Apparent Versus Intrinsic Processivity	80
5.	Concluding Remarks	89
	Acknowledgments	90
	References	90

6. **Distinguishing Xyloglucanase Activity in endo-β(1→4)glucanases**   97

Jens M. Eklöf, Marcus C. Ruda, and Harry Brumer

1.	Introduction	98
2.	Analytical Tools for Xyloglucan Oligo- and Polysaccharide Analysis	101
3.	Substrate Preparation	105
4.	Determination of Specific Xyloglucan Activity in a Glucanase	109
5.	Coda: Xyloglucan Hydrolysis by β-glucanases—Obvious but Often Overlooked	116
	Acknowledgments	116
	References	117

7. **Methods for Structural Characterization of the Products of Cellulose- and Xyloglucan-Hydrolyzing Enzymes** — 121
Maria J. Peña, Sami T. Tuomivaara, Breeanna R. Urbanowicz, Malcolm A. O'Neill, and William S. York

   1. Introduction — 122
   2. Preparation of Substrates — 124
   3. Purification of the Oligosaccharide Products — 124
   4. Chemical and Structural Analysis of the Reaction Products — 127
   Acknowledgments — 137
   References — 138

8. **The Crystallization and Structural Analysis of Cellulases (and Other Glycoside Hydrolases): Strategies and Tactics** — 141
Shirley M. Roberts and Gideon J. Davies

   1. General Crystallization Considerations — 144
   2. Optimization and Additive Screens — 149
   3. Cellulase- and Glycosidase-Specific Complex Tactics — 152
   4. Buffer and Cryo-protection Problems — 159
   5. Toward a Cellulase-Specific Approach? — 162
   Acknowledgments — 162
   References — 163

9. **Visualization of Cellobiohydrolase I from *Trichoderma reesei* Moving on Crystalline Cellulose Using High-Speed Atomic Force Microscopy** — 169
Kiyohiko Igarashi, Takayuki Uchihashi, Anu Koivula, Masahisa Wada, Satoshi Kimura, Merja Penttilä, Toshio Ando, and Masahiro Samejima

   1. Introduction — 170
   2. Sample Preparation for AFM Observations — 171
   3. Observation of Cellulase Molecules on Crystalline Cellulose — 174
   4. Image Analysis — 177
   5. Conclusion — 179
   Acknowledgments — 180
   References — 180

10. **Small-Angle X-ray Scattering and Crystallography: A Winning Combination for Exploring the Multimodular Organization of Cellulolytic Macromolecular Complexes** — 183
Mirjam Czjzek, Henri-Pierre Fierobe, and Véronique Receveur-Bréchot

    1. Introduction — 184
    2. Measuring and Analyzing SAXS Data — 186

3. Combining SAXS and Crystallography to Analyze Multimodular
Organization of Cellulolytic Enzymes and Complexes   191
4. Conclusions and Outlook   203
Acknowledgments   205
References   205

11. Quantitative Approaches to The Analysis of Carbohydrate-Binding Module Function   211

D. Wade Abbott and Alisdair B. Boraston

1. Introduction   212
2. Experimental Approaches to Investigating CBM Function   215
3. Summary and New Directions for CBM Research   228
References   229

12. *In Situ* Detection of Cellulose with Carbohydrate-Binding Modules   233

J. Paul Knox

1. Introduction   234
2. Preparation of Plant Materials   236
3. CBM-Labeling Procedures   240
4. Summary   244
Acknowledgment   244
References   244

13. Interactions Between Family 3 Carbohydrate Binding Modules (CBMs) and Cellulosomal Linker Peptides   247

Oren Yaniv, Felix Frolow, Maly Levy-Assraf, Raphael Lamed, and Edward A. Bayer

1. Introduction   248
2. Enzyme-Linked Immunosorbent Assay (ELISA) to Check Possible Interactions Between CBM3 and Cellulosomal Linkers   250
3. Isothermal Titration Calorimetry for Analysis of Interactions Between CBM3 and Cellulosomal Linkers   253
Acknowledgments   258
References   258

14. Approaches for Improving Thermostability Characteristics in Cellulases   261

Michael Anbar and Edward A. Bayer

1. Construction of Cellulase Libraries   262
2. Screening Endoglucanases for Enhanced Thermostability   265

	3. Screening β-Glucosidase for Enhanced Thermostability	267
	4. *In Vitro* Recombination Between Best Thermostable Mutants	269
	References	271

15. Thermophilic Glycosynthases for Oligosaccharides Synthesis 273

Beatrice Cobucci-Ponzano, Giuseppe Perugino, Andrea Strazzulli, Mosè Rossi, and Marco Moracci

	1. Introduction	274
	2. Thermophilic β-Glycosynthases from GH1	276
	3. Thermophilic α-L-Fucosyntases from GH29	285
	4. Thermophilic α-D-Galactosynthase from GH36	292
	5. Summary	295
	Acknowledgments	296
	References	296

16. Engineering Cellulase Activity into *Clostridium acetobutylicum* 301

Henri-Pierre Fierobe, Florence Mingardon, and Angélique Chanal

	1. Introduction	302
	2. Electrotransformation of *C. acetobutylicum* ATCC824 and Storage of Recombinant Strains	303
	3. Detection/Quantification of the Secreted Heterologous Cellulosomal Protein	306
	4. Recombinant Strain of *C. acetobutylicum* Secreting a Heterologous Minicellulosome	308
	5. Secretion Issue and the Use of "Carrier" Modules	311
	6. Outlook	314
	Acknowledgments	315
	References	315

17. Transformation of *Clostridium Thermocellum* by Electroporation 317

Daniel G. Olson and Lee R. Lynd

	1. Introduction	318
	2. Materials	318
	3. Methods	320
	4. Troubleshooting	328
	References	329

18. Genetic and Functional Genomic Approaches for the Study of Plant Cell Wall Degradation in *Cellvibrio japonicus* 331
Jeffrey G. Gardner and David H. Keating

1. Introduction 332
2. Growth and Maintenance of *C. japonicus* 333
3. Genetic Manipulation of *C. japonicus* 336
4. Construction of Mutations 338
5. Transcriptomic Analysis of *C. japonicus* 341
6. Global Expression Profiling 342
7. Summary and Future Directions 345
Acknowledgment 345
References 345

19. Methods for the Isolation of Cellulose-Degrading Microorganisms 349
James E. McDonald, David J. Rooks, and Alan J. McCarthy

1. Introduction 350
2. Cellulosic Substrates 351
3. Isolation and Cultivation of Cellulolytic Microorganisms 354
4. Methods for the Detection of Cellulose Degradation in Cultures 366
5. Future Prospects 368
Acknowledgments 369
References 369

20. Metagenomic Approaches to the Discovery of Cellulases 375
David J. Rooks, James E. McDonald, and Alan J. McCarthy

1. Introduction 376
2. Nucleic Acid Extraction from Environmental Samples 377
3. Metagenomic Libraries 379
4. Metatranscriptomics 385
5. Outlook 389
Acknowledgment 390
References 390

21. *Escherichia coli* Expression, Purification, Crystallization, and Structure Determination of Bacterial Cohesin–Dockerin Complexes 395
Joana L. A. Brás, Ana Luisa Carvalho, Aldino Viegas, Shabir Najmudin, Victor D. Alves, José A. M. Prates, Luís M. A. Ferreira, Maria J. Romão, Harry J. Gilbert, and Carlos M. G.A. Fontes

1. Introduction 396
2. Cloning of Cohesin and Dockerin Genes in Prokaryotic Expression Vectors 397

3.	Expression and Purification of Cohesin–Dockerin Complexes in *E. coli*	400
4.	The Dual Binding Mode and the Crystallization of Cohesin–Dockerin Complexes	402
5.	X-ray Crystallography of Cohesin–Dockerin Complexes	402
6.	Summary	413
	Acknowledgments	414
	References	414

22. Measurements of Relative Binding of Cohesin and Dockerin Mutants Using an Advanced ELISA Technique for High-Affinity Interactions 417

Michal Slutzki, Yoav Barak, Dan Reshef, Ora Schueler-Furman, Raphael Lamed, and Edward A. Bayer

1.	Introduction	418
2.	Preparation of Constructs for ELISA Procedure	420
3.	Detection of Free Dockerins Using iELISA	422
4.	Data Analysis	425
5.	Notes	426
6.	Summary	427
	References	427

23. Designer Cellulosomes for Enhanced Hydrolysis of Cellulosic Substrates 429

Yael Vazana, Sarah Moraïs, Yoav Barak, Raphael Lamed, and Edward A. Bayer

1.	Construction of Designer Cellulosomes	430
2.	Scaffoldin Testing and Analysis of Cohesin–Dockerin Interaction	433
3.	Chimeric Enzyme Testing	441
4.	Sugar Analysis	449
5.	Summary	450
	References	451

24. High-Throughput Screening of Cohesin Mutant Libraries on Cellulose Microarrays 453

Michal Slutzki, Vered Ruimy, Ely Morag, Yoav Barak, Rachel Haimovitz, Raphael Lamed, and Edward A. Bayer

1.	Introduction	454
2.	Cohesin Mutants: Library Preparation	456
3.	Screening Small Cohesin Libraries	456
4.	Screening Large Cohesin Libraries	458
5.	Microarray Probing	459

6. DNA Extraction	460
7. Notes	461
8. Summary	461
References	462

Author Index   *465*
Subject Index   *489*

Contributors

D. Wade Abbott
Lethbridge Research Station, Agriculture and Agri-Food Canada, Lethbridge, Alberta, Canada

Victor D. Alves
CIISA-Faculdade de Medicina Veterinária, Pólo Universitário do Alto da Ajuda, Avenida da Universidade Técnica, Lisboa, Portugal

Michael Anbar
Department of Biological Chemistry, The Weizmann Institute of Science, Rehovot, Israel

Toshio Ando
Department of Physics, Kanazawa University, Kanazawa, Japan

Yoav Barak
Chemical Research Support, The Weizmann Institute of Science, Rehovot, Israel

Edward A. Bayer
Department of Biological Chemistry, The Weizmann Institute of Science, Rehovot, and Department of Molecular Microbiology and Biotechnology, The Daniella Rich Institute for Structural Biology, Tel Aviv University, Ramat Aviv, Israel

Alisdair B. Boraston
Biochemistry & Microbiology, University of Victoria, P.O. Box 3055 STN CSC, Victoria, British Columbia, Canada

Joana Brás
CIISA-Faculdade de Medicina Veterinária, Pólo Universitário do Alto da Ajuda, Avenida da Universidade Técnica, Lisboa, Portugal

Jennifer R. Bromley
Department of Biochemistry, University of Cambridge, Cambridge, United Kingdom

Harry Brumer
Michael Smith Laboratories and Department of Chemistry, University of British Columbia, Vancouver, Canada, and Division of Glycoscience, School of Biotechnology, Royal Institute of Technology (KTH), AlbaNova University Center, Stockholm, Sweden

Marta Busse-Wicher
Department of Biochemistry, University of Cambridge, Cambridge, United Kingdom

Ana Luisa Carvalho
REQUIMTE/CQFB, Departamento de Química, Faculdade de Ciências e Tecnologia, Universidade Nova de Lisboa, Caparica, Portugal

Angélique Chanal
Total Gas and Power R & D Biotechnology Team, Paris, France

Beatrice Cobucci-Ponzano
Institute of Protein Biochemistry, National Research Council, Naples, Italy

Mirjam Czjzek
Université Pierre et Marie Curie, Paris 6, and Centre National de la Recherche Scientifique, Marine Plants and Biomolecules, UMR 7139, Station Biologique de Roscoff, Roscoff, France

Gideon J. Davies
York Structural Biology Laboratory, Department of Chemistry, The University of York, Heslington, York, YO10 5DD, United Kingdom

Anna Draga
Megazyme International Ireland, Bray Business Park, Bray, County Wicklow, Ireland

Paul Dupree
Department of Biochemistry, University of Cambridge, Cambridge, United Kingdom

Vincent G. H. Eijsink
Department of Chemistry, Biotechnology and Food Science, Norwegian University of Life Sciences, Aas, Norway

Jens M. Eklöf
Michael Smith Laboratories and Department of Chemistry, University of British Columbia, Vancouver, Canada, and Division of Glycoscience, School of Biotechnology, Royal Institute of Technology (KTH), AlbaNova University Center, Stockholm, Sweden

Luís M. A. Ferreira
CIISA-Faculdade de Medicina Veterinária, Pólo Universitário do Alto da Ajuda, Avenida da Universidade Técnica, Lisboa, Portugal

Henri-Pierre Fierobe
Laboratoire de Chimie Bactérienne, CNRS-UPR9043, CNRS IMM, Marseille, France

Carlos M. G. A. Fontes
CIISA-Faculdade de Medicina Veterinária, Pólo Universitário do Alto da Ajuda, Avenida da Universidade Técnica, Lisboa, Portugal

Felix Frolow
Department of Molecular Microbiology and Biotechnology, The Daniella Rich Institute for Structural Biology, Tel Aviv University, Ramat Aviv, Israel

Jeffrey G. Gardner
DOE Great Lakes Bioenergy Research Center, University of Wisconsin-Madison, Madison, Wisconsin, USA

Harry J. Gilbert
Institute for Cell and Molecular Biosciences Medical School, Newcastle University, Framlington Place, United Kingdom

Leonardo D. Gomez
CNAP, Department of Biology, University of York, Heslington, York, United Kingdom

Rachel Haimovitz
Department of Biological Chemistry, The Weizmann Institute of Science, Rehovot, Israel

Svein J. Horn
Department of Chemistry, Biotechnology and Food Science, Norwegian University of Life Sciences, Aas, Norway

Kiyohiko Igarashi
Department of Biomaterial Sciences, Graduate School of Agricultural and Life Sciences, The University of Tokyo, Bunkyo-ku, Tokyo, Japan

David H. Keating
DOE Great Lakes Bioenergy Research Center, University of Wisconsin-Madison, Madison, Wisconsin, USA

Satoshi Kimura
Department of Biomaterial Sciences, Graduate School of Agricultural and Life Sciences, The University of Tokyo, Bunkyo-ku, Tokyo, Japan, and College of Life Sciences, Kyung Hee University, Gyeonggi-do, Yongin-si, Republic of Korea

J. Paul Knox
Faculty of Biological Sciences, Centre for Plant Sciences, University of Leeds, Leeds, United Kingdom

Anu Koivula
VTT Technical Research Centre of Finland, P.O. Box 1000, VTT, Finland

Ondrej Kosik
Department of Biochemistry, University of Cambridge, Cambridge, United Kingdom

Raphael Lamed
Department of Molecular Microbiology and Biotechnology, The Daniella Rich Institute for Structural Biology, Tel Aviv University, Ramat Aviv, Israel

Maly Levy-Assraf
Department of Molecular Microbiology and Biotechnology, The Daniella Rich Institute for Structural Biology, Tel Aviv University, Ramat Aviv, Israel

Lee R. Lynd
Thayer School of Engineering at Dartmouth College; Mascoma Corporation, Lebanon; Department of Biology, Dartmouth College, Hanover, New Hampshire, and BioEnergy Science Center, Oak Ridge, Tennessee, USA

Marco Malten
Novozymes A/S, Bagsværd, Denmark

Alan J. McCarthy
Microbiology Research Group, Institute of Integrative Biology, Biosciences Building, University of Liverpool, Liverpool, United Kingdom

Barry V. McCleary
Megazyme International Ireland, Bray Business Park, Bray, County Wicklow, Ireland

James E. McDonald
School of Biological Sciences, Bangor University, Bangor, Gwynedd, Wales, United Kingdom

K. C. McFarland
Novozymes Inc., Davis, California, USA

Vincent McKie
Megazyme International Ireland, Bray Business Park, Bray, County Wicklow, Ireland

Simon J. McQueen-Mason
CNAP, Department of Biology, University of York, Heslington, York, United Kingdom

Florence Mingardon
Laboratoire de Chimie Bactérienne, UPR 9043, CNRS IMM, Marseille, France

Sarah Moraïs
Department of Biological Chemistry, The Weizmann Institute of Science, Rehovot, Israel

Marco Moracci
Institute of Protein Biochemistry, National Research Council, Naples, Italy

Ely Morag
Designer Energy, Rehovot Science Park, Rehovot, Israel

Shabir Najmudin
CIISA-Faculdade de Medicina Veterinária, Pólo Universitário do Alto da Ajuda, Avenida da Universidade Técnica, Lisboa, Portugal

Malcolm A. O'Neill
Complex Carbohydrate Research Center, University of Georgia, Athens, Georgia, USA

Daniel G. Olson
Thayer School of Engineering at Dartmouth College, Hanover, New Hampshire, and BioEnergy Science Center, Oak Ridge, Tennessee, USA

Maria J. Peña
Complex Carbohydrate Research Center, University of Georgia, Athens, Georgia, USA

Merja Penttilä
VTT Technical Research Centre of Finland, P.O. Box 1000, VTT, Finland

Giuseppe Perugino
Institute of Protein Biochemistry, National Research Council, Naples, Italy

José A. M. Prates
CIISA-Faculdade de Medicina Veterinária, Pólo Universitário do Alto da Ajuda, Avenida da Universidade Técnica, Lisboa, Portugal

Véronique Receveur-Bréchot
CRCM, UMR7258 CNRS, INSERM, Aix-Marseille University, IPC, Marseille cedex 9, France

Dan Reshef
Department of Microbiology and Molecular Genetics, Institute for Medical Research Israel-Canada, Hadassah Medical School, The Hebrew University, Jerusalem, Israel

Shirley M. Roberts
York Structural Biology Laboratory, Department of Chemistry, The University of York, Heslington, York, YO10 5DD, United Kingdom

Maria J. Romão
REQUIMTE/CQFB, Departamento de Química, Faculdade de Ciências e Tecnologia, Universidade Nova de Lisboa, Caparica, Portugal

David J. Rooks
Microbiology Research Group, Institute of Integrative Biology, Biosciences Building, University of Liverpool, Liverpool, United Kingdom

Mosè Rossi
Institute of Protein Biochemistry, National Research Council, Naples, Italy

Marcus C. Ruda
SweTree Technologies, P.O. Box 4095, Umeå, Sweden

Vered Ruimy
Department of Biological Chemistry, The Weizmann Institute of Science, Rehovot, Israel

Morten Sørlie
Department of Chemistry, Biotechnology and Food Science, Norwegian University of Life Sciences, Aas, Norway

Masahiro Samejima
Department of Biomaterial Sciences, Graduate School of Agricultural and Life Sciences, The University of Tokyo, Bunkyo-ku, Tokyo, Japan

Ora Schueler-Furman
Department of Microbiology and Molecular Genetics, Institute for Medical Research Israel-Canada, Hadassah Medical School, The Hebrew University, Jerusalem, Israel

Michal Slutzki
Department of Biological Chemistry, The Weizmann Institute of Science, Rehovot, Israel

Andrea Strazzulli
Institute of Protein Biochemistry, National Research Council, Naples, Italy

Marianne Toft
Novozymes A/S, Bagsværd, Denmark

Sami T. Tuomivaara
Complex Carbohydrate Research Center, and Department of Biochemistry and Molecular Biology, University of Georgia, Athens, Georgia, USA

Takayuki Uchihashi
Department of Physics, Kanazawa University, Kanazawa, Japan

Breeanna R. Urbanowicz
Complex Carbohydrate Research Center, University of Georgia, Athens, Georgia, USA

Kjell M. Vårum
Department of Biotechnology, Norwegian University of Science and Technology, Trondheim, Norway

Priit Väljamäe
Institute of Molecular and Cell Biology, University of Tartu, Tartu, Estonia

Yael Vazana
Department of Biological Chemistry, The Weizmann Institute of Science, Rehovot, Israel

Aldino Viegas
REQUIMTE/CQFB, Departamento de Química, Faculdade de Ciências e Tecnologia, Universidade Nova de Lisboa, Caparica, Portugal

Masahisa Wada
Department of Biomaterial Sciences, Graduate School of Agricultural and Life Sciences, The University of Tokyo, Bunkyo-ku, Tokyo, Japan, and College of Life Sciences, Kyung Hee University, Gyeonggi-do, Yongin-si, Republic of Korea

Caragh Whitehead
CNAP, Department of Biology, University of York, Heslington, York, United Kingdom

Bente Wischmann
Novozymes A/S, Bagsværd, Denmark

Oren Yaniv
Department of Molecular Microbiology and Biotechnology, The Daniella Rich Institute for Structural Biology, Tel Aviv University, Ramat Aviv, Israel

William S. York
Complex Carbohydrate Research Center, and Department of Biochemistry and Molecular Biology, University of Georgia, Athens, Georgia, USA

Zhinong Zhang
Department of Biochemistry, University of Cambridge, Cambridge, United Kingdom

Preface

Significant interest in cellulose degradation started in the 1950s when cotton-based uniforms of the U.S. military were being destroyed by microorganisms during the Korean War. At that time, the major focus for cellulase research was the inhibition of these enzymes to prevent the degradation of the uniforms. In the 1970s, which ushered in the first oil crisis due to instability in the Middle East, the interest in cellulases switched to the use of these enzymes to explore the potential of cellulosic biomass as a renewable, and thus sustainable, replacement of fossil-based sources of oil and its associated products. The use of cellulases in second-generation biofuels is now a major driver for the substantial worldwide interest in these enzymes. Over the past 60 years, cellulase research has evolved as new methods became available. Until the early 1980s, cellulases were purified from their endogenous host and then subjected to enzyme characterization. With the advent of gene cloning and gene sequencing, enzyme characterization from the endogenous host very rapidly became redundant, and for the following 10 years, cellulases were characterized from noncellulolytic hosts such as *Escherichia coli*. The sequence information of these enzymes led to their assignment to 12 discrete families providing evidence for convergent evolution. In 1983, Bayer and Lamed discovered that cellulase enzymes expressed by anaerobic organisms associate to form multienzyme complexes known as cellulosomes. In the late 1980s, it became evident that cellulases often displayed modular architectures in which the catalytic domain is linked to noncatalytic cellulose binding domains (now referred to as carbohydrate binding modules). Finally, in the early 1990s, and for 10 years thereafter, the crystal structure of a large number of cellulases were determined, which provided substantial insight into how the topology of the active site and distal substrate binding regions contribute to the mode of enzyme action. The new millennium brought additional advances in cellulase methodologies. These include the use of cellulases as biosynthetic enzymes through their conversion to glycosynthases; small angle X-ray scattering is now routinely used to study the topology of multimodular cellulases; it is now possible to visualize, in real time, the movement of cellulases across cellulose fibers using high-speed atomic force microscopy; cellulosomes assemblies can be visualized using X-ray crystallography and a suite of high-throughput technologies ranging from the rapid analysis of complex plant cell wall structures to the saccharification of plant biomass by cellulase cocktails have recently been developed.

This volume of *Methods in Enzymology* has attempted to inform readers of all the methods currently deployed in cellulase research. The volume is loosely grouped into six sections. Chapters 1–7 focus on cellulase assays that deal with activity measurements and product analysis; Chapters 8–10 describe methods used to characterize the structure of the enzymes and their interaction with substrate; Chapters 11–13 provide methods used to characterize cellulose-specific carbohydrate binding modules; Chapters 14–16 describe methods used to increase the industrial utility of enzymes and microbes; Chapters 17–20 target the methodology used to explore and isolate cellulolytic microorganisms, while the final four chapters, 21–24, describe methods used to study dockerin–cohesin interactions that lead to cellulosome assemblies. Each chapter provides a background to the topic described placing the enzymes, microorganisms, or analytical method in the context of cellulase research.

When working on cellulases that are attacking polysaccharides, particularly insoluble forms of cellulose, one must recognize that these enzymes depart from classical enzyme-catalyzed reactions. Thus, because of the polymeric nature of the substrate, target sites are not uniformly distributed in solution and thus it is not always easy to apply typical Michaelis–Menton kinetics. Maybe more significant are the problems associated with studying the action of cellulases on crystalline substrates. The reaction is slow and the substrate has several different substructures that vary in access to enzyme attack. Thus, simple kinetics (product release versus time) is rarely linear as substrate access will alter as the reaction proceeds. Typically, in the cellulase field, rather crude measurements of enzyme activity are used such as the time it takes for 80% of substrate to be degraded. Thus readers who come from a typical enzyme field should be aware of the problems presented with studying the action of cellulases against, particularly, insoluble substrates; there is no continuous assay, the initial rate has no relevance, and the degree of enzyme processivity (Chapter 5) needs to be carefully considered. I hope this volume will be particularly useful to scientists new to the cellulase field, while also providing an important methodological reference for multidisciplinary groups already working in the cellulase field.

Harry J. Gilbert

METHODS IN ENZYMOLOGY

VOLUME I. Preparation and Assay of Enzymes
Edited by SIDNEY P. COLOWICK AND NATHAN O. KAPLAN

VOLUME II. Preparation and Assay of Enzymes
Edited by SIDNEY P. COLOWICK AND NATHAN O. KAPLAN

VOLUME III. Preparation and Assay of Substrates
Edited by SIDNEY P. COLOWICK AND NATHAN O. KAPLAN

VOLUME IV. Special Techniques for the Enzymologist
Edited by SIDNEY P. COLOWICK AND NATHAN O. KAPLAN

VOLUME V. Preparation and Assay of Enzymes
Edited by SIDNEY P. COLOWICK AND NATHAN O. KAPLAN

VOLUME VI. Preparation and Assay of Enzymes *(Continued)*
Preparation and Assay of Substrates
Special Techniques
Edited by SIDNEY P. COLOWICK AND NATHAN O. KAPLAN

VOLUME VII. Cumulative Subject Index
Edited by SIDNEY P. COLOWICK AND NATHAN O. KAPLAN

VOLUME VIII. Complex Carbohydrates
Edited by ELIZABETH F. NEUFELD AND VICTOR GINSBURG

VOLUME IX. Carbohydrate Metabolism
Edited by WILLIS A. WOOD

VOLUME X. Oxidation and Phosphorylation
Edited by RONALD W. ESTABROOK AND MAYNARD E. PULLMAN

VOLUME XI. Enzyme Structure
Edited by C. H. W. HIRS

VOLUME XII. Nucleic Acids (Parts A and B)
Edited by LAWRENCE GROSSMAN AND KIVIE MOLDAVE

VOLUME XIII. Citric Acid Cycle
Edited by J. M. LOWENSTEIN

VOLUME XIV. Lipids
Edited by J. M. LOWENSTEIN

VOLUME XV. Steroids and Terpenoids
Edited by RAYMOND B. CLAYTON

VOLUME XVI. Fast Reactions
Edited by KENNETH KUSTIN

VOLUME XVII. Metabolism of Amino Acids and Amines (Parts A and B)
Edited by HERBERT TABOR AND CELIA WHITE TABOR

VOLUME XVIII. Vitamins and Coenzymes (Parts A, B, and C)
Edited by DONALD B. MCCORMICK AND LEMUEL D. WRIGHT

VOLUME XIX. Proteolytic Enzymes
Edited by GERTRUDE E. PERLMANN AND LASZLO LORAND

VOLUME XX. Nucleic Acids and Protein Synthesis (Part C)
Edited by KIVIE MOLDAVE AND LAWRENCE GROSSMAN

VOLUME XXI. Nucleic Acids (Part D)
Edited by LAWRENCE GROSSMAN AND KIVIE MOLDAVE

VOLUME XXII. Enzyme Purification and Related Techniques
Edited by WILLIAM B. JAKOBY

VOLUME XXIII. Photosynthesis (Part A)
Edited by ANTHONY SAN PIETRO

VOLUME XXIV. Photosynthesis and Nitrogen Fixation (Part B)
Edited by ANTHONY SAN PIETRO

VOLUME XXV. Enzyme Structure (Part B)
Edited by C. H. W. HIRS AND SERGE N. TIMASHEFF

VOLUME XXVI. Enzyme Structure (Part C)
Edited by C. H. W. HIRS AND SERGE N. TIMASHEFF

VOLUME XXVII. Enzyme Structure (Part D)
Edited by C. H. W. HIRS AND SERGE N. TIMASHEFF

VOLUME XXVIII. Complex Carbohydrates (Part B)
Edited by VICTOR GINSBURG

VOLUME XXIX. Nucleic Acids and Protein Synthesis (Part E)
Edited by LAWRENCE GROSSMAN AND KIVIE MOLDAVE

VOLUME XXX. Nucleic Acids and Protein Synthesis (Part F)
Edited by KIVIE MOLDAVE AND LAWRENCE GROSSMAN

VOLUME XXXI. Biomembranes (Part A)
Edited by SIDNEY FLEISCHER AND LESTER PACKER

VOLUME XXXII. Biomembranes (Part B)
Edited by SIDNEY FLEISCHER AND LESTER PACKER

VOLUME XXXIII. Cumulative Subject Index Volumes I–XXX
Edited by MARTHA G. DENNIS AND EDWARD A. DENNIS

VOLUME XXXIV. Affinity Techniques (Enzyme Purification: Part B)
Edited by WILLIAM B. JAKOBY AND MEIR WILCHEK

VOLUME XXXV. Lipids (Part B)
Edited by JOHN M. LOWENSTEIN

VOLUME XXXVI. Hormone Action (Part A: Steroid Hormones)
Edited by BERT W. O'MALLEY AND JOEL G. HARDMAN

VOLUME XXXVII. Hormone Action (Part B: Peptide Hormones)
Edited by BERT W. O'MALLEY AND JOEL G. HARDMAN

VOLUME XXXVIII. Hormone Action (Part C: Cyclic Nucleotides)
Edited by JOEL G. HARDMAN AND BERT W. O'MALLEY

VOLUME XXXIX. Hormone Action (Part D: Isolated Cells, Tissues, and Organ Systems)
Edited by JOEL G. HARDMAN AND BERT W. O'MALLEY

VOLUME XL. Hormone Action (Part E: Nuclear Structure and Function)
Edited by BERT W. O'MALLEY AND JOEL G. HARDMAN

VOLUME XLI. Carbohydrate Metabolism (Part B)
Edited by W. A. WOOD

VOLUME XLII. Carbohydrate Metabolism (Part C)
Edited by W. A. WOOD

VOLUME XLIII. Antibiotics
Edited by JOHN H. HASH

VOLUME XLIV. Immobilized Enzymes
Edited by KLAUS MOSBACH

VOLUME XLV. Proteolytic Enzymes (Part B)
Edited by LASZLO LORAND

VOLUME XLVI. Affinity Labeling
Edited by WILLIAM B. JAKOBY AND MEIR WILCHEK

VOLUME XLVII. Enzyme Structure (Part E)
Edited by C. H. W. HIRS AND SERGE N. TIMASHEFF

VOLUME XLVIII. Enzyme Structure (Part F)
Edited by C. H. W. HIRS AND SERGE N. TIMASHEFF

VOLUME XLIX. Enzyme Structure (Part G)
Edited by C. H. W. HIRS AND SERGE N. TIMASHEFF

VOLUME L. Complex Carbohydrates (Part C)
Edited by VICTOR GINSBURG

VOLUME LI. Purine and Pyrimidine Nucleotide Metabolism
Edited by PATRICIA A. HOFFEE AND MARY ELLEN JONES

VOLUME LII. Biomembranes (Part C: Biological Oxidations)
Edited by SIDNEY FLEISCHER AND LESTER PACKER

VOLUME LIII. Biomembranes (Part D: Biological Oxidations)
Edited by SIDNEY FLEISCHER AND LESTER PACKER

VOLUME LIV. Biomembranes (Part E: Biological Oxidations)
Edited by SIDNEY FLEISCHER AND LESTER PACKER

VOLUME LV. Biomembranes (Part F: Bioenergetics)
Edited by SIDNEY FLEISCHER AND LESTER PACKER

VOLUME LVI. Biomembranes (Part G: Bioenergetics)
Edited by SIDNEY FLEISCHER AND LESTER PACKER

VOLUME LVII. Bioluminescence and Chemiluminescence
Edited by MARLENE A. DELUCA

VOLUME LVIII. Cell Culture
Edited by WILLIAM B. JAKOBY AND IRA PASTAN

VOLUME LIX. Nucleic Acids and Protein Synthesis (Part G)
Edited by KIVIE MOLDAVE AND LAWRENCE GROSSMAN

VOLUME LX. Nucleic Acids and Protein Synthesis (Part H)
Edited by KIVIE MOLDAVE AND LAWRENCE GROSSMAN

VOLUME 61. Enzyme Structure (Part H)
Edited by C. H. W. HIRS AND SERGE N. TIMASHEFF

VOLUME 62. Vitamins and Coenzymes (Part D)
Edited by DONALD B. MCCORMICK AND LEMUEL D. WRIGHT

VOLUME 63. Enzyme Kinetics and Mechanism (Part A: Initial Rate and Inhibitor Methods)
Edited by DANIEL L. PURICH

VOLUME 64. Enzyme Kinetics and Mechanism
(Part B: Isotopic Probes and Complex Enzyme Systems)
Edited by DANIEL L. PURICH

VOLUME 65. Nucleic Acids (Part I)
Edited by LAWRENCE GROSSMAN AND KIVIE MOLDAVE

VOLUME 66. Vitamins and Coenzymes (Part E)
Edited by DONALD B. MCCORMICK AND LEMUEL D. WRIGHT

VOLUME 67. Vitamins and Coenzymes (Part F)
Edited by DONALD B. MCCORMICK AND LEMUEL D. WRIGHT

VOLUME 68. Recombinant DNA
Edited by RAY WU

VOLUME 69. Photosynthesis and Nitrogen Fixation (Part C)
Edited by ANTHONY SAN PIETRO

VOLUME 70. Immunochemical Techniques (Part A)
Edited by HELEN VAN VUNAKIS AND JOHN J. LANGONE

VOLUME 71. Lipids (Part C)
Edited by JOHN M. LOWENSTEIN

VOLUME 72. Lipids (Part D)
Edited by JOHN M. LOWENSTEIN

VOLUME 73. Immunochemical Techniques (Part B)
Edited by JOHN J. LANGONE AND HELEN VAN VUNAKIS

VOLUME 74. Immunochemical Techniques (Part C)
Edited by JOHN J. LANGONE AND HELEN VAN VUNAKIS

VOLUME 75. Cumulative Subject Index Volumes XXXI, XXXII, XXXIV–LX
Edited by EDWARD A. DENNIS AND MARTHA G. DENNIS

VOLUME 76. Hemoglobins
Edited by ERALDO ANTONINI, LUIGI ROSSI-BERNARDI, AND EMILIA CHIANCONE

VOLUME 77. Detoxication and Drug Metabolism
Edited by WILLIAM B. JAKOBY

VOLUME 78. Interferons (Part A)
Edited by SIDNEY PESTKA

VOLUME 79. Interferons (Part B)
Edited by SIDNEY PESTKA

VOLUME 80. Proteolytic Enzymes (Part C)
Edited by LASZLO LORAND

VOLUME 81. Biomembranes (Part H: Visual Pigments and Purple Membranes, I)
Edited by LESTER PACKER

VOLUME 82. Structural and Contractile Proteins (Part A: Extracellular Matrix)
Edited by LEON W. CUNNINGHAM AND DIXIE W. FREDERIKSEN

VOLUME 83. Complex Carbohydrates (Part D)
Edited by VICTOR GINSBURG

VOLUME 84. Immunochemical Techniques (Part D: Selected Immunoassays)
Edited by JOHN J. LANGONE AND HELEN VAN VUNAKIS

VOLUME 85. Structural and Contractile Proteins (Part B: The Contractile Apparatus and the Cytoskeleton)
Edited by DIXIE W. FREDERIKSEN AND LEON W. CUNNINGHAM

VOLUME 86. Prostaglandins and Arachidonate Metabolites
Edited by WILLIAM E. M. LANDS AND WILLIAM L. SMITH

VOLUME 87. Enzyme Kinetics and Mechanism (Part C: Intermediates, Stereo-chemistry, and Rate Studies)
Edited by DANIEL L. PURICH

VOLUME 88. Biomembranes (Part I: Visual Pigments and Purple Membranes, II)
Edited by LESTER PACKER

VOLUME 89. Carbohydrate Metabolism (Part D)
Edited by WILLIS A. WOOD

VOLUME 90. Carbohydrate Metabolism (Part E)
Edited by WILLIS A. WOOD

VOLUME 91. Enzyme Structure (Part I)
Edited by C. H. W. HIRS AND SERGE N. TIMASHEFF

VOLUME 92. Immunochemical Techniques (Part E: Monoclonal Antibodies and General Immunoassay Methods)
Edited by JOHN J. LANGONE AND HELEN VAN VUNAKIS

VOLUME 93. Immunochemical Techniques (Part F: Conventional Antibodies, Fc Receptors, and Cytotoxicity)
Edited by JOHN J. LANGONE AND HELEN VAN VUNAKIS

VOLUME 94. Polyamines
Edited by HERBERT TABOR AND CELIA WHITE TABOR

VOLUME 95. Cumulative Subject Index Volumes 61–74, 76–80
Edited by EDWARD A. DENNIS AND MARTHA G. DENNIS

VOLUME 96. Biomembranes [Part J: Membrane Biogenesis: Assembly and Targeting (General Methods; Eukaryotes)]
Edited by SIDNEY FLEISCHER AND BECCA FLEISCHER

VOLUME 97. Biomembranes [Part K: Membrane Biogenesis: Assembly and Targeting (Prokaryotes, Mitochondria, and Chloroplasts)]
Edited by SIDNEY FLEISCHER AND BECCA FLEISCHER

VOLUME 98. Biomembranes (Part L: Membrane Biogenesis: Processing and Recycling)
Edited by SIDNEY FLEISCHER AND BECCA FLEISCHER

VOLUME 99. Hormone Action (Part F: Protein Kinases)
Edited by JACKIE D. CORBIN AND JOEL G. HARDMAN

VOLUME 100. Recombinant DNA (Part B)
Edited by RAY WU, LAWRENCE GROSSMAN, AND KIVIE MOLDAVE

VOLUME 101. Recombinant DNA (Part C)
Edited by RAY WU, LAWRENCE GROSSMAN, AND KIVIE MOLDAVE

VOLUME 102. Hormone Action (Part G: Calmodulin and Calcium-Binding Proteins)
Edited by ANTHONY R. MEANS AND BERT W. O'MALLEY

VOLUME 103. Hormone Action (Part H: Neuroendocrine Peptides)
Edited by P. MICHAEL CONN

VOLUME 104. Enzyme Purification and Related Techniques (Part C)
Edited by WILLIAM B. JAKOBY

VOLUME 105. Oxygen Radicals in Biological Systems
Edited by LESTER PACKER

VOLUME 106. Posttranslational Modifications (Part A)
Edited by FINN WOLD AND KIVIE MOLDAVE

VOLUME 107. Posttranslational Modifications (Part B)
Edited by FINN WOLD AND KIVIE MOLDAVE

VOLUME 108. Immunochemical Techniques (Part G: Separation and Characterization of Lymphoid Cells)
Edited by GIOVANNI DI SABATO, JOHN J. LANGONE, AND HELEN VAN VUNAKIS

VOLUME 109. Hormone Action (Part I: Peptide Hormones)
Edited by LUTZ BIRNBAUMER AND BERT W. O'MALLEY

VOLUME 110. Steroids and Isoprenoids (Part A)
Edited by JOHN H. LAW AND HANS C. RILLING

VOLUME 111. Steroids and Isoprenoids (Part B)
Edited by JOHN H. LAW AND HANS C. RILLING

VOLUME 112. Drug and Enzyme Targeting (Part A)
Edited by KENNETH J. WIDDER AND RALPH GREEN

VOLUME 113. Glutamate, Glutamine, Glutathione, and Related Compounds
Edited by ALTON MEISTER

VOLUME 114. Diffraction Methods for Biological Macromolecules (Part A)
Edited by HAROLD W. WYCKOFF, C. H. W. HIRS, AND SERGE N. TIMASHEFF

VOLUME 115. Diffraction Methods for Biological Macromolecules (Part B)
Edited by HAROLD W. WYCKOFF, C. H. W. HIRS, AND SERGE N. TIMASHEFF

VOLUME 116. Immunochemical Techniques
(Part H: Effectors and Mediators of Lymphoid Cell Functions)
Edited by GIOVANNI DI SABATO, JOHN J. LANGONE, AND HELEN VAN VUNAKIS

VOLUME 117. Enzyme Structure (Part J)
Edited by C. H. W. HIRS AND SERGE N. TIMASHEFF

VOLUME 118. Plant Molecular Biology
Edited by ARTHUR WEISSBACH AND HERBERT WEISSBACH

VOLUME 119. Interferons (Part C)
Edited by SIDNEY PESTKA

VOLUME 120. Cumulative Subject Index Volumes 81–94, 96–101

VOLUME 121. Immunochemical Techniques (Part I: Hybridoma Technology and Monoclonal Antibodies)
Edited by JOHN J. LANGONE AND HELEN VAN VUNAKIS

VOLUME 122. Vitamins and Coenzymes (Part G)
Edited by FRANK CHYTIL AND DONALD B. MCCORMICK

VOLUME 123. Vitamins and Coenzymes (Part H)
Edited by FRANK CHYTIL AND DONALD B. MCCORMICK

VOLUME 124. Hormone Action (Part J: Neuroendocrine Peptides)
Edited by P. MICHAEL CONN

VOLUME 125. Biomembranes (Part M: Transport in Bacteria, Mitochondria, and Chloroplasts: General Approaches and Transport Systems)
Edited by SIDNEY FLEISCHER AND BECCA FLEISCHER

VOLUME 126. Biomembranes (Part N: Transport in Bacteria, Mitochondria, and Chloroplasts: Protonmotive Force)
Edited by SIDNEY FLEISCHER AND BECCA FLEISCHER

VOLUME 127. Biomembranes (Part O: Protons and Water: Structure and Translocation)
Edited by LESTER PACKER

VOLUME 128. Plasma Lipoproteins (Part A: Preparation, Structure, and Molecular Biology)
Edited by JERE P. SEGREST AND JOHN J. ALBERS

VOLUME 129. Plasma Lipoproteins (Part B: Characterization, Cell Biology, and Metabolism)
Edited by JOHN J. ALBERS AND JERE P. SEGREST

VOLUME 130. Enzyme Structure (Part K)
Edited by C. H. W. HIRS AND SERGE N. TIMASHEFF

VOLUME 131. Enzyme Structure (Part L)
Edited by C. H. W. HIRS AND SERGE N. TIMASHEFF

VOLUME 132. Immunochemical Techniques (Part J: Phagocytosis and Cell-Mediated Cytotoxicity)
Edited by GIOVANNI DI SABATO AND JOHANNES EVERSE

VOLUME 133. Bioluminescence and Chemiluminescence (Part B)
Edited by MARLENE DELUCA AND WILLIAM D. MCELROY

VOLUME 134. Structural and Contractile Proteins (Part C: The Contractile Apparatus and the Cytoskeleton)
Edited by RICHARD B. VALLEE

VOLUME 135. Immobilized Enzymes and Cells (Part B)
Edited by KLAUS MOSBACH

VOLUME 136. Immobilized Enzymes and Cells (Part C)
Edited by KLAUS MOSBACH

VOLUME 137. Immobilized Enzymes and Cells (Part D)
Edited by KLAUS MOSBACH

VOLUME 138. Complex Carbohydrates (Part E)
Edited by VICTOR GINSBURG

VOLUME 139. Cellular Regulators (Part A: Calcium- and Calmodulin-Binding Proteins)
Edited by ANTHONY R. MEANS AND P. MICHAEL CONN

VOLUME 140. Cumulative Subject Index Volumes 102–119, 121–134

VOLUME 141. Cellular Regulators (Part B: Calcium and Lipids)
Edited by P. MICHAEL CONN AND ANTHONY R. MEANS

VOLUME 142. Metabolism of Aromatic Amino Acids and Amines
Edited by SEYMOUR KAUFMAN

VOLUME 143. Sulfur and Sulfur Amino Acids
Edited by WILLIAM B. JAKOBY AND OWEN GRIFFITH

VOLUME 144. Structural and Contractile Proteins (Part D: Extracellular Matrix)
Edited by LEON W. CUNNINGHAM

VOLUME 145. Structural and Contractile Proteins (Part E: Extracellular Matrix)
Edited by LEON W. CUNNINGHAM

VOLUME 146. Peptide Growth Factors (Part A)
Edited by DAVID BARNES AND DAVID A. SIRBASKU

VOLUME 147. Peptide Growth Factors (Part B)
Edited by DAVID BARNES AND DAVID A. SIRBASKU

VOLUME 148. Plant Cell Membranes
Edited by LESTER PACKER AND ROLAND DOUCE

VOLUME 149. Drug and Enzyme Targeting (Part B)
Edited by RALPH GREEN AND KENNETH J. WIDDER

VOLUME 150. Immunochemical Techniques (Part K: *In Vitro* Models of B and T Cell Functions and Lymphoid Cell Receptors)
Edited by GIOVANNI DI SABATO

VOLUME 151. Molecular Genetics of Mammalian Cells
Edited by MICHAEL M. GOTTESMAN

VOLUME 152. Guide to Molecular Cloning Techniques
Edited by SHELBY L. BERGER AND ALAN R. KIMMEL

VOLUME 153. Recombinant DNA (Part D)
Edited by RAY WU AND LAWRENCE GROSSMAN

VOLUME 154. Recombinant DNA (Part E)
Edited by RAY WU AND LAWRENCE GROSSMAN

VOLUME 155. Recombinant DNA (Part F)
Edited by RAY WU

VOLUME 156. Biomembranes (Part P: ATP-Driven Pumps and Related Transport: The Na, K-Pump)
Edited by SIDNEY FLEISCHER AND BECCA FLEISCHER

VOLUME 157. Biomembranes (Part Q: ATP-Driven Pumps and Related Transport: Calcium, Proton, and Potassium Pumps)
Edited by SIDNEY FLEISCHER AND BECCA FLEISCHER

VOLUME 158. Metalloproteins (Part A)
Edited by JAMES F. RIORDAN AND BERT L. VALLEE

VOLUME 159. Initiation and Termination of Cyclic Nucleotide Action
Edited by JACKIE D. CORBIN AND ROGER A. JOHNSON

VOLUME 160. Biomass (Part A: Cellulose and Hemicellulose)
Edited by WILLIS A. WOOD AND SCOTT T. KELLOGG

VOLUME 161. Biomass (Part B: Lignin, Pectin, and Chitin)
Edited by WILLIS A. WOOD AND SCOTT T. KELLOGG

VOLUME 162. Immunochemical Techniques (Part L: Chemotaxis and Inflammation)
Edited by GIOVANNI DI SABATO

VOLUME 163. Immunochemical Techniques (Part M: Chemotaxis and Inflammation)
Edited by GIOVANNI DI SABATO

VOLUME 164. Ribosomes
Edited by HARRY F. NOLLER, JR., AND KIVIE MOLDAVE

VOLUME 165. Microbial Toxins: Tools for Enzymology
Edited by SIDNEY HARSHMAN

VOLUME 166. Branched-Chain Amino Acids
Edited by ROBERT HARRIS AND JOHN R. SOKATCH

VOLUME 167. Cyanobacteria
Edited by LESTER PACKER AND ALEXANDER N. GLAZER

VOLUME 168. Hormone Action (Part K: Neuroendocrine Peptides)
Edited by P. MICHAEL CONN

VOLUME 169. Platelets: Receptors, Adhesion, Secretion (Part A)
Edited by JACEK HAWIGER

VOLUME 170. Nucleosomes
Edited by PAUL M. WASSARMAN AND ROGER D. KORNBERG

VOLUME 171. Biomembranes (Part R: Transport Theory: Cells and Model Membranes)
Edited by SIDNEY FLEISCHER AND BECCA FLEISCHER

VOLUME 172. Biomembranes (Part S: Transport: Membrane Isolation and Characterization)
Edited by SIDNEY FLEISCHER AND BECCA FLEISCHER

VOLUME 173. Biomembranes [Part T: Cellular and Subcellular Transport: Eukaryotic (Nonepithelial) Cells]
Edited by SIDNEY FLEISCHER AND BECCA FLEISCHER

VOLUME 174. Biomembranes [Part U: Cellular and Subcellular Transport: Eukaryotic (Nonepithelial) Cells]
Edited by SIDNEY FLEISCHER AND BECCA FLEISCHER

VOLUME 175. Cumulative Subject Index Volumes 135–139, 141–167

VOLUME 176. Nuclear Magnetic Resonance (Part A: Spectral Techniques and Dynamics)
Edited by NORMAN J. OPPENHEIMER AND THOMAS L. JAMES

VOLUME 177. Nuclear Magnetic Resonance (Part B: Structure and Mechanism)
Edited by NORMAN J. OPPENHEIMER AND THOMAS L. JAMES

VOLUME 178. Antibodies, Antigens, and Molecular Mimicry
Edited by JOHN J. LANGONE

VOLUME 179. Complex Carbohydrates (Part F)
Edited by VICTOR GINSBURG

VOLUME 180. RNA Processing (Part A: General Methods)
Edited by JAMES E. DAHLBERG AND JOHN N. ABELSON

VOLUME 181. RNA Processing (Part B: Specific Methods)
Edited by JAMES E. DAHLBERG AND JOHN N. ABELSON

VOLUME 182. Guide to Protein Purification
Edited by MURRAY P. DEUTSCHER

VOLUME 183. Molecular Evolution: Computer Analysis of Protein and Nucleic Acid Sequences
Edited by RUSSELL F. DOOLITTLE

VOLUME 184. Avidin-Biotin Technology
Edited by MEIR WILCHEK AND EDWARD A. BAYER

VOLUME 185. Gene Expression Technology
Edited by DAVID V. GOEDDEL

VOLUME 186. Oxygen Radicals in Biological Systems (Part B: Oxygen Radicals and Antioxidants)
Edited by LESTER PACKER AND ALEXANDER N. GLAZER

VOLUME 187. Arachidonate Related Lipid Mediators
Edited by ROBERT C. MURPHY AND FRANK A. FITZPATRICK

VOLUME 188. Hydrocarbons and Methylotrophy
Edited by MARY E. LIDSTROM

VOLUME 189. Retinoids (Part A: Molecular and Metabolic Aspects)
Edited by LESTER PACKER

VOLUME 190. Retinoids (Part B: Cell Differentiation and Clinical Applications)
Edited by LESTER PACKER

VOLUME 191. Biomembranes (Part V: Cellular and Subcellular Transport: Epithelial Cells)
Edited by SIDNEY FLEISCHER AND BECCA FLEISCHER

VOLUME 192. Biomembranes (Part W: Cellular and Subcellular Transport: Epithelial Cells)
Edited by SIDNEY FLEISCHER AND BECCA FLEISCHER

VOLUME 193. Mass Spectrometry
Edited by JAMES A. MCCLOSKEY

VOLUME 194. Guide to Yeast Genetics and Molecular Biology
Edited by CHRISTINE GUTHRIE AND GERALD R. FINK

VOLUME 195. Adenylyl Cyclase, G Proteins, and Guanylyl Cyclase
Edited by ROGER A. JOHNSON AND JACKIE D. CORBIN

VOLUME 196. Molecular Motors and the Cytoskeleton
Edited by RICHARD B. VALLEE

VOLUME 197. Phospholipases
Edited by EDWARD A. DENNIS

VOLUME 198. Peptide Growth Factors (Part C)
Edited by DAVID BARNES, J. P. MATHER, AND GORDON H. SATO

VOLUME 199. Cumulative Subject Index Volumes 168–174, 176–194

VOLUME 200. Protein Phosphorylation (Part A: Protein Kinases: Assays, Purification, Antibodies, Functional Analysis, Cloning, and Expression)
Edited by TONY HUNTER AND BARTHOLOMEW M. SEFTON

VOLUME 201. Protein Phosphorylation (Part B: Analysis of Protein Phosphorylation, Protein Kinase Inhibitors, and Protein Phosphatases)
Edited by TONY HUNTER AND BARTHOLOMEW M. SEFTON

VOLUME 202. Molecular Design and Modeling: Concepts and Applications (Part A: Proteins, Peptides, and Enzymes)
Edited by JOHN J. LANGONE

VOLUME 203. Molecular Design and Modeling: Concepts and Applications (Part B: Antibodies and Antigens, Nucleic Acids, Polysaccharides, and Drugs)
Edited by JOHN J. LANGONE

VOLUME 204. Bacterial Genetic Systems
Edited by JEFFREY H. MILLER

VOLUME 205. Metallobiochemistry (Part B: Metallothionein and Related Molecules)
Edited by JAMES F. RIORDAN AND BERT L. VALLEE

VOLUME 206. Cytochrome P450
Edited by MICHAEL R. WATERMAN AND ERIC F. JOHNSON

VOLUME 207. Ion Channels
Edited by BERNARDO RUDY AND LINDA E. IVERSON

VOLUME 208. Protein–DNA Interactions
Edited by ROBERT T. SAUER

VOLUME 209. Phospholipid Biosynthesis
Edited by EDWARD A. DENNIS AND DENNIS E. VANCE

VOLUME 210. Numerical Computer Methods
Edited by LUDWIG BRAND AND MICHAEL L. JOHNSON

VOLUME 211. DNA Structures (Part A: Synthesis and Physical Analysis of DNA)
Edited by DAVID M. J. LILLEY AND JAMES E. DAHLBERG

VOLUME 212. DNA Structures (Part B: Chemical and Electrophoretic Analysis of DNA)
Edited by DAVID M. J. LILLEY AND JAMES E. DAHLBERG

VOLUME 213. Carotenoids (Part A: Chemistry, Separation, Quantitation, and Antioxidation)
Edited by LESTER PACKER

VOLUME 214. Carotenoids (Part B: Metabolism, Genetics, and Biosynthesis)
Edited by LESTER PACKER

VOLUME 215. Platelets: Receptors, Adhesion, Secretion (Part B)
Edited by JACEK J. HAWIGER

VOLUME 216. Recombinant DNA (Part G)
Edited by RAY WU

VOLUME 217. Recombinant DNA (Part H)
Edited by RAY WU

VOLUME 218. Recombinant DNA (Part I)
Edited by RAY WU

VOLUME 219. Reconstitution of Intracellular Transport
Edited by JAMES E. ROTHMAN

VOLUME 220. Membrane Fusion Techniques (Part A)
Edited by NEJAT DÜZGÜNEŞ

VOLUME 221. Membrane Fusion Techniques (Part B)
Edited by NEJAT DÜZGÜNEŞ

VOLUME 222. Proteolytic Enzymes in Coagulation, Fibrinolysis, and Complement Activation (Part A: Mammalian Blood Coagulation Factors and Inhibitors)
Edited by LASZLO LORAND AND KENNETH G. MANN

VOLUME 223. Proteolytic Enzymes in Coagulation, Fibrinolysis, and Complement Activation (Part B: Complement Activation, Fibrinolysis, and Nonmammalian Blood Coagulation Factors)
Edited by LASZLO LORAND AND KENNETH G. MANN

VOLUME 224. Molecular Evolution: Producing the Biochemical Data
Edited by ELIZABETH ANNE ZIMMER, THOMAS J. WHITE, REBECCA L. CANN, AND ALLAN C. WILSON

VOLUME 225. Guide to Techniques in Mouse Development
Edited by PAUL M. WASSARMAN AND MELVIN L. DEPAMPHILIS

VOLUME 226. Metallobiochemistry (Part C: Spectroscopic and Physical Methods for Probing Metal Ion Environments in Metalloenzymes and Metalloproteins)
Edited by JAMES F. RIORDAN AND BERT L. VALLEE

VOLUME 227. Metallobiochemistry (Part D: Physical and Spectroscopic Methods for Probing Metal Ion Environments in Metalloproteins)
Edited by JAMES F. RIORDAN AND BERT L. VALLEE

VOLUME 228. Aqueous Two-Phase Systems
Edited by HARRY WALTER AND GÖTE JOHANSSON

VOLUME 229. Cumulative Subject Index Volumes 195–198, 200–227

VOLUME 230. Guide to Techniques in Glycobiology
Edited by WILLIAM J. LENNARZ AND GERALD W. HART

VOLUME 231. Hemoglobins (Part B: Biochemical and Analytical Methods)
Edited by JOHANNES EVERSE, KIM D. VANDEGRIFF, AND ROBERT M. WINSLOW

VOLUME 232. Hemoglobins (Part C: Biophysical Methods)
Edited by JOHANNES EVERSE, KIM D. VANDEGRIFF, AND ROBERT M. WINSLOW

VOLUME 233. Oxygen Radicals in Biological Systems (Part C)
Edited by LESTER PACKER

VOLUME 234. Oxygen Radicals in Biological Systems (Part D)
Edited by LESTER PACKER

VOLUME 235. Bacterial Pathogenesis (Part A: Identification and Regulation of Virulence Factors)
Edited by VIRGINIA L. CLARK AND PATRIK M. BAVOIL

VOLUME 236. Bacterial Pathogenesis (Part B: Integration of Pathogenic Bacteria with Host Cells)
Edited by VIRGINIA L. CLARK AND PATRIK M. BAVOIL

VOLUME 237. Heterotrimeric G Proteins
Edited by RAVI IYENGAR

VOLUME 238. Heterotrimeric G-Protein Effectors
Edited by RAVI IYENGAR

VOLUME 239. Nuclear Magnetic Resonance (Part C)
Edited by THOMAS L. JAMES AND NORMAN J. OPPENHEIMER

VOLUME 240. Numerical Computer Methods (Part B)
Edited by MICHAEL L. JOHNSON AND LUDWIG BRAND

VOLUME 241. Retroviral Proteases
Edited by LAWRENCE C. KUO AND JULES A. SHAFER

VOLUME 242. Neoglycoconjugates (Part A)
Edited by Y. C. LEE AND REIKO T. LEE

VOLUME 243. Inorganic Microbial Sulfur Metabolism
Edited by HARRY D. PECK, JR., AND JEAN LEGALL

VOLUME 244. Proteolytic Enzymes: Serine and Cysteine Peptidases
Edited by ALAN J. BARRETT

VOLUME 245. Extracellular Matrix Components
Edited by E. RUOSLAHTI AND E. ENGVALL

VOLUME 246. Biochemical Spectroscopy
Edited by KENNETH SAUER

VOLUME 247. Neoglycoconjugates (Part B: Biomedical Applications)
Edited by Y. C. LEE AND REIKO T. LEE

VOLUME 248. Proteolytic Enzymes: Aspartic and Metallo Peptidases
Edited by ALAN J. BARRETT

VOLUME 249. Enzyme Kinetics and Mechanism (Part D: Developments in Enzyme Dynamics)
Edited by DANIEL L. PURICH

VOLUME 250. Lipid Modifications of Proteins
Edited by PATRICK J. CASEY AND JANICE E. BUSS

VOLUME 251. Biothiols (Part A: Monothiols and Dithiols, Protein Thiols, and Thiyl Radicals)
Edited by LESTER PACKER

VOLUME 252. Biothiols (Part B: Glutathione and Thioredoxin; Thiols in Signal Transduction and Gene Regulation)
Edited by LESTER PACKER

VOLUME 253. Adhesion of Microbial Pathogens
Edited by RON J. DOYLE AND ITZHAK OFEK

VOLUME 254. Oncogene Techniques
Edited by PETER K. VOGT AND INDER M. VERMA

VOLUME 255. Small GTPases and Their Regulators (Part A: Ras Family)
Edited by W. E. BALCH, CHANNING J. DER, AND ALAN HALL

VOLUME 256. Small GTPases and Their Regulators (Part B: Rho Family)
Edited by W. E. BALCH, CHANNING J. DER, AND ALAN HALL

VOLUME 257. Small GTPases and Their Regulators (Part C: Proteins Involved in Transport)
Edited by W. E. BALCH, CHANNING J. DER, AND ALAN HALL

VOLUME 258. Redox-Active Amino Acids in Biology
Edited by JUDITH P. KLINMAN

VOLUME 259. Energetics of Biological Macromolecules
Edited by MICHAEL L. JOHNSON AND GARY K. ACKERS

VOLUME 260. Mitochondrial Biogenesis and Genetics (Part A)
Edited by GIUSEPPE M. ATTARDI AND ANNE CHOMYN

VOLUME 261. Nuclear Magnetic Resonance and Nucleic Acids
Edited by THOMAS L. JAMES

VOLUME 262. DNA Replication
Edited by JUDITH L. CAMPBELL

VOLUME 263. Plasma Lipoproteins (Part C: Quantitation)
Edited by WILLIAM A. BRADLEY, SANDRA H. GIANTURCO, AND JERE P. SEGREST

VOLUME 264. Mitochondrial Biogenesis and Genetics (Part B)
Edited by GIUSEPPE M. ATTARDI AND ANNE CHOMYN

VOLUME 265. Cumulative Subject Index Volumes 228, 230–262

VOLUME 266. Computer Methods for Macromolecular Sequence Analysis
Edited by RUSSELL F. DOOLITTLE

VOLUME 267. Combinatorial Chemistry
Edited by JOHN N. ABELSON

VOLUME 268. Nitric Oxide (Part A: Sources and Detection of NO; NO Synthase)
Edited by LESTER PACKER

VOLUME 269. Nitric Oxide (Part B: Physiological and Pathological Processes)
Edited by LESTER PACKER

VOLUME 270. High Resolution Separation and Analysis of Biological Macromolecules (Part A: Fundamentals)
Edited by BARRY L. KARGER AND WILLIAM S. HANCOCK

VOLUME 271. High Resolution Separation and Analysis of Biological Macromolecules (Part B: Applications)
Edited by BARRY L. KARGER AND WILLIAM S. HANCOCK

VOLUME 272. Cytochrome P450 (Part B)
Edited by ERIC F. JOHNSON AND MICHAEL R. WATERMAN

VOLUME 273. RNA Polymerase and Associated Factors (Part A)
Edited by SANKAR ADHYA

VOLUME 274. RNA Polymerase and Associated Factors (Part B)
Edited by SANKAR ADHYA

VOLUME 275. Viral Polymerases and Related Proteins
Edited by LAWRENCE C. KUO, DAVID B. OLSEN, AND STEVEN S. CARROLL

VOLUME 276. Macromolecular Crystallography (Part A)
Edited by CHARLES W. CARTER, JR., AND ROBERT M. SWEET

VOLUME 277. Macromolecular Crystallography (Part B)
Edited by CHARLES W. CARTER, JR., AND ROBERT M. SWEET

VOLUME 278. Fluorescence Spectroscopy
Edited by LUDWIG BRAND AND MICHAEL L. JOHNSON

VOLUME 279. Vitamins and Coenzymes (Part I)
Edited by DONALD B. MCCORMICK, JOHN W. SUTTIE, AND CONRAD WAGNER

VOLUME 280. Vitamins and Coenzymes (Part J)
Edited by DONALD B. MCCORMICK, JOHN W. SUTTIE, AND CONRAD WAGNER

VOLUME 281. Vitamins and Coenzymes (Part K)
Edited by DONALD B. MCCORMICK, JOHN W. SUTTIE, AND CONRAD WAGNER

VOLUME 282. Vitamins and Coenzymes (Part L)
Edited by DONALD B. MCCORMICK, JOHN W. SUTTIE, AND CONRAD WAGNER

VOLUME 283. Cell Cycle Control
Edited by WILLIAM G. DUNPHY

VOLUME 284. Lipases (Part A: Biotechnology)
Edited by BYRON RUBIN AND EDWARD A. DENNIS

VOLUME 285. Cumulative Subject Index Volumes 263, 264, 266–284, 286–289

VOLUME 286. Lipases (Part B: Enzyme Characterization and Utilization)
Edited by BYRON RUBIN AND EDWARD A. DENNIS

VOLUME 287. Chemokines
Edited by RICHARD HORUK

VOLUME 288. Chemokine Receptors
Edited by RICHARD HORUK

VOLUME 289. Solid Phase Peptide Synthesis
Edited by GREGG B. FIELDS

VOLUME 290. Molecular Chaperones
Edited by GEORGE H. LORIMER AND THOMAS BALDWIN

VOLUME 291. Caged Compounds
Edited by GERARD MARRIOTT

VOLUME 292. ABC Transporters: Biochemical, Cellular, and Molecular Aspects
Edited by SURESH V. AMBUDKAR AND MICHAEL M. GOTTESMAN

VOLUME 293. Ion Channels (Part B)
Edited by P. MICHAEL CONN

VOLUME 294. Ion Channels (Part C)
Edited by P. MICHAEL CONN

VOLUME 295. Energetics of Biological Macromolecules (Part B)
Edited by GARY K. ACKERS AND MICHAEL L. JOHNSON

VOLUME 296. Neurotransmitter Transporters
Edited by SUSAN G. AMARA

VOLUME 297. Photosynthesis: Molecular Biology of Energy Capture
Edited by LEE MCINTOSH

VOLUME 298. Molecular Motors and the Cytoskeleton (Part B)
Edited by RICHARD B. VALLEE

VOLUME 299. Oxidants and Antioxidants (Part A)
Edited by LESTER PACKER

VOLUME 300. Oxidants and Antioxidants (Part B)
Edited by LESTER PACKER

VOLUME 301. Nitric Oxide: Biological and Antioxidant Activities (Part C)
Edited by LESTER PACKER

VOLUME 302. Green Fluorescent Protein
Edited by P. MICHAEL CONN

VOLUME 303. cDNA Preparation and Display
Edited by SHERMAN M. WEISSMAN

VOLUME 304. Chromatin
Edited by PAUL M. WASSARMAN AND ALAN P. WOLFFE

VOLUME 305. Bioluminescence and Chemiluminescence (Part C)
Edited by THOMAS O. BALDWIN AND MIRIAM M. ZIEGLER

VOLUME 306. Expression of Recombinant Genes in Eukaryotic Systems
Edited by JOSEPH C. GLORIOSO AND MARTIN C. SCHMIDT

VOLUME 307. Confocal Microscopy
Edited by P. MICHAEL CONN

VOLUME 308. Enzyme Kinetics and Mechanism (Part E: Energetics of Enzyme Catalysis)
Edited by DANIEL L. PURICH AND VERN L. SCHRAMM

VOLUME 309. Amyloid, Prions, and Other Protein Aggregates
Edited by RONALD WETZEL

VOLUME 310. Biofilms
Edited by RON J. DOYLE

VOLUME 311. Sphingolipid Metabolism and Cell Signaling (Part A)
Edited by ALFRED H. MERRILL, JR., AND YUSUF A. HANNUN

VOLUME 312. Sphingolipid Metabolism and Cell Signaling (Part B)
Edited by ALFRED H. MERRILL, JR., AND YUSUF A. HANNUN

VOLUME 313. Antisense Technology
(Part A: General Methods, Methods of Delivery, and RNA Studies)
Edited by M. IAN PHILLIPS

VOLUME 314. Antisense Technology (Part B: Applications)
Edited by M. IAN PHILLIPS

VOLUME 315. Vertebrate Phototransduction and the Visual Cycle (Part A)
Edited by KRZYSZTOF PALCZEWSKI

VOLUME 316. Vertebrate Phototransduction and the Visual Cycle (Part B)
Edited by KRZYSZTOF PALCZEWSKI

VOLUME 317. RNA–Ligand Interactions (Part A: Structural Biology Methods)
Edited by DANIEL W. CELANDER AND JOHN N. ABELSON

VOLUME 318. RNA–Ligand Interactions (Part B: Molecular Biology Methods)
Edited by DANIEL W. CELANDER AND JOHN N. ABELSON

VOLUME 319. Singlet Oxygen, UV-A, and Ozone
Edited by LESTER PACKER AND HELMUT SIES

VOLUME 320. Cumulative Subject Index Volumes 290–319

VOLUME 321. Numerical Computer Methods (Part C)
Edited by MICHAEL L. JOHNSON AND LUDWIG BRAND

VOLUME 322. Apoptosis
Edited by JOHN C. REED

VOLUME 323. Energetics of Biological Macromolecules (Part C)
Edited by MICHAEL L. JOHNSON AND GARY K. ACKERS

VOLUME 324. Branched-Chain Amino Acids (Part B)
Edited by ROBERT A. HARRIS AND JOHN R. SOKATCH

VOLUME 325. Regulators and Effectors of Small GTPases
(Part D: Rho Family)
Edited by W. E. BALCH, CHANNING J. DER, AND ALAN HALL

VOLUME 326. Applications of Chimeric Genes and Hybrid Proteins
(Part A: Gene Expression and Protein Purification)
Edited by JEREMY THORNER, SCOTT D. EMR, AND JOHN N. ABELSON

VOLUME 327. Applications of Chimeric Genes and Hybrid Proteins
(Part B: Cell Biology and Physiology)
Edited by JEREMY THORNER, SCOTT D. EMR, AND JOHN N. ABELSON

VOLUME 328. Applications of Chimeric Genes and Hybrid Proteins (Part C: Protein–Protein Interactions and Genomics)
Edited by JEREMY THORNER, SCOTT D. EMR, AND JOHN N. ABELSON

VOLUME 329. Regulators and Effectors of Small GTPases (Part E: GTPases Involved in Vesicular Traffic)
Edited by W. E. BALCH, CHANNING J. DER, AND ALAN HALL

VOLUME 330. Hyperthermophilic Enzymes (Part A)
Edited by MICHAEL W. W. ADAMS AND ROBERT M. KELLY

VOLUME 331. Hyperthermophilic Enzymes (Part B)
Edited by MICHAEL W. W. ADAMS AND ROBERT M. KELLY

VOLUME 332. Regulators and Effectors of Small GTPases (Part F: Ras Family I)
Edited by W. E. BALCH, CHANNING J. DER, AND ALAN HALL

VOLUME 333. Regulators and Effectors of Small GTPases (Part G: Ras Family II)
Edited by W. E. BALCH, CHANNING J. DER, AND ALAN HALL

VOLUME 334. Hyperthermophilic Enzymes (Part C)
Edited by MICHAEL W. W. ADAMS AND ROBERT M. KELLY

VOLUME 335. Flavonoids and Other Polyphenols
Edited by LESTER PACKER

VOLUME 336. Microbial Growth in Biofilms (Part A: Developmental and Molecular Biological Aspects)
Edited by RON J. DOYLE

VOLUME 337. Microbial Growth in Biofilms (Part B: Special Environments and Physicochemical Aspects)
Edited by RON J. DOYLE

VOLUME 338. Nuclear Magnetic Resonance of Biological Macromolecules (Part A)
Edited by THOMAS L. JAMES, VOLKER DÖTSCH, AND ULI SCHMITZ

VOLUME 339. Nuclear Magnetic Resonance of Biological Macromolecules (Part B)
Edited by THOMAS L. JAMES, VOLKER DÖTSCH, AND ULI SCHMITZ

VOLUME 340. Drug–Nucleic Acid Interactions
Edited by JONATHAN B. CHAIRES AND MICHAEL J. WARING

VOLUME 341. Ribonucleases (Part A)
Edited by ALLEN W. NICHOLSON

VOLUME 342. Ribonucleases (Part B)
Edited by ALLEN W. NICHOLSON

VOLUME 343. G Protein Pathways (Part A: Receptors)
Edited by RAVI IYENGAR AND JOHN D. HILDEBRANDT

VOLUME 344. G Protein Pathways (Part B: G Proteins and Their Regulators)
Edited by RAVI IYENGAR AND JOHN D. HILDEBRANDT

VOLUME 345. G Protein Pathways (Part C: Effector Mechanisms)
Edited by RAVI IYENGAR AND JOHN D. HILDEBRANDT

VOLUME 346. Gene Therapy Methods
Edited by M. IAN PHILLIPS

VOLUME 347. Protein Sensors and Reactive Oxygen Species (Part A: Selenoproteins and Thioredoxin)
Edited by HELMUT SIES AND LESTER PACKER

VOLUME 348. Protein Sensors and Reactive Oxygen Species (Part B: Thiol Enzymes and Proteins)
Edited by HELMUT SIES AND LESTER PACKER

VOLUME 349. Superoxide Dismutase
Edited by LESTER PACKER

VOLUME 350. Guide to Yeast Genetics and Molecular and Cell Biology (Part B)
Edited by CHRISTINE GUTHRIE AND GERALD R. FINK

VOLUME 351. Guide to Yeast Genetics and Molecular and Cell Biology (Part C)
Edited by CHRISTINE GUTHRIE AND GERALD R. FINK

VOLUME 352. Redox Cell Biology and Genetics (Part A)
Edited by CHANDAN K. SEN AND LESTER PACKER

VOLUME 353. Redox Cell Biology and Genetics (Part B)
Edited by CHANDAN K. SEN AND LESTER PACKER

VOLUME 354. Enzyme Kinetics and Mechanisms (Part F: Detection and Characterization of Enzyme Reaction Intermediates)
Edited by DANIEL L. PURICH

VOLUME 355. Cumulative Subject Index Volumes 321–354

VOLUME 356. Laser Capture Microscopy and Microdissection
Edited by P. MICHAEL CONN

VOLUME 357. Cytochrome P450, Part C
Edited by ERIC F. JOHNSON AND MICHAEL R. WATERMAN

VOLUME 358. Bacterial Pathogenesis (Part C: Identification, Regulation, and Function of Virulence Factors)
Edited by VIRGINIA L. CLARK AND PATRIK M. BAVOIL

VOLUME 359. Nitric Oxide (Part D)
Edited by ENRIQUE CADENAS AND LESTER PACKER

VOLUME 360. Biophotonics (Part A)
Edited by GERARD MARRIOTT AND IAN PARKER

VOLUME 361. Biophotonics (Part B)
Edited by GERARD MARRIOTT AND IAN PARKER

VOLUME 362. Recognition of Carbohydrates in Biological Systems (Part A)
Edited by YUAN C. LEE AND REIKO T. LEE

VOLUME 363. Recognition of Carbohydrates in Biological Systems (Part B)
Edited by YUAN C. LEE AND REIKO T. LEE

VOLUME 364. Nuclear Receptors
Edited by DAVID W. RUSSELL AND DAVID J. MANGELSDORF

VOLUME 365. Differentiation of Embryonic Stem Cells
Edited by PAUL M. WASSAUMAN AND GORDON M. KELLER

VOLUME 366. Protein Phosphatases
Edited by SUSANNE KLUMPP AND JOSEF KRIEGLSTEIN

VOLUME 367. Liposomes (Part A)
Edited by NEJAT DÜZGÜNEŞ

VOLUME 368. Macromolecular Crystallography (Part C)
Edited by CHARLES W. CARTER, JR., AND ROBERT M. SWEET

VOLUME 369. Combinational Chemistry (Part B)
Edited by GUILLERMO A. MORALES AND BARRY A. BUNIN

VOLUME 370. RNA Polymerases and Associated Factors (Part C)
Edited by SANKAR L. ADHYA AND SUSAN GARGES

VOLUME 371. RNA Polymerases and Associated Factors (Part D)
Edited by SANKAR L. ADHYA AND SUSAN GARGES

VOLUME 372. Liposomes (Part B)
Edited by NEJAT DÜZGÜNEŞ

VOLUME 373. Liposomes (Part C)
Edited by NEJAT DÜZGÜNEŞ

VOLUME 374. Macromolecular Crystallography (Part D)
Edited by CHARLES W. CARTER, JR., AND ROBERT W. SWEET

VOLUME 375. Chromatin and Chromatin Remodeling Enzymes (Part A)
Edited by C. DAVID ALLIS AND CARL WU

VOLUME 376. Chromatin and Chromatin Remodeling Enzymes (Part B)
Edited by C. DAVID ALLIS AND CARL WU

VOLUME 377. Chromatin and Chromatin Remodeling Enzymes (Part C)
Edited by C. DAVID ALLIS AND CARL WU

VOLUME 378. Quinones and Quinone Enzymes (Part A)
Edited by HELMUT SIES AND LESTER PACKER

VOLUME 379. Energetics of Biological Macromolecules (Part D)
Edited by JO M. HOLT, MICHAEL L. JOHNSON, AND GARY K. ACKERS

VOLUME 380. Energetics of Biological Macromolecules (Part E)
Edited by JO M. HOLT, MICHAEL L. JOHNSON, AND GARY K. ACKERS

VOLUME 381. Oxygen Sensing
Edited by CHANDAN K. SEN AND GREGG L. SEMENZA

VOLUME 382. Quinones and Quinone Enzymes (Part B)
Edited by HELMUT SIES AND LESTER PACKER

VOLUME 383. Numerical Computer Methods (Part D)
Edited by LUDWIG BRAND AND MICHAEL L. JOHNSON

VOLUME 384. Numerical Computer Methods (Part E)
Edited by LUDWIG BRAND AND MICHAEL L. JOHNSON

VOLUME 385. Imaging in Biological Research (Part A)
Edited by P. MICHAEL CONN

VOLUME 386. Imaging in Biological Research (Part B)
Edited by P. MICHAEL CONN

VOLUME 387. Liposomes (Part D)
Edited by NEJAT DÜZGÜNEŞ

VOLUME 388. Protein Engineering
Edited by DAN E. ROBERTSON AND JOSEPH P. NOEL

VOLUME 389. Regulators of G-Protein Signaling (Part A)
Edited by DAVID P. SIDEROVSKI

VOLUME 390. Regulators of G-Protein Signaling (Part B)
Edited by DAVID P. SIDEROVSKI

VOLUME 391. Liposomes (Part E)
Edited by NEJAT DÜZGÜNEŞ

VOLUME 392. RNA Interference
Edited by ENGELKE ROSSI

VOLUME 393. Circadian Rhythms
Edited by MICHAEL W. YOUNG

VOLUME 394. Nuclear Magnetic Resonance of Biological Macromolecules (Part C)
Edited by THOMAS L. JAMES

VOLUME 395. Producing the Biochemical Data (Part B)
Edited by ELIZABETH A. ZIMMER AND ERIC H. ROALSON

VOLUME 396. Nitric Oxide (Part E)
Edited by LESTER PACKER AND ENRIQUE CADENAS

VOLUME 397. Environmental Microbiology
Edited by JARED R. LEADBETTER

VOLUME 398. Ubiquitin and Protein Degradation (Part A)
Edited by RAYMOND J. DESHAIES

VOLUME 399. Ubiquitin and Protein Degradation (Part B)
Edited by RAYMOND J. DESHAIES

VOLUME 400. Phase II Conjugation Enzymes and Transport Systems
Edited by HELMUT SIES AND LESTER PACKER

VOLUME 401. Glutathione Transferases and Gamma Glutamyl Transpeptidases
Edited by HELMUT SIES AND LESTER PACKER

VOLUME 402. Biological Mass Spectrometry
Edited by A. L. BURLINGAME

VOLUME 403. GTPases Regulating Membrane Targeting and Fusion
Edited by WILLIAM E. BALCH, CHANNING J. DER, AND ALAN HALL

VOLUME 404. GTPases Regulating Membrane Dynamics
Edited by WILLIAM E. BALCH, CHANNING J. DER, AND ALAN HALL

VOLUME 405. Mass Spectrometry: Modified Proteins and Glycoconjugates
Edited by A. L. BURLINGAME

VOLUME 406. Regulators and Effectors of Small GTPases: Rho Family
Edited by WILLIAM E. BALCH, CHANNING J. DER, AND ALAN HALL

VOLUME 407. Regulators and Effectors of Small GTPases: Ras Family
Edited by WILLIAM E. BALCH, CHANNING J. DER, AND ALAN HALL

VOLUME 408. DNA Repair (Part A)
Edited by JUDITH L. CAMPBELL AND PAUL MODRICH

VOLUME 409. DNA Repair (Part B)
Edited by JUDITH L. CAMPBELL AND PAUL MODRICH

VOLUME 410. DNA Microarrays (Part A: Array Platforms and Web-Bench Protocols)
Edited by ALAN KIMMEL AND BRIAN OLIVER

VOLUME 411. DNA Microarrays (Part B: Databases and Statistics)
Edited by ALAN KIMMEL AND BRIAN OLIVER

VOLUME 412. Amyloid, Prions, and Other Protein Aggregates (Part B)
Edited by INDU KHETERPAL AND RONALD WETZEL

VOLUME 413. Amyloid, Prions, and Other Protein Aggregates (Part C)
Edited by INDU KHETERPAL AND RONALD WETZEL

VOLUME 414. Measuring Biological Responses with Automated Microscopy
Edited by JAMES INGLESE

VOLUME 415. Glycobiology
Edited by MINORU FUKUDA

VOLUME 416. Glycomics
Edited by MINORU FUKUDA

VOLUME 417. Functional Glycomics
Edited by MINORU FUKUDA

VOLUME 418. Embryonic Stem Cells
Edited by IRINA KLIMANSKAYA AND ROBERT LANZA

VOLUME 419. Adult Stem Cells
Edited by IRINA KLIMANSKAYA AND ROBERT LANZA

VOLUME 420. Stem Cell Tools and Other Experimental Protocols
Edited by IRINA KLIMANSKAYA AND ROBERT LANZA

VOLUME 421. Advanced Bacterial Genetics: Use of Transposons and Phage for Genomic Engineering
Edited by KELLY T. HUGHES

VOLUME 422. Two-Component Signaling Systems, Part A
Edited by MELVIN I. SIMON, BRIAN R. CRANE, AND ALEXANDRINE CRANE

VOLUME 423. Two-Component Signaling Systems, Part B
Edited by MELVIN I. SIMON, BRIAN R. CRANE, AND ALEXANDRINE CRANE

VOLUME 424. RNA Editing
Edited by JONATHA M. GOTT

VOLUME 425. RNA Modification
Edited by JONATHA M. GOTT

VOLUME 426. Integrins
Edited by DAVID CHERESH

VOLUME 427. MicroRNA Methods
Edited by JOHN J. ROSSI

VOLUME 428. Osmosensing and Osmosignaling
Edited by HELMUT SIES AND DIETER HAUSSINGER

VOLUME 429. Translation Initiation: Extract Systems and Molecular Genetics
Edited by JON LORSCH

VOLUME 430. Translation Initiation: Reconstituted Systems and Biophysical Methods
Edited by JON LORSCH

VOLUME 431. Translation Initiation: Cell Biology, High-Throughput and Chemical-Based Approaches
Edited by JON LORSCH

VOLUME 432. Lipidomics and Bioactive Lipids: Mass-Spectrometry–Based Lipid Analysis
Edited by H. ALEX BROWN

VOLUME 433. Lipidomics and Bioactive Lipids: Specialized Analytical Methods and Lipids in Disease
Edited by H. ALEX BROWN

VOLUME 434. Lipidomics and Bioactive Lipids: Lipids and Cell Signaling
Edited by H. ALEX BROWN

VOLUME 435. Oxygen Biology and Hypoxia
Edited by HELMUT SIES AND BERNHARD BRÜNE

VOLUME 436. Globins and Other Nitric Oxide-Reactive Protiens (Part A)
Edited by ROBERT K. POOLE

VOLUME 437. Globins and Other Nitric Oxide-Reactive Protiens (Part B)
Edited by ROBERT K. POOLE

VOLUME 438. Small GTPases in Disease (Part A)
Edited by WILLIAM E. BALCH, CHANNING J. DER, AND ALAN HALL

VOLUME 439. Small GTPases in Disease (Part B)
Edited by WILLIAM E. BALCH, CHANNING J. DER, AND ALAN HALL

VOLUME 440. Nitric Oxide, Part F Oxidative and Nitrosative Stress in Redox Regulation of Cell Signaling
Edited by ENRIQUE CADENAS AND LESTER PACKER

VOLUME 441. Nitric Oxide, Part G Oxidative and Nitrosative Stress in Redox Regulation of Cell Signaling
Edited by ENRIQUE CADENAS AND LESTER PACKER

VOLUME 442. Programmed Cell Death, General Principles for Studying Cell Death (Part A)
Edited by ROYA KHOSRAVI-FAR, ZAHRA ZAKERI, RICHARD A. LOCKSHIN, AND MAURO PIACENTINI

VOLUME 443. Angiogenesis: *In Vitro* Systems
Edited by DAVID A. CHERESH

VOLUME 444. Angiogenesis: *In Vivo* Systems (Part A)
Edited by DAVID A. CHERESH

VOLUME 445. Angiogenesis: *In Vivo* Systems (Part B)
Edited by DAVID A. CHERESH

VOLUME 446. Programmed Cell Death, The Biology and Therapeutic Implications of Cell Death (Part B)
Edited by ROYA KHOSRAVI-FAR, ZAHRA ZAKERI, RICHARD A. LOCKSHIN, AND MAURO PIACENTINI

VOLUME 447. RNA Turnover in Bacteria, Archaea and Organelles
Edited by LYNNE E. MAQUAT AND CECILIA M. ARRAIANO

VOLUME 448. RNA Turnover in Eukaryotes: Nucleases, Pathways
and Analysis of mRNA Decay
Edited by LYNNE E. MAQUAT AND MEGERDITCH KILEDJIAN

VOLUME 449. RNA Turnover in Eukaryotes: Analysis of Specialized and Quality
Control RNA Decay Pathways
Edited by LYNNE E. MAQUAT AND MEGERDITCH KILEDJIAN

VOLUME 450. Fluorescence Spectroscopy
Edited by LUDWIG BRAND AND MICHAEL L. JOHNSON

VOLUME 451. Autophagy: Lower Eukaryotes and Non-Mammalian Systems (Part A)
Edited by DANIEL J. KLIONSKY

VOLUME 452. Autophagy in Mammalian Systems (Part B)
Edited by DANIEL J. KLIONSKY

VOLUME 453. Autophagy in Disease and Clinical Applications (Part C)
Edited by DANIEL J. KLIONSKY

VOLUME 454. Computer Methods (Part A)
Edited by MICHAEL L. JOHNSON AND LUDWIG BRAND

VOLUME 455. Biothermodynamics (Part A)
Edited by MICHAEL L. JOHNSON, JO M. HOLT, AND GARY K. ACKERS (RETIRED)

VOLUME 456. Mitochondrial Function, Part A: Mitochondrial Electron Transport
Complexes and Reactive Oxygen Species
Edited by WILLIAM S. ALLISON AND IMMO E. SCHEFFLER

VOLUME 457. Mitochondrial Function, Part B: Mitochondrial Protein Kinases,
Protein Phosphatases and Mitochondrial Diseases
Edited by WILLIAM S. ALLISON AND ANNE N. MURPHY

VOLUME 458. Complex Enzymes in Microbial Natural Product Biosynthesis,
Part A: Overview Articles and Peptides
Edited by DAVID A. HOPWOOD

VOLUME 459. Complex Enzymes in Microbial Natural Product Biosynthesis,
Part B: Polyketides, Aminocoumarins and Carbohydrates
Edited by DAVID A. HOPWOOD

VOLUME 460. Chemokines, Part A
Edited by TRACY M. HANDEL AND DAMON J. HAMEL

VOLUME 461. Chemokines, Part B
Edited by TRACY M. HANDEL AND DAMON J. HAMEL

VOLUME 462. Non-Natural Amino Acids
Edited by TOM W. MUIR AND JOHN N. ABELSON

VOLUME 463. Guide to Protein Purification, 2nd Edition
Edited by RICHARD R. BURGESS AND MURRAY P. DEUTSCHER

VOLUME 464. Liposomes, Part F
Edited by NEJAT DÜZGÜNEŞ

VOLUME 465. Liposomes, Part G
Edited by NEJAT DÜZGÜNEŞ

VOLUME 466. Biothermodynamics, Part B
Edited by MICHAEL L. JOHNSON, GARY K. ACKERS, AND JO M. HOLT

VOLUME 467. Computer Methods Part B
Edited by MICHAEL L. JOHNSON AND LUDWIG BRAND

VOLUME 468. Biophysical, Chemical, and Functional Probes of RNA Structure, Interactions and Folding: Part A
Edited by DANIEL HERSCHLAG

VOLUME 469. Biophysical, Chemical, and Functional Probes of RNA Structure, Interactions and Folding: Part B
Edited by DANIEL HERSCHLAG

VOLUME 470. Guide to Yeast Genetics: Functional Genomics, Proteomics, and Other Systems Analysis, 2nd Edition
Edited by GERALD FINK, JONATHAN WEISSMAN, AND CHRISTINE GUTHRIE

VOLUME 471. Two-Component Signaling Systems, Part C
Edited by MELVIN I. SIMON, BRIAN R. CRANE, AND ALEXANDRINE CRANE

VOLUME 472. Single Molecule Tools, Part A: Fluorescence Based Approaches
Edited by NILS G. WALTER

VOLUME 473. Thiol Redox Transitions in Cell Signaling, Part A Chemistry and Biochemistry of Low Molecular Weight and Protein Thiols
Edited by ENRIQUE CADENAS AND LESTER PACKER

VOLUME 474. Thiol Redox Transitions in Cell Signaling, Part B Cellular Localization and Signaling
Edited by ENRIQUE CADENAS AND LESTER PACKER

VOLUME 475. Single Molecule Tools, Part B: Super-Resolution, Particle Tracking, Multiparameter, and Force Based Methods
Edited by NILS G. WALTER

VOLUME 476. Guide to Techniques in Mouse Development, Part A Mice, Embryos, and Cells, 2nd Edition
Edited by PAUL M. WASSARMAN AND PHILIPPE M. SORIANO

VOLUME 477. Guide to Techniques in Mouse Development, Part B Mouse Molecular Genetics, 2nd Edition
Edited by PAUL M. WASSARMAN AND PHILIPPE M. SORIANO

VOLUME 478. Glycomics
Edited by MINORU FUKUDA

VOLUME 479. Functional Glycomics
Edited by MINORU FUKUDA

VOLUME 480. Glycobiology
Edited by MINORU FUKUDA

VOLUME 481. Cryo-EM, Part A: Sample Preparation and Data Collection
Edited by GRANT J. JENSEN

VOLUME 482. Cryo-EM, Part B: 3-D Reconstruction
Edited by GRANT J. JENSEN

VOLUME 483. Cryo-EM, Part C: Analyses, Interpretation, and Case Studies
Edited by GRANT J. JENSEN

VOLUME 484. Constitutive Activity in Receptors and Other Proteins, Part A
Edited by P. MICHAEL CONN

VOLUME 485. Constitutive Activity in Receptors and Other Proteins, Part B
Edited by P. MICHAEL CONN

VOLUME 486. Research on Nitrification and Related Processes, Part A
Edited by MARTIN G. KLOTZ

VOLUME 487. Computer Methods, Part C
Edited by MICHAEL L. JOHNSON AND LUDWIG BRAND

VOLUME 488. Biothermodynamics, Part C
Edited by MICHAEL L. JOHNSON, JO M. HOLT, AND GARY K. ACKERS

VOLUME 489. The Unfolded Protein Response and Cellular Stress, Part A
Edited by P. MICHAEL CONN

VOLUME 490. The Unfolded Protein Response and Cellular Stress, Part B
Edited by P. MICHAEL CONN

VOLUME 491. The Unfolded Protein Response and Cellular Stress, Part C
Edited by P. MICHAEL CONN

VOLUME 492. Biothermodynamics, Part D
Edited by MICHAEL L. JOHNSON, JO M. HOLT, AND GARY K. ACKERS

VOLUME 493. Fragment-Based Drug Design
Tools, Practical Approaches, and Examples
Edited by LAWRENCE C. KUO

VOLUME 494. Methods in Methane Metabolism, Part A
Methanogenesis
Edited by AMY C. ROSENZWEIG AND STEPHEN W. RAGSDALE

VOLUME 495. Methods in Methane Metabolism, Part B
Methanotrophy
Edited by AMY C. ROSENZWEIG AND STEPHEN W. RAGSDALE

VOLUME 496. Research on Nitrification and Related Processes, Part B
Edited by MARTIN G. KLOTZ AND LISA Y. STEIN

VOLUME 497. Synthetic Biology, Part A
Methods for Part/Device Characterization and Chassis Engineering
Edited by CHRISTOPHER VOIGT

VOLUME 498. Synthetic Biology, Part B
Computer Aided Design and DNA Assembly
Edited by CHRISTOPHER VOIGT

VOLUME 499. Biology of Serpins
Edited by JAMES C. WHISSTOCK AND PHILLIP I. BIRD

VOLUME 500. Methods in Systems Biology
Edited by DANIEL JAMESON, MALKHEY VERMA, AND HANS V. WESTERHOFF

VOLUME 501. Serpin Structure and Evolution
Edited by JAMES C. WHISSTOCK AND PHILLIP I. BIRD

VOLUME 502. Protein Engineering for Therapeutics, Part A
Edited by K. DANE WITTRUP AND GREGORY L. VERDINE

VOLUME 503. Protein Engineering for Therapeutics, Part B
Edited by K. DANE WITTRUP AND GREGORY L. VERDINE

VOLUME 504. Imaging and Spectroscopic Analysis of Living Cells
Optical and Spectroscopic Techniques
Edited by P. MICHAEL CONN

VOLUME 505. Imaging and Spectroscopic Analysis of Living Cells
Live Cell Imaging of Cellular Elements and Functions
Edited by P. MICHAEL CONN

VOLUME 506. Imaging and Spectroscopic Analysis of Living Cells
Imaging Live Cells in Health and Disease
Edited by P. MICHAEL CONN

VOLUME 507. Gene Transfer Vectors for Clinical Application
Edited by THEODORE FRIEDMANN

VOLUME 508. Nanomedicine
Cancer, Diabetes, and Cardiovascular, Central Nervous System, Pulmonary and Inflammatory Diseases
Edited by NEJAT DÜZGÜNEŞ

VOLUME 509. Nanomedicine
Infectious Diseases, Immunotherapy, Diagnostics, Antifibrotics, Toxicology and Gene Medicine
Edited by NEJAT DÜZGÜNEŞ

VOLUME 510. Cellulases
Edited by HARRY J. GILBERT

CHAPTER ONE

MEASUREMENT OF *ENDO*-1,4-β-GLUCANASE

Barry V. McCleary, Vincent McKie, *and* Anna Draga

Contents

1. Introduction	2
2. Purity of EG Enzymes Included in the Study	2
3. Polysaccharide and Oligosaccharide Substrates	3
4. Viscometric Assay of EG Activity	4
4.1. Substrate preparation	4
4.2. Assay of enzyme activity	5
4.3. Analytical results	5
5. Reducing-Sugar Assays for EG Activity	6
5.1. Nelson–Somogyi reducing-sugar assay (slightly modified)	6
5.2. PAHBAH reducing-sugar assay (slightly modified)	7
5.3. Analytical results	8
6. Soluble Chromogenic Substrates	10
6.1. Substrate preparation	10
6.2. Precipitant solutions	12
6.3. Assay of enzyme activity	12
7. Insoluble Chromogenic Substrates	14
7.1. Substrate preparation	14
7.2. Assay of enzyme activity	14
8. Summary	15
References	16

Abstract

Several procedures are available for the measurement of *endo*-1,4-β-glucanase (EG). Primary methods employ defined oligosaccharides or highly purified polysaccharides and measure the rate of hydrolysis of glycosidic bonds using a reducing-sugar method. However, these primary methods are not suitable for the measurement of EG in crude fermentation broths due to the presence of reducing sugars and other enzymes active on these substrates. In such cases, dyed soluble or insoluble substrates are preferred as they are specific,

Megazyme International Ireland, Bray Business Park, Bray, County Wicklow, Ireland

sensitive, easy to use, and are not affected by other components, such as reducing sugars, in the enzyme preparation.

1. Introduction

Enzymic dissolution of crystalline and amorphous cellulose appears to require the concerted action of endo-1,4-β-glucanase (EG) and exo-cellobiohydrolases (CBH; Canevascini and Gattlen, 1981; Sharrock, 1988). exo-Glucosidases may also be involved. Lower degree of polymerization (DP) cello-oligosaccharides are hydrolyzed to D-glucose by β-glucosidase, which may remove product inhibition of the former two enzymes. Many methods have been proposed for the measurement of CBH, but most are only applicable to analysis of the purified enzyme. A procedure proposed by Deshpande et al. (1984), which employed p-nitrophenyl cellobiose and p-nitrophenyl lactoside, would appear to allow selective measurement of exo-1,4-β-glucanases in the presence of cellulolytic enzymes. β-Glucosidase is routinely assayed with an aryl-β-glucoside substrate such as p-nitrophenyl-β-glucoside, 6-bromo-2-naphthyl-β-glucoside, and methylumbelliferyl-β-glucoside. However, action on an aryl-β-glucoside does not necessarily mean that the enzyme will hydrolyze cellobiose, thus care should be applied in using aryl-β-glucosides in screening for β-glucosidase enzymes. Theoretically, with exo-glucosidases, the rate of hydrolysis of cello-oligosaccharides would increase as the DP increases from 2 to 6. This would allow them to be distinguished from β-glucosidase where the rate of hydrolysis remains much the same as the DP increases (McCleary and Harrington, 1988).

In this chapter, methods for the measurement of endo-1,4-β-glucanase (EG) will be described, with some discussion on the strengths and limitations of the various methods employed.

2. Purity of EG Enzymes Included in the Study

The EG included in this study were:

(a) The major EG from *Trichoderma longibrachiatum*. Purified from Genencor Laminex BG by conventional chromatographic procedures. Appears as a single major band on isoelectric focusing (pI = 4.7) and SDS gel electrophoresis (MW = 57,200; Fig. 1.1). Contamination with β-glucosidase less than 0.001% (on an activity basis).
(b) The major EG purified from *Aspergillus niger* industrial preparation. Crystalline and appears as a single band on isoelectric focusing (pI = 4.55) and SDS gel electrophoresis (MW = 27,000). Contamination with β-glucosidase less than 0.00001% (on an activity basis).

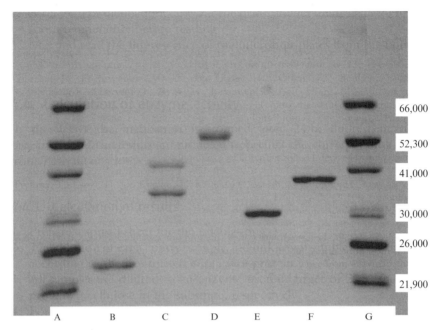

Figure 1.1 SDS Gel electrophoresis of pure endo-1,4-β-glucanases from: B. *Aspergillus niger*. C. *Talaromyces emersonii*. D. *Trichoderma longibrachiatum*. E. *Bacillus amyloliquifaciens* (recombinant), F. *Thermotoga maritima* (recombinant). A and G are molecular weight standards. (For the color version of this figure, the reader is referred to the Web version of this chapter.)

(c) The major EG purified from a *Talaromyces emersonii* industrial preparation. Two major bands on isoelectric focusing (pI = 3.4 and 3.6). Two bands on SDS gel electrophoresis (EG component with MW = 37,000). Contamination with β-glucosidase less than 0.006% (on an activity basis).
(d) EG from *Bacillus amyloliquefaciens* (recombinant). Single band on isoelectric focusing (pI = 6.1) and SDS gel electrophoresis (MW = 34,300). Contamination with β-glucosidase less than 0.005% (on an activity basis).
(e) EG from *Thermotoga maritima* (recombinant). Single band on isoelectric focusing (pI = 6.2) and on SDS gel electrophoresis (MW = 38,200). Contamination with β-glucosidase less than 0.007% (on an activity basis).

3. POLYSACCHARIDE AND OLIGOSACCHARIDE SUBSTRATES

A range of native and chemically modified polysaccharides are susceptible to hydrolysis by EG, and different EG have varying abilities to hydrolyze these polymers. None of the available polysaccharides could be considered to be an ideal substrate for EG. Native cellulose is insoluble and EG has very limited ability to hydrolyze the amorphous and crystalline regions in the substrate. Solubility is imparted by chemical modification such as carboxymethylation or hydroxypropylation. However, overmodification interferes with the ability of the EG to hydrolyze the 1,4-β-D-glucan backbone. For example, CM-cellulose 7M [degree of substitution (DS) of approximately 0.7] is completely soluble, but is resistant to hydrolysis by some EG. CM-cellulose 4M (DS approximately 0.4), in contrast, is only partially soluble in water, but is readily hydrolyzed by most EG.

Barley or oat 1,3:1,4-β-glucan is a very useful substrate for the assay of EG. It is water soluble and readily hydrolyzed by all EG. The major structural features are cellotriosyl and cellotetraosyl units joined in an irregular pattern through 1,3-β-D-linkages (Varghese et al., 1994). β-Glucan can be extracted and purified to greater than 96% purity from barley or oat flour. On storage of a 10 mg/ml solution at room temperature over several days, the β-glucan will tend to precipitate/gel from solution; however, this can be readily redissolved by heating the solution at approximately 80 °C for 5–10 min.

Xyloglucan from *Tamarindus indica* seed consists of a main chain of β-(1-4)-linked D-glucosyl residues to which D-xylosyl groups are attached α-(1-6) to three out of every four main-chain residues. D-Galactosyl groups are attached β-(1-2) to some of the D-xylosyl groups (Matheson and McCleary, 1985). The combined yield of hepta-, octa-, and nona-saccharides on hydrolysis of the xyloglucan by certain cellulases, and the composition of the oligosaccharides, indicate that there is an average structural unit made up of one-unsubstituted and three-substituted D-glucosyl residues in the main chain. Consequently, tamarind xyloglucan serves as an interesting substrate to distinguish EG that can act on highly branched substrates.

Konjac glucomannan has a backbone of β-(1-4)-linked D-glucosyl and D-mannosyl residues in a random or nonregular distribution. The ratio of D-glucose to D-mannose is 2:3 (Matheson and McCleary, 1985). Evidence for the presence of both isolated and contiguous D-glucosyl residues has been confirmed from the structures of the oligosaccharides produced on both EG and β-mannanase hydrolysis. Partial acetylation of the polysaccharide imparts solubility. Konjac glucomannan serves as an interesting substrate to distinguish EG based on their ability to bind D-mannose as well as D-glucose in the enzyme active site.

4. Viscometric Assay of EG Activity

4.1. Substrate preparation

CM-cellulose 7M (Sigma cat. no. C5013-500G; 5 mg/ml): Carefully sprinkle 1 g of CMC (7M) into 180 ml of vigorously stirring buffer (100 mM; desired pH) and heat on a hot plate stirrer to approximately 80 °C until the polysaccharide completely dissolves. While heating, loosely cover the beaker with aluminum foil to prevent the formation of a "skin" on the surface. Allow the solution to cool to room temperature and adjust the volume to 200 ml. Store in a 250 ml Duran bottle at room temperature. Add two drops of toluene to prevent microbial contamination.

Barley β-glucan (Megazyme cat. no. P-BGBM; 5 mg/ml): Weigh 1 g of pure high viscosity barley β-glucan into a 500 ml Pyrex beaker and wet with a few milliliters of ethanol. Add 180 ml of buffer (100 mM, desired pH) and stir and heat to approximately 80 °C (until the β-glucan completely dissolves). While heating, loosely cover the beaker with aluminum foil to prevent the formation of a "skin" on the surface. Allow the solution to cool to room temperature and adjust the volume to 200 ml. Store in a 250 ml Duran bottle at room temperature. Add two drops of toluene to prevent microbial contamination.

4.2. Assay of enzyme activity

Transfer 12 ml of substrate solution (CM-cellulose 7M or barley β-glucan) to a C-Type, U-tube viscometer incubated in a water bath at 40 °C. Allow the solution to equilibrate to assay temperature over 5 min. Start the reaction by adding 0.2 ml of enzyme solution with thorough mixing. Record the time of addition of the enzyme as time zero. Immediately record the time of flow of solution between the two marks on the viscometer. Take further readings at various time intervals over 1 h. Record the flow times in seconds.

The activity of the enzyme preparation can be simply shown graphically as the rate of decrease in the viscosity of the polysaccharide solution with time of incubation.

4.3. Analytical results

Viscometry gives a specific measurement of EG. Results for a typical viscometric assay of EG are shown in Fig. 1.2. *endo*-Hydrolysis of the polysaccharide causes a rapid decrease in viscosity. For EG, several polysaccharide substrates can be used, including high viscosity CM-cellulose 7M, barley β-glucan, tamarind xyloglucan (Megazyme cat. no. P-XYGLN), and Konjac glucomannan (Megazyme cat. no. P-GLCMH). CM-cellulose 4M (Megazyme cat. no. P-CMC4M) cannot be used because it does not completely dissolve, and would thus block the capillary tube in the viscometer. While *exo*-acting enzymes can remove glucose from the reducing end of the polysaccharide, extensive hydrolysis is required before there is any effect on viscosity. Also, with chemically modified polysaccharides such as CM-cellulose 7M, hydrolysis by *exo*-acting enzymes is blocked by the carboxymethyl groups. Viscometric assays also offer high sensitivity; cleavage of a single bond per molecule can result in a dramatic decrease in viscosity. Assays based on this

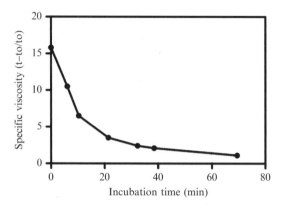

Figure 1.2 Hydrolysis of high viscosity barley 1,3:1,4-β-glucan by *Trichoderma longibrachiatum* EG. Barley β-glucan (12 ml, 0.5%, w/v) in 100 mM sodium acetate buffer, pH 4.5 incubated with 0.5 ml (60 mU) of EG at 40 °C.

principle are very useful in detecting trace amounts of *endo*-activity in *exo*-acting enzymes. Another advantage offered by viscometric assays is that they can be performed on crude fermentation broths. Sugars and other components in the broth do not interfere with the assay. Such assays are limited by the fact that it is hard to define activity in conventional enzyme units (micromoles of substrate hydrolyzed or product produced per minute). In the brewing industry, this problem was partially resolved for the analysis of β-glucanase in malt by performing inverse reciprocal plots of the data to obtain inverse reciprocal viscosity units (IRVU; Bathgate, 1979). Such assays require a pure β-glucan substrate with viscosities in a well defined range.

5. Reducing-Sugar Assays for EG Activity

5.1. Nelson–Somogyi reducing-sugar assay (slightly modified)

5.1.1. Preparation of reagents

Reagent A: Dissolve 25 g of anhydrous sodium carbonate, 25 g of sodium potassium tartrate, and 200 g of sodium sulfate in 800 ml of distilled water. Dilute to 1 l and filter if necessary (Nelson, 1944; Somogyi, 1952). Store at room temperature.

Reagent B: Dissolve 30 g of copper sulfate pentahydrate in 200 ml of distilled water containing four drops concentrated sulfuric acid. Store at room temperature.

Reagent C: Dissolve 50 g of ammonium molybdate in 900 ml of distilled water and add 42 ml of concentrated sulfuric acid. Separately dissolve 6 g of sodium arsenate heptahydrate in 50 ml of water and add to the above solution. Dilute the whole to 1 l. If necessary, warm the solution to 55 °C to obtain complete dissolution. Store at room temperature.

Reagent D: Add 1 ml of Reagent B to 25 ml of Reagent A and mix. Store at room temperature.

Reagent E: Dilute an aliquot of solution C fivefold (e.g., 50–250 ml) with distilled water before use (stable at 4 °C for approximately 1 week). Store at room temperature.

5.1.1. Assay of enzyme activity

Transfer 0.5 ml of polysaccharide substrate solution (10 mg/ml) [or 0.2 ml of oligosaccharide substrate solution (10 mM)] in 100 mM buffer (of the required pH) to the bottom of five glass test tubes (16 × 120 mm) and preincubate at 40 °C for 5 min. Preincubate enzyme solution (approximately 5 ml) in the same buffer at 40 °C for 5 min. Transfer 0.2 ml of enzyme solution to each of the test tubes containing substrate, mix well and incubate at 40 °C for 3, 6, 9, and 12 min. Terminate the reaction by adding 0.5 ml of Nelson–Somogyi solution D with vigorous stirring on a vortex mixer. Prepare the zero time incubation by adding Nelson–Somogyi solution D to the substrate solution before addition of the enzyme. On addition of the enzyme, stir the tube contents vigorously.

Prepare a D-glucose sugar standard curve by adding 0.2 ml of D-glucose standard (125 and 250 mg/ml) in duplicate to the bottom of glass test tubes. Add 0.5 ml of the polysaccharide substrate solution (or 0.2 ml of oligosaccharide substrate solution) as used in the assay followed by 0.5 ml of Nelson–Somogyi solution D and mix the tube contents vigorously. Prepare a substrate blank by mixing 0.2 ml of distilled water, 0.5 ml of polysaccharide substrate (or 0.2 ml of oligosaccharide), and 0.5 ml of Nelson–Somogyi solution D.

Incubate all solutions in a boiling water bath for 20 min. Remove the tubes and add 3.0 ml of Nelson–Somogyi solution D with vigorous stirring. Allow the tubes to stand at room temperature for 10 min and mix the contents again. If a turbidity forms, centrifuge the tubes at 3000 rpm for 10 min in a bench centrifuge.

Measure the absorbance of the reaction solutions and standards against the substrate blank in a spectrophotometer set at 520 nm.

Calculate enzyme activity as micromoles of reducing-sugar equivalents produced per minute under the defined assay conditions.

5.2. PAHBAH reducing-sugar assay (slightly modified)

5.2.1. Preparation of reagents

Reagent A: Add 10 g of *p*-hydroxy benzoic acid hydrazide (PAHBAH) to 60 ml of distilled water and slurry. Add 10 ml of concentrated hydrochloric acid to obtain complete dissolution and adjust the volume to 200 ml (Lever, 1972). Store at room temperature.

Reagent B: Add 24.9 g of trisodium citrate to 500 ml of water and dissolve with stirring. Add 2.2 g of calcium chloride and dissolve with stirring. Then, add 40.0 g of sodium hydroxide and dissolve. Adjust the volume to 2 l. The solution should be clear. Store at room temperature.

Reagent C: Immediately before use, add 10 ml of Reagent A to 90 ml of Reagent B. Store on ice between uses.

5.2.2. Assay of enzyme activity

Transfer 0.5 ml of polysaccharide substrate solution (10 mg/ml) [or 0.2 ml of oligosaccharide substrate solution (10 mM)] in 100 mM buffer (of the required pH) to the bottom of five glass test tubes (16 × 120 mm) and preincubate at 40 °C for 5 min. Preincubate enzyme solution (approximately 5 ml) in the same buffer at 40 °C for 5 min. Transfer 0.2 ml of enzyme solution to each of the test tubes containing substrate, mix well and incubate at 40 °C for 3, 6, 9, and 12 min. Terminate the reaction by adding 5 ml of PAHBAH solution C with mixing.

Prepare a D-glucose sugar standard curve by adding 0.2 ml of D-glucose standard (125 and 250 µg/ml) in duplicate to the bottom of glass test tubes. Add 0.5 ml of the polysaccharide substrate solution as used in the assay (or 0.2 ml of oligosaccharide substrate solution) followed by 5 ml of PAHBAH solution C with thorough mixing. Prepare a substrate blank by mixing 0.2 ml of distilled water, 0.5 ml of polysaccharide substrate solution (or 0.2 ml of oligosaccharide solution), and 5 ml of PAHBAH solution C. Incubate all solutions in a boiling water bath for exactly 6 min. Remove the rack containing the tubes and place it in an ice-water bath to cool. If a turbidity is present, centrifuge the tubes at 3000 rpm for 10 min in a bench centrifuge. Within 15 min, measure the absorbance of the reaction solutions and standards against the substrate blank in a spectrophotometer set at 490 nm.

Calculate enzyme activity as micromoles of reducing-sugar equivalents produced per minute under the defined assay conditions.

5.3. Analytical results

In the measurement of EG, the ultimate or primary unit of activity is micromoles of glycosidic bonds cleaved in the substrate per minute under defined assay conditions. This is determined using one of many reducing-sugar assays. In industry, the most commonly used reducing-sugar assay is the dinitrosalicylic acid (DNSA) method (Ghose, 1987; Miller, 1959) which utilizes quite alkaline reagents. Alternatives which are more applicable to laboratory situations are the Nelson–Somogyi (Nelson, 1944; Somogyi, 1952) reducing-sugar method and the PAHBAH reducing-sugar method

(Lever, 1972). The latter method, like the DNSA procedure, employs very alkaline reagents. Under highly alkaline conditions, the 1,3-β-linkage in barley β-glucan is susceptible to cleavage, resulting in very high assay blanks. However, the reagent causes no problems with 1,4-β-glucan substrates. The Nelson–Somogyi reagent does not cause such problems; however, several polysaccharides tend to partially precipitate in the presence of the reagents, so centrifugation of the reaction mixture is essential before the absorbance of assay solutions can be measured.

Since hydrolysis of polysaccharides releases an array of oligosaccharides and polysaccharides of varying DP, it is important that the reducing color formed is essentially independent of the DP of the fragment size, that is, that 1 μmol of cellobiose gives essentially the same color formation as 1 μmol of cellotriose, etc. The color formed by 1 μmol of cellobiose, cellotriose, etc., with the Nelson–Somogyi and PAHBAH methods are shown in Table 1.1. Clearly, for both methods, there is a slight increase in the color response for equimolar amounts of the oligosaccharides as the DP increases. To minimize such effects in enzyme assays, cellobiose is used to standardize the assay instead of D-glucose.

The relative rates of hydrolysis of a range of highly purified polysaccharides by several EG are shown in Table 1.2. Some EG act on a broad range of polysaccharides, while others have limited ability to hydrolyze highly branched polysaccharides such as xyloglucan or CM-cellulose 7M (compared to CM-cellulose 4M; Matheson and McCleary, 1985). In screening for EG, it is important to employ substrates that will detect all types of EG. However, in selecting for EG with particular activity patterns, it is useful to have substrates that select for different types of specificities. In the latter case, comparisons of the rates of hydrolysis of CM-cellulose 4M, xyloglucan, and glucomannan can be useful. This can also be useful when checking enzyme preparations for adulteration. In wine manufacture, EG from *A. niger* are

Table 1.1 Absorbance values obtained for 0.1 μmol of glucose, cellobiose, cellotriose, cellotetraose, cellopentaose, and cellohexaose with the Nelson–Somogyi and PAHBAH reducing-sugar methods

Saccharide	Absorbance at 520 or 490 nm	
	Nelson–Somogyi (at 520 nm)	PAHBAH (at 490 nm)
Glucose	0.20	0.19
Cellobiose	0.25	0.22
Cellotriose	0.27	0.23
Cellotetraose	0.30	0.25
Cellopentaose	0.32	0.27
Cellohexaose	0.33	0.28

Table 1.2 Relative initial rates of hydrolysis of polysaccharides by various *endo*-1,4-β-glucanases

Source of *endo*-1,4-β-glucanase	CMC4M	CMC7M	Barley β-glucan	Tamarind xyloglucan	Konjac glucomannan
T. longibrachiatum	100	100	43	38	8
T. emersonii	100	100	89	55	0.6
A. niger	100	45	135	n.d.	0.08
Bacillus amyloliquefaciens	100	58	109	4.3	n.d.
Thermotoga maritima	100	79	133	n.d.	n.d.

n.d.: not detectible. Assays performed as described in the text.

allowed, whereas those from *Trichoderma* sp. are not. *A. niger* EG is unable to hydrolyze tamarind xyloglucan, whereas *Trichoderma* sp. EG readily hydrolyze this substrate. Consequently, ability to hydrolyze tamarind xyloglucan indicates the presence of *Trichoderma* sp. EG in a particular enzyme preparation (e.g., one from *A. niger*).

EG can also be assayed using cello-oligosaccharides or borohydride-reduced cello-oligosaccharides. The former cannot be used in reducing-sugar assays due to the high blank value derived from the substrate. Borohydride reduction removes this. A comparison of the rates of hydrolysis of reduced cellobiose to cellohexaose gives an insight into the substrate binding requirements of the particular EG. Relative rates of hydrolysis of borohydride reduced cello-oligosaccharides by several EG are shown in Fig. 1.3 and Table 1.3. Clearly, there are differences in action patterns and binding requirements of the EG studied. *Trichoderma* sp. EG rapidly hydrolyzes borohydride-reduced cellotriose, whereas the *A. niger* EG requires a chain length of five or six glucose residues for maximum hydrolysis rate. This difference in ability to hydrolyze small DP cello-oligosaccharides explains the ability of the *Trichoderma* sp. EG to hydrolyze the highly substituted tamarind xyloglucan.

6. Soluble Chromogenic Substrates

6.1. Substrate preparation

Substrates for the measurement of EG have been prepared by dyeing a range of modified celluloses with various dyes (Biely *et al.*, 1985; Huang and Tang, 1976; McCleary, 1980). The substrates discussed in this report are

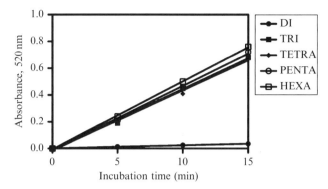

Figure 1.3 Time course of hydrolysis of borohydride reduced cello-oligosaccharides by *Trichoderma longibrachiatum* EG. An aliquot (0.2 ml) of reduced cello-oligosaccharide (10 mM) incubated with 0.2 ml of EG (30 mU) at 40 °C. The reaction was stopped at 0, 5, 10, and 15 min by addition of Nelson–Somogyi reagent D and color developed.

Table 1.3 Relative initial rates of hydrolysis of borohydride reduced cello-oligosaccharides by various *endo*-1,4-β-glucanases

Substrate	*T. longibrachiatum*	*T. emersonii*	*A. niger*	*B. amyloliquefaciens*	*T. maritima*
CMC4M	108	42	76	91	74
Cellobiitol	4.7	n.d.	n.d.	<2	<5
Cellotriitol	89	6	n.d.	<2	<5
Cellotetritol	90	24	17	85	62
Cellopentaitol	95	80	66	98	99
Cellohexaitol	100	100	100	100	100

n.d.: not detectible. Assays performed as described in the text.

polysaccharides (CM-cellulose 4M, barley β-glucan, and tamarind xyloglucan) that have been dyed with Remazol Brilliant Blue dye to a dye content of approximately one dye molecule per 15–20 anhydrohexose units and are available commercially from Megazyme. To dissolve, carefully sprinkle 1 or 2 g of dyed polysaccharide (see Figures 1.4 and 1.5) into 90 ml of vigorously stirring water or buffer (100 mM; required pH). Continue stirring and heating (to about 80 °C) until the polysaccharide completely dissolves. Allow the solution to cool to room temperature and check the pH and adjust if necessary. Adjust the volume to 100 ml with buffer and store the solution in a 100 ml Duran bottle at 4 °C. Add two drops of toluene to prevent microbial infection. Before use, warm the solution to approximately 30–40 °C and shake the bottle contents vigorously.

6.2. Precipitant solutions

6.2.1. Precipitant solution A (for Azo-CMC assay)
Dissolve 40 g of sodium acetate trihydrate and 4 g of zinc acetate in 150 ml of distilled water. Adjust the pH to 5.0 with 5 M HCl and the volume to 200 ml with distilled water. Add 200 ml of this solution to 800 ml of industrial methylated spirit (95%) or ethanol (95%), mix well and store at room temperature in a well-sealed bottle.

6.2.2. Precipitant solution B (for Azo-barley glucan assay)
Dissolve 30.0 g sodium acetate ($CH_3COONa \cdot 3H_2O$) and 3.0 g zinc acetate in 250 ml of distilled water. Adjust the pH to 5.0 with concentrated HCl, adjust the volume to 300 ml. Add 700 ml of industrial methylated spirit (95%) or ethanol and mix well. Store in a well-sealed glass bottle at room temperature.

6.3. Assay of enzyme activity
Add 0.5 ml of appropriately diluted and pre-equilibrated enzyme solution to 0.5 ml of pre-equilibrated substrate solution (10 mg/ml; buffered or unbuffered) and incubate at 40 °C for 10 min. Add ethanolic precipitant solution (2.5 or 3.0 ml; see Figs. 1.4 and 1.5) and vigorously stir the tube and

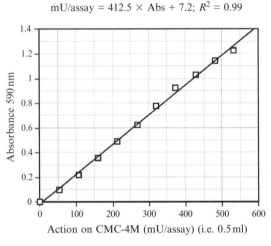

Figure 1.4 Standard curve for the action of *Trichoderma longibrachiatum* EG on Azo-CM-cellulose (Lot 90504). An aliquot (0.5 ml) of Azo-CM-cellulose (2%, w/v) in 100 mM sodium acetate buffer (pH 4.5) was incubated with 0.5 ml of EG in 100 mM sodium acetate buffer (pH 4.5) at a range of concentrations (0–540 mU/assay) for 10 min at 40 °C. The reaction was terminated and high molecular weight dyed CM-cellulose precipitated by the addition of 2.5 ml of Precipitant A. Tube contents were mixed thoroughly, centrifuged, and the absorbance of the supernatant solution read against a substrate blank at 590 nm.

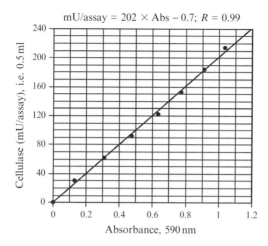

Figure 1.5 Standard curve for the action of *Trichoderma longibrachiatum* EG on Azo-barley glucan (Lot 60602). An aliquot (0.5 ml) of Azo-barley glucan (1% w/v in distilled water) was incubated with 0.5 ml of EG at a range of concentrations (0–230 mU/assay) in sodium acetate buffer (100 mM, pH 4.5) for 10 min at 40 °C. The reaction was terminated and high molecular weight dyed β-glucan precipitated by the addition of 3 ml of Precipitant B. Tube contents were mixed thoroughly, centrifuged, and the absorbance of the supernatant solution read against a substrate blank at 590 nm.

contents for 15–20 s. This terminates the reaction and precipitates the nonhydrolyzed, higher molecular weight dyed polysaccharide substrate. After 5 min, mix the tube contents vigorously on a vortex mixer and centrifuge the reaction tubes at 3000 rpm for 10 min in a bench centrifuge. Read the absorbance of the supernatant solution against a substrate blank. Prepare this blank by mixing 0.5 ml of buffer with 0.5 ml of substrate solution, followed by precipitant solution (2.5 or 3.0 ml). Mix the contents vigorously and centrifuge the tubes at 3000 rpm for 10 min.

Calculate enzyme activity by reference to a standard curve pre-prepared for the particular substrate and *endo*-1,4-β-glucanase. Standard curves will vary by as much as 10% between substrate lots and between different types of EG.

A typical standard curve for the activity of *T. longibrachiatum* EG on Azo-CM-cellulose is shown in Fig. 1.4 and on Azo-barley glucan is shown in Fig. 1.5. These curves are created by first standardizing the particular EG activity on CM-cellulose 4M (or barley β-glucan) using the Nelson–Somogyi reducing-sugar procedure. The enzyme at a range of concentrations is then run in the assay employing the dyed substrate under defined assay conditions (usually, 40 °C and 10 min incubation). The concentration of unknown preparations of EG can then be standardized using this assay and by reference to the standard curve.

7. Insoluble Chromogenic Substrates

7.1. Substrate preparation

These substrates are usually prepared by dyeing and crosslinking soluble polysaccharides such as barley β-glucan, tamarind xyloglucan, or hydroxypropyl cellulose (McCleary, 2001). Optimization of the degree of dyeing and crosslinking is determined empirically, with the aim of producing a substrate with the highest possible sensitivity and a standard curve that can be used over an absorbance range of 2 absorbance units. The substrate consists of gelatinous dyed particles of the substrate that rapidly hydrate in water or buffer, but are insoluble. These particles are readily hydrolyzed by the target enzyme, releasing soluble, dyed polysaccharide fragments into solution. The rate of release of dyed fragments (increase in absorbance at 590 nm) is related to milli-Units of enzyme activity through a standard curve. The substrates can be used in a powder form (azurine, crosslinked polysaccharides) or the substrate can be incorporated into tablets. The powder form of the substrate is useful for screening for enzymes in agar plates or in acrylamide gels. Tablet forms of the substrate (e.g., Cellazyme C, Cellazyme T, and Beta-Glucazyme) are useful for assaying enzyme activity in solutions.

7.2. Assay of enzyme activity

Add 0.5 ml of appropriately diluted and buffered enzyme solution to the bottom of a 16 × 120 mm glass test tube and equilibrate at 40 °C for 5 min. Using forceps, add a Cellazyme C tablet (or Cellazyme T or Beta-Glucazyme tablet) and allow the reaction to proceed for 10 min (do not stir the tube contents). The tablet rapidly hydrates and absorbs the buffered enzyme solution. After 10 min add 10 ml of 2% (w/v) Trizma Base (pH ~ 9) or 2% trisodium phosphate (pH 11) and vigorously stir the tube contents for 10 s. This terminates the reaction. Allow the tubes to stand at room temperature for 5 min and then stir the contents again and filter the suspension through a Whatman no. 1 (9 cm) filter circle. Measure the absorbance of the filtrate at 590 nm against a substrate/enzyme blank. Prepare the substrate/enzyme blank by adding Trizma Base to the enzyme solution before the addition of the Cellazyme C tablet. A single blank is required for each set of determinations and this is used to zero the spectrophotometer.

Calculate enzyme activity by reference to a standard curve pre-prepared for the particular substrate and *endo*-1,4-β-glucanase. Standard curves will vary by as much as 10% between substrate lots.

Typical standard curves for the hydrolysis of Cellazyme C tablets (containing dyed and crosslinked HP-cellulose) and Beta-Glucazyme tablets (containing dyed and crosslinked barley β-glucan) are shown in Figs. 1.6

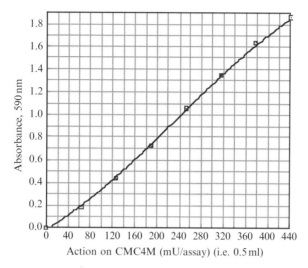

Figure 1.6 Standard curve for the action of *Trichoderma longibrachiatum* EG on Cellazyme C tablets (Lot 80201). An aliquot (0.5 ml) of EG at a range of concentrations (0–440 mU/assay) in sodium acetate buffer (25 mM, pH 4.5) was preincubated at 40 °C for 5 min. A Cellazyme C tablet was then added (without stirring) and the reaction allowed to proceed for exactly 10 min. The reaction was terminated by the addition of 10 ml of 2% Trizma Base (pH ~ 9) and the tube contents mixed vigorously on a vortex stirrer. After 5 min at room temperature, the tube contents were mixed again and then filtered through Whatman no. 1 (9 cm) filter papers. The absorbance of the filtrate was read against that of a substrate blank at 590 nm.

and 1.7. To use these standard curves to determine the concentration of EG in unknown enzyme preparations, assays must be performed exactly as described (10 min incubation at 40 °C).

8. Summary

A range of methods are available for the measurement of EG. The usefulness of these methods depends, in large part, on the purity of the enzyme being analyzed. Assays based on the measurement of glycosidic bonds cleaved in a defined substrate such as CM-cellulose 4M or a borohydride reduced cello-oligosaccharide give activity in well-defined International Units (micromoles of bonds cleaved per minute), but such assays are of limited use for the measurement of EG in crude fermentation broths. In such matrices, sugars present interfere with the assay and the substrates are susceptible to hydrolysis by enzymes other than EG; for example, *exo*-cellulase, cellobiohydrolase, and β-glucosidase. In spite of the limitations

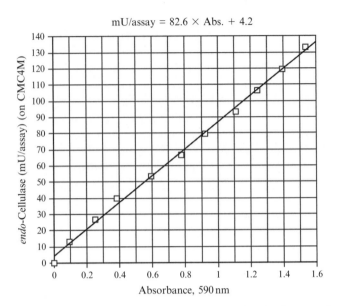

Figure 1.7 Standard curve for the action of *Trichoderma longibrachiatum* EG on Cellazyme T tablets (Lot 00601). The assay was performed exactly as for Cellazyme C (Fig. 1.6) with a range of enzyme concentrations (0–133 mU/assay).

of reducing-sugar assays, since they measure the rate of cleavage of glycosidic bonds, they are the best primary methods. Viscometric assays employing high viscosity polysaccharides such as barley β-glucan or CM-cellulose 7M are very useful because of their specificity for EG. However, determination of units of activity and translation of these "arbitrary" units back to International Units is difficult. Perhaps, the most convenient methods for the specific measurement of EG are those based on the use of soluble or insoluble (dyed, crosslinked) polysaccharides. These give specific measurement of EG and they can be used to assay activity in crude fermentation broths. They are secondary methods, so they must be standardized using a primary method such as the Nelson–Somogyi reducing-sugar method.

REFERENCES

Bathgate, G. N. (1979). The determination of *endo*-glucanase activity in malt. *J. Inst. Brew.* **85,** 92–94.
Biely, P., Markovič, O., and Mislovičová, D. (1985). Sensitive detection of *endo*-1,4-β-glucanases and endo-1,4-β-xylanases in gels. *Anal. Biochem.* **144,** 147–151.
Canevascini, G., and Gattlen, C. (1981). A comparative investigation of various cellulase assay procedures. *Biotechnol. Bioeng.* **XXIII,** 1573–1590.

Deshpande, M. V., Eriksson, K.-E., and Pettersson, G. (1984). An assay for selective determination of exo-1,4,-β-glucanases in a mixture of cellulolytic enzymes. *Anal. Biochem.* **138**, 481–487.

Ghose, T. M. (1987). Measurement of cellulase activities. *Pure Appl. Chem.* **59**, 257–268.

Huang, J. S., and Tang, J. (1976). Sensitive assay for cellulase and dextranase. *Anal. Biochem.* **73**, 369–377.

Lever, M. (1972). A new reaction for colorimetric determination of carbohydrates. *Anal. Biochem.* **47**, 273–279.

Matheson, N. K., and McCleary, B. V. (1985). Enzymes metabolising polysaccharides and their application to the analysis of the structure and function of glucans. *In* "The Polysaccharides," (G. O. Aspinall, ed.), pp. 1–105. Academic Press, New York. Vol. 3.

McCleary, B. V. (1980). New chromogenic substrates for the assay of α-amylase and β-1,4-glucanase. *Carbohydr. Res.* **86**, 97–104.

McCleary, B. V. (2001). Analysis of feed enzymes. *In* "Enzymes in Farm Animal Nutrition," (M. Bedford and G. Partridge, eds.), pp. 85–107. CAB International, Wallingford, UK.

McCleary, B. V., and Harrington, J. (1988). Purification of β-D-glucosidase from *Aspergillus niger*. *Methods Enzymol.* **160**, 575–582.

Miller, G. L. (1959). Use of dinitrosalicylic acid reagent for determination of reducing sugar. *Anal. Chem.* **31**, 426–428.

Nelson, N. (1944). A photometric adaption of the Somogyi method for the determination of glucose. *J. Biol. Chem.* **153**, 375–380.

Sharrock, K. R. (1988). Cellulase assay methods. *J. Biochem. Biophys. Methods* **17**, 81–106.

Somogyi, M. (1952). Note on sugar determination. *J. Biol. Chem.* **195**, 19–23.

Varghese, J. N., Garrett, T. P. J., Colman, P. M., Chen, L., Hoj, P. B., and Fincher, G. B. (1994). Three-dimensional structures of two plant 1,3-glucan endohydrolases with distinct substrate specificities. *Proc. Natl. Acad. Sci. USA* **91**, 2785–2789.

CHAPTER TWO

Biomass Conversion Determined via Fluorescent Cellulose Decay Assay

Bente Wischmann,[*] Marianne Toft,[*] Marco Malten,[*] and K. C. McFarland[†]

Contents

1. Introduction	20
2. Correlation Between Cellulose Decay Monitored by HPLC and by Fluorescence	21
3. Assay Development and Modeling of Response Using Statistics	24
3.1. Development of plate layout	25
3.2. Reduction of assay time	28
4. Final Assay Procedure	32
4.1. Substrate preparation	32
4.2. Protein determination, sample dilution, and loading of microtiter plates	33
4.3. Incubation and fluorescence reading of FCD assay plates	34
4.4. Calculation of enzyme activity	35
Acknowledgment	36
References	36

Abstract

An example of a rapid microtiter plate assay (fluorescence cellulose decay, FCD) that determines the conversion of cellulose in a washed biomass substrate is reported. The conversion, as verified by HPLC, is shown to correlate to the monitored FCD in the assay. The FCD assay activity correlates to the performance of multicomponent enzyme mixtures and is thus useful for the biomass industry. The development of an optimized setup of the 96-well microtiter plate is described, and is used to test a model that shortens the assay incubation time from 72 to 24 h. A step-by-step procedure of the final assay is described.

[*] Novozymes A/S, Bagsværd, Denmark
[†] Novozymes Inc., Davis, California, USA

Methods in Enzymology, Volume 510
ISSN 0076-6879, DOI: 10.1016/B978-0-12-415931-0.00002-1

© 2012 Elsevier Inc.
All rights reserved.

1. Introduction

One of the big challenges for improving the efficiency of enzyme-catalyzed reactions in the biomass industry is that multicomponent enzyme mixtures are needed to provide the required performance. Hence, there are several different enzymatic activities in a single product such as Novozymes Cellic® CTec, and there are synergies between the activities. The consequence is that there is not one single activity measurement that can monitor the performance of the enzyme system used in a specific application. Optimally, a true application-relevant method must be used. Synthetic substrates including Avicel, carboxymethyl cellulose (CMC), or AZO CMC are only able to reveal a few of the enzyme activities present in complex biomass enzyme products and cannot show their synergistic interactions.

Therefore, in order to properly assess multicomponent biomass enzyme systems, the selected substrate has to be a natural insoluble plant cell wall material. Plant cell walls are composed of a mixture of interlocking polymers that include carbohydrates and lignin. Only multicomponent enzymes, acting synergistically, are therefore able to fully break down natural plant cell wall material.

As the method of choice, a performance correlation method has been developed (Malten and McFarland, 2011). It is a microtiter plate-based method that detects cellulose hydrolysis through a fluorescence enhancer bound to cellulose, and is defined as the fluorescent cellulose decay (FCD) assay. The substrate is pretreated corn stover, which was washed, ground, and sieved (WGS-PCS).

The substrate is mixed with a fluorescence enhancer (Calcofluor White, FB28) and the enzymatic hydrolysis of the substrate is monitored as a decrease in fluorescence emission intensity. FB28 is a compound that binds to cellulose and, when cellulose is hydrolyzed, FB28 is released. Binding to cellulose increases the quantum yield of the FB28 (Ruchel et al., 2001). Consequently, the emission response decreases upon hydrolysis of cellulose.

Initially, the method was set up with an incubation time of 72 h (Malten et al., 2012) at 50 °C with a goal of 80% conversion of cellulose to glucose, in alignment with possible commercial process timelines, and efficiency targets for hydrolysis. In the final assay described here, this incubation time could be reduced to 24 h using a model obtained by comparing 24- and 72-h conversion, allowing a faster analytical response time. This model can be used for predicting the 72-h conversion for unknown enzyme mixtures of similar composition to those used for building the model. For unknown enzyme samples with very different composition, the model can be used as an initial estimate. In the final assay, the substrate is incubated with enzyme at 50 °C, pH 5.0 for 24 h after which the results are normalized and calculated relative to the content of cellulose in the substrate and the result converted to that of a 72-h assay. The result is thus expressed as the predicted enzyme load (in mg protein/g cellulose) necessary for the conversion of 80% of the cellulose in the substrate after incubation at pH 5.0 and 50 °C for 72 h.

Firstly, this chapter describes the correlation between cellulose decay monitored by high-pressure liquid chromatography (HPLC) and by FCD. Secondly, it demonstrates the use of modeling to shorten the assay incubation time from 72 to 24 h, but in a way that can predict the performance of an enzyme sample under commercial process conditions. Thirdly, the final assay procedure is explained as a step-by-step procedure.

2. Correlation Between Cellulose Decay Monitored by HPLC and by Fluorescence

The aim of enzyme hydrolysis of biomass substrates is to break down sugar polymers into fermentable monosaccharides. HPLC enables quantification of such generated monosaccharides, which can be expressed as a percent (%) conversion of the total cellulose present in the substrate. The total concentration of sugar polymers, typically cellulose, in the substrate with 3% total solids was determined by HPLC after complete hydrolysis with a high enzyme load. The completion of the hydrolysis was verified by comparison to data obtained by compositional analysis.

In the experiments described here, 50 mg enzyme protein/g cellulose was well above the amount sufficient to reach the target of 80% cellulose conversion after 72-h incubation at 50 °C, and approximates the maximal conversion of enzyme-accessible cellulose to glucose (∼98% of theoretically achievable glucose from cellulose, based on composition for a typical WGS-PCS). The conversion achieved at 50 mg control enzyme/g cellulose is set to 100% in the calculations.

During hydrolysis of the cellulose in the WGS-PCS, a reduction of the fluorescent brightener 28 (FB28)-based fluorescence emission signal was observed, which was essentially linear to the produced glucose. In order to obtain a % conversion from the fluorescence emission signal, the following normalization was used (Eq. 2.1):

$$\text{FCD estimated conversion} = \frac{AU_{72h,0mg} - AU_{72h,Xmg}}{AU_{72h,0mg} - AU_{72h,50mg}} \quad (2.1)$$

where AU is the arbitrary unit of fluorescence emission intensity at the measured time, 72 h. $AU_{0\,mg}$ is the signal from the wells without enzyme and $AU_{50\,mg}$ is the signal from the wells of 50 mg sample protein sufficient to ensure complete hydrolysis of enzyme available cellulose.

By this normalization, the value is also independent of the measuring apparatus, which is important because actual measured values of the fluorescence emission intensity can vary between fluorescence readers.

By applying Eq. (2.1), it is possible directly to compare the FCD assay-estimated % conversion, obtained from reading of fluorescence, with the HPLC measured conversion as shown in Fig. 2.1. In Fig. 2.1, the FCD assay was used to estimate the conversion of two batches of WGS-PCS substrate (batch no. 81 and batch no. 96) using Eq. (2.1), and the result plotted against the HPLC measured conversion. In the range between 20% and 80% conversion, the correlation is linear.

An enzyme dose–response curve of the HPLC measured conversion shows that the WGS-PCS from batch 96 needs a higher amount of enzyme for 80% conversion than batch 81 (see Fig. 2.2A and B). At higher and lower conversions, the curves merge again (see Fig. 2.2C). The figures show that the FCD assay can be used to rank the hydrolysis performance of different enzyme doses on the same WGS-PCS substrate, and can be used to show the differences in hydrolysis of different substrate batches.

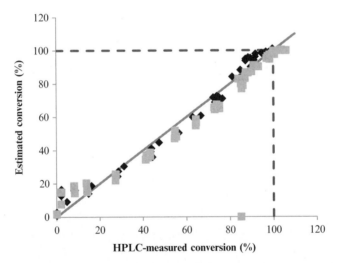

Figure 2.1 FCD assay of cellulose conversion (%) versus HPLC-measured cellulose conversion (%) after 72 h hydrolysis of two different WGS-PCS substrate batches, 81 (dark diamonds) and 96 (light squares), using varied doses of enzyme (Novozymes Cellic® CTec). Below 20%, the prediction of HPLC-measured conversion from FCD conversion is invalid. From 20% to 85%, the prediction for both biomass batches is slightly low. Above 85%, the assay predicts very well for batch 96 and gives slightly low values for batch 81. In the range above 20%, the average deviation of the FCD and HPLC assays is 5.6% for batch 81 and 8.8% for batch 96.

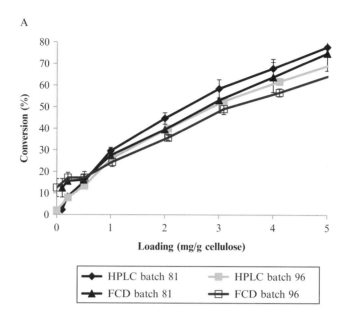

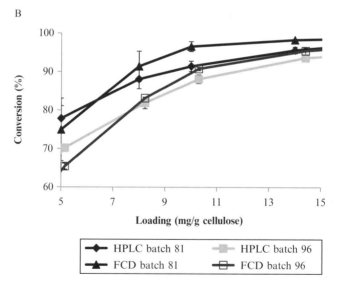

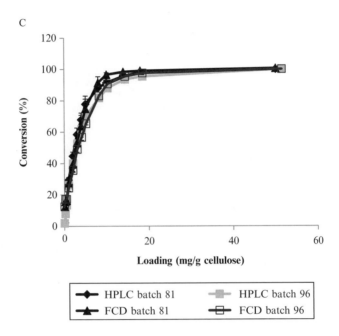

Figure 2.2 Dose–response curves of HPLC-measured conversion (%) and FCD-measured conversion (%) for varying concentrations of enzyme mixes (mg/g cellulose) using a 72-h incubation at 50 °C and pH 5.0. Even though discrepancy between the HPLC and the FCD responses are seen, the FCD assay is excellent for comparing enzyme dosing on the same biomass substrate batch, as well as comparing different biomasses. (A) Response profile for an enzyme load of up to 5 mg/g cellulose. (B) Dose–response curve for the enzyme loads of between 5 and 15 mg/g cellulose. (C) The profile of the whole range of dose–response curves up to 100% conversion.

3. Assay Development and Modeling of Response Using Statistics

After demonstrating the potential application of the FCD assay, further optimization was needed to ensure good reproducibility and reduce the time of the assay. It was also important to take into account that the assay should be as easy as possible to perform. Hence, several sets of data were collected and used for statistical investigations and optimization. For statistical analysis, JMP ver. 8.0.2 (SAS Institute, Cary, NC, USA) was used. Furthermore, the final output of the assay should not be an estimated conversion, but the enzyme load needed to reach a defined target such as 80% cellulose conversion after 72 h hydrolysis. This allows different enzyme mixtures to be compared even when the dose–response curves are nonlinear. In order to obtain this, each enzyme sample was analyzed at three different enzyme loadings bracketing the target conversion. Another requirement for the assay was to include a control sample to calculate the normalized conversion based on values of 0% and 100% hydrolysis of cellulose.

3.1. Development of plate layout

In a 96-well microtiter plate, the temperature and fluorescence measurements may vary between the wells, for example, between the wells lying on the border of the plate and the wells inside. Hence, for optimization of the assay, the plate layout was analyzed by replicating a single enzyme dose with a single enzyme mixture (the control enzyme) over the whole plate. This made it possible to determine the variation between rows and between columns in order to optimize the plate layout for the samples to be tested. In addition, the control enzyme was assayed in a series of seven dilutions to check for the expected FCD profile. It was also determined whether one dilution series of the control enzyme was sufficient, or whether two dilution series were needed due to large variation between the wells.

Five microtiter plates were filled with two dilution series (0, 0, 0, 2, 5, 6, 8, 10, 20, 50, 50, and 50 mg protein/g cellulose) of the control enzyme in rows D and E (two different randomized orders of concentrations) as shown in Fig. 2.3. The remaining wells were filled with 8 mg protein/g cellulose control enzyme. Fluorescence was read at 0, 24, and 72 h.

First, the effect of the well position in the microtiter plate was investigated excluding rows D and E wells which contain the dilution series of the control enzyme. The signal output data expressed in arbitrary units (AU) of fluorescence emission intensity at the measuring time from rows A to C and F to H was analyzed using the model: Result (AU) = μ + Row no. + Column no. + Plate no., where μ is the overall average. The effects were

	1	2	3	4	5	6	7	8	9	10	11	12
A	8	8	8	8	8	8	8	8	8	8	8	8
B	8	8	8	8	8	8	8	8	8	8	8	8
C	8	8	8	8	8	8	8	8	8	8	8	8
D	10	0	0	50	7	6	50	2	20	0	50	5
E	20	0	6	50	50	10	2	5	0	7	0	50
F	8	8	8	8	8	8	8	8	8	8	8	8
G	8	8	8	8	8	8	8	8	8	8	8	8
H	8	8	8	8	8	8	8	8	8	8	8	8

Figure 2.3 Plate layout for the 96-well microtiter plates in experiment 1. Columns 1–12 and rows A–H are indicated as well as the protein load in mg/g cellulose in the substrate.

first included as fixed effects to obtain prediction profiles (see Fig. 2.4) and afterward as random effects in order to get estimates of the individual contribution to the error (Table 2.1).

It is seen in Table 2.1 and Fig. 2.4 that the variation from column number is larger than the variation from row number. There is also a clear structure in how the result varies over the columns with smallest results near the edges and largest results near the center. However, the variation across all plates, rows, and columns corresponds to a coefficient of variation (standard deviation divided by the average, CV) of only 2.3% and 5.3% for 24 and 72 h, respectively. Therefore, it was decided to accept this source of variation as part of the assay, but use the knowledge in the specific setup of the plate design.

Since column number accounts for the largest amount of variation, the wells used for the three enzyme doses (dilutions) of each unknown enzyme sample are selected across columns and not within columns. This also gives the possibility of analyzing more samples on each microtiter plate than if selected within columns. However, if the wells are selected as 1-2-3 for one sample and 4-5-6 for the next etc., then there is a risk of a consistent but wrong result for some samples as the wells consistently increase or decrease within the set of three wells. On the other hand, it would be very impractical in a manual assay if the wells were not selected as three wells next to each other. It was, therefore, decided to make two measurements (two sets of

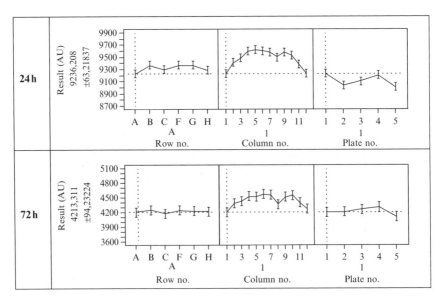

Figure 2.4 Prediction profiles at two incubation times: 24 and 72 h, excluding rows D and E wells which contain the dilution series of the enzyme. For each incubation time, the parameters row number, column number, and plate number, are shown.

Table 2.1 Estimates of the individual contributions to the error (expressed as variance components)

Component	Variance components estimates	% of total	CV
Response result (AU) at $T=24$ h			
Row no. and random	2906.29	6.032	0.6
Column no. and random	18313.85	38.008	1.4
Plate no. and random	9255.53	19.209	1.0
Residual	17707.90	36.751	1.4
Total	48183.57	100.000	2.3
Response result (AU) at $T=72$			
Row no. and random	152.36	0.265	0.3
Column no. and random	13088.56	22.762	2.5
Plate no. and random	4915.86	8.549	1.6
Residual	39344.09	68.423	4.4
Total	57500.86	100.000	5.3

three wells) in different positions on the microtiter plate for each dilution set of unknown enzyme sample.

To calculate the % conversion (as in Eq. 2.1), the values of 0 and 50 mg of the dilution series of the control enzyme at 72 h are used. The CV of a single determination was determined to be 5.2% for 0 mg and to 2.2% for 50 mg (using variance components). Using the average of three determinations (from one row of dilution series of control enzyme), the CV is reduced to 3.6% and 1.3%, respectively. Using the average of six determinations (from two rows of dilution series of the control enzyme), the CV is reduced to 3.1% and 1.0%, respectively. From these values and the CV for sample readings (shown in Table 2.1), the CV of the % conversion can be calculated. At 72 h, the CV of % conversion is reduced from 9.5% to 9.1% using two rows of control enzyme instead of only one. This is considered a small reduction and thus it was decided to use only one row for the control enzyme.

Based on this study, a final plate layout is chosen in which the dilution series of the control enzyme is placed in a nonrandom way over row D. Each unknown enzyme sample (sample) is weighed out twice and analyzed in three dilutions. A total of 14 samples can be analyzed in one "run." Samples 1–14 are placed across the plate, first in row A, then B, and so on. The second weighing of each sample is then applied starting with sample 1 in positions E7–9 and so on. The plate layout is shown in Fig. 2.5. A copy of this plate is made using the dilutions made previously (i.e., not new weighings) in order to reduce the error coming from plate-to-plate variation. Thus, four results will be given for each sample dilution. The initial experiment (see Fig. 2.4) showed that the values varied over the plate,

	1	2	3	4	5	6	7	8	9	10	11	12
A	\multicolumn{3}{c}{Sample 1a}									
A	2	5	8	2	5	8	2	5	8	2	5	8
B	Sample 5a			Sample 6a			Sample 7a			Sample 8a		
B	2	5	8	2	5	8	2	5	8	2	5	8
C	Sample 9a			Sample 10a			Sample 11a			Sample 12a		
C	2	5	8	2	5	8	2	5	8	2	5	8
D	Control enzyme											
D	0	0	0	2	5	6	8	10	20	50	50	50
E	Sample 13a			Sample 14a			Sample 1b			Sample 2b		
E	2	5	8	2	5	8	2	5	8	2	5	8
F	Sample 3b			Sample 4b			Sample 5b			Sample 6b		
F	2	5	8	2	5	8	2	5	8	2	5	8
G	Sample 7b			Sample 8b			Sample 9b			Sample 10b		
G	2	5	8	2	5	8	2	5	8	2	5	8
H	Sample 11b			Sample 12b			Sample 13b			Sample 14b		
H	2	5	8	2	5	8	2	5	8	2	5	8

Figure 2.5 Microtiter plate layout. The microtiter plate layout shows how to distribute the samples in the 96-wells from A1 to H12. A total of 14 samples may be analyzed in one run (i.e., two plates). Fourteen unknown enzyme mixtures (samples) 1–14 are weighed out twice (a and b) and diluted into three enzyme protein concentrations (2, 5, and 8 mg/g cellulose). The control enzyme is placed in row D (D1–D12) in a series of dilutions 0, 2, 5, 6, 8, 10, 20, and 50 mg/g cellulose. 0 and 50 mg/g cellulose are placed in three wells on each microtiter plate.

both across rows and across columns. This is a variation that cannot be avoided, but with this plate setup, the influence of this variation on the final result (average) is reduced as each sample dilution is analyzed twice in two different rows and two different columns. Furthermore, the influence of variation coming from the plates is reduced as each sample is analyzed on two different plates.

3.2. Reduction of assay time

A number of samples were analyzed in order to establish a preliminary model for the relationship between values measured after 24 and 72 h of incubation. This investigation led to decisions about how to transform the data before statistical modeling, and to a proof of principle for the idea of predicting the result after 72 h from data read after 24 h.

The aim of the assay was to be able to give the results that *would have been* obtained after 72 h of hydrolysis (Y), but from data acquired after only 24 h (X). The raw data acquired after 72 h are transformed relative to the 0- and 50-mg control sample results as described in Eq. (2.1) and the results lie between 0% and 100% conversion. In Fig. 2.6, an example is shown that visualizes patterns of hydrolysis after incubation of 72 h at 50 °C. In this example (Fig. 2.6), the protein load needed to reach 80% conversion is between 4.4 and 18.0 mg protein/g cellulose, dependent on

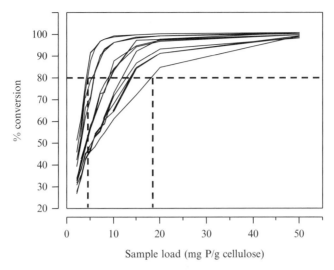

Figure 2.6 The % conversion of substrate after 72 h of incubation is shown as a function of sample load in mg protein/g cellulose. The curves are derived from 16 different samples at 12 different levels of sample load from 0 to 50 mg protein/g cellulose. The dashed lines indicate how the Y is found as the protein load per g cellulose needed to achieve 80% conversion for two samples of extreme performance. The sample to the left reaches 80% conversion with a protein load of 4.4 mg protein/g cellulose at 50 °C after 72 h of incubation (a very strong enzyme sample). The right-most sample reaches 80% conversion with a protein load of 18.0 mg/g cellulose (a very weak enzyme sample).

the sample. This is the desired output of the assay (the result or the Y value).

In the experimental setup, seven plates are filled with control enzyme in row D and seven different samples in rows A–C and E–H, in all 42 enzyme samples. Concentrations for the samples were 0, 2, 3, 4, 5, 6, 7, 8, 10, 15, 20, and 50 mg protein/g cellulose placed randomly over the row. The plates were read at 0, 24, and 72 h.

The optimal format of the data read at 24 h was obtained after several modifications of the raw fluorescence readings (AU). The modifications included: no transformation, subtracting time zero value, and normalizing to time zero value. The best result (in terms of ability to model the result as a function of concentration) was found with normalizing to time zero value for each well as seen in Eq. (2.2):

$$\text{Rel0h} = \frac{\text{AU}_{0h} - \text{AU}_{24h}}{\text{AU}_{0h}} \qquad (2.2)$$

where AU is the arbitrary unit of fluorescence emission intensity at the measured time.

This normalization gives a final format which is similar to the one used for the 72-h incubation which is expressed in % conversion in the range between 0% and 100% (Eq. 2.1).

3.2.1. Expressing the data showing the actual result

From the data acquired at 72-h incubation, Y was determined in principle as shown in Fig. 2.6. In more detail, the natural logarithm was taken to both concentration and % conversion resulting in a linear relationship. Then the result (Y) was found using linear interpolation between the two protein loads that gave just above and just below 80% conversion.

3.2.2. The final model for predicted result

A model was then built between Rel0h data from concentrations 2–10 mg protein/g cellulose read at 24 h (X) and the result (Y) determined from data read at 72 h. Both X and Y were logarithm transformed. The regression model was built using PLS regression (Wold et al., 2001) as the X data are highly correlated. This model was compared with a model built only on concentrations 2, 5, and 8 mg protein/g cellulose as X. Both models performed well, and thus the concentrations 2, 5, and 8 mg protein/g cellulose are sufficient (data now shown) and will therefore be used in the final assay setup.

The final model is thus:

$$\log(Y) = \beta_0 + \beta_1 \times \log(\text{Rel0h}_{2\,\text{mg}}) + \beta_2 \times \log(\text{Rel0h}_{5\,\text{mg}}) + \beta_3 \times \log(\text{Rel0h}_{8\,\text{mg}}) \quad (2.3)$$

This model is fitted using multiple linear regression (MLR) as the three remaining Xs are less strongly correlated.

3.2.3. Final proof of concept

The natural logarithm to the actual results ($\log(Y)$ Actual) are plotted against the predicted results derived by using Eq. (2.3) ($\log(Y)$ Predicted) in Fig. 2.7. In Fig. 2.7A, all samples are shown. It is clear that the high values are badly predicted (the dots should be lying close to the straight line). As this high region (corresponding to very weak samples) is of less interest, it was decided to leave out samples with Y value higher than 9.0. This corresponds to 2.2 in the plot as $\log(9.0) = 2.2$. A model of the remaining samples is shown in Fig. 2.7B. This model has a good agreement between actual and predicted values with a maximal difference of 0.36 in original values. The range in original values is from 3.5 to 9.0 mg protein/g cellulose. The proof of concept is thus achieved.

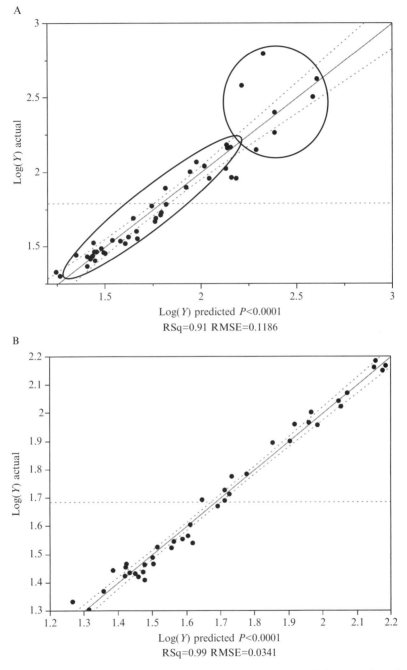

Figure 2.7 The actual values of the results (Y) derived from 72-h data are plotted on the ordinate axis. The predicted results (Y) derived from 24-h data using Eq. (2.3) are plotted on the abscissa. (A) Original model based on all 42 samples. (B) Reduced model based on samples with Y determined to 3.5–9.9 mg protein/g cellulose.

3.2.4. Validation of the model

A study with seven new enzyme samples not previously tested was used to validate the model. The study was made using the final plate setup with two plates and each sample weighed out twice on each plate. The study also included two technicians.

The results showed good agreement between the actual Y (protein load needed for 80% conversion) determined at 72 h and the predicted Y calculated from data collected after 24 h of incubation.

The CV of the result (Y) was estimated to be 5.8%. This intermediate precision includes variation from days, plates, position on the plate, and technicians and represents the CV for *a single determination*. The final result for an unknown enzyme sample is the average of four single determinations (each sample being weighed out twice on each of two plates) and the CV of this average is estimated to be 4.9%.

4. FINAL ASSAY PROCEDURE

4.1. Substrate preparation

The substrate is dilute sulfuric acid pretreated corn stover provided by the National Renewable Energy Laboratory (NREL, Golden, CO, USA; Schell *et al.*, 2003) produced at conditions of 190 °C, 1 min residence time, 0.05 g acid/g dry biomass, and at a 30% total solid concentration in the NREL vertical pretreatment reactor system. It is further modified by 6 h of wet grinding, sieved through a 425 μm mesh, and subsequently washed thoroughly in distilled water. The substrate is referred to as WGS-PCS (i.e., washed, ground, sieved, pretreated corn stover). To the substrate is added sodium acetate (NaOAc) and manganese sulfate ($MnSO_4$) to a final concentration of 60 and 1.2 mM, respectively with adjustment to pH 5.0. The FB28 (Sigma-Aldrich F3543, also called Calcofluor White M2R by other suppliers) is added to a concentration of 180 μM. Insoluble dry matter (TIS) from WGS-PCS is 3.7% in the substrate stock. The cellulose content of the WGS-PCS in this study is 52.9% of the insoluble solids (IS), as determined by NREL (method of Sluiter *et al.*, 2010). Other components of the IS include xylan 2.43%, arabinan 0.17%, galactan 0.03%, lignin 31.8%, acetyl 0.70%, and ash 8.00%. The substrate concentration in the final assay is 3% total solid.

4.1.1. Substrate loading

1. The substrate is stored refrigerated in beakers.
2. Pour an aliquot of the substrate into a beaker and stir intensively on a magnetic stirrer.

3. While the substrate is stirred, aliquots are withdrawn and 250 µl/well is loaded into the polypropylene flat-bottomed microtiter plates (Costar 96-well microplate cf. Costar 734-1551). Use, for example, Eppendorf multipette® Xstream no. #613-0618 with combi-tip 25 ml for loading the substrate into the Costar plates (MT Plates). Be aware that the substrate settles quickly.
4. Prepare two microtiter plates per sample and mark them with date and plate ID.
5. The plates are now ready to load samples.
6. Plates may be loaded with substrate, heat sealed and stored refrigerated for later use.

4.2. Protein determination, sample dilution, and loading of microtiter plates

The protein concentration in enzyme samples was determined by the micro BCA assay (Thermo Scientific) using the manufacturer's method relative to a BSA standard.

1. Make sure that the enzyme protein concentration is determined for the samples and control sample. Also ensure that the cellulose concentration of the substrate is determined, for example, as described in Sluiter *et al.* (2010). The dilution of the samples is done to three target protein concentrations calculated relative to the cellulose concentration of the substrate (mg enzyme/g cellulose).
2. Weigh out each sample twice and dissolve in Milli Q water (resistivity ≤ 18.2 MΩ cm at 25 °C) in 25-ml volumetric flasks. Dilute each weighing to the three different target concentrations: 2, 5, and 8 mg/g cellulose.
3. It is suggested to prepare the sample dilutions in deep well plates following the plate layout in Fig. 2.5.
4. Transfer 50 µl of the diluted samples from this dilution plate to each of two identical substrate-preloaded plates by use of a multichannel pipette, for example, a Biohit 8-channel e-line pipette.

4.2.1. Control sample

1. Weigh out the control sample (Cellic® CTec2, Novozymes) with an accuracy of ±0.05% and transfer to a 25-ml volumetric flask.
2. Fill up with Milli Q water and stir the solution for between 15 and 30 min. (The same diluted sample is used for two microtiter plates.)
3. Dilute further to the following target concentrations of enzyme protein relative to the concentration of cellulose in the substrate: 2, 5, 6, 8, 10, 20, 50 mg protein/g cellulose.

4. Follow the plate layout shown in Fig. 2.5.
5. Milli Q water is used as blank (i.e., 0 mg protein/g cellulose).
6. Place 50 µl of each dilution in the wells of each of two identical substrate-preloaded microtiter plates.

4.2.2. Assay samples

1. Weigh out each sample twice with an accuracy of ±0.05% and transfer to a 25-ml volumetric flask.
2. Fill up with Milli Q water and stir the solution for between 15 and 30 min. (The same diluted sample is used for two microtiter plates.)
3. Dilute further to the following target concentrations of enzyme protein relative to the concentration of cellulose in the substrate: 2, 5, and 8 mg/g cellulose.
4. Place 50 µl of each dilution in the wells of each of two identical substrate-preloaded microtiter plates.

4.2.3. Plate sealing and mixing

1. Heat seal the plates with, for example, Thermo Scientific cat# AB0745 (Easy Peel for manual Heat Sealer model 50V). Ensure that the sealing time, for example, 2.0 s at 168 °C is sufficient to ensure tight closure of the plate in order to avoid evaporation. This can be done by sealing an empty plate and removing the seal, which shall then show even and complete attachment marks around each well. Furthermore, plates can be weighed before and after incubation to check that no evaporation occurs.
2. Mix the sample with substrate manually by upside down thorough mixing or by use of a table shaker at 1200 rpm for 10 s. Bubbles in the wells are visible from the bottom of the inverted plate when the mixing is sufficient.

4.3. Incubation and fluorescence reading of FCD assay plates

The assay measures the decay of cellulose hydrolysis in a substrate mixed with the FB28 fluorescence enhancer. The cellulose hydrolysis is monitored as a decrease of fluorescence (excitation/emission: 360 nm/460 nm). It is important to read the fluorescence through the bottom of the plate. This way the plate can remain sealed; a top read was shown not to be useful (Malten and McFarland, 2011).

4.3.1. Incubation, mixing, and reading

1. Read the fluorescence of the microtiter plates from the bottom, $T=0$ h.

2. After 24 h of incubation, take out the plates and let them stand 20 min at room temperature before reading.
3. Mix and read the fluorescence of the microtiter plates from the bottom, $T = 24$ h.

4.4. Calculation of enzyme activity

In the assay, the incubation is stopped after 24 h, fluorescent signal measured, and the results are modeled to predict the corresponding results from 72-h incubation.

4.4.1. Calculation of results

1. Prepare a spread sheet for the calculations.
2. The dilution of each control sample and unknown samples is determined according to their estimated content of protein.
3. All samples are diluted to an enzyme protein target of 2, 5, and 8 mg protein/g cellulose in the substrate, respectively.
4. For each enzyme load, the normalized relative response in AU is calculated: Normalized response = $(AU_{0\,h} - AU_{24\,h})/AU_{0\,h}$.
5. By use of a model (Eq. 2.3) that correlates the AU at $T = 0$, 24, and 72 h, the results are calculated using the data read after 24 h of incubation. The normalized result for the three enzyme loads of each sample is entered into Eq. (2.3). And the final output of the assay is expressed as the enzyme load needed to reach after 72 h an 80% cellulose conversion.

4.4.2. Quality control and approval of results

Each sample is weighed out twice and analyzed in three dilutions on the plates. Two identical plates are prepared in every analytical run. The three dilutions give rise to one result. The two times two results of each sample are tested statistically for outliers, the average is calculated and presented as the final result of the given sample. The result is given as the enzyme load (in mg) needed to obtain an 80% conversion of cellulose in the substrate.

4.4.3. Approval criteria

1. *Weight*: Both control enzyme and unknown enzyme samples should be weighed out with an accuracy of $\pm 0.05\%$.
2. *Sample*: The lower r^2 limit for the three sample dilutions is 0.98.
3. *Limits of results*: The lower and upper limits of results are 3.8 and 9.0 mg/g cellulose, respectively, for the enzyme mixtures tested.

ACKNOWLEDGMENT

National Renewable Energy Laboratory (NREL) is a sponsor of this work. The NREL (1617 Cole Blvd., Golden, CO 80401) is a national laboratory of the U.S. Department of Energy managed by the Midwest Research Institute for the U.S. Department of Energy under Contract Number DE-AC36-08GO28308.

REFERENCES

Malten, M., and McFarland, K. C. (2011). A method of analyzing cellulose decay in cellulosic material hydrolysis. WIPO Patent application WO2011/008785 (A2).

Malten, M., Wischmann, B., Jensen, H. M., Kramer, R., and McFarland, K. C. (2012). Fluorescent cellulose decay assay for rapid measurement of enzymatic pretreated corn stover hydrolysis. (In preparation).

Ruchel, R., Behe, M., Torp, B., and Laatsch, H. (2001). Usefulness of optical brighteners in medical mycology. *Rev. Iberoam. Micol.* **18,** 147–149.

Schell, D., Farmer, J., Newman, M., and McMillan, J. (2003). Dilute-sulfuric acid pretreatment of corn stover in pilot-scale reactor. *Appl. Biochem. Biotechnol.* **105–108,** 69–85.

Sluiter, J., Ruiz, R., Scarlata, C., Sluiter, A., and Templeton, D. (2010). Compositional analysis of lignocellulosic feedstocks. 1. Review and description of methods. *J. Agric. Food Chem.* **58**(16), 9043–9053.

Wold, S., Sjöström, M., and Eriksson, L. (2001). PLS-regression: A basic tool of chemometrics. *Chemom. Intell. Lab. Sys.* **58,** 109–130.

CHAPTER THREE

The Analysis of Saccharification in Biomass Using an Automated High-Throughput Method

Caragh Whitehead, Leonardo D. Gomez, *and* Simon J. McQueen-Mason

Contents

1. Introduction	38
2. Automatic Saccharification Platform	39
2.1. Grinding and weighing robot	39
2.2. Liquid handling robot	42
3. Enzyme Characterization	44
3.1. Analysis of the time course in enzyme activity	44
3.2. Determination of enzyme efficiency	45
3.3. Monosaccharide analysis	46
3.4. Determination of enzyme filter paper units	48
4. Concluding Remarks	49
Acknowledgments	49
References	49

Abstract

The recalcitrance of the cell wall to enzymatic hydrolysis represents one of the greatest challenges for using biomass to replace the petroleum as a feedstock for fuels and chemicals. Cell walls are complex in architecture and composition, posing a biochemical challenge for the development of efficient enzymes to release the sugars from the polysaccharide components. The complex composition of the polymers that constitute the cell wall requires a mixture of enzymes to hydrolyze the different glycosidic bonds present in biomass. The improvement of the properties of biomass, in turn, requires the screening of large populations of plants in order to identify markers associated with saccharification potential or pinpoint the genes that regulate recalcitrance. The improvement of both, enzymes and biomass together, requires the capacity to deal with large numbers of variables in a combinatorial approach. We have developed a high-throughput system that allows the determination of cellulolytic activity in a 96-well plate format by automatically handling biomass materials, carrying out hydrolytic reactions, and determining the release of reducing sugars. This platform consists of a purpose-made robot that grinds, formats, and dispenses precise amounts of solids into 96-well plates, and a liquid-handling station specifically designed to carry out pretreatments, hydrolysis, and the determination of released reducing sugar equivalents using a colorimetric assay. These modules can be used individually or in combination according to the function needed. Here we show some examples of the capabilities of the platforms in terms of enzyme and biomass evaluation, as well as combining the robot with off-line analytical tools.

1. INTRODUCTION

In recent years, plant biomass has been considered as a relatively cheap, abundant, and renewable source to replace our reliance on fossil fuels for liquid transportation fuels and other chemicals. Approximately 75% of lignocellulosic biomass contains energy-rich polysaccharides that, when broken down to its sugars by enzymatic hydrolysis, can be fermented to produce useful commodities such as biofuels, bioplastics, chemicals, as well as food and feed ingredients (Gomez et al., 2008).

Possibly the toughest obstacle to overcome in using biomass as an industrial feedstock for biofuel production is the natural properties of plant cell walls that make them resistant to digestion. The main factor impeding the digestion of lignocellulose is its structure, as the cellulose microfibrils are embedded in a matrix of hemicelluloses such as arabinoxylans and xylans. These polysaccharides are also associated with lignin, a phenolic polymer that effectively seals the polysaccharides into a resistant macromolecular structure that is recalcitrant to digestion (Gomez et al., 2010). Other factors that have been shown to contribute to indigestibility are the degree of cellulose crystallinity and the heterogeneous nature of cell wall composition between cell type and tissues, as well as different plant lineages (King et al., 2011; Santoro et al., 2010).

During industrial saccharification, fermentable sugars are released in a two-step process from lignocellulosic biomass. The first step involves using physical and/or chemical pretreatments to make the polysaccharides more accessible for enzymatic hydrolysis. This is followed by the second step wherein the exposed cellulose is treated with commercial cellulolytic enzyme cocktails that are usually produced from filamentous fungi such as *Trichoderma reesei*, which then degrade the cellulose and hemicelluloses into simple sugars (Navarro et al., 2010).

The hydrolysis of cellulose in these cocktails is generally effected by a consortium of endoglucanases, cellobiohydrolases, and β-glucosidases (Gilbert, 2010). However, more recently, a new class of cellulose oxidases, belonging to family GH61, have been shown to play a key role in this process by attacking crystalline areas of the cellulose (Quinlan *et al.*, 2011). Various industries such as paper recycling and cotton processing also use cellulases, which makes them, by dollar volume, the third largest industrially produced enzyme (Wilson, 2009).

The saccharification of plant biomass into fermentable sugars represents the biggest barrier to developing cost-competitive cellulosic-derived biofuels. These processing costs might be effectively reduced by producing either more effective enzymes or more digestible plant biomass. In order to identify plant genes that have an impact on biomass digestibility, a robust and reliable system for the high-throughput analysis of plant materials is needed. Various high-throughput methods have previously been used to screen for changes in plant biochemistry and phenotype, leading to the identification of gene and enzyme function within a number of pathways (Colbert *et al.*, 2001; Sessions *et al.*, 2002). However, current methods used for the analysis of cell wall digestion are too labor intensive and time consuming to be used for screening large populations of plants. A high-throughput method that is automated, sensitive, and reliable could be used to identify differences in digestibility and identify genetic loci that could play a role in improving the quality of plant biomass as well as revealing combinations of pretreatments and enzymes that lead to improved saccharification (Gomez *et al.*, 2010; Santoro *et al.*, 2010).

Much work is being carried out to improve the enzymes that are used in biofuel production. The areas being studied are very varied and range from matching specific enzyme cocktails to particular plant materials (King *et al.*, 2011), to the creation of enzyme cocktails that have optimal efficiency depending on the pretreatment used (Gao *et al.*, 2011) as well as the identification of novel enzymes with higher activity from organisms which have not previously been studied (Wilson, 2009). Here we describe a semiautomated high-throughput platform, which can perform saccharification analyses in a 96-well system, to analyze the digestibility of different plant materials or the effectiveness of different enzymes. This high-throughput platform was originally described for the analysis of saccharification potential in differing plant materials (Gomez *et al.*, 2010). Here we demonstrate that this high-throughput platform can also be used to assess the relative efficiency of different enzymes to digest different plant biomass.

2. Automatic Saccharification Platform

2.1. Grinding and weighing robot

A high-throughput system to improve the performance of enzymes as well as the quality of biomass requires the handling of small amounts of materials in an automated manner. Designing this high-throughput system involves the miniaturization of the desired process. The scale that we required for a 96-well plate-based protocol meant that we needed to work with milligram amounts of biomass. This proved to be a challenge as lignocellulosic plant material is a heterogeneous and insoluble material (Gomez et al., 2010). To overcome this, we commissioned a grinding and weighing platform (Labman Automation, Stokesley, North Yorkshire, UK) to format the material into 96-well plates that allowed us to work in a high-throughput manner (Fig. 3.1A). This platform dispenses biomass from 2 ml vials that have been filled with plant material into the wells in precise amounts. The materials are precut to fit inside the vials and three ball bearings are added. Each individual vial is collected by the robotic arm from a rack and moved to the grinding station (Fig. 3.1B) where it is shaken at $\sim$5000 rpm for a specified length of time depending on the type of material present in the tube. It needs to be remembered that the grinding itself can be seen as a form of pretreatment if the particle size of the material is too small. However, it has been observed the material that has been milled to a mesh particle size of 20–80 µm does not show an affected digestibility (Decker et al., 2009). Therefore, we kept the particle size above this threshold to prevent it from possibly increasing the saccharification of the samples analyzed.

After grinding, the vial is moved to the declogging station where it is inverted four times to loosen any compacted material. In the next step, the vial is moved to the piercing station where a small hole is pierced in the bottom of the vial by a needle. The size of the hole in the vial can be varied by using a different size needle. At the tube holder, the vial is moved from the gripper to the vibration tool on the robotic arm. The robotic arm moves over the designated well of a 96-deep well plate that is sitting on top of a balance. When it is in position, it dispenses the material by vibrating at variable speeds until it reaches the predetermined weight of 4 mg. The final weight, which has an accuracy of 0.1 mg, is recorded. The plates were formatted so that they contained 16 repetitions of each sample (Fig. 3.1C); however, the plate layout as well as the amount of sample dispensed can be adjusted to the needs of the experiment. The last two columns of the plate are always left empty as they are used for the addition of glucose standards during the saccharification assay. When the plate has been completed, it is sealed with a silicon mat to prevent contamination during storage, as well as evaporation during the saccharification assay.

In this work, we used *Miscanthus* and *Arabidopsis* biomass to test our platform. *Miscanthus sinensis* was grown in Wageningen under field conditions and harvested at the end of the growing season, after senescence. The plants were ground using a hammer mill and the particle size selected using a 1-mm sieve to retain large pieces of stems. *Arabidopsis thaliana* ecotype Columbia, was grown using an 8-h light regime for 6 weeks, and subsequently moved to 16 h light. After senescence, the main bolts were harvested and ground using a ball mill to a particle size ranging between 0.5 and 1 mm.

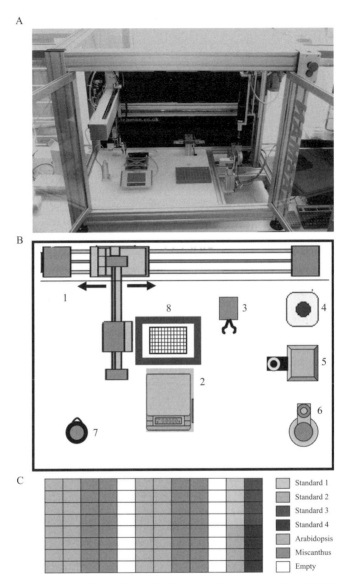

Figure 3.1 Grinding and weighing platform for biomass samples. (A) View of the robot. (B) Plan of the biomass dispensing robot showing the position of the robotic arm (1); balance (2); tube holder (3); grinding station (4); declogging station (5); piercing station (6); arm cleaning position (7); and 96-well plate (8). (C) Example of plate layout showing the standards in column 11 and 12 and different material formatted across the plates. (For the color version of this figure, the reader is referred to the Web version of this chapter.)

2.2. Liquid handling robot

The deep well plate that contains the formatted samples is placed on the liquid handling robot (Tecan Evo 200; Tecan Group Ltd., Männedorf, Switzerland; Fig. 3.2), where it then undergoes the saccharification assay (Fig. 3.3). First a mild pretreatment, 0.5 N of NaOH was added to the wells and the plates were heated on a temperature-controlled block (Torrey Pines Scientific, Carlsbad, CA, USA) at 90 °C for 30 min and cooled to 14 °C for 5 min. An alkaline pretreatment was chosen based on previous work that we had done as well as work published by Pedersen et al. (2011), which indicated that the amount of glucose and xylose released enzymatically is generally higher after an alkaline pretreatment compared with acid especially when performed at a low temperature (< 140 °C). The pretreatment conditions can be adjusted according to the type of plant material being used; however, it is limited to maximum temperature of 100 °C.

After the pretreatment, the samples were washed five times with 500 μl of 25 mM sodium acetate buffer, pH 4.5, allowing time between each wash for

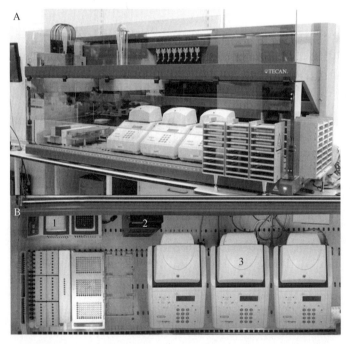

Figure 3.2 Liquid handling station for the automated saccharification assay. (A) General view of the liquid handling station. (B) Work table showing the pretreatment heating block (1), incubation oven for hydrolysis (2), and thermocyclers (3). (For the color version of this figure, the reader is referred to the Web version of this chapter.)

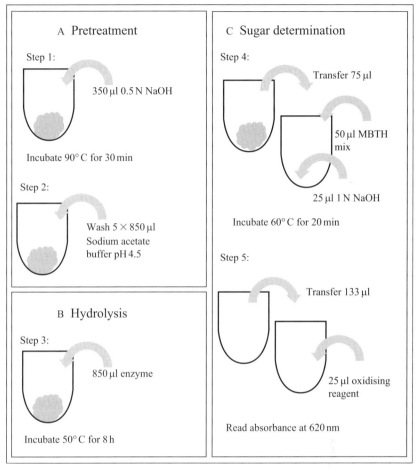

Figure 3.3 Diagram of the saccharification assay showing the volumes and incubation conditions during pretreatment, hydrolysis, and determination of reducing sugars. (For the color version of this figure, the reader is referred to the Web version of this chapter.)

the solids in the sample to settle. This was done to remove the pretreatment and to get the sample ready for enzymatic hydrolysis by bringing it to the correct pH. Next, 850 μl of enzyme was added to each well and the plate was moved into the shaking 50 °C incubator (Tecan Group Ltd.) for 8 h. The standard incubation time can be adjusted depending on the experiment.

The volumes of the reaction were equalized by adding 25 mM sodium acetate buffer, pH 4.5, to a final volume of 850 μl. Aliquots (75 μl) of the hydrolysis reaction were transferred to three independent 96-well semiskirted plates, and glucose standards of 50, 100, and 150 nmol were added to the empty columns. The modified 3-methyl-2-benzothiazolinone hydrazone

(MBTH) method was used for sugar determination (Anthon and Barrett, 2002; Gomez et al., 2010); 50 μl of a solution containing 0.43 mg/ml MBTH and 0.14 mg/ml dithiothreitol (DTT) as well as 25 μl of 1 N NaOH were added to the plate. After incubation at 60 °C for 20 min in three thermocyclers with motorized lids (Bio-Rad Laboratories Ltd., Hemel Hempstead, UK), 133 μl of the solution is transferred to an optical plate and 89 μl oxidizing reagent (0.5% $FeNH_4(SO_4)_2$, 0.5% sulfamic acid, and 0.25 N HCl) was added. This incubation is preformed in thermocyclers to provide constant heating across all wells. The absorbance of the plates was read at 620 nm using a Sunrise plate reader (Tecan Group Ltd.).

3. Enzyme Characterization

The high-throughput platform has been designed to be sufficiently flexible to serve different functions in biomass and enzyme characterization. We are able to screen a number of large populations of different biomass to identify the difference in saccharification potential within different genotypes. Similarly, we can perform characterization of enzymes by automatically determining parameters such as affinity, time courses, and activity.

3.1. Analysis of the time course in enzyme activity

Two different enzyme mixtures from Novozymes (Bagsvaerd, Denmark) were used to test the automated method for saccharification. These enzymes were specifically developed for hydrolyzing lignocellulosic materials. The first mixture was an enzyme cocktail of a 4:1 ratio of Celluclast and Novozyme 188 (CN188) and the second was Cellic CTec2 (CEL). The commercial preparations were filtered using a Hi-Trap desalting column (GE Healthcare, Little Chalfont, Buckinghamshire, UK) using 25 mM sodium acetate buffer, pH 4.5, to equilibrate the column. The proteins in the purified enzymes were quantified using Bradford reagent. The absorbance of the reaction and the BSA standards were read at 595 nm.

We used the automated platform to establish the saccharification of the CN188 enzyme mixture over 18 h (Fig. 3.4). The same saccharification analysis conditions were used during the time course experiment, namely a mild pretreatment of 0.5 N NaOH at 90 °C for 30 min followed by the addition of the enzyme. However, this time the plate was incubated in the shaking oven at 50 °C for 2, 6, 12, and 18 h with aliquots taken for sugar determination at each of these time points. We tested the enzyme's activity on *Arabidopsis* and *Miscanthus* as these represent a monocot and grass species, respectively. After 2 h of incubation at 50 °C, there was no difference between the amount of reducing sugars released from

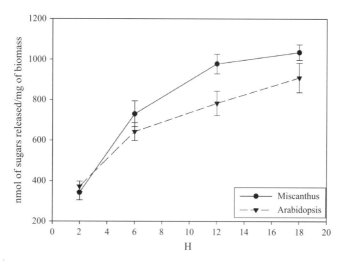

Figure 3.4 Automated determination of the time course for the saccharification of *Arabidopsis* and *Miscanthus*. Four-milligram stem samples of *Arabidopsis thaliana* Col 0 or *Miscanthus sinensis* were digested with a mixture of Celluclast and Novozyme 188 for 2, 6, 12, and 18 h. The resulting hydrolyzates were assayed for reducing sugars. Bars represent values ± SD of three experiments.

Arabidopsis and *Miscanthus* (Fig. 3.5). However, at 6 h, a difference can be observed, with *Miscanthus* releasing more sugar than *Arabidopsis*. This difference increases further at 12h of incubation. For *Miscanthus*, the saccharification is very efficient during the first 12 h of enzyme hydrolysis and then it slows down until it almost plateaus between 12 and 18 h. *Arabidopsis* shows a lower sugar release during the first 12 h and, then, the saccharification proceeds at lower speed. These types of experiments are used to determine the initial conditions to assay the saccharification potential in combinations of biomasses and enzymes.

3.2. Determination of enzyme efficiency

The standard saccharification analysis method (Fig. 3.3) was used on the liquid handling robot to compare the saccharification efficiency of the CN188 enzyme mixture with the Cellic CTec2 (CEL) enzyme on two different types of plant material, *Arabidopsis* and *Miscanthus*. The results after 8 h of enzyme hydrolysis at 50 °C show that the amount of reducing sugar released by CEL from *Arabidopsis* is 16% higher than that by CN188. This difference, however, is greater in *Miscanthus*, where CEL releases almost 50% more reducing sugar than CN188 during the same incubation time. This would indicate that the CEL enzyme is more effective in these saccharification conditions for *Miscanthus* biomass.

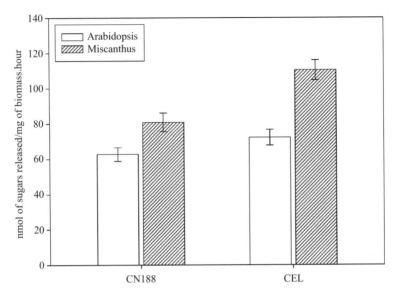

Figure 3.5 Comparison of the saccharification of *Arabidopsis* and *Miscanthus* performed with two different enzyme mixtures. *Arabidopsis* and *Miscanthus* samples were digested with Celluclast and Novozyme 188 (CN188) and Cellic2 (CEL) enzyme preparations. The results represent the amount of reducing sugars released after 8 h of saccharification at 50 °C.

These determinations can be carried out using up to 10 different enzymes and 20 biomass samples in one run.

3.3. Monosaccharide analysis

The high-throughput system can be linked to detailed analysis in a lower throughput approach such as the analysis of the monosaccharide released by enzymatic hydrolysis determined by high-performance anion exchange chromatography (HPAEC). Following saccharification in the robotic platform, the hydrolysates were subjected to chemical hydrolysis using 2 M trifluoroacetic acid. The resulting monosaccharides were washed twice with isopropyl alcohol, the isopropanol was evaporated, and resuspended in water. The HPAEC was performed using a Dionex ICS-3000, with a Carbopac P20 column (Dionex, Camberley, Surrey, UK). The chromatography was performed following the procedure of Jones et al. (2003).

Figure 3.6A shows that the monosaccharide profile released by CEL is richer in xylose than the hydrolysis using CN188 in *Arabidopsis* saccharification. When these enzyme preparations are used in *Miscanthus*, a higher release of glucose is observed using the CEL enzyme, compared with the CN188 enzyme (Fig. 3.6B), indicating a possible increase in glucose release from cellulose. The monosaccharide profile present in the hydrolyzate of

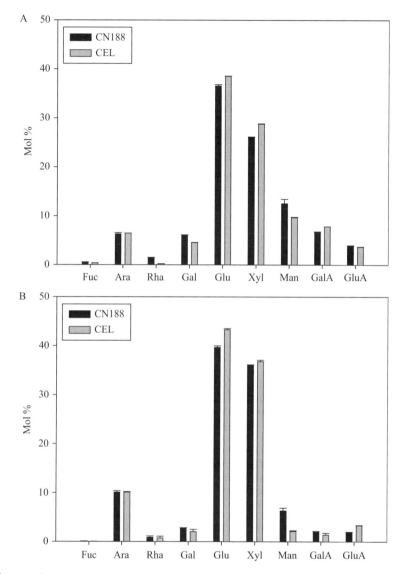

Figure 3.6 Monosaccharides released from *Arabidopsis* and *Miscanthus* by two enzyme preparations. Stem samples from *Arabidopsis* (A) and *Miscanthus* (B) were put though the saccharification system using two enzyme mixtures (CN188: Celluclast and Novozyme 188, 4:1; CEL: Cellic2). The resulting hydrolyzate was analyzed using HPAEC.

Arabidopsis suggests the degradation of pectic polysaccharides, as well as the hemicellulosic and cellulosic fractions. The hydrolyzate released by the saccharification of *Miscanthus* is composed mainly of glucose, xylose, and arabinose, suggesting the efficient degradation of arabinoxylans.

3.4. Determination of enzyme filter paper units

The automated platform offers the possibility of multiple determinations of activity parameters in enzymes and enzyme mixtures in an automated format. We routinely perform these determinations in order to standardize the enzyme loadings in saccharification assays.

A single filter paper disk (Whatman No. 1) was placed into each well of a 96-deep well plate leaving the last two columns free for standards and sealed with a silicone mat. The liquid handling robot was used for the determination of filter paper units (FPUs) of the two enzyme mixtures, CN188 and CEL. In this case, the paper discs were pretreated with 350 µl 1% H_2SO_4 per well and various amounts (0, 50, 100, 150, and 200 µl) of the two enzymes were added to the plates and left to hydrolyze for 2 h at 50 °C in a shaking incubator.

The results obtained (Fig. 3.7) indicate a higher slope for the activity of CEL using increasing enzyme amounts under excess substrate. The FPU values obtained in this experiment were 500.7 and 403.3 U/ml for CEL and CN188, respectively.

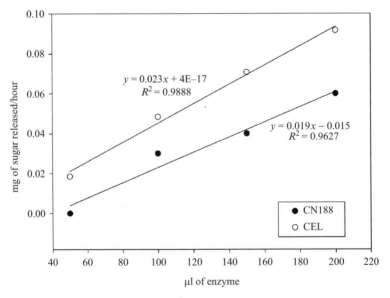

Figure 3.7 Automated determination of cellulose activity. Sugars released from filter paper discs using different loading of two enzyme preparations (CN188: Celluclast and Novozyme 188, 4:1; CEL: Cellic2).

4. Concluding Remarks

The need for enzyme discovery to meet the challenge of obtaining renewable products requires a high-throughput approach to enzyme characterization. Biofuels and biorenewables require the deconstruction of a wide range of biomass and the utilization of complex enzyme mixtures. This situation implies that the number of combinations required for optimizing the enzyme mixture with specific substrates is high. The high-throughput platform presented here is an example of a dedicated tool designed to assist in enzyme discovery programs, as well as the improvement of biomass sources. By using these platforms, we can reliably determine the activity of enzyme combinations in biomass deconstruction, while also assessing the saccharification potential of different sources of biomass. Here, we show that we can automatically determine the efficiency of the saccharification of several biomasses with different enzyme preparations, determine the activity of enzyme preparations, and integrate the platform with detailed downstream analyses of the products released from different biomasses.

ACKNOWLEDGMENTS

We are very grateful to Oene Dolstra and Luisa Trinidade for the generous gift of *Miscanthus* biomass. The enzyme mixtures were a kind gift from Anders Viksø-Nielson at Novozymes, Copenhagen, Denmark. This work is supported by FP7 programmes RENEWALL and SUNLIBB, and by BBSRC projects BB/G016178 and BB/G016194.

REFERENCES

Anthon, G. E., and Barrett, D. M. (2002). Determination of reducing sugars with 3-methyl-2-benzothiazolinonehydrazone. *Anal. Biochem.* **305,** 287–289.

Colbert, T., Till, B. J., Tompa, R., Reynolds, S., Steine, M. N., Yeung, A. T., McCallum, C. M., Comai, L., and Henikoff, S. (2001). High-throughput screening for induced point mutations. *Plant Physiol.* **126,** 480–484.

Decker, S., Brunecky, R., Tucker, M., Himmel, M., and Selig, M. (2009). High-throughput screening techniques for biomass conversion. *Bioenergy Res.* **2,** 179–192.

Gao, D., Upppugundla, N., Chundawat, S. P. S., Yu, X., Hermanson, S., Gowda, K., Brumm, P., Mead, D., Balan, V., and Dale, D. E. (2011). Hemicellulases and auxiliary enzymes for improved conversion of lignocellulosic biomass to monosaccharides. *Biotechnol. Biofuels* **4,** 5.

Gilbert, H. J. (2010). The biochemistry and structural biology of plant cell wall deconstruction. *Plant Physiol.* **153,** 444–455.

Gomez, L. D., Steele-King, C. G., and McQueen-Mason, S. J. (2008). Sustainable liquid biofuels from biomass: The writing's on the walls. *New Phytol.* **178,** 473–485.

Gomez, L. D., Whitehead, C., Barakate, A., Halpin, C., and McQueen-Mason, S. J. (2010). Automated saccharification assay for determination of digestibility in plant material. *Biotechnol. Biofuels* **3,** 23.

Jones, L., Milne, J. L., Ashford, D., and McQueen-Mason, S. J. (2003). Cell wall arabinan is essential for guard cell function. *Proc. Natl. Acad. Sci. USA* **100,** 11783–11788.

King, B. C., Waxman, K. D., Nenni, N. V., Walker, L. P., Bergstrom, G. C., and Gibson, D. M. (2011). Arsenal of plant cell wall degrading enzymes reflects host preference among plant pathogenic fungi. *Biotechnol. Biofuels* **4,** 4.

Navarro, D., Couturier, M., da Silva, G. G. D., Berrin, J.-G., Rouau, X., Asther, M., and Bignon, C. (2010). Automated assay for screening the enzymatic release of reducing sugars from micronized biomass. *Microb. Cell Fact.* **9,** 58.

Pedersen, M., Johansen, K. S., and Meyer, A. S. (2011). Low temperature lignocelluloses pretreatment: Effects and interactions of pretreatment pH are critical for maximizing enzymatic monosaccharide yields from wheat straw. *Biotechnol. Biofuels* **4,** 11.

Quinlan, R. J., Sweeney, M. D., Lo Leggio, L., Otten, H., Poulsen, J. C., Johansen, K. S., Krogh, K. B., Jorgensen, C. I., Tovborg, M., Anthonsen, A., Tryfona, T., Walter, C. P., *et al*. (2011). Insights into the oxidative degradation of cellulose by a copper metalloenzyme that exploits biomass components. *Proc. Natl. Acad. Sci. USA* **108,** 15079–15084.

Santoro, N., Cantu, S. L., Tornqvist, C.-E., Falbel, T. G., Bolivar, J. L., Patterson, S. E., Pauly, M., and Walton, J. D. (2010). A high-throughput platform for screening milligram quantities of plant biomass for lignocellulose digestion. *Bioenergy Res.* **3,** 93–102.

Sessions, A., Burke, E., Presting, G., Aux, G., McElver, J., Patton, D., Dietrich, B., Ho, P., Bacwaden, J., Ko, C., Clarke, J. D., Cotton, D., *et al*. (2002). A high-throughput *Arabidopsis* reverse genetics system. *Plant Cell* **14,** 2985–2994.

Wilson, D. B. (2009). Cellulases and biofuels. *Curr. Opin. Biotechnol.* **20,** 295–299.

CHAPTER FOUR

Studies of Enzymatic Cleavage of Cellulose Using Polysaccharide Analysis by Carbohydrate gel Electrophoresis (PACE)

Ondrej Kosik,[1] Jennifer R. Bromley,[1] Marta Busse-Wicher,[1] Zhinong Zhang, and Paul Dupree

Contents

1. Introduction	52
2. The Principle of PACE	54
2.1. Poly-, oligo-, and monosaccharide sample preparation	55
2.2. Derivatization of poly-, oligo-, and monosaccharides with ANTS	56
2.3. PACE gel preparation, running, and interpretation	56
2.4. Cellulolytic hydrolysis of PASC	59
2.5. Substrate disappearance assay for kinetic constant determination	60
2.6. Mechanism of hydrolysis	63
3. Discussion: Other Possible Applications	64
4. Summary	65
Acknowledgments	65
References	65

Abstract

With the advent of fast genome analysis, many genes encoding novel putative cellulolytic enzymes are being identified in diverse bacterial and fungal genomes. The discovery of these genes calls for quick, robust, and reliable methods for qualitative and quantitative characterization of the enzymatic activities of the encoded proteins. Here, we describe the use of the polysaccharide analysis by carbohydrate gel electrophoresis (PACE) method, which was previously used, among other applications, to characterize various hemicellulose degrading enzymes; for structural elucidation of these carbohydrates; and for analysis

Department of Biochemistry, University of Cambridge, Cambridge, United Kingdom
[1] These authors contributed equally to the work.

of products resulting from enzymatic cleavage of cellulose. PACE relies on fluorescent labeling of mono-, oligo-, and polysaccharides at their reducing end and separation of the labeled carbohydrates by polyacrylamide gel electrophoresis. Labeling can be carried out before or after enzymatic digestion. PACE is very sensitive and allows analysis of both substrate specificities and kinetic properties of cellulolytic enzymes.

ABBREVIATIONS

AIR	alcohol insoluble residue
AMAC	2-aminoacridone
ANTS	8-aminonaphthalene-1,3,6-trisulfonic acid
DMSO	dimethyl sulfoxide
DP	degree of polymerization
GH	glycoside hydrolase
HPLC	high-performance liquid chromatography
MALDI	matrix-assisted laser desorption/ionization
MS/MS	tandem mass spectrometry
MWCO	molecular weight cut off
NMR	nuclear magnetic resonance
NP-LC	normal-phase liquid chromatography
PACE	polysaccharide analysis by carbohydrate gel electrophoresis
PASC	phosphoric acid swollen cellulose
TOF/TOF	tandem flight mass spectrometry
UV	ultraviolet (light)

1. INTRODUCTION

Cellulose, the most abundant polysaccharide of plant cell walls, consists of β-1,4-linked glucosyl units. Despite this simple biochemical composition, the cellulose chains in the microfibrils may adopt a variety of crystalline or amorphous topologies (Klemm et al., 2005). To degrade cellulose completely, in nature or in industrial processes, several cellulolytic activities acting in synergy are needed. The glucan chains of cellulose are cleaved into shorter chains by endocellulases (also referred to as endoglucanases) and exo-acting processive enzymes, defined as cellobiohydrolases, which may act from the reducing or nonreducing end of cellulose chains. The recently described GH61 enzymes may also oxidatively cleave crystalline cellulose. Finally, β-glucosidases catalyze the digestion of soluble oligosaccharides and cellobiose into glucose (Bayer et al., 1998; Cantarel et al., 2009; Quinlan et al., 2011).

Additionally, in the plant cell wall, cellulose is networked with other components of the wall such as hemicelluloses, pectins, and lignins. This structural complexity increases the resistance of cellulose to biological decomposition. It is, therefore, not surprising that organisms which utilize cellulose as a source of carbon express an extensive repertoire of enzymes that act in synergy to degrade the plant cell wall. These enzymes are of great interest to the textile, detergent, food, paper, and pulp industries (Kirk et al., 2002).

As biofuels derived from lignocellulosic biomass are expected to become a major transport fuel (Wilson, 2009), cellulose-degrading enzymes are likely to become the largest volume of industrial enzymes. The role of cellulases is to generate glucose from cellulosic biomass, which can then be fermented by microbes to generate biofuels such as ethanol and butanol. In the hunt for new enzymes, genomic sequencing of cellulolytic organisms and metagenomic enzyme prospecting is now commonplace, resulting in a growing list of newly annotated genes encoding putative glycoside hydrolases (GHs) (Cantarel et al., 2009).

The products generated by GHs acting on native polymers or oligosaccharide standards are often analyzed by a suite of techniques including chromatography, capillary electrophoresis, HPLC, MALDI-TOF/TOF mass spectrometry, NMR, and Fourier transform infrared spectroscopy, or by simple chemical reducing sugar and total sugar assays (Fagard et al., 2000; Fry, 2000; James and Jenkins, 1997; Jarvis and McCann, 2000; Monsarrat et al., 1999; Mort and Chen, 1996). These techniques have proved to be successful in assigning enzyme activity and in the determination of product structure; however, they often have limitations, such as the need for a relatively large amount of a pure compound for analysis, and they may not always provide specific information on the site of hydrolysis. As the discovery of increasing numbers of GHs continues at a rapid rate, simple, sensitive, and relatively high-throughput techniques are required for the characterization of their hydrolytic specificities.

Polysaccharide analysis by carbohydrate gel electrophoresis (PACE) is a quantitative separation technique, which achieves sensitive and reproducible analyses of the products of enzyme hydrolysis (Goubet et al., 2002; Jackson, 1990). PACE relies on the derivatization of the reducing ends of mono-, oligo-, and polysaccharides with a fluorophore, and this allows separation through the gel matrix under optimized conditions and detection through sensitive fluorescence imaging of the polyacrylamide gel. 8-aminonaphthalene-1,3,6-trisulfonic acid (ANTS) is often used as the fluorophore giving neutral poly- and oligomers a charge, and, thus, allowing them to migrate through the polyacrylamide gel in an electric field. Thus, PACE is flexible in its application in enzyme product (or substrate) analysis across a range of polysaccharides. Using combinations of GHs of known function, PACE has been successfully applied to the characterization of hemicelluloses (Brown *et al.*, 2007, 2009; Goubet *et al.*, 2009; Mortimer *et al.*, 2010) and pectins (Barton *et al.*, 2006). When coupled with other analytical techniques such as NP-LC and MALDI-TOF/TOF-MS/MS, PACE has also been used in the structural characterization of complex carbohydrates (Tryfona *et al.*, 2010).

Moreover, PACE is a useful tool in the characterization of the hydrolytic enzymes themselves and can be used to qualitatively and quantitatively assay the specificity and kinetics of novel GHs (Hogg *et al.*, 2003; Palackal *et al.*, 2004). The change in intensities and number of oligosaccharide bands over a time course reflects the increase in number of lower oligomer bands and decrease in higher oligomer bands, due to hydrolysis of glycosidic linkages. This enzyme-mediated change can be quantified, allowing comparisons between hydrolysis conditions, be it pH, temperature, buffer, or substrate (Goubet *et al.*, 2002). PACE has been relatively little used for studies on cellulose degradation. However, PACE has recently been deployed in the analysis of the degradation of cellulose by a GH61 copper-dependent oxidase from *Thermoascus aurantiacus* (Quinlan *et al.*, 2011).

PACE has undoubtedly proved to be a useful technique in the study of many aspects of polysaccharide biosynthesis and hydrolysis. This medium throughput technique requires only small quantities of material for analysis, with sensitivities of 500 fmol (Goubet *et al.*, 2002) and little requirement for specialized equipment. This makes PACE ideal for the broad analyses of novel enzyme activities including product profile analysis, and in the identification of optimal reaction conditions. This chapter describes the protocols for analysis of products of enzyme-mediated cleavage of cello-oligosaccharides and cellulosic polysaccharides.

2. THE PRINCIPLE OF PACE

PACE is an analytical technique for the separation of fluorophore-labeled oligosaccharide products of enzymatic digestions, based on their size and charge (Fig. 4.1). Mono- and oligosaccharides with a free reducing end can react with a fluorophore containing a primary amino group, when the reducing end sugar ring is open, to form a Schiff base. After reduction by a suitable reducing agent, the coupling to a fluorophore becomes irreversible. Two fluorophores have been widely used for this derivatization: ANTS and 2-aminoacridone. The former of these molecules is charged, while the latter is uncharged, allowing differentiation of charged and uncharged products of hydrolysis. For products of cellulolytic hydrolases, ANTS is the preferred fluorophore since products are uncharged, thus derivatization with a charged reporter is required to facilitate migration into the gel matrix.

Analysis of Cellulose Hydrolysis Products by PACE

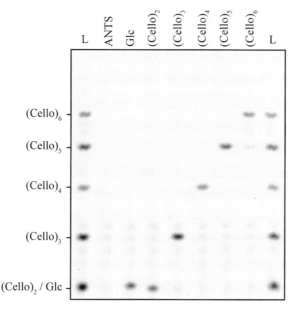

Figure 4.1 PACE gel showing separation of cello-oligosaccharides (DP 2–6 and glucose) derivatized with ANTS (giving them identical charge) by their size. Lanes from left to right: L—mixture of cello-oligosaccharides (DP 2–6) and glucose; ANTS—8-aminonaphtalene-1,3,6-trisulphonic acid only (blank); Glc—glucose; $(Cello)_2$—cellobiose; $(Cello)_3$—cellotriose; $(Cello)_4$—cellotetraose; $(Cello)_5$—cellopentaose; $(Cello)_6$—cellohexaose. Shorter oligosaccharides migrate faster through the mesh of polyacrylamide gel. Note that under standard conditions, glucose and cellobiose closely comigrate on a PACE gel making them hard to distinguish.

Migration is carried out in a direct electrical field through the mesh of a polyacrylamide gel, alongside a set of known oligosaccharide standards that allow quantitation and identification through comigration.

2.1. Poly-, oligo-, and monosaccharide sample preparation

Oligosaccharides obtained from many sources may be analyzed by PACE. Here, commercially available cello-oligosaccharides including cellohexaose (Megazyme) and phosphoric acid swollen cellulose (PASC), obtained from Avicel (Sigma-Aldrich) according to the method of Zhang et al. (2006), were used. It is not always necessary to use pure substrates, and it is also possible to use a cell wall extract, such as an alcohol insoluble residue (AIR) derived from plant material, generated by the method of Mortimer et al. (2010). It is important to note that when analyzing AIR, an alkali pretreatment step prior to hydrolysis is often required for analysis of cell wall polysaccharides to improve enzyme access. Similarly, to measure enzyme activity on cellulose, it may be necessary to loosen the structure, for example, by use of phosphoric acid.

Purified cellohexaose was solubilized to 0.5 mg ml^{-1} in water and divided into 20 μg aliquots, while PASC was prepared as a 5 mg ml^{-1} suspension in water and divided into 500 μg aliquots for analysis. PASC or purified oligosaccharides can be derivatized with ANTS before enzymatic hydrolysis to follow a single reducing end through the hydrolytic process, or can be hydrolyzed first and the newly formed reducing ends derivatized to track the products of a reaction. Specificities of cellulolytic enzymes classified into four GH families (GH3, GH7, GH45, and GH61) were analyzed on cellulose and cello-oligosaccharides. By combining these two approaches, it is possible to assay for cellulolytic activity on polysaccharides and cello-oligosaccharides, to understand substrate preference, exo- or endohydrolytic activity, and kinetic constant determination.

2.2. Derivatization of poly-, oligo-, and monosaccharides with ANTS

1. Prepare samples and standards of 1–5 pmol of cello-oligosaccharides degree of polymerization (DP) 2–6 mixture and glucose for derivatization with ANTS (Invitrogen) by addition of 10 μl dimethyl sulfoxide (DMSO) buffer (3:17:20 acetic acid:water:DMSO, v/v/v), 5 μl 0.2 M ANTS (dissolved in water: acetic acid, 17:3, v/v), and 5 μl 1 M NaCNBH$_3$ (in DMSO).
2. Following a brief vortex and brief centrifugation to collect the solution, samples are incubated at 37 °C overnight.
3. Following incubation, a brief centrifugation is required to collect the condensate. Samples are ventilated in a fume hood for 30 min to allow escape of remaining HCN gas before drying *in vacuo* at 45 °C.
4. Dried samples are resuspended in 100 μl of 3 M urea, whereas standards are resuspended in 100 μl of 6 M urea. Both can then be stored medium term at −20 °C. An overnight period at −20 °C is encouraged prior to separation by polyacrylamide gel electrophoresis, as it improves resolubilization of the saccharides.

2.3. PACE gel preparation, running, and interpretation

PACE gels, consisting of stacking and resolving parts that differ in acrylamide content, and thus mesh size, need to be poured 1 day in advance, to allow the gel to polymerize fully. During this period of polymerization, gels should be kept moist by overlaying with wet paper towels and stored in polythene bags at 4 °C. A maximum of 5 μl of ANTS-labeled samples are loaded into the wells of vertical gels and are run in an electric field at two different voltages for 1.5–2 h. The initial lower voltage is to ensure all oligosaccharides have entered the resolving gel and the second higher voltage separates oligosaccharides through the gel mesh. Gels are visualized using an ultraviolet (UV) trans-illuminator at 365 nm. Images are recorded using a camera equipped with the appropriate filters and are interpreted and quantified with suitable software.

2.3.1. Preparation of PACE gels

1. PACE gel separation is performed using Hoefer SE660 vertical gel electrophoresis equipment with 25 cm glass plates with 0.75 mm spacers. The gel cassette should be cleaned thoroughly with 96% (v/v) ethanol using lint-free tissue and handled with powder free gloves, prior to assembly of the gel casting apparatus, to ensure strongly fluorescing particulate matter is removed.
2. The resolving gel (22 cm high) consists of 19.7% (w/v) polyacrylamide containing, 0.04% (v/v) N,N,N',N'-tetramethylethylenediamine (TEMED), 0.04% (w/v) ammonium persulphate in 0.1 M Tris-borate buffer, pH 8.2. After thorough mixing of the acrylamide solution, the gel is poured between 25 cm high glass plates and overlaid with isopropanol and allowed to polymerize.
3. After 45–60 min, isopropanol is removed with blotting paper.
4. Within 20 min of polymerization of the resolving gel, a 2 cm high stacking gel consisting of 11.2% (w/v) polyacrylamide containing 0.08% (w/v) TEMED, 0.04% (w/v) ammonium persulphate in 0.1 M Tris-borate buffer (pH 8.2) is overlaid on the previously poured and polymerized resolving gel and the 28 well gel-comb should be positioned before the gel begins to polymerize. The timing of pouring the stacking gel is key to ensure that good contact is made between the resolving and stacking gels.
5. After 45–60 min, the top of the gel, including the gel-comb, is covered with damp tissue and the whole casting apparatus covered with a polythene bag to prevent the gel from drying out. The gel should be stored at 4 °C overnight to allow for complete polymerization.

2.3.2. Running and analyzing the samples on PACE gels

1. The gel-comb is carefully removed prior to electrophoresis, and the wells filled with 0.1 M Tris-borate buffer (pH 8.2).
2. One to two microliters of sample or standards is loaded into every other well using a Hamilton syringe. The first and last 3 wells of the gel should be left blank to allow for splaying of lanes at the edges of the gel.
3. Place the upper anode chamber on top of the loaded PACE gel and place the assembled apparatus into the lower, cooled cathode chamber.

Both chambers are filled with 0.1 M Tris-borate running buffer (pH 8.2). The lower chamber should be cooled to 10 °C. It is key to ensure even cooling throughout electrophoresis as uneven cooling can result in loss of resolution and artefactual oligosaccharide doublets have been reported.
4. Electrophoresis is carried out for 30 min at 200 V to allow samples to move evenly into the stacking gel before raising the voltage to 1000 V for 90 min.
5. PACE gels are visualized using a Genebox (Syngene) equipped with a UV-transilluminator with long-wave tubes (emitting at 365 nm) and a camera equipped with short-pass (500–600 nm) filter. In initial checks of migration during electrophoresis, illumination periods of the gel in the plates should be kept brief, to around 400 ms, to ensure cross-linking of oligosaccharides to the gel is avoided. After electrophoresis, the gel is removed from the glass plates for imaging. Free ANTS migrates ahead of derivatized monosaccharides, and this should be cut from the gel before the final exposures for analysis are made.
6. Images of PACE gels are taken using GeneSnap (Syngene) software and quantified using GeneTools (Syngene) software.

2.3.3. PACE gel interpretation and quantification

The interpretation of PACE gels is assisted by comparing the digestion pattern of the products to migration standards (mono- and oligosaccharides of known structure, similar to analyzed polymers). In order to make the PACE gels quantifiable, three different concentrations (e.g., 10, 20 and 50 pmol) of derivatized standards need to be included on each PACE gel. Using mono- and disaccharides is preferable for quantification as these are commercially available at the highest purity. The fluorescence intensity of the standard is independent of DP of the oligosaccharide (Goubet et al., 2002). The calibration curve produced using analytical GeneTools software and the quantity of each hydrolytic product (according to the specific band on the PACE gel) is determined from the fluorescence intensity emitted by the fluorophore and recalculated according to the molecular mass of given oligosaccharide.

In contrast to noncellulosic mono- and disaccharides (Brown et al., 2007; Goubet et al., 2009, 2002; Mortimer et al., 2010), glucose and cellobiose very closely comigrate on a PACE gel under standard conditions (Fig. 4.1) and this makes them hard to distinguish. Therefore, cellobiohydrolase and β-glucosidase activities are not easily distinguishable. It is possible that by altering acrylamide concentrations or electrophoresis buffer, conditions might be found to separate these saccharides. Longer oligosaccharides are easily separated into distinct bands and oligosaccharides up to a 20-mer can be quantitatively resolved by PACE under standard conditions.

2.4. Cellulolytic hydrolysis of PASC

Due to the crystalline nature of cellulose, PASC is prepared by solubilisation of microcrystalline cellulose (Avicel, Sigma) in concentrated phosphoric acid according to the method of Zhang *et al.* (2006) to provide a substrate that is susceptible to enzyme attack.

1. Prepared PASC is incubated with an excess of enzyme at 50 °C for up to 52 h in 500 μl reaction volume. To ensure efficient hydrolysis, appropriate conditions should be used according to the GH. For GH3, GH7, and GH45 enzymes, the reaction was carried out in 0.1 M ammonium acetate buffer, pH 5.5, according to the method of Andersen *et al.* (2008); while for GH61 digests, the reaction was performed in 10 mM ascorbate, 2.5 mM triethylammonium acetate according to the method of Quinlan *et al.* (2011). To ensure exhaustive digestion, a second volume of fresh enzyme is added following the initial 52 h incubation for a further hour. The initial addition of enzyme is intended to be in excess of that required to fully digest the enzyme accessible substrate, but this second addition is sometimes necessary to cleave any incompletely digested products.
2. Samples may also be taken at regular intervals during hydrolysis and terminated by boiling to monitor the time course of reaction.
3. Termination of hydrolysis is performed by boiling for 10 min.
4. Samples need to be taken to dryness *in vacuo* at 45 °C before reducing ends are derivatized with ANTS (Section 2.2), separated by PACE and visualized (Section 2.3).

Cello-oligosaccharides of different sizes were released and detected as products of enzymatic digestions of PASC by GH3, GH7, GH45 (Fig. 4.2), and GH61 (Fig. 4.3) enzymes. GH3 or GH7 digestion resulted in production of glucose and cellobiose, respectively, consistent with their known activities as β-glucosidase (Krogh *et al.*, 2010) and cellobiohydrolase, prepared from heterologously expressed protein according to the principle of Shoemaker *et al.* (1983). GH45 digest of PASC led to the production of several oligosaccharides, predominantly DP 1–3, consistent with its endo-glucanase activity (Davies *et al.*, 1993). Interestingly, some other products not comigrating with standard cello-oligosaccharide standards were generated by the GH7, and these may be minor side products of PASC hydrolysis. When PASC was digested with mixture of the GH3, GH7, and GH45 enzymes, substantial digestion of the polymer to small oligosaccharides occurred (Fig. 4.2). The GH61 digestion of PASC was carried out as a time course, where reactions were stopped at different time points. Increasing amounts of the oligosaccharide products were observed (Fig. 4.3), confirming the results of Quinlan *et al.* (2011). GH61 produces a mixture of products, and only some have a reducing end available for derivatization

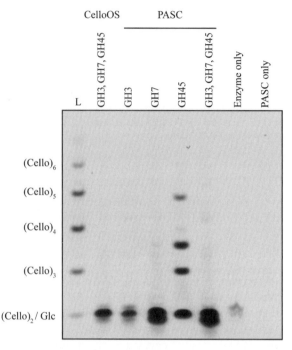

Figure 4.2 GH3, GH7, and GH45 hydrolysis of cello-oligosaccharides and PASC. Lanes from left to right: L—mixture of cello-oligosaccharides (DP 2–6); cello-oligosaccharides hydrolyzed with GH3, GH7, and GH45 enzymes; PASC hydrolyzed with GH3, GH7, GH45, or all three enzymes; enzymes incubated without substrate addition; PASC incubation with no addition of enzyme. Cellohexaose and/or PASC are digested to completeness with the mixture of all three enzymes. GH45 hydrolase produces a ladder of short oligosaccharides when applied on PASC. GH3 and GH7 digestion of PASC result in glucose and/or cellobiose.

whereas others have an oxidized reducing end. It is important to note that only the GH61 products that have non-oxidized reducing ends are visualized by this technique.

2.5. Substrate disappearance assay for kinetic constant determination

To follow the activity of an enzyme on a given material, or to follow activity at the first enzyme subsite following the reducing end, the initial substrate can be derivatized prior to hydrolysis. However, as derivatization is carried out in the presence of an excess of ANTS, the possibility of secondary product labeling must be avoided by reaction quenching and free ANTS removal prior to enzyme hydrolysis. In this example, cellohexaose ((Cello)$_6$)

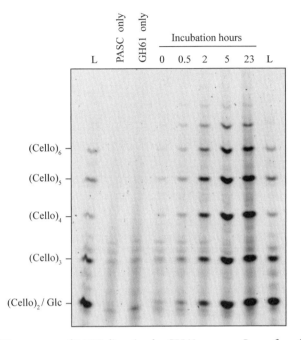

Figure 4.3 Time course of PASC digestion by GH61 enzyme. Lanes from left to right: L—mixture of cello-oligosaccharides (DP 2–6) and glucose; PASC only—PASC with no addition of enzyme; GH61 only—enzyme incubation without substrate addition; 0, 0.5, 2, 5, and 23—enzymatic reactions stopped at different time points (h) by boiling. The increasing band intensity with the longer duration of incubation indicates the increasing release of oligosaccharides from polymeric PASC substrate.

was derivatized first and hydrolysis of this substrate tracked over time using a PACE-based substrate disappearance assay. The number of reported reducing ends remains constant throughout, allowing kinetic constants to be determined for individual hydrolases (Fig. 4.4).

2.5.1. Substrate derivatization and quenching of reducing conditions

1. Twenty micrograms of dry (Cello)$_6$ is derivatized overnight with ANTS as described above (Section 2.2) and, after a brief centrifugation to collect any condensate, vented.
2. Quench the remaining NaCNBH$_3$ by addition of an equimolar amount of 1 M HCl. Return the sample to neutral pH by addition of NaOH.
3. Dialyze in 500–1000 Da molecular weight cut off (MWCO) dialysis tubing against DI H$_2$O for 2 days, changing the water at least twice per day. *N.B.* (Cello)$_6$, labeled with ANTS—$M_r = 1371.22$ Da, (Cello)$_5$

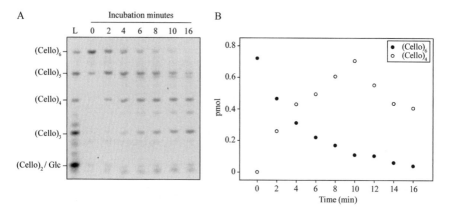

Figure 4.4 GH3 enzyme substrate disappearance assay. (Cello)$_6$ was first derivatized with ANTS following hydrolysis over time. (A) PACE showing the decrease of hydrolyzed ANTS-labeled (Cello)$_6$ substrate. Lanes from left to right: L—mixture of cellooligosaccharides (DP 2–6) and glucose; 0, 2, 4, 6, 8, 10, and 16—fractions removed and reaction terminated by ethanol addition at given times (min). (B) Representative plot showing the decrease of ANTS-labeled (Cello)$_6$ in the time course. Also shown is the initial increase followed by the subsequent decrease of the (Cello)$_4$-ANTS intermediate product of the GH3 hydrolytic reaction.

labeled with ANTS—$M_r = 1191.16$. For smaller oligosaccharides, a lower MWCO is required.
4. Following dialysis, take samples to dryness either *in vacuo* at 45 °C or by freeze drying. Samples should be transferred to 1.5 ml tubes for hydrolysis.

2.5.2. Digestion of labeled oligosaccharides over a time course for kinetic analysis

1. Following derivatization, quenching, and fluorophore removal, oligosaccharides are hydrolyzed by addition of 0.8 μg of hydrolase in 300 μl 0.1 M ammonium acetate. For GH3 hydrolysis, 0.1 M ammonium acetate, pH 5.5, is used as a buffer as the pH optimum has previously been determined to be in the region of pH 4–6 (Krogh *et al.*, 2010).
2. Hydrolysis may be monitored over a time course by removal of 2 μg substrate aliquots and addition of 2 volumes of 100% (v/v) ethanol to precipitate proteins (Fig. 4.4). To retain solubility of longer oligosaccharides, the reaction may also be terminated by addition of formic acid to 40% (v/v) or simply by boiling. The latter of these three techniques is less desirable since the enzyme continues to hydrolyze substrates at altered rates before being denatured by high temperature.

3. Following hydrolysis and termination of reactions, samples are dried and resuspended in 3 M urea for separation and interpretation as described previously (Section 2.3). GeneTools software is used to determine the intensity of each oligosaccharide product generated.

GH3 hydrolysis of ANTS-labeled (Cello)$_6$ was followed over a 16-min time course to determine the initial velocity (V_0) for a provided substrate mass of 2 µg, equivalent to 0.72 pmol (Cello)$_6$ of 90% purity (Megazyme). The 0 time sample contains known quantity of (Cello)$_6$, which allows quantitation of the rate of disappearance of substrate and appearance of products, shown in Fig. 4.4B for (Cello)$_6$ and cellotetraose ((Cello)$_4$). GH3 is known to follow Michaelis Menten kinetics when hydrolyzing cellobiose (Krogh et al., 2010). Use of this type of assay to study V_0 over a range of enzyme and substrate concentrations could extend our understanding of enzyme preference for cello-oligosaccharides of specific lengths.

2.6. Mechanism of hydrolysis

Use of the labeling strategies, before and after hydrolysis, as described in Sections 2.4 and 2.5, provides insight into the mechanism of hydrolysis of cello-oligosaccharides. Many exo-acting GHs have been characterized to access their substrates from either the reducing or nonreducing end, such as GH6 and GH7 cellobiohydrolases, acting synergistically in the hydrolysis of cellulose (Boisset et al., 2000). Using PACE to analyze the hydrolysis of reducing end labeled and unlabeled oligosaccharides, the end preference of exo-acting hydrolases can be determined.

1. (Cello)$_6$ is divided into 20 µg aliquots and taken to dryness *in vacuo* at 45 °C.
2. For each hydrolase, 1 aliquot is reducing end labeled with ANTS as described above (Section 2.5.1) while a second remains unlabeled.
3. In appropriate conditions for digestion by GH3, GH7, and GH45, 500 µl 0.1 M ammonium acetate, pH 5.5; an excess of enzyme is added to the labeled and unlabeled (Cello)$_6$ aliquots and incubated overnight at 21 °C.
4. Following overnight incubation, a second volume of enzyme is added to ensure exhaustive digest of the oligosaccharide mixture before terminating the reaction by boiling for 10 min.
5. Samples are taken to dryness *in vacuo* at 45 °C. The labeled sample is simply resuspended in 100 µl 3 M urea, while the unlabeled sample is derivatized as described above (Section 2.2), before resuspending in the same volume of 3 M urea.
6. Samples should be separated as described above (Section 2.3) and the oligosaccharide ladder interpreted.

Through this analysis, it can be confirmed that GH3 β-glucosidase has exohydrolytic activity acting at the nonreducing end since both labeled and unlabeled substrates are digested to cellobiose or glucose (Fig. 4.5). GH7, as a second example of exo-cellulolytic activity, proceeds from the reducing end of its substrate since the reducing end labeled substrate could not be accommodated by this enzyme, while the unlabeled substrate was digested to completion. GH45 was confirmed to have endohydrolytic activity since the enzyme was able to digest both labeled and unlabeled (Cello)$_6$, resulting in the production of a heterogeneous mixture of cello-oligosaccharide products (Fig. 4.5).

3. Discussion: Other Possible Applications

The presented experiments are only a few examples of possible applications of PACE. For example, given that cellulose adopts both amorphous and crystalline forms, a range of substrates may be easily compared for

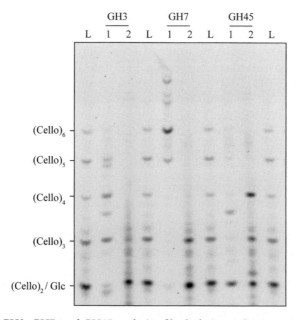

Figure 4.5 GH3, GH7, and GH45 analysis of hydrolytic mechanism. Lanes from left to right: L—mixture of cello-oligosaccharides (DP 2–6) and glucose; 1—ANTS-labeled (Cello)$_6$ prior to enzymatic hydrolysis; 2—products of hydrolysis labeled with ANTS after the reaction was terminated. Whereas GH3 and GH45 hydrolases can process reducing end-labeled (Cello)$_6$ and produce a number of products, GH7 hydrolase is not active on this modified substrate. All three enzymes can process native (Cello)$_6$ resulting into number of different products. These data are consistent with the view that the GH7 hydrolase processes the substrate from reducing end, whereas GH3 and GH45 hydrolases act from the nonreducing end.

different enzymes. Labeled products of one enzymatic reaction may be tested as a substrate for another enzyme. By analyzing the products of partial digests, the mechanism of enzyme action may be better understood. The kinetics may be analyzed both by observing the increase in product or disappearance of the substrate. Minor products can be identified owing to the sensitivity of the method. By using sequential digestions with combinations of various specific hydrolases, the structure of complex oligosaccharides and polysaccharides can also be revealed (Tryfona et al., 2010).

4. Summary

PACE has proven to be extremely useful in the analysis of complex polysaccharides, including hemicelluloses and pectins, as well as their corresponding degrading enzymes. Here, we demonstrate the potential of PACE in the characterization of cellulolytic enzymes. The versatility of PACE allows for the identification of both substrate preferences of the cellulolytic enzymes and the kinetic properties of the hydrolytic reactions.

ACKNOWLEDGMENTS

PD and ZZ were supported by BBSRC Grant BB/G016240/1 (BBSRC Sustainable Bioenergy Centre). We thank Katja Salomon Johansen and Kristian Bertel Rømer Mørkeberg Krogh (Novozymes A/S) for the enzymes used in this study.

REFERENCES

Andersen, N., Johansen, K. S., Michelsen, M., Stenby, E. H., Krogh, K. B. R. M., and Olsson, L. (2008). Hydrolysis of cellulose using mono-component enzymes shows synergy during hydrolysis of phosphoric acid swollen cellulose (PASC), but competition on Avicel. *Enzyme Microb. Technol.* **42,** 362–370.

Barton, C. J., Tailford, L. E., Welchman, H., Zhang, Z., Gilbert, H. J., Dupree, P., and Goubet, F. (2006). Enzymatic fingerprinting of Arabidopsis pectic polysaccharides using polysaccharide analysis by carbohydrate gel electrophoresis (PACE). *Planta* **224,** 163–174.

Bayer, E. A., Chanzy, H., Lamed, R., and Shoham, Y. (1998). Cellulose, cellulases and cellulosomes. *Curr. Opin. Struct. Biol.* **8,** 548–557.

Boisset, C., Fraschini, C., Schulein, M., Henrissat, B., and Chanzy, H. (2000). Imaging the enzymatic digestion of bacterial cellulose ribbons reveals the endo character of the cellobiohydrolase Cel6A from *Humicola insolens* and its mode of synergy with cellobiohydrolase Cel7A. *Appl. Environ. Microbiol.* **66,** 1444–1452.

Brown, D. M., Goubet, F., Vicky, W. W. A., Goodacre, R., Stephens, E., Dupree, P., and Turner, S. R. (2007). Comparison of five xylan synthesis mutants reveals new insight into the mechanisms of xylan synthesis. *Plant J.* **52,** 1154–1168.

Brown, D. M., Zhang, Z. N., Stephens, E., Dupree, P., and Turner, S. R. (2009). Characterization of IRX10 and IRX10-like reveals an essential role in glucuronoxylan biosynthesis in Arabidopsis. *Plant J.* **57,** 732–746.

Cantarel, B. L., Coutinho, P. M., Rancurel, C., Bernard, T., Lombard, V., and Henrissat, B. (2009). The Carbohydrate-Active EnZymes database (CAZy): an expert resource for Glycogenomics. *Nucleic Acids Res.* **37,** D233–D238.

Davies, G. J., Dodson, G. G., Hubbard, R. E., Tolley, S. P., Dauter, Z., Wilson, K. S., Hjort, C., Mikkelsen, J. M., Rasmussen, G., and Schülein, M. (1993). Structure and function of endoglucanase V. *Nature* **365,** 362–364.

Fagard, M., Desnos, T., Desprez, T., Goubet, F., Refregier, G., Mouille, G., McCann, M., Rayon, C., Vernhettes, S., and Höfte, H. (2000). PROCUSTE1 encodes a cellulose synthase required for normal cell elongation specifically in roots and dark-grown hypocotyls of Arabidopsis. *Plant Cell* **12,** 2409–2423.

Fry, S. C. (2000). The Growing Plant Cell Wall: Chemical and Metabolic Analysis. The Blackburn Press, Caldwell.

Goubet, F., Jackson, P., Deery, M. J., and Dupree, P. (2002). Polysaccharide analysis using carbohydrate gel electrophoresis: A method to study plant cell wall polysaccharides and polysaccharide hydrolases. *Anal. Biochem.* **300,** 53–68.

Goubet, F., Barton, C. J., Mortimer, J. C., Yu, X., Zhang, Z., Miles, G. P., Richens, J., Liepman, A. H., Seffen, K., and Dupree, P. (2009). Cell wall glucomannan in Arabidopsis is synthesised by CSLA glycosyltransferases, and influences the progression of embryogenesis. *Plant J.* **60,** 527–538.

Hogg, D., Pell, G., Dupree, P., Goubet, F., Martin-Orue, S. M., Armand, S., and Gilbert, H. J. (2003). The modular architecture of *Cellvibrio japonicus* mannnases in glycoside hydrolase families 5 and 26 point to differences in their role in mannan degradation. *Biochem. J.* **371,** 1027–1043.

Jackson, P. (1990). The use of polyacrylamide-gel electrophoresis for the high-resolution separation of reducing saccharides labelled with the fluorophore 8-aminonaphthalene-1,3,6-trisulphonic acid. Detection of picomolar quantities by an imaging system based on a cooled charge-coupled device. *Biochem. J.* **270,** 705–713.

James, D. C., and Jenkins, N. (1997). Analysis of N-glycans by matrix-assisted laser desorption/ionization mass spectrometry. *In* "A Laboratory Guide to Glycoconjugate Analysis," (P. Jackson and J. T. Gallagher, eds.), Vol. 9, pp. 91–112. Birkhäuser Verlag, Basel.

Jarvis, M. C., and McCann, M. C. (2000). Macromolecular biophysics of the plant cell wall: Concepts and methodology. *Plant Physiol. Biochem.* **38,** 1–13.

Kirk, O., Borchert, T. V., and Fuglsang, C. C. (2002). Industrial enzyme applications. *Curr. Opin. Biotechnol.* **13,** 345–351.

Klemm, D., Heublein, B., Fink, H.-P., and Bohn, A. (2005). Cellulose: Fascinating biopolymer and sustainable raw material. *Angew. Chem. Int. Ed Engl.* **44,** 3358–3393.

Krogh, K., Harris, P., Olsen, C., Johansen, K., Hojer-Pedersen, J., Borjesson, J., and Olsson, L. (2010). Characterization and kinetic analysis of a thermostable GH3 β-glucosidase from *Penicillium brasilianum*. *Appl. Microbiol. Biotechnol.* **86,** 143–154.

Monsarrat, B., Brando, T., Condouret, P., Nigou, J., and Puzo, G. (1999). Characterization of mannooligosaccharide caps in mycobacterial lipoarabinomannan by capillary electrophoresis/electrospray mass spectrometry. *Glycobiology* **9,** 335–342.

Mort, A. J., and Chen, E. M. W. (1996). Separation of 8-aminonaphthalene-1,3,6-trisulfonate (ANTS)-labeled oligomers containing galacturonic acid by capillary electrophoresis: Application to determining the substrate specificity of endopolygalacturonases. *Electrophoresis* **17,** 379–383.

Mortimer, J. C., Miles, G. P., Brown, D. M., Zhang, Z., Segura, M. P., Weimar, T., Yu, X., Seffen, K. A., Stephens, E., Turner, S. R., and Dupree, P. (2010). Absence of branches from xylan in Arabidopsis *gux* mutants reveals potential for simplification of lignocellulosic biomass. *Proc. Natl. Acad. Sci. USA* **107,** 17409–17414.

Palackal, N., Brennan, Y., Callen, N. C., Dupree, P., Frey, G., Goubet, F., Hazlewood, G. P., Healey, S., Kang, Y. E., Kretz, K. A., Lee, E., Tan, X., *et al.* (2004). An evolutionary route to xylanase process fitness. *Protein Sci.* **13,** 494–503.

Quinlan, R. J., Sweeney, M. D., Lo Leggio, L., Otten, H., Poulsen, J.-C. N., Johansen, K. S., Krogh, K. B. R. M., Jørgensen, C. I., Tovborg, M., Anthonsen, A., Tryfona, T., Walter, C. P., *et al.* (2011). Insights into the oxidative degradation of cellulose by a copper metalloenzyme that exploits biomass components. *Proc. Natl. Acad. Sci. USA* **108,** 15079–15084.

Shoemaker, S., Schweickart, V., Ladner, M., Gelfand, D., Kwok, S., Myambo, K., and Innis, M. (1983). Molecular cloning of exo-cellobiohydrolase I derived from *Trichoderma reesei* strain L27. *Nat. Biotechnol.* **1,** 691–696.

Tryfona, T., Liang, H.-C., Kotake, T., Kaneko, S., Marsh, J., Ichinose, H., Lovegrove, A., Tsumuraya, Y., Shewry, P. R., Stephens, E., and Dupree, P. (2010). Carbohydrate structural analysis of wheat flour arabinogalactan protein. *Carbohydr. Res.* **345,** 2648–2656.

Wilson, D. B. (2009). Cellulases and biofuels. *Curr. Opin. Biotechnol.* **20,** 295–299.

Zhang, Y. H. P., Cui, J., Lynd, L. R., and Kuang, L. R. (2006). A transition from cellulose swelling to cellulose dissolution by o-phosphoric acid: Evidence from enzymatic hydrolysis and supramolecular structure. *Biomacromolecules* **7,** 644–648.

CHAPTER FIVE

MEASURING PROCESSIVITY

Svein J. Horn,* Morten Sørlie,* Kjell M. Vårum,† Priit Väljamäe,‡ and Vincent G. H. Eijsink*

Contents

1. Introduction	70
2. Classical Methods for Measuring Processivity	71
2.1. Measuring product ratios	72
2.2. Measuring soluble- versus insoluble-reducing ends	74
3. Learning from Family 18 Chitinases	76
3.1. Measuring processivity with chitosan	76
3.2. The structural basis of processivity and directionality	78
4. Methods Based on Labeling—Apparent Versus Intrinsic Processivity	80
4.1. Measurement of P^{app} using a single-hit approach	81
4.2. Measurement of P^{app} using single-turnover approach	82
4.3. Intrinsic processivity	86
5. Concluding Remarks	89
Acknowledgments	90
References	90

Abstract

Natural cellulolytic enzyme systems as well as leading commercial cellulase cocktails are dominated by enzymes that degrade cellulose chains in a processive manner. Despite the abundance of processivity among natural cellulases, the molecular basis as well as the biotechnological implications of this mechanism are only partly understood. One of the major limitations lies in the fact that it is not straightforward to measure and quantify processivity in what essentially are biphasic experimental systems. Here, we describe and discuss both well-established methods and newer methods for measuring cellulase processivity. In addition, we discuss recent insights from studies on chitinases that may help direct further studies on processivity in cellulases.

* Department of Chemistry, Biotechnology and Food Science, Norwegian University of Life Sciences, Aas, Norway
† Department of Biotechnology, Norwegian University of Science and Technology, Trondheim, Norway
‡ Institute of Molecular and Cell Biology, University of Tartu, Tartu, Estonia

1. Introduction

Glycoside hydrolases acting on cellulose face a major challenge in gaining access to the substrate. For substrate binding to be productive, these enzymes need to bind several consecutive sugar moieties, and extracting such a stretch of sugars from a crystalline or otherwise insoluble substrate seems a daunting task (Beckham et al., 2011). Thus, it is no surprise that natural cellulolytic systems often are dominated by cellobiohydrolases, which appear to be optimized to bind to the (more accessible) chain ends of cellulose chains (Merino and Cherry, 2007). In the classic view on enzymatic cellulose degradation, endoglucanases help generate such chain ends by cleaving randomly in the more amorphous regions of the substrate. Recent work has shown that also enzymes currently classified as CBM33 and GH61 (see Cantarel et al., 2009 for classification) may contribute to generating chain ends by introducing chain breaks in the more crystalline regions (Forsberg et al., 2011; Quinlan et al., 2011; Vaaje-Kolstad et al., 2010; Westereng et al., 2011).

Another crucial property of cellobiohydrolases is their processivity, that is, the ability to remain attached to the substrate in between subsequent hydrolytic reactions (Davies and Henrissat, 1995; Robyt and French, 1967; Rouvinen et al., 1990; Teeri, 1997). Such a mechanism is thought to be beneficial for the degradation of crystalline substrates because the enzyme remains closely associated with the detached single polymer chain in between subsequent hydrolytic steps (Harjunpää et al., 1996; Teeri, 1997; von Ossowski et al., 2003). Because of the 180° rotation between consecutive sugar units, processive degradation of cellulose yields disaccharides (Davies and Henrissat, 1995; Rouvinen et al., 1990). Processive cellulases from *Trichoderma reesei* have been extensively studied (Divne et al., 1998; Harjunpää et al., 1996; Igarashi et al., 2009, 2011; Imai et al., 1998; Jalak and Väljamäe, 2010; Kipper et al., 2005; Koivula et al., 1998; Kurašin and Väljamäe, 2011; von Ossowski et al., 2003; Zou et al., 1999) and so have processive cellulases from *Humicola insolens* (Varrot et al., 1999, 2003), *Thermobifida fusca* (Li et al., 2007; Vuong and Wilson, 2009; Zhou et al., 2004), and *Clostridium cellulolyticum* (Mandelman et al., 2003; Parsiegla et al., 1998, 2000, 2008). At the macroscopic level, processivity may be observed as a unidirectional sharpening of substrate crystals (Boisset et al., 2000; Chanzy and Henrissat, 1985; Imai et al., 1998).

Despite this massive body of experimental work, the molecular mechanism of processivity and the implication of processivity on biomass conversion efficiencies are only partly understood. This is due to the intrinsic complexity of the biphasic experimental system and to nonstandard kinetic complexities of the enzymatic reaction, especially when it comes to the description and quantification of the product "release" process. In a

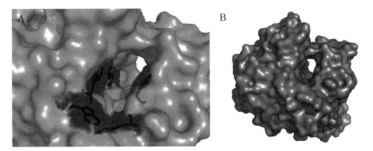

Figure 5.1 The catalytic domain of Cel6A (formerly called CBH II) from *Trichoderma reesei*. Cel6A, a processive cellobiohydrolase, is the first cellulase for which a three-dimensional structure could be determined (Rouvinen *et al.*, 1990; Zou *et al.*, 1999). Panels A and B show different views. The view in panel B highlights the tunnel character of the substrate-binding cleft. Panel A depicts several aromatic residues (shown in black) found in the tunnel. The catalytic domains of processive glycoside hydrolases are generally characterized by deep substrate-binding grooves that sometimes have closed "roofs", such as in the case of Cel6A. Note that these active sites are adapted to bind a single chain of polysaccharide. Single chains are threaded through these clefts or tunnels while consecutive disaccharides are being cleaved off at the catalytic center. See Varrot *et al.* (2003) for a detailed crystallographic study of this process.

processive reaction, one of the products, a dimer, is indeed released, whereas the other product, the shortened polymer, somehow will "slide" through the enzyme (Fig. 5.1).

In principle, further insight into the structural basis, molecular mechanism, and biotechnological implications of cellulase processivity could be obtained from detailed studies of natural enzymes with varying properties, as well as from structure–function studies (protein engineering). This requires good methods for measuring and quantifying processivity which, unfortunately, represents a major challenge, primarily due to the fact that insoluble substrates are not amenable to straightforward biochemical analysis. Several of the currently available methods are described and discussed below.

2. Classical Methods for Measuring Processivity

Figure 5.2 shows a scheme for processive degradation of a polysaccharide. It is important to note that each productive binding of a processive enzyme, be it an endo-binding or an exo-binding enzyme, to a highly polymeric substrate chain leads to production of maximally one product with an odd number of sugars (normally a trimer or a monomer; see legend to Fig. 5.2), whereas all other products resulting from the same initial

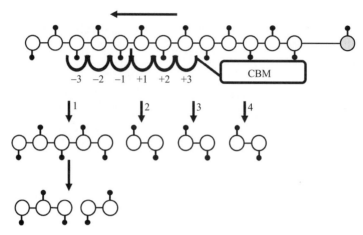

Figure 5.2 Essential features of processivity, illustrated for an endo-binding chitinase with a carbohydrate-binding module (CBM) and six subsites in its catalytic domain (−3 to +3). In chitin and cellulose, consecutive sugars are rotated by 180° with the consequence that only every second sugar will have a productive configuration when the polymer is threaded through the active site of the enzyme. The small black dots indicate the N-acetyl groups in chitin and underpin the periodicity in the substrate. The reducing end sugar is colored gray and the direction of the processive movement of the polymer chain is indicated by an arrow. The first hydrolysis (indicated by 1) will produce a pentamer, whereas all further cleavages (2–4) resulting from the same initial enzyme–substrate association will produce dimers. The pentamer (or any other odd-numbered oligosaccharide originating from the first cut) will eventually be degraded to disaccharides and a trisaccharide (which some enzymes may convert to a monomer and a dimer). Picture reproduced from Eijsink *et al.* (2008).

productive enzyme–substrate association will be dimers. This may be exploited for quantification of processivity, as discussed in Section 2.1. Furthermore, all products generated by a processive enzyme are soluble, except for the first product where initial binding is endo (internal). Thus, processive enzymes create much more soluble reducing ends (i.e., the reducing ends of the generated dimers) than, for example, endoglucanases which will mainly generate nonsoluble reducing ends, because higher DP cellodextrins are hardly soluble and remain attached to the substrate. This may also be exploited to measure processivity, as discussed in Section 2.2.

2.1. Measuring product ratios

As illustrated by Fig. 5.2, processive cellulases will convert cellulose chains to cellobiose and minor fractions of other products. For most enzymes, these other products, such as the pentamer initially generated in the degradation scheme of Fig. 5.2, will eventually end up as cellotriose, cellobiose, and glucose (Fox et al., 2012; Medve et al., 1998; Teeri, 1997). Trimeric and monomeric products are indicative of initial binding, whereas dimeric products are primarily, but not exclusively, generated by processive hydrolytic steps. Therefore, the ratio between these products provides an indication of the degree of processivity.

This approach to measuring processivity is attractive, because it is based on quantitative analysis of soluble hydrolytic products (monomers to trimers) which can easily be achieved by standard HPLC methods that are available in most laboratories. Analysis of glucose and short glucose oligomers is typically based on the use of Aminex columns for separation and refractive index monitoring for product detection (e.g., Koivula et al., 1998).

One way of interpreting the outcome of a degradation reaction in terms of the degree of processivity, P, is to use the following formula (Vuong and Wilson, 2009):

$$P = ([\text{dimers}] - [\text{monomers}])/([\text{trimers}] + [\text{monomers}]). \qquad (5.1)$$

The term [dimers] − [monomers] reflects the total number of consecutive cuts, where the concentration of dimers is corrected for dimers that result from conversion of the trimer to a monomer and a dimer (it is assumed that monomers are not released directly from the polymeric substrate). The term [trimers] + [monomers] represents the number of initial cuts. The value P reflects the number of cuts following the initial cut; the number of cuts per productive enzyme–substrate association equals $P+1$.

While this method is attractive due to its experimental and theoretical simplicity, it has several pitfalls that need consideration. First, there are no *a priori* reasons to assume that initial binding, be it endo- or exo-, preferentially yields an odd-numbered product, be it a trimer or a pentamer as in Fig. 5.2, or a longer polymeric fragment with an odd-numbered DP. If one assumes that 50% of the initial cleavages do not yield an odd-numbered product, Eq. (5.1) leads to an overestimation of processivity, primarily because the number of initial cuts is underestimated. An exo-acting non-processive enzyme with a strong preference for cleaving off dimers would seem processive when using Eq. (5.1) for interpretation of product profiles.

Another complication may arise from the formation of intermediate products during the reaction that are not covered by Eq. (5.1) (tetramers, pentamers, etc.). In such cases, one would run reactions to completion in order to end up with the simplest possible product profile (normally only DP1–DP3). Still, the generation of intermediate soluble products will create problems due to the major impact of the preferred binding modes used during further degradation of these products. For example, a hexameric intermediate product may be converted to two trimers (resulting in lower apparent processivity) or to three dimers (resulting in higher apparent processivity), depending on processivity-independent binding preferences.

In attempts to handle these pitfalls, several variants of Eq. (5.1) have been used in the literature (e.g., Fox et al., 2012; von Ossowski et al., 2003; Zhang et al., 2010), but none of these are perfect. Typical processivity values found using these approaches are in the range of 5–25 (Kurašin and Väljamäe, 2011 and references therein). Interestingly, the methods based on labeling that are discussed in Section 4 generally give higher values.

Despite its pitfalls, this product ratio method can be very useful, for example, for semiquantitative determination of mutational effects. In support of the validity of the method, Vuong and Wilson (2009) showed a reasonable correlation between processivity values determined according to Eq. (5.1) and values determined using the method described in Section 2.2. In the mutational analysis of chitinase processivity discussed in Section 3, there was a clear correlation between the degree of processivity (as assessed by the method described in Section 3.1) and P values determined using Eq. (5.1). Most importantly though, illustrating the problems discussed above, chitinase variants that by all other means were deemed to no longer be processive still had a P of 4.

2.2. Measuring soluble- versus insoluble-reducing ends

Another well-known method for the determination of processivity is based on the simultaneous determination of soluble- and insoluble-reducing ends (i.e., determination of the amount of reducing ends among the soluble sugars and in the nonsoluble substrate; Zhang et al., 2000). Endo-acting nonprocessive enzymes will mainly create insoluble-reducing ends. Any processive enzyme, be it endo- or exo-acting, will yield a higher fraction of soluble reducing ends because each initial cut is followed by repetitive production of soluble dimers. The key problem with this method is that nonprocessive enzymes that preferably bind in an exo-mode (i.e., at chain ends) will yield the same result as processive enzymes. Thus, mutational effects on the soluble/nonsoluble ratio may reflect changes in processivity as well as changes in the ratio of endo- versus exo-binding. From a practical point of view, the type and amount of substrate used play an important role because the number of binding sites for exo-attack (i.e., chain ends) may vary quite considerably (the DPs of various cellulose substrates differ a lot; Zhang and Lynd, 2005).

This method has proven useful to study mutational effects on processivity (Li et al., 2007; Vuong and Wilson, 2009; Zhang et al., 2000; Zhou et al., 2004). Furthermore, Vuong and Wilson (2009) showed that use of this method yields the same approximate trends among mutants as the methods described in Section 2.1. More generally, processivity values for cellulases determined with this method are in the same range (up to 25) as those found with the method of Section 2.1. Similar conclusions may be drawn from work on chitinases (S. J. Horn and V. G. H. Eijsink, unpublished observations).

Comparative studies have shown that this method is quite suitable for discriminating between true nonprocessive endoglucanases and enzymes that are processive and/or exo-acting (Irwin *et al.*, 1993). Nonprocessive endoglucanases stand out by their low production of soluble reducing ends. To discriminate between exo-action and endo-action, one may carry out activity assays on carboxymethylcellulose (or in the case of chitin, chitosan; see Section 3), where enzyme activity leads to a reduction of viscosity that depends on the exo- (almost no reduction of viscosity) or endo-mode (rapid reduction of viscosity) of the enzyme (see Sikorski *et al.*, 2006; Watson *et al.*, 2009).

To determine the ratio between soluble and insoluble reducing groups (SRGs and IRGs) (Doner and Irwin, 1992; Irwin *et al.*, 1998; Watson *et al.*, 2009; Zhang *et al.*, 2000), the supernatant and the substrate are separated, and reducing groups in the supernatant are determined by common methods for this purpose such as the DNS assay. Before determining reducing groups in the remaining substrate, for practical reasons often paper discs, bound protein is removed by washing with 6 M guanidine hydrochloride. After reconditioning the substrate in an appropriate buffer, reducing ends may be determined using a modified bicinchoninate assay as described by Doner and Irwin (1992), for example, using the Pierce microBCA reagent kit.

As indicated above, the interpretation of these experiments depends very much on knowing whether the enzyme binds in an endo- or an exo-fashion. The latter is not only dependent on the enzyme but also on the substrate, where ends are much more accessible but normally much less abundant than "internal"-binding sites (the latter, of course, depends on the DP of the substrate used). Processive enzymes tend to have long and deep, sometimes "tunnel-like", substrate-binding clefts (Fig. 5.1) that enclose their substrates to different extents, and this structural property is associated with processivity and/or exo-binding. There is evidence in the literature that the "walls" and "roofs" of tunnel-like processive glycoside hydrolases, including cellobiohydrolases, are quite flexible and that these enzymes do have the ability to bind to the substrate in an endo-fashion (Boisset *et al.*, 2000; Kurašin and Väljamäe, 2011; Michel *et al.*, 2003; Sikorski *et al.*, 2006; Varrot *et al.*, 1999; Zou *et al.*, 1999). If initial endo-attack were to be the default, the methods based on measuring soluble- versus insoluble-reducing ends would be fine. However, most experimental data for true crystalline substrates such as cellulose (e.g., Imai *et al.*, 1998) and chitin (e.g., Hult *et al.*, 2005) indicate that initial exo-binding is more prominent. If exo-binding is preferred, the present method leads to overestimation of the degree of processivity.

The limitations of the methods discussed so far can be overcome by using the labeling techniques discussed in Section 4. Original reducing ends in the substrate can be labeled before the enzymatic reaction, whereas reducing ends emerging in the soluble substrate may be labeled after the enzymatic reaction. One of the "labeling" techniques is to reduce the cellulose prior to hydrolysis.

3. Learning from Family 18 Chitinases

Chitin is a crystalline polymer of β(1–4)-linked N-acetylglucosamine occurring in the exo-skeletons of organisms with an outer skeleton, such as insects and crustaceans. Chitosan is a family of water-soluble chitin derivatives which can be prepared with varying degrees of acetylation and chain lengths. In the studies referred to here, the chitosans are produced from chitin by a homogeneous process where the acetyl groups are randomly distributed, that is, according to a Bernoullian distribution. These water-soluble chitosans may still be highly acetylated (up to about 65%), meaning that they still resemble chitin (in the studies discussed here, 65% randomly acetylated chitosans were used). Because catalysis in family 18 chitinases is substrate-assisted and depends on the acetamido group of the (acetylated) sugar bound in the -1 subsite (Tews et al., 1997; van Aalten et al., 2001), these enzymes can bind nonproductively to chitosans. As originally pointed out by Sørbotten et al. (2005) and explained further below, processive family 18 chitinases acting on chitosans will yield diagnostic product patterns that are dominated by oligomers comprising an even number of sugars. This provides a tool to study processivity.

3.1. Measuring processivity with chitosan

Figure 5.3 shows diagnostic product patterns obtained after degradation of a water-soluble highly acetylated chitosan with processive or nonprocessive chitinases. The legend to Fig. 5.3 explains in detail why processive family 18 chitinases yield these patterns.

It is important to prepare chitosans that have a random (Bernoullian) distribution of the acetylated units (Vårum et al., 1991a,b), which can be achieved through the homogenous deacetylation of chitin (Sannan et al., 1976). Such high-molecular weight and fully water-soluble chitosans are then used as substrates for the enzymes. Degraded chitosan samples are prepared by incubating the enzyme with the chitosan substrate for an appropriate time, and reactions are stopped by boiling the samples. The products of the enzymatic reactions are analyzed by size-exclusion chromatography (Horn et al., 2006b; Sørbotten et al., 2005). In order to obtain the required resolution of the oligomeric products, three analytical columns packed with Superdex 30, coupled in series, were used (overall dimensions: 2.60 (i.d.) × 180 cm). The mobile phase was 0.15 M ammonium acetate (pH 4.5) at a flow rate of 0.8 mL/min. A refractive index detector was used to monitor the products.

These chitosan-based methods for monitoring processivity provide an excellent tool for qualitative monitoring of processivity (Fig. 5.3B) and

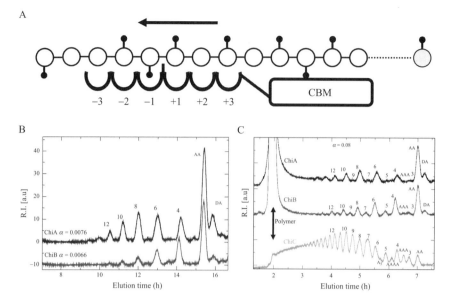

Figure 5.3 Observing processivity in family 18 chitinases. Panels B and C show size-exclusion chromatograms of product mixtures obtained upon incubating various family 18 chitinases with 65% acetylated chitosan. Product peaks are labeled by the DP of the oligomer they contain or, in some cases, by the sequence of the oligomer (A, acetylated sugar unit; D, deacetylated sugar unit). α indicates the degree of scission expressed as the fraction of cleaved glycosidic bonds. The void peak, containing material with DP > 40, is only shown in panel C. Under these conditions, all three enzymes initially bind to the chitosan substrate in an endo-fashion (Sikorski et al., 2006). Panel B shows the product profile obtained for chitobiohydrolases ChiA and ChiB at very low α-values, when rebinding and cleavage of released intermediate products have, most probably, not yet occurred. Panel C shows product profiles obtained later during the reaction ($\alpha = 0.08$) and includes a product profile for a nonprocessive family 18 endo-chitinase, ChiC, for comparison. Panels B and C show the diagnostic dominance of even-numbered products in the early stages of the reactions with ChiA and ChiB, which is indicative of processivity. These product profiles may be explained by Panel A, which is similar to Fig. 5.2 but the substrate now is water-soluble chitosan where several acetyl groups, indicated by the black dots, are lacking. In the situation shown, the first product would be a pentamer with sequence ADDAA. If the enzyme is processive, the next product would normally be a dimer, but in the present situation, this does not happen because the sugar then bound in the −1 subsite (i.e., the sugar bound in +2 in the picture) is lacking the acetyl group. If the enzyme is not processive or if the formation of a nonproductive "processive complex" would lead to enzyme–substrate dissociation, one would observe no clear dominance of even-numbered products, as for ChiC in panel B. However, if processive "movement" continues until a next productive complex emerges, longer even-numbered products will be formed; in the situation shown in panel C, the next product would be a hexamer with sequence ADADAA. So, if the enzyme is processive, every product resulting from the same initial enzyme–substrate association will be even-numbered, except for the first, leading to the diagnostic product patterns seen for ChiA and ChiB. The extent to which these longer oligomers can rebind and be cleaved depends on the binding preferences of the various subsites of the enzyme in question (Horn et al., 2006b; Sørbotten et al., 2005). In the present example, the hexamer with sequence ADADAA will eventually be converted to two trimers, ADA and DAA (Sørbotten et al., 2005). Panels B and C were adapted from Sikorski et al. (2006).

this tool has recently been used to obtain novel insight into enzyme functionality (see Section 3.2). Quantitative analysis of the degree of processivity is quite straightforward in this monophasic system, as described in detail in Sikorski et al. (2006). Degradation experiments are conducted in a capillary viscometer that allows continuous monitoring of relative viscosity, a parameter that is sensitive only to changes in the high-molecular weight fraction of the substrate and that thus reflects the number of initial (endo-) cuts (referred to as α_{pol}). By determining the total amount of reducing chain ends created in the enzymatic reaction using a reducing end assay that is insensitive for the length and composition of the products (Horn and Eijsink, 2004), one obtains values for the total amount of cleaved glycoside bonds (referred to as α). By comparing α with α_{pol}, one obtains a value for processivity, which was found to be in the order of 10 and 4 for ChiA and ChiB, respectively (Sikorski et al., 2006).

Obviously, the quantitative approach comes in handy when, for example, comparing mutants, but the absolute values should be handled with great care. Chitosan is not a natural substrate and the need for the substrate to move by (on average) a multitude of two sugar residues in between consecutive hydrolytic steps is not natural neither. It is quite likely that the processivity values coming out of this approach are underestimated, as the chance of full enzyme–substrate dissociation occurring is likely to become larger with the length of the "sliding path" (which corresponds to a dimer for chitin, but is, on average, longer for chitosan). This might be especially true for enzymes such as ChiB, whose substrate binding cleft contains a "porch loop" (van Aalten et al., 2001) that may create sterical hindrance for sliding. In fact, judged by dimer/trimer ratios (Section 2.1) obtained when degrading chitin, ChiB is more processive than ChiA, in contrast with the results from the chitosan experiments.

3.2. The structural basis of processivity and directionality

While the chitosan method is based on a nonnatural substrate, cannot be used for cellulases, and has its quantitative limitations, it has been extremely useful for unraveling some fundamental aspects of processivity. In particular, studies on the action of the two chitobiohydrolases from *Serratia marcescens*, ChiA (Zakariassen et al., 2009) and ChiB (Horn et al., 2006a), on chitosan have revealed the exceptional role of aromatic residues in processivity.

The substrate-binding clefts of processive glycoside hydrolases are often lined with aromatic residues, in particular tryptophan residues, and it is well known that these residues are important for enzyme function (e.g., Koivula et al., 1998; Uchiyama et al., 2001; Fig. 5.4). Hydrophobic stacking interactions may enable processive enzymes to remain attached to their

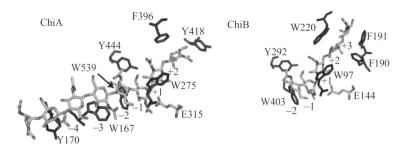

Figure 5.4 Substrate binding in processive chitobiohydrolases ChiA and ChiB. The pictures show details of the crystal structures of ChiA in complex with an octameric substrate (Papanikolau et al., 2001) and ChiB in complex with a pentameric substrate (van Aalten et al., 2001). In addition to the substrate, the pictures show the side chains of the catalytic acids (E315 in ChiA, E144 in ChiB) and the side chains of aromatic residues interacting with the substrate. In ChiA, the $+1$ and $+2$ subsites are the "product" sites, where dimeric products are released. In ChiB, these product subsites are -1 and -2. Picture adapted from Zakariassen et al. (2009). (See Color Insert).

substrates, while retaining the ability to "slide" during the processive mode of action. Compared to, for example, hydrogen bonds, hydrophobic stacking interactions are nonspecific and involve larger interaction surfaces (Sørlie et al., 2012). Mutational analysis of aromatic residues in the substrate-binding clefts of ChiA and ChiB, using the chitosan method for qualitative inspection of processivity (Fig. 5.5), revealed that specific aromatic residues are crucial for processivity. In fact, it was possible to almost completely abolish processivity by single point mutations of aromatic residues in subsites interacting with the polymeric part of the substrate ($-$ subsites in ChiA and $+$ subsites in ChiB). In ChiB, mutation of Trp97 (Fig. 5.5) or Trp220 in subsites $+1$ and $+2$ yielded nonprocessive mutants (Horn et al., 2006a). In ChiA, only mutation of Trp167 (-3 subsite) led to abolished processivity. Mutation of Trp275 and Phe396 (analogous to Trp97 and Trp220 in ChiB) did affect enzyme efficiency but had only minor effects on processivity (Zakariassen et al., 2009). So, using the chitosan method, it was possible to pinpoint structural determinants of processivity and its directionality.

From a biotechnological point of view, it is interesting to note that these nonprocessive mutants, expectedly, showed reduced activity on chitin, but that they showed strongly increased activity on chitosan. Thus, it may seem that processivity is unfavorable if the substrate is soluble and well accessible. For further information on the structural basis of processivity in ChiA and ChiB, as well as on some kinetic and thermodynamic aspects of this process, the reader is referred to Zakariassen et al. (2010) and to a recent review by Sørlie et al. (2012).

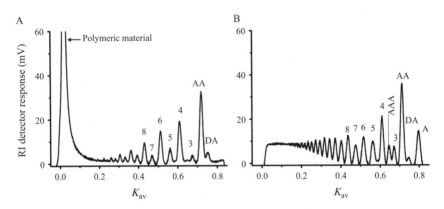

Figure 5.5 Product profiles upon degradation of 65% acetylated chitosan with ChiB (A) and its nonprocessive W97A mutant (B). See Fig. 5.3 for explanation of the annotation. The extent of degradation is approximately the same for both reactions ($\alpha = 0.14$). K_{av} denotes partition coefficient. Picture is taken from Horn et al. (2006a).

4. METHODS BASED ON LABELING—APPARENT VERSUS INTRINSIC PROCESSIVITY

In a totally different approach, compared to the approaches described in Section 2, one may measure and quantify processivity by direct measurement under single-hit or single-turnover conditions. Such approaches have recently been pioneered by Väljamäe and coworkers. Like the methods described in Section 2, these approaches may be used to determine the apparent processivity (P^{app}). It should be noted, though, that processivity may be limited by the substrate (Igarashi et al., 2011; Kurašin and Väljamäe, 2011) and that values for P^{app} thus may differ between substrates. Thus, determination of apparent processivities does not necessarily tell the whole story and additional approaches may be needed. In an alternative approach, also recently described by Väljamäe and coworkers, one may determine the intrinsic processivity of an enzyme by determination of underlying rate constants.

In an early study, processivity for cellobiohydrolase Cel7A from *T. reesei* was assessed by studying the hydrolysis of cellulose labeled at its reducing end with anthranilic acid under single-hit conditions (Kipper et al., 2005). Processivity was calculated by determining the ratio between fluorescent soluble products, released by initial cuts at the reducing end, and nonfluorescent products released by subsequent processive steps. This analysis yielded high values for processivity (up to 88 on bacterial cellulose), which, in part, may be due to the fact that the occurrence of endo-attacks leads to overestimation of processivity. In later work, discussed below, methods were developed to circumvent these problems.

4.1. Measurement of P^{app} using a single-hit approach

Apparent processivity (P^{app}) equals the average number of catalytic acts (N_{catal}) that an enzyme performs per one initiation of a processive run (N_{init}).

$$P^{app} = \frac{N_{catal}}{N_{init}}. \tag{5.2}$$

The experimental challenge for the quantification of P^{app} is the determination of N_{init}. In the following, we will describe two approaches to measure P^{app}: (i) a single-hit and (ii) a single-turnover approach.

The single-hit approach relies on the simultaneous measurement of enzyme-generated SRGs and IRGs under conditions that minimize chances of the same chain being hit twice. Such conditions may be obtained by using low enzyme to substrate ratios and by working at very low percentages of total conversion. The amount of released SRGs accounts for N_{catal}. The amount of generated IRGs accounts for the number of endo-initiations (N_{init})$_{Endo}$ (Irwin et al., 1993). If reduced cellulose is used as a substrate, then IRGs account for the sum of (N_{init})$_{Endo}$ and reducing end exo-initiations ((N_{init})$_{R-exo}$) (Kurašin and Väljamäe, 2011). Currently, it is not possible to quantify the exo-initiations from the nonreducing ends and their presence will lead to the overestimation of P^{app}. When using reduced cellulose, N_{catal} equals the total amount of generated reducing groups, RG_{Tot}, which obviously equals SRGs plus IRGs.

In order to measure the generation of SRGs and IRGs under single-hit conditions, a sensitive analytical procedure for their detection must be available. While the sensitivity of absorbance-based methods, like the bicinchoninic acid (BCA) method, is usually sufficient for determination of SRGs, fluorescence labeling techniques must be employed for IRGs. So far, anthranilic acid (AA) (Kipper et al., 2005; Velleste et al., 2010) and 2,6-diaminopyridine (DAP) (Kurašin and Väljamäe, 2011) have been used for fluorescence labeling of IRGs on cellulose. In both cases, the labeling involves reductive amination. In reduced cellulose, all aldehyde groups in the former reducing ends are reduced to corresponding alcohols which do not react in reductive amination. Upon the action of cellulases, an aldehyde is generated and visualized by reductive amination. Reduced cellulose used as substrate must be pure cellulose without noncellulosic components that may interfere with labeling or later fluorescence measurements. Reduced celluloses with high degree of polymerization (DP), such as reduced bacterial cellulose (rBC), must be used as substrates to avoid the limitation of P^{app} by the DP. For preparation of reduced cellulose, see Velleste et al. (2010). As even low amounts of endoglucanase (EG) will produce significant amounts of IRGs, cellobiohydrolase (CBH) preparations with very high purity must be used to measure processivity (Kurašin and Väljamäe, 2011).

In the experiments, reduced cellulose (1–5 g/L) is incubated with a cellulase (usually below 50 nM) for different times between 10 and 60 min. The reaction is ended by separating the cellulose from the solution by centrifugation and by stopping residual enzyme activity in the pellet by adding 0.2 M NaOH. The pellet is then used to measure IRGs, whereas the supernatant is used to determine SRGs. It is important to remove all bound cellulases with alkali and subsequent proteinase K treatment before the labeling of IRGs. Because of the low concentrations of SRGs and IRGs, care must be taken to prepare adequate zero time points. For detailed protocols of AA and DAP labeling of IRGs, see Velleste et al. (2010) and Kurašin and Väljamäe (2011). P^{app} equals to the slope of the line obtained when plotting [RG$_{Tot}$] versus [IRG] (Fig. 5.6) If single-hit conditions are met, this plot should give a straight line with the intercept approaching zero, as has indeed been observed (Fig. 5.6). For Cel7A acting on rBC, this method yielded a P^{app} of 61.

4.2. Measurement of P^{app} using single-turnover approach

This method relies on the hydrolysis of uniformly ^{14}C-labeled cellulose. Cellulases are mixed with labeled cellulose to start the hydrolysis. After a short interval, an excess of nonlabeled substrate is added to trap all unbound cellulases. Under these conditions, each bound cellulase is allowed to perform only one processive run on labeled cellulose. ^{14}C-labeled bacterial cellulose (^{14}C-BC) may be prepared by cultivating *Acetobacter xylinum* (ATCC 53582) in the presence of ^{14}C-glucose as a carbon source. Preparation of BC from this organism is described in Velleste et al. (2010).

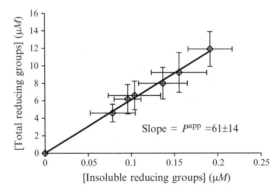

Figure 5.6 Single-hit approach for determination of apparent processivity (P^{app}). Reduced bacterial cellulose (1.0 g/L) was incubated with 20 nM *Trichoderma reesei* Cel7A in 50 mM sodium acetate, pH 5.0 at 30 °C for 15–120 min. Soluble-reducing groups and insoluble-reducing groups were determined using the BCA method and DAP labeling, respectively. Data are adopted from Kurašin and Väljamäe (2011).

Regenerated amorphous cellulose (RAC) with high binding capacity for cellulases is prepared from Avicel (Zhang et al., 2006) and used as the trap cellulose (RAC trap).

In the experiment, ^{14}C-BC (0.5 g/L; specific activity 450,000 DPM/mg) is incubated with a cellulase (in the present example Cel7A from T. reesei) and Aspergillus β-glucosidase (0.12 µM) in 50 mM sodium acetate buffer, pH 5.0 containing BSA (0.1 g/L) at 25 °C. After 30 s, an equal volume of RAC trap is added (10 g/L; note that the cellulose concentration is 20 times higher than for the ^{14}C-BC and that the RAC cellulose is much more enzyme accessible than the BC). At selected times, 0.2 mL aliquots of the reaction mixture are withdrawn and added to 20-µL 1 M NaOH to stop the hydrolysis. After separation of cellulose by centrifugation, the radioactivity in the supernatant is quantified using a liquid scintillation counter. To enable accurate handling of samples, the 0 and 30 s (time that corresponds to the addition of trap) time points were prepared in parallel reactions, using exactly the same conditions. Figure 5.7A shows the release of ^{14}C-cellobiose (^{14}CB, calculated from the released radioactivity assuming that CB is the only product) in hydrolysis of ^{14}C-BC with Cel7A. After addition of trap, the release of ^{14}CB continues only a little and levels off (Fig. 5.7A). If the RAC trap is replaced by the equal amount of buffer, the production of ^{14}CB continues for the whole experimental period (Fig. 5.7A).

The time course of ^{14}CB formation during a single processive run follows Eq. (5.3)

$$[^{14}\text{CB}] = [^{14}\text{CB}]_{\max}(1 - e^{-kt}) \quad (5.3)$$

where $[^{14}\text{CB}]_{\max}$ is the leveling-off value of $[^{14}\text{CB}]$ and k is the pseudo first-order rate constant for passing through one processive run (see below). The value of $[^{14}\text{CB}]_{\max}$ depends on (i) the concentration of productive cellulase–cellulose complexes at the moment of the trap addition ($[\text{ES}]_{\text{trap}}$) and (ii) the average number of ^{14}CB units released during a single run (i.e., the P^{app}). Therefore, in order to find the P^{app} from the values of $[^{14}\text{CB}]_{\max}$, the values of $[\text{ES}]_{\text{trap}}$ must be available. For Cel7A, a method for the quantification of the cellulase population with the active site occupied by the cellulose chain ($[\text{ES}]_{\text{OA}}$; OA for occupied active site) can be used to find the values of $[\text{ES}]_{\text{trap}}$ (Jalak and Väljamäe, 2010). The method is based on the competition between a low molecular weight model substrate (reporter molecule) and cellulose chain. As the trap must be added shortly after the initiation of the hydrolysis (usually after 10–30 s), a sensitive quantification of $[\text{ES}]_{\text{trap}}$ is necessary. Here we used fluorescent methylumbelliferyl lactoside (MUL) as the reporter molecule. If the concentration of MUL is far below its K_M (5 µM MUL was used and the K_M of Cel7A for MUL is about 300 µM;

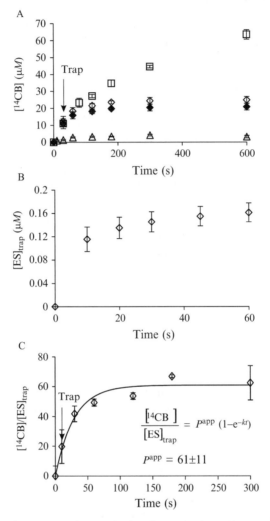

Figure 5.7 Single-turnover approach for determination of apparent processivity (P^{app}). (A) ^{14}C labeled bacterial cellulose (^{14}C-BC, 0.5 g/L) was incubated with 1.0 μM *T. reesei* Cel7A and 0.12 μM β-glucosidase in 50 mM sodium acetate buffer, pH 5.0 at 25 °C. After 30 s, an equal volume of RAC trap (10 g/L) was added. ^{14}C-cellobiose (^{14}CB) was quantified from the released radioactivity. (◇), ^{14}CB release in the standard experiment, (△), "leakage of trap" found in the control experiment where RAC trap was added before cellulases, (◆), data corrected for the "leakage of trap", (□), control experiment without addition of the RAC trap (instead an equal amount of buffer was added). (B) The concentration of Cel7A (total 0.25 μM) with the active site occupied by a cellulose chain ($[ES]_{OA}$) was found from the hydrolysis of BC (0.5 g/L) in the presence of β-glucosidase (0.85 μM) and methylumbelliferyl lactoside (5 μM) as reporter molecule. $[ES]_{OA}$ was used to account for the concentration of enzyme–substrate complexes at the moment of the trap addition ($[ES]_{trap}$). (C) ^{14}C-BC, (0.5 g/L) was incubated with 0.25 μM Cel7A and 0.12 μM β-glucosidase. After 10 s, an equal volume of RAC trap (10 g/L) was added. Released ^{14}CB was divided by $[ES]_{trap}$ after 10 s (from Panel B). Values of P^{app} and k were found from nonlinear regression according to Eq. (5.5).

Voutilainen et al., 2008), the rate of MUL hydrolysis (v_{MUL}) is given by

$$v_{MUL} = \left(\frac{k_{cat}}{K_M}\right)_{MUL} [E]_{FA}[MUL]. \tag{5.4}$$

$[E]_{FA}$ ("FA" for free active site) is the concentration of cellulase with the active site free from cellulose chain, that is, free to hydrolyze MUL. $(k_{cat}/K_M)_{MUL}$ can be found from the calibration of MUL hydrolysis in the absence of cellulose, in which case $[E]_{FA}$ equals $[E]_{Total}$. If $[E]_{FA}$ is measured from the hydrolysis of MUL in the presence of cellulose, the $[ES]_{OA}$ is available from $[ES]_{OA} = [E]_{Total} - [E]_{FA}$. It is important to supplement the hydrolysis mixtures with excess β-glucosidase activity so that the MUL hydrolysis is not inhibited by the CB released from cellulose (Jalak and Väljamäe, 2010).

Figure 5.7B shows the buildup of the population of $[ES]_{trap}$ in time. After dividing the released [^{14}CB] with corresponding $[ES]_{trap}$, Eq. (5.3) can now be expressed in the form of

$$\frac{[^{14}CB]}{[ES]_{trap}} = \left(\frac{[^{14}CB]}{[ES]_{trap}}\right)_{max} (1 - e^{-kt}) = P^{app}(1 - e^{-kt}). \tag{5.5}$$

Fitting of the experimental data to Eq. (5.5) now allows the values of P^{app} and k to be determined. Using this approach, a P^{app} value of 61 ± 11 was found for Cel7A (Fig. 5.7C). This figure is remarkably similar to the P^{app} of 61 ± 14 found for Cel7A on rBC using the single-hit approach (Fig. 5.6). Obviously, the fact that the match is almost perfect may be a matter of coincidence, but it does indicate that both methods are similarly valid.

The rate constant for passing through one processive run (k) in Eqs. (5.3) and (5.5) is related to the rate constant (k_{cat}), leading from one productive complex to another productive complex one CB unit further on the cellulose chain, as shown in Eq. (5.6):

$$k = \frac{k_{cat}}{P^{app}}. \tag{5.6}$$

The k value for Cel7A on ^{14}C-BC was 0.034 ± 0.010 s^{-1} (Fig. 5.7C). From Eq. (5.6) it follows that the k_{cat} is 2.1 ± 0.6 s^{-1}, a figure that is well in line with earlier estimates of k_{cat} for Cel7A on cellulose (Fig. 5.8A; Gruno et al., 2004; Igarashi et al., 2011; Kurašin and Väljamäe, 2011).

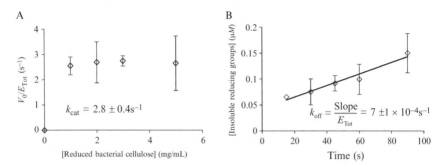

Figure 5.8 Intrinsic processivity (P^{Intr}) can be found from catalytic constant (k_{cat}) and dissociation rate constant (k_{off}). (A) v_0 is the rate of the formation of total reducing groups (measured after 10 s of hydrolysis) in hydrolysis of reduced bacterial cellulose (rBC) with 0.1 μM T. reesei Cel7A in 50 mM sodium acetate buffer, pH 5.0 at 30 °C. Under these conditions, the enzyme was saturated already in the case of lowest substrate concentration used and k_{cat} was found as an average of all $v_0/[E]_{Tot}$ values. (B) Formation of insoluble-reducing groups (IRGs) in hydrolysis of rBC (1.0 g/L) with Cel7A (20 nM) in 50-mM sodium acetate buffer, pH 5.0 at 30 °C. If enzyme recruitment is limited by dissociation, the k_{off} is available from the slope of the line. Data are adopted from Kurašin and Väljamäe (2011).

4.3. Intrinsic processivity

As the P^{app} of cellulases depends on the substrate, methods have been developed to reveal the "processivity potential", that is, the true or intrinsic processivity (P^{Intr}). P^{Intr} can be described in two equivalent ways. For enzymes active against nucleic acids, such as polymerases or helicases, a probability (p) that the enzyme will proceed to make a catalytic step rather than dissociate has often been used as a measure of processivity (McClure and Chow, 1980). Numerical values of p fall between 0 and 1 (a hypothetical enzyme with infinite processivity has a p value of 1). P^{Intr} can also be described as an average number of consecutive catalytic acts performed before dissociation. In this case, P^{Intr} is given by Eq. (5.7) (Lucius et al., 2003)

$$P^{Intr} = -\frac{1}{\ln(1-P_d)} \approx \frac{1}{P_d} \approx \frac{k_{cat}}{k_{off}}. \quad (5.7)$$

In Eq. (5.7), a dissociation probability (P_d) was used instead of p (note that $p = 1 - P_d$). P_d is governed by the rate constants for two processes: (i) the dissociation of the enzyme–substrate complex (k_{off}) and (ii) the catalytic rate (k_{cat}). P_d is related to k_{cat} and k_{off} according to Eq. (5.8)

$$P_d = \frac{k_{off}}{k_{off} + k_{cat}}. \quad (5.8)$$

In the case of processive enzymes, $P_d \ll 1$ and P^{Intr} approximate to the rightmost term of Eq. (5.7), k_{cat}/k_{off}. Thus, to find the value of P^{Intr}, we need to know the values of k_{cat} and k_{off}. It must be noted that P^{Intr} equals P^{app} (Eq. 5.2) only on an ideal polymer, where P_d is independent of the position of the enzyme on the polymer and where the enzymes meet no obstacles or chain ends. Using values for k_{cat} and k_{off} determined as described below and shown in Fig. 5.8, we found that P^{Intr} for Cel7A on rBC is ~ 4000 (Kurašin and Väljamäe, 2011).

4.3.1. Measuring k_{cat}

Measurement of k_{cat} values for cellulases acting on cellulose is complicated by the heterogeneity of the substrate, nonlinear kinetics, and by the fact that most enzymes have more than one binding mode. Although it is possible to achieve apparent saturating conditions by increasing the cellulose concentration, the assumption that, in this case, all enzymes are in productive complexes with cellulose may not hold. Cellulases are often multidomain enzymes consisting of catalytic domain (CD) and one or more carbohydrate-binding modules (CBMs) where both domains can bind individually to cellulose (Boraston et al., 2004). Binding through the CBM only results in nonproductive binding (Ståhlberg et al., 1991). Although the relative contribution of different binding modes (Jalak and Väljamäe, 2010; Kostylev and Wilson, 2011) is usually not known, it is expected that the fraction of nonproductively bound enzymes increases with increasing enzyme to substrate ratio and therefore better estimates for k_{cat} can be achieved at low enzyme to substrate ratios (Igarashi et al., 2006).

Cellulase action on cellulose commonly shows nonlinear kinetics, and both enzyme- and substrate-related limitations may be responsible for the phenomenon (Bansal et al., 2009; Zhang and Lynd, 2004). It has been suggested that, in the case of processive CBHs, the rate of product formation is governed by k_{cat} only at the initial pre-steady-state of the hydrolysis (first 10 s) (Jalak and Väljamäe, 2010). Therefore, the measurements of k_{cat} must rely on the initial "burst phase" and not on the steady state (Kipper et al., 2005; Kurašin and Väljamäe, 2011; Murphy et al., 2012; Praestgaard et al., 2011).

For the determination of k_{cat}, a cellulose suspension is treated with cellulase for a short time (usually between 5 and 30 s). The concentration of cellulose must be as high as possible and the enzyme concentration should be as low as reliable. Short hydrolysis times mean that reactions need to be stopped effectively; for cellulases, the addition of alkali seems most appropriate. After stopping the reaction, the enzyme-generated reducing ends are determined using the most sensitive of available methods whose response does not depend on the DP of the reducing sugar. To the best of our knowledge, the greatest sensitivity and reliability can be achieved with the modified BCA method (Zhang and Lynd, 2005). In this case, the reaction can be stopped by the addition of an equal volume of BCA reagent to the hydrolysis mixture (Kurašin and Väljamäe, 2011). At this point, there are two possibilities: (i) if RG_{Tot} is measured, the stopped hydrolysis mixture is heated at 75 °C for 30 min. After cooling to room temperature, cellulose is separated by centrifugation and the absorbance at 560 nm in the supernatant is recorded; (ii) if only SRGs are measured, the stopped hydrolysis mixture is centrifuged, supernatant is removed and heated at 75 °C for 30 min followed by the absorbance measurements. The latter approach is more suitable for CBHs that produce mainly SRGs. For EGs that produce significant amounts of IRGs, the first approach must be used. When measuring IRGs, it is essential to use reduced cellulose (Kongruang et al., 2004; Sakamoto et al., 1989; Velleste et al., 2010) to minimize the (high) background from the IRGs in the starting material. As the BCA method is sensitive to protein, it is very important to make adequate zero time point measurements. To do so, the BCA reagent is first mixed with cellulose followed by the addition of enzyme after which the samples are treated as described above. An example of the measurement of the k_{cat} of Cel7A from *T. reesei* on rBC is given in Fig. 5.8A.

4.3.2. Measuring the k_{off} value

Direct measurement of k_{off} values for cellulases is complicated because of the insoluble and heterogeneous substrate and different binding modes of enzymes. Furthermore, it has been reported that the binding of cellulase to cellulose may be fully or partially irreversible (Bothwell et al., 1997; Kyriacou et al., 1989; Ma et al., 2008; Palonen et al., 1999; Zhu et al., 2009). By measuring fluorescence recovery after photobleaching, k_{off} values in the range of 10^{-2} to 10^{-3} s^{-1} have been determined for processive cellulases from *T. fusca* (Moran-Mirabal et al., 2011). Off-rates have also been measured following the establishment of new binding equilibrium after disturbance (e.g., dilution; Carrard and Linder, 1999; Linder and Teeri, 1996). Fitting of progress curves to kinetic models has also been used to derive k_{off} values (Harjunpää et al., 1996; Praestgaard et al., 2011).

k_{off} values may also be estimated from the rate of initiations in the steady state. This approach assumes that the rate of enzyme recruitment is limited by dissociation of the enzyme, which is plausible assumption for processive CBHs (Kurašin and Väljamäe, 2011). If the k_{off} is to be used for calculating P^{Intr}, it is imperative that it represents the dissociation of a cellulose chain from the active site of the enzyme. The presence of CBMs may complicate this analysis because they may keep the enzyme attached to the substrate in a nonproductive manner (i.e., the active site may be empty). It is thus advisable to (also) measure k_{off} on truncated enzyme versions containing the CD only. Available data indicate that for *T. reesei* Cel7A, the dissociation of the CBM is not rate limiting for the dissociation of the full-length enzyme but this assumption cannot be generalized (Carrard and Linder,

1999). An example of the measurement of k_{off} using DAP labeling of Cel7A generated IRGs on rBC is given in Fig. 5.8B. As the nonreducing end lacks specific chemistry, the initiations from that end cannot be determined. Thus, this method for determination of k_{off} can only be used for enzymes that employ endo- or reducing end exo-mode initiation.

5. Concluding Remarks

Processivity is important for the functionality of cellulolytic enzymes and it is thus important to understand this phenomenon and to be able to measure it. In fact, taking into account the discussion about endo- versus exo-action, and on the basis of our own published and unpublished observations and considerations, we would like to propose that the degree of processivity, in fact, is a major functional discriminator between cellulases. The question whether an initial attack is endo- or exo- has received much attention in the literature, but the question whether or not an initial attack is followed by processive action may, in fact, be more important.

An interesting topic, not addressed in detail in this chapter, concerns the kinetics and thermodynamics of the processive step. This step includes release of a short product (usually a dimer) from the product binding sites as well as displacement of a longer product (the rest of the polymer) from one position of the enzyme to another (Fig. 5.2). Describing the latter displacement in kinetic terms and understanding of the driving forces behind it are major challenges. The work on chitinases described in Section 3 clearly shows that processive movement of the polymeric chain through the enzyme does not depend on the glycosidic bonds being broken (hence the formation of, e.g., octamers shown in Fig. 5.3A). It would seem that the best way to view processive movement is as a process of limited diffusion, where the polymeric chain is released in an environment that is optimized to promote rebinding rather than full release. Indeed, the active sites of processive glycoside hydrolases show several features that seem to restrict the polymer from diffusing away and/or promote rebinding. This includes the presence of carefully positioned aromatic residues (see Section 3.2) as well as, in some but far from all processive enzymes, the presence of loops that cover the polymeric part of the substrate (e.g., von Ossowski et al., 2003; Fig. 5.1). It is possible that the presence of CBMs also contributes to limiting the diffusional freedom of the enzyme-bound polymer.

As to the quantification of processivity, it is very important to emphasize the fact that processivity is affected by the substrate and that values for P^{app} thus differ between cellulose variants. Several of the reports cited above, addressing most of the methods described above, clearly show that the degree of processivity depends on the type of cellulose used. These substrate effects have been ascribed to the occurrence of obstacles on the enzyme path (Igarashi et al., 2011; Kurašin and Väljamäe, 2011).

Unfortunately, there are no straightforward ways to measure processivity accurately. The "classical" methods discussed in Section 2 are inaccurate and provide limited insight; the methods used for chitinases (Section 3) have been of major importance for our understanding of the process but are not amenable to cellulases, and the labeling methods described in Section 4 are accurate, but rather difficult to implement. Still, recent work does show that it is possible to get to grips with processivity, especially if one combines the various analytical methods discussed above with extensive site-directed mutagenesis work and crystallography. Modeling and molecular dynamics studies of cellulase and chitinase action have shown promising results and may add further to our understanding of processivity (Beckham and Crowley, 2011; Beckham et al., 2011; Payne et al., 2011). More advanced modeling of the process itself based on experimental progress curves may also help (Nutt et al., 1998; Praestgaard et al., 2011; Sikorski et al., 2005). When it comes to the analytical methods, it is imperative to have a thorough understanding of what these methods actually show and where the possible pitfalls are.

ACKNOWLEDGMENTS

P. V.'s work was supported by the European Commission (FP7/2007–2013, Grant 213139). V. G. H. E., K. M. V., M. S., and S. J. H. were supported by grants 164653, 178428, 177542, and 196885 from the Norwegian Research Council. We thank Bjørge Westereng and Gustav Vaaje-Kolstad for helpful discussions and help in preparing some of the figures.

REFERENCES

Bansal, P., Hall, M., Realff, M. J., Lee, J. H., and Bommarius, A. S. (2009). Modeling cellulase kinetics on lignocellulosic substrates. *Biotechnol. Adv.* **27,** 833–848.

Beckham, G. T., and Crowley, M. F. (2011). Examination of the α-chitin structure and decrystallization thermodynamics at the nanoscale. *J. Phys. Chem. B* **115,** 4516–4522.

Beckham, G. T., Bomble, Y. J., Bayer, E. A., Himmel, M. E., and Crowley, M. F. (2011). Applications of computational science for understanding enzymatic deconstruction of cellulose. *Curr. Opin. Biotechnol.* **22,** 231–238.

Boisset, C., Fraschini, C., Schulein, M., Henrissat, B., and Chanzy, H. (2000). Imaging the enzymatic digestion of bacterial cellulose ribbons reveals the endo character of the cellobiohydrolase Cel6A from *Humicola insolens* and its mode of synergy with cellobiohydrolase Cel7A. *Appl. Environ. Microbiol.* **66,** 1444–1452.

Boraston, A. B., Bolam, D. N., Gilbert, H. J., and Davies, G. J. (2004). Carbohydrate-binding modules: Fine-tuning polysaccharide recognition. *Biochem. J.* **382,** 769–781.

Bothwell, M. K., Wilson, D. B., Irwin, D. C., and Walker, L. P. (1997). Binding reversibility and surface exchange of *Thermomonospora fusca* E-3 and E-5 and Trichoderma reesei CBHI. *Enzyme Microb. Technol.* **20,** 411–417.

Cantarel, B. L., Coutinho, P. M., Rancurel, C., Bernard, T., Lombard, V., and Henrissat, B. (2009). The Carbohydrate-Active EnZymes database (CAZy): An expert resource for Glycogenomics. *Nucleic Acids Res.* **37,** D233–D238.
Carrard, G., and Linder, M. (1999). Widely different off rates of two closely related cellulose-binding domains from Trichoderma reesei. *Eur. J. Biochem.* **262,** 637–643.
Chanzy, H., and Henrissat, B. (1985). Unidirectional degradation of *valonia* cellulose microcrystals subjected to cellulase action. *FEBS Lett.* **184,** 285–288.
Davies, G., and Henrissat, B. (1995). Structures and mechanisms of glycosyl hydrolases. *Structure* **3,** 853–859.
Divne, C., Ståhlberg, J., Teeri, T. T., and Jones, T. A. (1998). High-resolution crystal structures reveal how a cellulose chain is bound in the 50 angstrom long tunnel of cellobiohydrolase I from *Trichoderma reesei. J. Mol. Biol.* **275,** 309–325.
Doner, L. W., and Irwin, P. L. (1992). Assay of reducing end-groups in oligosaccharide homologues with 2,2′-bicinchoninate. *Anal. Biochem.* **202,** 50–53.
Eijsink, V. G. H., Vaaje-Kolstad, G., Vårum, K. M., and Horn, S. J. (2008). Towards new enzymes for biofuels: Lessons from chitinase research. *Trends Biotechnol.* **26,** 228–235.
Forsberg, Z., Vaaje-Kolstad, G., Westereng, B., Bunæs, A. C., Stenstrøm, Y., MacKenzie, A., Sørlie, M., Horn, S. J., and Eijsink, V. G. H. (2011). Cleavage of cellulose by a CBM33 protein. *Protein Sci.* **20,** 1479–1483.
Fox, J. M., Levine, S. E., Clark, D. S., and Blanch, H. W. (2012). Initial- and processive-cut products reveal cellobiohydrolase rate limitations and the role of companion enzymes. *Biochemistry* **51,** 442–452.
Gruno, M., Väljamäe, P., Pettersson, G., and Johansson, G. (2004). Inhibition of the *Trichoderma reesei* cellulases by cellobiose is strongly dependent on the nature of the substrate. *Biotechnol. Bioeng.* **86,** 503–511.
Harjunpää, V., Teleman, A., Koivula, A., Ruohonen, L., Teeri, T. T., Teleman, O., and Drakenberg, T. (1996). Cello-oligosaccharide hydrolysis by cellobiohydrolase II from *Trichoderma reesei*: Association and rate constants derived from an analysis of progress curves. *Eur. J. Biochem.* **240,** 584–591.
Horn, S. J., and Eijsink, V. G. H. (2004). A reliable reducing end assay for chito-oligosaccharides. *Carbohydr. Polym.* **56,** 35–39.
Horn, S. J., Sikorski, P., Cederkvist, J. B., Vaaje-Kolstad, G., Sørlie, M., Synstad, B., Vriend, G., Vårum, K. M., and Eijsink, V. G. H. (2006a). Costs and benefits of processivity in enzymatic degradation of recalcitrant polysaccharides. *Proc. Natl. Acad. Sci. USA* **103,** 18089–18094.
Horn, S. J., Sørbotten, A., Synstad, B., Sikorski, P., Sørlie, M., Vårum, K. M., and Eijsink, V. G. H. (2006b). Endo/exo mechanism and processivity of family 18 chitinases produced by Serratia marcescens. *FEBS J.* **273,** 491–503.
Hult, E. L., Katouno, F., Uchiyama, T., Watanabe, T., and Sugiyama, J. (2005). Molecular directionality in crystalline beta-chitin: Hydrolysis by chitinases A and B from *Serratia marcescens* 2170. *Biochem. J.* **388,** 851–856.
Igarashi, K., Wada, M., Hori, R., and Samejima, M. (2006). Surface density of cellobiohydrolase on crystalline celluloses—A critical parameter to evaluate enzymatic kinetics at a solid–liquid interface. *FEBS J.* **273,** 2869–2878.
Igarashi, K., Koivula, A., Wada, M., Kimura, S., Penttila, M., and Samejima, M. (2009). High speed atomic force microscopy visualizes processive movement of *Trichoderma reesei* cellobiohydrolase I on crystalline cellulose. *J. Biol. Chem.* **284,** 36186–36190.
Igarashi, K., Uchihashi, T., Koivula, A., Wada, M., Kimura, S., Okamoto, T., Penttila, M., Ando, T., and Samejima, M. (2011). Traffic jams reduce hydrolytic efficiency of cellulase on cellulose surface. *Science* **333,** 1279–1282.
Imai, T., Boisset, C., Samejima, M., Igarashi, K., and Sugiyama, J. (1998). Unidirectional processive action of cellobiohydrolase Cel7A on *Valonia* cellulose microcrystals. *FEBS Lett.* **432,** 113–116.

Irwin, D. C., Spezio, M., Walker, L. P., and Wilson, D. B. (1993). Activity studies of eight purified cellulases: Specificity, synergism, and binding domain effects. *Biotechnol. Bioeng.* **42,** 1002–1013.

Irwin, D., Shin, D. H., Zhang, S., Barr, B. K., Sakon, J., Karplus, P. A., and Wilson, D. B. (1998). Roles of the catalytic domain and two cellulose binding domains of *Thermomonospora fusca* E4 in cellulose hydrolysis. *J. Bacteriol.* **180,** 1709–1714.

Jalak, J., and Väljamäe, P. (2010). Mechanism of initial rapid rate retardation in cellobiohydrolase catalyzed cellulose hydrolysis. *Biotechnol. Bioeng.* **106,** 871–883.

Kipper, K., Väljamäe, P., and Johansson, G. (2005). Processive action of cellobiohydrolase Cel7A from *Trichoderma reesei* is revealed as 'burst' kinetics on fluorescent polymeric model substrates. *Biochem. J.* **385,** 527–535.

Koivula, A., Kinnari, T., Harjunpää, V., Ruohonen, L., Teleman, A., Drakenberg, T., Rouvinen, J., Jones, T. A., and Teeri, T. T. (1998). Tryptophan 272: An essential determinant of crystalline cellulose degradation by *Trichoderma reesei* cellobiohydrolase Cel6A. *FEBS Lett.* **429,** 341–346.

Kongruang, S., Han, M. J., Breton, C. I. G., and Penner, M. H. (2004). Quantitative analysis of cellulose-reducing ends. *Appl. Biochem. Biotechnol.* **113,** 213–231.

Kostylev, M., and Wilson, D. B. (2011). Determination of the catalytic base in family 48 glycosyl hydrolases. *Appl. Environ. Microbiol.* **77,** 6274–6276.

Kurašin, M., and Väljamäe, P. (2011). Processivity of cellobiohydrolases is limited by the substrate. *J. Biol. Chem.* **286,** 169–177.

Kyriacou, A., Neufeld, R. J., and Mackenzie, C. R. (1989). Reversibility and competition in the adsorption of *Trichoderma reesei* cellulase components. *Biotechnol. Bioeng.* **33,** 631–637.

Li, Y. C., Irwin, D. C., and Wilson, D. B. (2007). Processivity, substrate binding, and mechanism of cellulose hydrolysis by *Thermobifida fusca* Cel9A. *Appl. Environ. Microbiol.* **73,** 3165–3172.

Linder, M., and Teeri, T. T. (1996). The cellulose-binding domain of the major cellobiohydrolase of Trichoderma reesei exhibits true reversibility and a high exchange rate on crystalline cellulose. *Proc. Natl. Acad. Sci. USA* **93,** 12251–12255.

Lucius, A. L., Maluf, N. K., Fischer, C. J., and Lohman, T. M. (2003). General methods for analysis of sequential "n-step" kinetic mechanisms: Application to single turnover kinetics of helicase-catalyzed DNA unwinding. *Biophys. J.* **85,** 2224–2239.

Ma, A. Z., Hu, Q., Qu, Y. B., Bai, Z. H., Liu, W. F., and Zhuang, G. Q. (2008). The enzymatic hydrolysis rate of cellulose decreases with irreversible adsorption of cellobiohydrolase I. *Enzyme Microb. Technol.* **42,** 543–547.

Mandelman, D., Belaich, A., Belaich, J. P., Aghajari, N., Driguez, H., and Haser, R. (2003). X-ray crystal structure of the multidomain endoglucanase Cel9G from *Clostridium cellulolyticum* complexed with natural and synthetic cello-oligosaccharides. *J. Bacteriol.* **185,** 4127–4135.

McClure, W. R., and Chow, Y. (1980). The kinetics and processivity of nucleic acid polymerases. *In* "Methods in Enzymology," (L. P. Daniel, ed.), pp. 277–297. Academic Press, San Diego, CA.

Medve, J., Karlsson, J., Lee, D., and Tjerneld, F. (1998). Hydrolysis of microcrystalline cellulose by cellobiohydrolase I and endoglucanase II from *Trichoderma reesei*: Adsorption, sugar production pattern, and synergism of the enzymes. *Biotechnol. Bioeng.* **59,** 621–634.

Merino, S., and Cherry, J. (2007). Progress and challenges in enzyme development for biomass utilization. *In* "Biofuels," (L. Olsson, ed.), pp. 95–120. Springer, Berlin/Heidelberg.

Michel, G., Helbert, W., Kahn, R., Dideberg, O., and Kloareg, B. (2003). The structural bases of the processive degradation of iota-carrageenan, a main cell wall polysaccharide of red algae. *J. Mol. Biol.* **334,** 421–433.

Moran-Mirabal, J. M., Bolewski, J. C., and Walker, L. P. (2011). Reversibility and binding kinetics of *Thermobifida fusca* cellulases studied through fluorescence recovery after photobleaching microscopy. *Biophys. Chem.* **155,** 20–28.

Murphy, L., Cruys-Bagger, N., Damgaard, H. D., Baumann, M. J., Olsen, S. N., Borch, K., Lassen, S. F., Sweeney, M., Tatsumi, H., and Westh, P. (2012). Origin of initial burst in activity for *Trichoderma reesei* endo-glucanases hydrolyzing insoluble cellulose. *J. Biol. Chem.* **287,** 1252–1260.

Nutt, A., Sild, V., Pettersson, G., and Johansson, G. (1998). Progress curves—A mean for functional classification of cellulases. *Eur. J. Biochem.* **258,** 200–206.

Palonen, H., Tenkanen, M., and Linder, M. (1999). Dynamic interaction of *Trichoderma reesei* cellobiohydrolases Ce16A and Ce17A and cellulose at equilibrium and during hydrolysis. *Appl. Environ. Microbiol.* **65,** 5229–5233.

Papanikolau, Y., Prag, G., Tavlas, G., Vorgias, C. E., Oppenheim, A. B., and Petratos, K. (2001). High resolution structural analyses of mutant chitinase A complexes with substrates provide new insight into the mechanism of catalysis. *Biochemistry* **40,** 11338–11343.

Parsiegla, G., Juy, M., Reverbel-Leroy, C., Tardif, C., Belaich, J. P., Driguez, H., and Haser, R. (1998). The crystal structure of the processive endocellulase CelF of Clostridium cellulolyticum in complex with a thiooligosaccharide inhibitor at 2.0 angstrom resolution. *EMBO J.* **17,** 5551–5562.

Parsiegla, G., Reverbel-Leroy, C., Tardif, C., Belaich, J. P., Driguez, H., and Haser, R. (2000). Crystal structures of the cellulase Ce148F in complex with inhibitors and substrates give insights into its processive action. *Biochemistry* **39,** 11238–11246.

Parsiegla, G., Reverbel, C., Tardif, C., Driguez, H., and Haser, R. (2008). Structures of mutants of cellulase Ce148F of *Clostridium cellulolyticum* in complex with long hemithiocellooligosaccharides give rise to a new view of the substrate pathway during processive action. *J. Mol. Biol.* **375,** 499–510.

Payne, C. M., Bomble, Y. J., Taylor, C. B., McCabe, C., Himmel, M. E., Crowley, M. F., and Beckham, G. T. (2011). Multiple functions of aromatic–carbohydrate interactions in a processive cellulase examined with molecular simulation. *J. Biol. Chem.* **286,** 41028–41035.

Praestgaard, E., Elmerdahl, J., Murphy, L., Nymand, S., McFarland, K. C., Borch, K., and Westh, P. (2011). A kinetic model for the burst phase of processive cellulases. *FEBS J.* **278,** 1547–1560.

Quinlan, R. J., Sweeney, M. D., Lo Leggio, L., Otten, H., Poulsen, J. C. N., Johansen, K. S., Krogh, K., Jorgensen, C. I., Tovborg, M., Anthonsen, A., Tryfona, T., Walter, C. P., et al. (2011). Insights into the oxidative degradation of cellulose by a copper metalloenzyme that exploits biomass components. *Proc. Natl. Acad. Sci. USA* **108,** 15079–15084.

Robyt, J. F., and French, D. (1967). Multiple attach hypothesis of alpha-amylase action: Action of porcine pancreatic, human salivary, and Aspergillus oryzae alpha-amylases. *Arch. Biochem. Biophys.* **122,** 8–16.

Rouvinen, J., Bergfors, T., Teeri, T., Knowles, J. K. C., and Jones, T. A. (1990). Three-dimensional structure of cellobiohydrolase II from *Trichoderma reesei*. *Science* **249,** 380–386.

Sakamoto, R., Arai, M., and Murao, S. (1989). Reduced cellulose as a substrate of cellulases. *Agric. Biol. Chem.* **53,** 1407–1409.

Sannan, T., Kurita, K., and Iwakura, Y. (1976). Studies on chitin. 2. Effect of deacetylation on solubility. *Macromol. Chem. Phys.* **177,** 3589–3600.

Sikorski, P., Stokke, B. T., Sørbotten, A., Vårum, K. M., Horn, S. J., and Eijsink, V. G. H. (2005). Development and application of a model for chitosan hydrolysis by a family 18 chitinase. *Biopolymers* **77,** 273–285.

Sikorski, P., Sørbotten, A., Horn, S. J., Eijsink, V. G. H., and Vårum, K. M. (2006). *Serratia marcescens* chitinases with tunnel-shaped substrate-binding grooves show endo activity and different degrees of processivity during enzymatic hydrolysis of chitosan. *Biochemistry* **45**, 9566–9574.

Sørbotten, A., Horn, S. J., Eijsink, V. G. H., and Vårum, K. M. (2005). Degradation of chitosans with chitinase B from *Serratia marcescens*—Production of chito-oligosaccharides and insight into enzyme processivity. *FEBS J.* **272**, 538–549.

Sørlie, M., Zakariassen, H., Norberg, A. L., and Eijsink, V. G. H. (2012). Processivity and substrate-binding in family 18 chitinases. *Biocat. Biotrans.* Doi:10.3109/10242422.2012.676282.

Ståhlberg, J., Johansson, G., and Pettersson, G. (1991). A new model for enzymatic hydrolysis of cellulose based on the two-domain structure of cellobiohydrolase I. *Biotechnology* **9**, 286–290.

Teeri, T. T. (1997). Crystalline cellulose degradation: New insight into the function of cellobiohydrolases. *Trends Biotechnol.* **15**, 160–167.

Tews, I., van Scheltinga, A. C. T., Perrakis, A., Wilson, K. S., and Dijkstra, B. W. (1997). Substrate-assisted catalysis unifies two families of chitinolytic enzymes. *J. Am. Chem. Soc.* **119**, 7954–7959.

Uchiyama, T., Katouno, F., Nikaidou, N., Nonaka, T., Sugiyama, J., and Watanabe, T. (2001). Roles of the exposed aromatic residues in crystalline chitin hydrolysis by chitinase a from *Serratia marcescens* 2170. *J. Biol. Chem.* **276**, 41343–41349.

Vaaje-Kolstad, G., Westereng, B., Horn, S. J., Liu, Z. L., Zhai, H., Sørlie, M., and Eijsink, V. G. H. (2010). An oxidative enzyme boosting the enzymatic conversion of recalcitrant polysaccharides. *Science* **330**, 219–222.

van Aalten, D. M. F., Komander, D., Synstad, B., Gåseidnes, S., Peter, M. G., and Eijsink, V. G. H. (2001). Structural insights into the catalytic mechanism of a family 18 exo-chitinase. *Proc. Natl. Acad. Sci. USA* **98**, 8979–8984.

Varrot, A., Hastrup, S., Schulein, M., and Davies, G. J. (1999). Crystal structure of the catalytic core domain of the family 6 cellobiohydrolase II, Cel6A, from *Humicola insolens*, at 1.92 angstrom resolution. *Biochem. J.* **337**, 297–304.

Varrot, A., Frandsen, T. P., von Ossowski, I., Boyer, V., Cottaz, S., Driguez, H., Schulein, M., and Davies, G. J. (2003). Structural basis for ligand binding and processivity in cellobiohydrolase Cel6A from *Humicola insolens*. *Structure* **11**, 855–864.

Vårum, K. M., Anthonsen, M. W., Grasdalen, H., and Smidsrød, O. (1991a). High-field NMR-spectroscopy of partially n-deacetylated chitins (chitosans). 1. Determination of the degree of n-acetylation and the distribution of n-acetyl groups in partially n-deacetylated chitins (chitosans) by high-field nmr-spectroscopy. *Carbohydr. Res.* **211**, 17–23.

Vårum, K. M., Anthonsen, M. W., Grasdalen, H., and Smidsrød, O. (1991b). High-field NMR-spectroscopy of partially n-deacetylated chitins (chitosans). 3. C-13-NMR studies of the aceyltation sequences in partially n-deacetylated chitins (chitosans). *Carbohydr. Res.* **217**, 19–27.

Velleste, R., Teugjas, H., and Väljamäe, P. (2010). Reducing end-specific fluorescence labeled celluloses for cellulase mode of action. *Cellulose* **17**, 125–138.

von Ossowski, I., Ståhlberg, J., Koivula, A., Piens, K., Becker, D., Boer, H., Harle, R., Harris, M., Divne, C., Mahdi, S., Zhao, Y. X., Driguez, H., et al. (2003). Engineering the exo-loop of *Trichoderma reesei* cellobiohydrolase, Ce17A. A comparison with *Phanerochaete chrysosporium* Cel7D. *J. Mol. Biol.* **333**, 817–829.

Voutilainen, S. P., Puranen, T., Siika-Aho, M., Lappalainen, A., Alapuranen, M., Kallio, J., Hooman, S., Viikri, L., Vehmaanpera, J., and Koivula, A. (2008). Cloning, expression, and characterization of novel thermostable family 7 cellobiohydrolases. *Biotechnol. Bioeng.* **101**, 515–528.

Vuong, T. V., and Wilson, D. B. (2009). Processivity, synergism, and substrate specificity of *Thermobifida fusca* Cel6B. *Appl. Environ. Microbiol.* **75,** 6655–6661.
Watson, B. J., Zhang, H. T., Longmire, A. G., Moon, Y. H., and Hutcheson, S. W. (2009). Processive endoglucanases mediate degradation of cellulose by *Saccharophagus degradans*. *J. Bacteriol.* **191,** 5697–5705.
Westereng, B., Ishida, T., Vaaje-Kolstad, G., Wu, M., Eijsink, V. G. H., Igarashi, K., Samejima, M., Ståhlberg, J., Horn, S. J., and Sandgren, M. (2011). The putative endoglucanase PcGH61D from *Phanerochaete chrysosporium* is a metal-dependent oxidative enzyme that cleaves cellulose. *PLoS One* **6,** e27807. 10.1371/journal.pone.0027807.
Zakariassen, H., Aam, B. B., Horn, S. J., Vårum, K. M., Sørlie, M., and Eijsink, V. G. H. (2009). Aromatic residues in the catalytic center of chitinase A from *Serratia marcescens* affect processivity, enzyme activity, and biomass converting efficiency. *J. Biol. Chem.* **284,** 10610–10617.
Zakariassen, H., Eijsink, V. G. H., and Sørlie, M. (2010). Signatures of activation parameters reveal substrate-dependent rate determining steps in polysaccharide turnover by a family 18 chitinase. *Carbohydr. Polym.* **81,** 14–20.
Zhang, Y. H. P., and Lynd, L. R. (2004). Toward an aggregated understanding of enzymatic hydrolysis of cellulose: Noncomplexed cellulase systems. *Biotechnol. Bioeng.* **88,** 797–824.
Zhang, Y. H. P., and Lynd, L. R. (2005). Determination of the number-average degree of polymerization of cellodextrins and cellulose with application to enzymatic hydrolysis. *Biomacromolecules* **6,** 1510–1515.
Zhang, S., Irwin, D. C., and Wilson, D. B. (2000). Site-directed mutation of noncatalytic residues of *Thermobifida fusca* exocellulase Cel6B. *Eur. J. Biochem.* **267,** 3101–3115.
Zhang, Y. H. P., Cui, J. B., Lynd, L. R., and Kuang, L. R. (2006). A transition from cellulose swelling to cellulose dissolution by o-phosphoric acid: Evidence from enzymatic hydrolysis and supramolecular structure. *Biomacromolecules* **7,** 644–648.
Zhang, X. Z., Zhang, Z. M., Zhu, Z. G., Sathitsuksanoh, N., Yang, Y. F., and Zhang, Y. H. P. (2010). The noncellulosomal family 48 cellobiohydrolase from *Clostridium phytofermentans* ISDg: Heterologous expression, characterization, and processivity. *Appl. Microbiol. Biotechnol.* **86,** 525–533.
Zhou, W. L., Irwin, D. C., Escovar-Kousen, J., and Wilson, D. B. (2004). Kinetic studies of *Thermobifida fusca* Cel9A active site mutant enzymes. *Biochemistry* **43,** 9655–9663.
Zhu, Z. G., Sathitsuksanoh, N., and Zhang, Y. H. P. (2009). Direct quantitative determination of adsorbed cellulase on lignocellulosic biomass with its application to study cellulase desorption for potential recycling. *Analyst* **134,** 2267–2272.
Zou, J. Y., Kleywegt, G. J., Ståhlberg, J., Driguez, H., Nerinckx, W., Claeyssens, M., Koivula, A., Teerii, T. T., and Jones, T. A. (1999). Crystallographic evidence for substrate ring distortion and protein conformational changes during catalysis in cellobiohydrolase Cel6A from *Trichoderma reesei*. *Structure* **7,** 1035–1045.

CHAPTER SIX

Distinguishing Xyloglucanase Activity in *endo*-β(1 → 4)glucanases

Jens M. Eklöf,*,† Marcus C. Ruda,‡ *and* Harry Brumer*,†

Contents

1. Introduction	98
2. Analytical Tools for Xyloglucan Oligo- and Polysaccharide Analysis	101
2.1. Thin layer chromatography	101
2.2. High-performance liquid chromatography	101
2.3. Mass spectrometry	103
2.4. Nuclear magnetic resonance spectroscopy	105
3. Substrate Preparation	105
3.1. Xyloglucan solubilization	106
3.2. Preparation of a mixture of Glc_4-based xylogluco-oligosaccharides	106
3.3. Isolation of individual Glc_4-based XGOs	107
3.4. Preparation of longer xylogluco-oligosaccharides	108
4. Determination of Specific Xyloglucan Activity in a Glucanase	109
4.1. Use of the bicinchoninic acid reducing-sugar assay to measure enzymatic xyloglucan hydrolysis	109
4.2. Determination of the *endo*- versus *exo*-cleavage preference of xyloglucanases using GPC	110
4.3. Determination of bond cleavage specificity of xyloglucanases by HPAEC-PAD analysis of polysaccharide and XXXGXXXG limit-digestion products	111
4.4. Determination of the extent of xyloglucan transglycosylation versus hydrolysis using Glc_8-based XGOs and HPAEC-PAD	112
4.5. Use of aryl β-glycosides of xyloglucan oligosaccharides to determine the contribution of individual enzyme subsites to catalysis	114
5. Coda: Xyloglucan Hydrolysis by β-glucanases—Obvious but Often Overlooked	116

* Michael Smith Laboratories and Department of Chemistry, University of British Columbia, Vancouver, Canada
† Division of Glycoscience, School of Biotechnology, Royal Institute of Technology (KTH), AlbaNova University Center, Stockholm, Sweden
‡ SweTree Technologies, P.O. Box 4095, Umeå, Sweden

Acknowledgments 116
References 117

Abstract

The ability of β-glucanases to cleave xyloglucans, a family of highly decorated β-glucans ubiquitous in plant biomass, has traditionally been overlooked in functional biochemical studies. An emerging body of data indicates, however, that a spectrum of xyloglucan specificity resides in diverse glycoside hydrolases from a range of carbohydrate-active enzyme families—including classic "cellulase" families. This chapter outlines a series of enzyme kinetic and product analysis methods to establish degrees of xyloglucan specificity and modes of action of glycosidases emerging from enzyme discovery projects.

1. INTRODUCTION

The practicing biochemist will appreciate that the substrate specificity of individual enzymes within any class, glycoside hydrolases (GHs) included, ranges from broad to strict. Indeed, the name given to an enzyme is purely operational, based upon the maximal activity observed (occasionally, "mixed-function" enzymes are described), which is, in turn, dependent upon the panel of potential substrates tested. *endo*-Cellulases—*endo*-β(1 → 4)-glucanases, EC 3.2.1.4—are often defined as such solely on the basis of activity on water-soluble (and thus easily assayed) artificial derivatives such as hydroxyethyl cellulose (HEC) and carboxymethyl cellulose (CMC), rather than the experimentally challenging, crystalline natural substrate. And while diverse soluble polysaccharides, such as β(1 → 4)xylans, β(1 → 4)mannans, and mixed-linkage β(1 → 3)/β(1 → 4)glucans, are often employed in parallel assays to establish a degree of "cellulase" specificity, nature's water-soluble cellulose (β(1 → 4) glucan) derivatives, the xyloglucans, have, historically, rarely been included in such comparisons. We would argue that this is a rather important oversight, given the ubiquity of xyloglucans in biomass, and the likely frequency with which these "cellulases" might encounter these polysaccharides.

The xyloglucans are heavily α(1 → 6)xylose-substituted β(1 → 4)glucans (Fig. 6.1), which are present in essentially all land plants (Popper *et al.*, 2011). In dicotyledons and in nongrassy monocots, specific xyloglucans can comprise as much as 20% of the dry weight of the primary cell wall, where they interact tightly with cellulose to confer specific structural and mechanical properties to the wall (Cosgrove, 2005; Pauly *et al.*, 1999; Vincken *et al.*, 1997b). Gramineous monocots, secondary cell walls, and reaction tissues all have trace levels of xyloglucans, while in the seeds of certain species, xyloglucans can, as principal storage polysaccharides, comprise a remarkable 50% of the dry weight (Buckeridge *et al.*, 2000; Nishikubo *et al.*, 2007, 2011; Vogel, 2008).

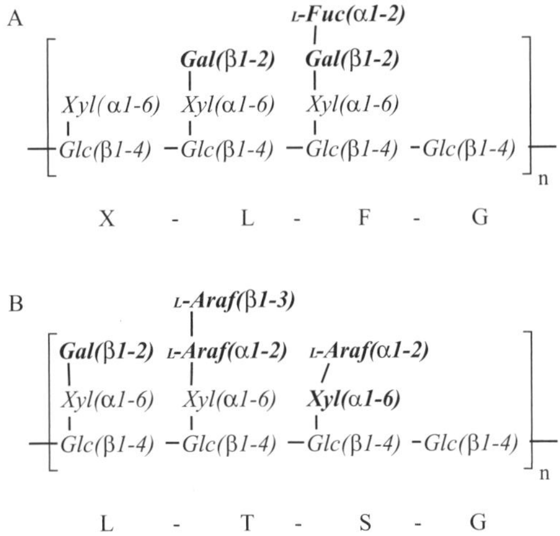

A

$$\begin{bmatrix} & & \textit{L-Fuc}(\alpha\textit{1-2}) & \\ & & | & \\ & \textit{Gal}(\beta\textit{1-2}) & \textit{Gal}(\beta\textit{1-2}) & \\ & | & | & \\ Xyl(\alpha 1\text{-}6) & Xyl(\alpha 1\text{-}6) & Xyl(\alpha 1\text{-}6) & \\ | & | & | & \\ -Glc(\beta 1\text{-}4) & -Glc(\beta 1\text{-}4) & -Glc(\beta 1\text{-}4) & -Glc(\beta 1\text{-}4) \end{bmatrix}_n$$

X - L - F - G

B

$$\begin{bmatrix} & \textit{L-Araf}(\beta\textit{1-3}) & & \\ & | & & \\ \textit{Gal}(\beta\textit{1-2}) & \textit{L-Araf}(\alpha\textit{1-2}) & \textit{L-Araf}(\alpha\textit{1-2}) & \\ | & | & / & \\ Xyl(\alpha 1\text{-}6) & Xyl(\alpha 1\text{-}6) & Xyl(\alpha 1\text{-}6) & \\ | & | & | & \\ -Glc(\beta 1\text{-}4) & -Glc(\beta 1\text{-}4) & -Glc(\beta 1\text{-}4) & -Glc(\beta 1\text{-}4) \end{bmatrix}_n$$

L - T - S - G

Figure 6.1 Common xyloglucan repeating structures highlighting variable side chain substituents in bold font (Vincken et al., 1997b; Hoffman et al., 2005; Hilz et al., 2007; Hsieh and Harris, 2009). (A) "XXXG"-type xyloglucans, based on a Glc_4Xyl_3 repeat, found in most angiosperms. (B) "XXGG"-type xyloglucans, based on a Glc_4Xyl_2 repeat, found in Solanaceae. Substructures are denoted by the commonly used single letter notation (Fry et al., 1993).

Structurally, the xyloglucans can be divided into two major classes, those based on either the "XXGG" or the "XXXG" repeats (Fig. 6.1), where "G" represents an unbranched $\beta(1 \rightarrow 4)$-linked backbone glucosyl residue and "X" represents a [Xyl$\alpha(1 \rightarrow 6)$]Glc$\beta(1 \rightarrow 4)$ moiety (Vincken et al., 1997b). The abundant, commercially available tamarind seed xyloglucan bears variable $\beta(1 \rightarrow 2)$-linked galactosyl extensions on the two "internal" Xyl residues of the XXXG repeat ("L" units, Fig. 6.1A), while primary cell wall xyloglucans are further distinguished by the regiospecific addition of an $\alpha(1 \rightarrow 2)$-linked fucosyl residue to the first Gal residue in XLLG (and XXLG) to give the "F" unit (Fig. 6.1A). Xyloglucans from the Solanaceae (e.g., paprika, chili pepper, potato, tomato, eggplant, and tobacco) are typically built upon the XXGG repeat, with the addition of arabinofuranosyl residues giving rise to the "S" and "T" units (Fig. 6.1B). A full catalogue of species-specific xyloglucan structures is beyond the scope of this chapter (see Hilz et al., 2007; Hoffman et al., 2005; Hsieh and Harris, 2009; Vincken et al., 1997b for a comprehensive summary), but it is essential to bear in mind the rich side chain diversity of xyloglucans when assaying the specificity of individual endo-$\beta(1 \rightarrow 4)$glucanases. Throughout this chapter, mixtures of xyloglucan oligosaccharides (XGOs) with variable branching, obtained by enzymatic digestion of native polysaccharides, will be referred to by the length of the glucan backbone, for example, XGO_{Glc4}, XGO_{Glc8}, XGO_{Glc12}, etc.

The potential of endo-glucanases to cleave xyloglucan was first underscored in the seminal work of Vincken et al. (1994, 1997a) late in the past century. Presently, a diversity of xyloglucan-specific hydrolytic activities have been demonstrated in members of GH families 5, 7, 9, 12, 16, 44, and 74 (Ariza *et al.*, 2011; Benkö *et al.*, 2008; Gilbert *et al.*, 2008; Irwin *et al.*, 2003; Vlasenko *et al.*, 2010; York and Hawkins, 2000 and references therein):

- "xyloglucan-specific *endo*-$\beta(1\rightarrow 4)$-glucanase" activity (EC 3.2.1.151, hereafter "*endo*-xyloglucanase" activity),
- xyloglucan-specific *exo*-$\beta(1\rightarrow 4)$-glucanase activity (EC 3.2.1.155, hereafter "*exo*-xyloglucanase" activity),
- "oligoxyloglucan reducing-end-specific cellobiohydrolase" activity (EC 3.2.150).

Many of these families are classic "cellulase" families (reviewed in Henrissat, 1997), and it is not unreasonable to expect that xyloglucanase activity may be found in other known $\beta(1\rightarrow 4)$-glucanase-containing GH families, as well as in novel, emerging families (Vlasenko *et al.*, 2010).

Currently, demonstrated xyloglucanases range from those that are very specific for the polysaccharide by productively harnessing binding of side chains (Baumann *et al.*, 2007; Gloster *et al.*, 2007; Irwin *et al.*, 2003; Master *et al.*, 2008; Wong *et al.*, 2010) to those that passively accommodate, or actively legislate against, certain backbone branching patterns (Ariza *et al.*, 2011; Gloster *et al.*, 2007; Irwin *et al.*, 2003; Martinez-Fleites *et al.*, 2006). Again by analogy with cellulases and other polysaccharidases, certain xyloglucanases demonstrate degrees of *exo*- versus *endo*-activity (Grishutin *et al.*, 2004), specifically due to the presence of active-site cleft "blocking loops" (Desmet *et al.*, 2007; Yaoi *et al.*, 2007). However, likely reflecting the branched nature of the substrate, the active site of xyloglucanases does not display the tunnel-like topology evident in cellobiohydrolases. A further level of complexity is brought by some xyloglucan backbone-cleaving enzymes, most notably plant members of GH16, that demonstrate significant or even exclusive xyloglucan *endo*-transglycosylase (xyloglucan:xyloglucosyltransferase) activity (EC 2.4.1.207; see Eklöf and Brumer, 2010 for a recent review).

In this context, this chapter outlines a basic experimental framework to distinguish modes of xyloglucan activity in a protein. Given the inherent complexity and diversity of xyloglucans as substrates, such an analysis can clearly expand to become quite involved, especially if one is interested in, for example, mapping the energetic contributions of individual enzymeside chain interactions to catalysis (Ariza *et al.*, 2011; Ibatullin *et al.*, 2008; Saura-Valls *et al.*, 2008). We restrict ourselves here to methods for substrate preparation (for assays and structural biology), enzyme kinetic analysis, and reaction product analysis, which can be performed in a reasonably wellequipped (bio)chemistry laboratory or department, using the commercially available tamarind xyloglucan. Where useful, literature references to more involved methods, for example, those for advanced substrate preparation, are included for those wishing to delve deeper.

2. Analytical Tools for Xyloglucan Oligo- and Polysaccharide Analysis

The analysis of xyloglucan oligo- and polysaccharides is an essential component for the preparation of substrates for xyloglucan hydrolase and xyloglucan transglycosylase analysis, as well as the determination of enzymatic reaction products. While not meant to be restrictive, the following sections highlight routine analytical methods for xyloglucan saccharide analysis, including sample preparation suggestions, some of which have been further optimized in our laboratory. Basic familiarity with each method is assumed, so only very brief descriptions of key technical parameters are presented. The methods are listed in approximate order of increasing requirements for equipment resources and operator skill.

2.1. Thin layer chromatography

XGO based on a four glucosyl unit backbone (XGO_{Glc4}) from tamarind kernel power can be conveniently resolved according to their monosaccharide stoichiometry by TLC on normal-phase silica using 2:1 acetonitrile/water as the eluent (R_f values: XXXG, 0.14; XXLG/XLXG, 0.17; XLLG, 0.22). Longer XGOs (i.e., XGO_{Glc8} and above) remain on the baseline under these conditions. Paper chromatography offers an alternative to silica TLC (Fry, 1986).

2.2. High-performance liquid chromatography

The wide range of molar masses encountered in the analysis of XGO (e.g., XXXG, 1062 Da) to polysaccharides (up to 500 kDa) necessitates the use of multiple high-performance liquid chromatography (HPLC) methods. The high-performance anion exchange chromatography (HPAEC; widely known as "Dionex" carbohydrate chromatography) and gel permeation chromatography (GPC) methods described in the following sections provide convenient overlap in the medium molar-mass range (XGO_{Glc8}–XGO_{Glc20}), thus allowing analysis across all oligo- and polysaccharide distributions.

2.2.1. High-performance anion exchange chromatography with pulsed amperometric detection

High-performance anion exchange chromatography with pulsed amperometric detection (HPAEC-PAD) run at elevated pH values is, in our experience, superior to other methods for routine analytical XGO separation. A number of different methods using Dionex columns such as CarboPac™ PA1 (Marry et al., 2003; McDougall and Fry, 1991) and PA100 (Baumann et al., 2007) have been described, while we have found the newer-generation CarboPac™ PA200 column superior for separation of both oligosaccharides smaller than

XGO$_{Glc4}$ and for separation of XGO$_{Glc4}$ to longer XGOs (potentially XGO$_{Glc40}$; Fig. 6.2). The following eluent gradients have been optimized for use on the Dionex ICS-3000 with a ternary solvent system to resolve XGOs of different lengths. Eluent reservoirs A, B, and C contain ultrapure water, 1 M NaOH, and 1 M NaOAc (sodium acetate), respectively, and mixtures are balanced to 100% with A (water).

- *Gradient 1, for separation of XGO$_{Glc4}$ or shorter:* 0–5 min, 10% B+2% C; 5–12 min, 10% B+2–30% (linear gradient) C; 12–12.1 min, 50% B+50% C; 12.1–13 min, a curved gradient (Dionex Chromeleon software profile 1) of B and C back to initial conditions; 13–17 min, equilibration at initial conditions.
- *Gradient 2, for separation of XGO$_{Glc4}$ or longer:* 0–4 min, 10% B+6% C; 4–17 min, 10% B+6–25% (linear gradient) C; 17–18 min, 50% B and 50% C; 18–22 min, equilibration at initial conditions.

Representative HPAEC chromatograms of enzymatically hydrolyzed xyloglucan are shown in Fig. 6.2.

2.2.2. Gel permeation chromatography a.k.a. size-exclusion chromatography

Many columns with suitable molar-mass ranges are available for GPC of xyloglucan oligo- and polysaccharides in either water or DMSO. We prefer the use of 100% DMSO as an eluent to minimize polysaccharide chain

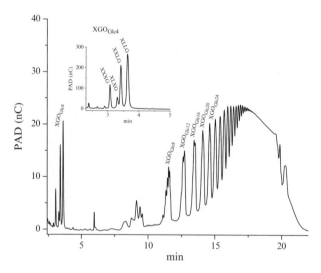

Figure 6.2 Separation of tamarind xyloglucan oligosaccharides (XGOs) produced by the controlled action of an *endo*-xyloglucanase from GH5 (Gloster *et al.*, 2007). The inset shows the identity of the limit-digestion products (Glc$_4$-based XGOs) and their relative amounts.

aggregation, despite the drawback that samples must first be lyophilized from water and redissolved in DMSO. This is usually trivial for shorter oligosaccharides but may require gentle heating and intensive mixing for longer polysaccharide samples.

GPC is conveniently performed with a single-pump system operating at a flow rate of 1 mL/min, with two Tosoh TSKgel columns connected in series (5000 and 3000 H_{HR}, both 7.8 × 300 mm) and maintained at 70 °C with a column oven; run times are typically 22 min. The choice of these columns with two different matrix porosities provides a good upper molar mass limit ($>1 \times 10^6$) together with resolution of low molar mass XGO repeats. Analyte detection is performed with an online Polymer Labs (PL) evaporative light-scattering 1000 detector, whose high-operating temperature is particularly suited for evaporation of the high-boiling eluent (*Settings*: airflow, 1.5 bar; evaporator, 170 °C; nebulizer, 150 °C). The system can be conveniently calibrated with pullulan molar mass standards over the range 180–788,000 Da (Polymer Laboratories). Representative GPC chromatograms of enzymatically hydrolyzed xyloglucan are shown in Fig. 6.3.

If a GPC system capable of handling DMSO is not available, the following water-based system is broadly useful. Columns, connected in series: 1 PL aquagel-OH MIXED 8 μm 50 × 7.5 mm guard column, 2 PL aquagel-OH MIXED 8 μm 300 × 7.5 mm, and 1 PL aquagel-OH 20 5 μm 300 × 7.5 mm; eluent: 100 mM NaCO$_3$ and 50 mM NH$_4$OAc in ultrapure water, flow rate 0.5 mL/min, run time 85 min, refractive index detection.

2.3. Mass spectrometry

Intact XGO$_{Glc4}$ are conveniently ionized by the biomolecule-friendly techniques electrospray ionization (ESI) and matrix-assisted laser desorption/ionization (MALDI); longer XGOs may also be analyzed by mass spectrometry (MS) but suffer from decreasing ionization and detection efficiency, dependent, to some extent, on the equipment used. As with HPLC, a range of MS systems can be employed, which combine either of these ionization sources with a range of mass analyzers (e.g., quadrupoles, time-of-flight (TOF) tubes, and ion traps; see Harvey, 2011; Hilz *et al.*, 2007; Hsieh and Harris, 2009 and references therein). Historically, fast-atom bombardment ionization of XGOs has also been used (York *et al.*, 1990).

2.3.1. ESI

For ESI, XGO$_{Glc4}$ (ca. 10 μM) are routinely sprayed from 1:1 methanol/water solution. The addition of 0.5 mM NaCl (final concentration) in this solution facilitates the generation of sodiated adducts. Varying the cone voltage in the range (35–130 V) can be effectively used to favor [M + Na]$^+$

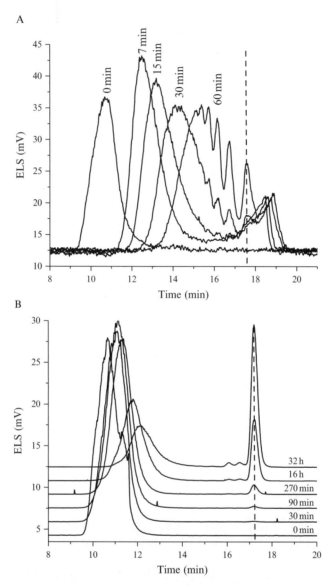

Figure 6.3 Representative GPC chromatograms from *endo*- and *exo*-acting xyloglucanases. (A) *Endo*-hydrolytic activity of a *Paenibacillus pabuli* GH5 xyloglucanase shifts the molar mass of native tamarind polysaccharide toward lower values, without the appearance of Glc_4-based XGOs (dotted line) until the end of the depolymerization. (The peak at 18.5–19 min is due to buffer. Data from Gloster et al., 2007.) (B) The *exo*-hydrolytic activity of a GH74 xyloglucanase from *Chrysosporium lucknowense* (Grishutin et al., 2004) produces XGO_{Glc4} (dotted line) at early stages of the reaction without a dramatic shift in the molar mass of the polysaccharide (Eklöf and Brumer, unpublished data; *C. lucknowense* GH74 xyloglucanase purchased from Dyadic NL).

or $[M+2Na]^{2+}$ adducts, while also improving signal overall, at least with the Micromass/Waters Z-spray ion source (Martinez-Fleites et al., 2006). Variation of the cone voltage can also be used to improve the signal of XGO_{Glc8} and longer XGOs, which can be detected as multiple-charged sodiated adducts, although signal intensity rapidly drops off with increasing molar mass. With ESI and high-resolution mass analyzers (e.g., orthogonal acceleration TOF), monoisotopic masses of the $[M+Na]^+$ adducts of XXXG, XLXG and XXLG (isobaric), and XLLG are observed at mass-to-charge ratio (m/z) values 1085.37, 1247.42, and 1409.48, respectively.

2.3.2. Matrix-assisted laser desorption/ionization

For UV-MALDI-TOF analysis, the common matrix 2,5-dihydroxybenzoic acid (DHB) provides good ionization of XGO_{Glc4} (Harvey, 2011), while, as with ESI, longer XGO repeats can be observed with decreasing sensitivity. Monosodiated ($[M+Na]^+$) adducts are typically the most abundant, with $[M+K]^+$ adducts often observed unless a deionization step (e.g., using ion-exchange resin) is included prior to MALDI sample preparation. Admixture of the sample and matrix solution prior to spotting is most convenient: Mix equal volumes of DHB (10 g/L) and oligosaccharide(s) (typically 0.5 mM in water), deposit a droplet (microliter volume) on a MALDI plate, and allow to air-dry (a gentle stream of air can speed the process) before inserting into the spectrometer for analysis.

2.4. Nuclear magnetic resonance spectroscopy

XGO comprised up to ca. 10 monosaccharides is readily amenable to nuclear magnetic resonance (NMR) analysis in solution, due in part to high water solubility. Acquisition of NMR data requires significant equipment resources (often available institutionally) and a high level of analyst skill in data interpretation. For the nonspecialist, a significant body of literature, beginning with seminal studies of tamarind seed XGOs in particular (York et al., 1990, 1993), already exists for comparative spectral analysis. These data include individual peak assignments from one- and two-dimensional experiments on diverse XGOs (see Hilz et al., 2007; Hoffman et al., 2005 and references therein).

3. SUBSTRATE PREPARATION

Xyloglucans from diverse plant sources can be isolated in milligram- to gram-scale quantities using comparatively straightforward protocols that involve extraction under basic conditions and purification by selective precipitation, most often with alcohol. Tamarind seed xyloglucan, which is comprised of XXXG, XLXG, XXLG, and XLLG repeats (Fig. 6.1) in the approximate ratio 15:10:30:55 (York et al., 1990), is by far the most widely used xyloglucan for experimental work. This xyloglucan comprises ca. 50–60% of tamarind kernel powder (itself available in vast quantities as an agricultural by-product; Mishra and Malhotra, 2009), is easily isolated, and, indeed, can be purchased in a highly purified form from Megazyme International (Ireland) at reasonable cost for laboratory-scale work. The following sections are thus focused on methods for the preparation of solutions of this xyloglucan and subsequent digestion to oligosaccharides, which are broadly general. The reader is referred to the detailed primary literature for methods for the extraction of specific xyloglucans, as well as oligosaccharide compositions obtained following subsequent *endo*-(xylo)glucanase digestion (see Hilz *et al.*, 2007; Hoffman *et al.*, 2005; Hsieh and Harris, 2009 and references therein).

3.1. Xyloglucan solubilization

Concentrations of native, high molar mass tamarind seed xyloglucan of up to 50 g/L may be achieved, but due to increasingly high viscosity, which complicates pipetting, a stock concentration of 10 g/L is most practical for enzyme assays.

1. Preheat water (deionized, if desired) to 50–60 °C. A large volume (hundreds of mL to L) in a suitably sized vessel to accommodate vigorous stirring is essential.
2. Add xyloglucan powder (from *Tamarindus indica*, Megazyme cat. no. P-XYGLN) portion-wise under constant, vigorous stirring. *NB*: Slow, portion-wise addition of the powder is essential to prevent clumping, which will greatly impede dissolution.
3. Maintain stirring at 50–60 °C until completely dissolved (may take several hours).

Xyloglucan solutions can be stored at 4 °C for shorter time periods, but for long-term storage, snap-freezing aliquots by immersion in liquid nitrogen are recommended to avoid microbial growth.

3.2. Preparation of a mixture of Glc_4-based xylogluco-oligosaccharides

The limit-digestion products of a number of *endo*-(xylo)glucanases are the minimal repeating units of the xyloglucan polysaccharide (Hoffman *et al.*, 2005; Hsieh and Harris, 2009). In the case of XXXG-type xyloglucans, such as that from tamarind seed, these repeats are based on a Glc_4 backbone (Fig. 6.1). The oligosaccharide mixture so obtained is directly useful to obtain product complexes for crystallography (Ariza *et al.*, 2011; Gloster *et al.*, 2007; Johansson *et al.*, 2004; Mark *et al.*, 2009; Martinez-Fleites *et al.*, 2006), as a mixture of acceptor substrates for transglycosylation assays (Baumann *et al.*,

2007; Sulová et al., 1995), or for further substrate synthesis (Faure et al., 2006; Ibatullin et al., 2008).

1. Make 500 mL of a 20 g/L tamarind xyloglucan solution. *Note*: This method may be scaled down to limit xyloglucan consumption.
2. Add 5 mL of 1 M NH$_4$OAc pH 4.5 and add 0.15% (w/v) *Trichoderma reesei* "cellulase" (Sigma-Aldrich C8546), and incubate at 30 °C. *Note*: This reagent is not a pure enzyme, but rather likely a mixture of extracellular enzymes, one or more of which has *endo*-xyloglucanase activity.
3. Check reaction progress by TLC (Section 2.1) or HPAEC-PAD (Gradient 2, Section 2.2.1). The digestion typically takes about 16 h.
4. When the reaction has gone to completion, add 1 mL NH$_3$ (37% in H$_2$O) and pump the solution over a Q Sepharose column (GE 5 × 2.6 cm) to remove the enzymes. *NB*: Removal of residual endo-xyloglucanase activity from the preparation is essential for further assay work. In our hands, we have found this resin-based method more effective in removing robust fungal enzymes than thermal inactivation.
5. Freeze-dry the flow-through after freezing the solution rapidly at dry-ice temperature or lower. Yields are nearly quantitative.

Subsequent separation, also in harness with *exo*-glycosidase treatment, can be used to produce individually pure oligosaccharides. For example, XXXG has been produced in large quantities using β-galactosidase treatment of mixed tamarind XGOs (Greffe et al., 2005; York et al., 1993) (*Caution*: Commonly used commercial preparations of *Aspergillus* β-galactosidase contain low-level side activities, which can completely degrade XGOs (Desmet et al., 2007). Careful reaction monitoring is thus necessary). Likewise, nonreducing-end dexylosylated oligosaccharides GXXG and GLLG can be produced using specific α-xylosidase treatment (Larsbrink et al., 2011).

3.3. Isolation of individual Glc$_4$-based XGOs

Individually, pure fractions of XXXG, XLXG/XXLG, and XLLG may be obtained in milligram-scale by careful fractionation by size-exclusion chromatography (SEC), for example, on a Bio-Gel P-6 matrix. Bio-Gel P-6 has a pore size that is optimal for diverse XGO fractionation (see Section 3.4), although historically Bio-Gel P-2 has been used effectively (York et al., 1990, 1993), also in tandem with normal-phase HPLC (McDougall and Fry, 1991). For researchers with access to basic organic/carbohydrate chemistry facilities, a more scalable method involves per-O-acetylation of the XGOs with acetic anhydride in pyridine, followed by "flash" normal-phase silica gel chromatography using toluene/acetone mixtures as eluents (Faure et al., 2006; Greffe et al., 2005).

3.4. Preparation of longer xylogluco-oligosaccharides

Longer oligosaccharides derived from xyloglucans, in particular those based on a Glc_8 backbone (XGO dimers), are useful as minimal substrates for assays and ligands for crystallography with inactive enzyme variants, as they contain a single internal unbranched Glc residue, which is the typical cleavage point for xyloglucan hydrolases and transglycosylases (Ariza et al., 2011; Baumann et al., 2007; Powlowski et al., 2009; Saura-Valls et al., 2008). Limiting the degree of hydrolysis by controlling the reaction time and enzyme addition in the method described in Section 3.2 results in a distribution of longer oligosaccharides, which must then be fractionated by preparative GPC. By carefully monitoring the hydrolysis step, the product distribution can be biased in favor of XGO_{Glc8}, or higher order XGOs (e.g., XGO_{Glc12} and XGO_{Glc16}), which may also be isolated (Martinez-Fleites et al., 2006).

1. Make 500 mL of 20 g/L tamarind xyloglucan solution. *Note*: This method may be scaled down to limit xyloglucan consumption.
2. Add 5 mL of 1 M NH_4OAc pH 4.5, followed by 0.05% (w/v) *T. reesei* cellulase (Sigma-Aldrich C8546), and incubate at 30 °C. *Note*: This reagent is not a pure enzyme, but rather likely a mixture of extracellular enzymes, one or more of which has *endo*-xyloglucanase activity.
3. Check reaction progress by HPAEC-PAD (Gradient 2, Section 2.2.1). The digest usually takes about 16 h.
4. Stop the digest by boiling for 30 min, cool to room temperature and filter through a 0.45 μm filter.
5. Concentrate the solution *in vacuo* to 20 mL or less (avoiding XGO precipitation).
6. Load 5 mL of this solution onto two water-jacketed columns (2 × 90 cm, 2.6 cm in diameter), connected in series and maintained at 60 °C, containing Bio-Gel P-6 (Bio-Rad, California) equilibrated with pure water. Elute with a flow rate of 0.3 mL/min.
7. Check fractions (5 mL) with HPAEC-PAD (Gradient 2, Section 2.2.1) to determine fraction composition.
8. Repeat steps 6 and 7 until all of the crude XGO solution has been fractionated.
9. Pool and freeze-dry desired fractions.

As with shorter XGOs, the XGO_{Glc8} mixture can be similarly treated with a β-galactosidase to produce the pure substrate XXXGXXXG following a second chromatography step to remove free galactose (Martinez-Fleites et al., 2006; but see *Caution* in Section 3.2). Selectively xylosylated variants of XXXGXXXG, suitable for enzyme subsite mapping studies, have been produced using elegant, but involved, chemo-enzymatic syntheses employing glycosynthase technology (Faure et al., 2006).

4. DETERMINATION OF SPECIFIC XYLOGLUCAN ACTIVITY IN A GLUCANASE

Paralleling the classical methods for characterizing the specificity and mode(s) of action of cellulases, a first-level approach to delineating the degree of xyloglucan specificity in a GH has two principal components:

1. Quantifying the specific activity of the enzyme for one or more xyloglucans versus other polysaccharides.
2. Analyzing the products of the enzymatic reaction to determine the mode of action (*endo-* vs. *exo-*, hydrolysis vs. transglycosylation) and the regiospecificity of bond cleavage.

Obtaining basic xyloglucan activity data are well within reach of most (bio) chemical laboratories using commercially available xyloglucan, straightforward reducing-sugar assays, and the aforementioned chromatographic methods (in harness with MS, if available). In-depth analysis of cleavage regiospecificity and determination of the relative rates of hydrolysis and transglycosylation can be achieved using the Glc_8-based "XGO dimers" produced as described above, while the contributions of specific branching residues to catalysis can be delineated using more complex, synthetic substrates.

4.1. Use of the bicinchoninic acid reducing-sugar assay to measure enzymatic xyloglucan hydrolysis

Hydrolytic activity on polysaccharides, xyloglucan included, is best quantified using assays which assay the generation of new reducing-end groups. Many such assays have been developed, each with its own merits, for example, the DNSA assay (Noelting and Bernfeld), the Nelson–Somogyi assay (Nelson, 1944), the MBTH assay (Anthon and Barrett, 2002), and the PAHBAH assay (Lever, 1972). We prefer the bicinchoninic acid (BCA) assay (McFeeters, 1980), in part due to its high sensitivity. The BCA assay is based on the reduction of Cu^{2+} to Cu^{1+} by the hemiacetal/aldehyde "reducing-end" of saccharides, which leads to the formation of a purple complex between Cu^{1+} and two BCA molecules (Smith *et al.*, 1985). A drawback associated with this assay is interference from diverse reducing substances, notably proteins (which are also routinely quantified using BCA), that can lead to impractically high backgrounds when assaying crude enzyme preparations. However, this is rarely a problem where the detailed kinetic analysis of a suitably diluted, purified enzyme is concerned.

Quantitation of enzyme (specific) activity with the BCA assay requires a standard curve based on hydrolysis product standards. While these are in principal comparatively easily obtained for xyloglucan (see Section 3.2), this is often not the case for most polysaccharides, which are irregularly substituted. Thus, glucose is typically used as a standard for all assays, and our experience shows that glucose gives an identical molar response to the Glc_4-based XGO mixture from tamarind within 5% or better. Using the protocol below, the linear working range of the BCA assay is 1–75 μM glucose with fresh solutions. In this context, it is worth remembering the rule of thumb that initial-rate enzyme kinetic measurements must be performed with less than 10% substrate conversion. For example, a 1 g/L xyloglucan solution contains just below 1 mM cleavage sites (assuming an approximate molar mass of an XGO repeat of 1000 Da, and cleavage at each unbranched Glc, typical for most *endo*-xyloglucanases), which means that initial rates should be measured at less than 100 μM produced reducing ends or lower.

The BCA assay is suitable for comparative assays with diverse potential polysaccharide substrates and can be used either for specific activity (e.g., kat/mg or $v_0/[E_t]$) measurements at a single substrate concentration or for full Michaelis–Menten analyses ($v_0/[E_t]$ vs. [S]), as well as pH-rate profile determinations. The performance of the BCA solutions will deteriorate over time but these are usually stable up to a year (Olson and Markwell, 2007). The condition of the BCA stock solutions should be validated each day by generating a standard curve.

1. Prepare solution A (54.28 g/L Na_2CO_3, 24.2 g/L $NaHCO_3$, and 1.942 g/L disodium 2,2′-bicinchoninate) and store at room temperature.
2. Prepare solution B (1.248 g/L $CuSO_4:5H_2O$, 1.262 g/L L-serine) and store at 4 °C.
3. Mix solution A and solution B 1:1. *Note*: Prepare this mixture fresh on the day of use.
4. Prepare at least six (6) standard samples containing glucose in the concentration range of 1–75 μM.
5. Add 250 μL BCA solution to 250 μL of glucose standard or enzyme reaction.
6. Develop color at 80 °C for 30 min, allow to cool to room temperature, and measure A_{560} in a spectrophotometer.
7. Determine enzyme activity from a plot of the standard curve (mol Glc produced) and the enzyme reaction time.

4.2. Determination of the *endo*- versus *exo*-cleavage preference of xyloglucanases using GPC

GPC provides a direct means of assessing whether an enzyme depolymerizes high molar mass xyloglucan (and indeed any polysaccharide) in *endo*-hydrolytic fashion, involving random cleavage at multiple sites along the polymer, or in an end-specific *exo*-fashion. As with cellulases, the presence of polypeptide loops in the active-site cleft can close-off one end of the cleft to various extents, resulting in more or less *exo-/endo*-specificity in individual enzymes. Time-course experiments on enzymes displaying *endo*-activity give rise to a series of chromatograms in which the molar mass of the xyloglucan peak decreases smoothly without the accumulation of low molar mass XGOs until only very late stages of the reaction, that is, at high degrees of hydrolysis (e.g., Fig. 6.3A). A strictly *exo*-acting xyloglucanase (or an *endo*-acting/processive enzyme), on the other hand, would produce low molar mass directly in the initial stages of the reaction, with a correspondingly slow shift of the molar mass of the polysaccharide (e.g., Fig. 6.3B).

1. Calibrate the DMSO–GPC system by injecting 100 µL of each pullulan standard (1 g/L) in the molar-mass range 180–788,000 g/mol (Section 2.2.2).
2. Make a reaction solution of >1200 µL (for eight time points) containing 2.5 g/L of xyloglucan and 5–25 mM of a buffer with a pH suitable for the enzyme under study. *Note:* The use of volatile buffer systems, e.g., ammonium formate, ammonium acetate, ammonium carbonate, etc., is strongly recommended to ensure that smaller products (e.g., XGO_{Glc4}) are not occluded by the lower molar mass salt peak.
3. Add enzyme and incubate at a temperature suitable for the enzyme under study. Trial and error will be required to find a suitable enzyme addition level.
4. Withdraw 150 µL of reaction solution per time-point and boil for 10 min to inactivate the enzyme.
5. Snap-freeze at dry-ice temperature or lower and lyophilize (freeze-dry) the samples. *Note*: The use of a freeze-drier is crucial to yield a powder allowing easier dissolution of the polysaccharide in DMSO prior to GPC analysis. Evaporation of water from liquid samples (e.g., in a "Speed-vac") yields intractable gums.
6. Dissolve the freeze-dried samples in 250 µL DMSO by heating to ca. 65 °C, with occasional vortex mixing, to facilitate dissolution.
7. Inject 100 µL per sample in the GPC system (Section 2.2.2).

While more limited in terms of information provided, due to the lack of resolution and insensitivity to high molar mass xyloglucan, HPAEC-PAD can provide complementary information to GPC analysis for the delineation of *endo*- versus *exo*-activity (Fig. 6.2 cf. Fig. 6.3). However, the greater strength of HPAEC-PAD is in analysis of low molar mass XGOs, limit-digest products, as described below.

4.3. Determination of bond cleavage specificity of xyloglucanases by HPAEC-PAD analysis of polysaccharide and XXXGXXXG limit-digestion products

Analysis of the so-called "limit-digest" products obtained at the endpoint of enzymatic xyloglucan hydrolysis is an essential part of xyloglucanase characterization, which can provide crucial insight into the bond cleavage specificity. In the most common case, a xyloglucanase may cleave the polysaccharide backbone exclusively at the glycosidic bond of unbranched Glc (i.e., at "G–X" linkages), thus producing XXXG, XLXG, XXLG, and XLLG from tamarind seed xyloglucan (as used preparatively in Section 3.2; Gloster et al., 2007; Martinez-Fleites et al., 2006; Vincken et al., 1997b; York et al., 1993). However, some xyloglucanases accept or prefer branched glucosyl residues in the -1 subsite (Ariza et al., 2011; Yaoi et al., 2007), thereby cleaving "X–X" or "X–G" linkages. A simple fingerprint of the limit-digest products is obtained using HPAEC-PAD (see below).

A further degree of insight into the cleavage mode(s) of a xyloglucanase can be readily gained by analyzing the hydrolysis of the Glc_8-based tetradecasaccharide, XXXGXXXG by HPAEC-PAD. This long model substrate, prepared by controlled *endo*-xyloglucanase and β-galactosidase hydrolysis of tamarind xyloglucan (see Section 3.4), is readily water soluble (whereas cellooctaose is not) and provides an internal unbranched glucosyl unit, hydrolysis at which yields 2 equivalents of the heptasaccharide XXXG. Hydrolysis at other positions generates unique reducing- and nonreducing-end fragments, which are discernible by HPAEC-PAD and can be further confirmed by MS (Ariza et al., 2011; Desmet et al., 2007; Yaoi et al., 2007).

1. Set up a 100 µL reaction containing 1 g/L of xyloglucan or 1 mM XXXGXXXG in 10–50 mM buffer at a pH suitable for the enzyme under study.
2. Add enough enzyme to degrade the samples into their respective limit-digest products overnight; trial and error will be required to determine a suitable enzyme loading.
3. After overnight incubation, inject 10 µL on the HPAEC-PAD system (Gradient 2, Section 2.2.1).

If limit-digestion of tamarind seed xyloglucan gives the product profile seen in the inset in Fig. 6.2, or if a single product is seen in the XXXGXXXG-digest, the enzyme cleaves xyloglucan exclusively at unbranched glucosyl residues. MS analysis (Section 2.3) provides a useful method to verify the reaction products anticipated from HPAEC analysis. In some cases, a diverse library of XGOs, as well as time-course experiments, may be required to fully disentangle atypical cleavage modes.

4.4. Determination of the extent of xyloglucan transglycosylation versus hydrolysis using Glc_8-based XGOs and HPAEC-PAD

The potential to perform substrate transglycosylation under kinetic reaction conditions is a common feature of all GHs that use the canonical "retaining mechanism" of hydrolysis. When transglycosylation occurs, the covalent glycosyl-enzyme intermediate is intercepted by a saccharide as an alternate acceptor substrate to the hydrolytic water molecule (see Claeyssens et al., 1990 for an early example, and Eklöf and Brumer, 2010; Gilbert et al., 2008 for general reviews of the mechanism). Indeed, xyloglucan endo-transglycosylation (EC 2.4.1.207) is the dominant—arguably, exclusive—mode of action of many plant GH16 members (Eklöf and Brumer, 2010). A particular difficulty in measuring the rate of transglycosylation is that no new polysaccharide chain ends are formed, thus confounding standard reducing-sugar assays. Drawbacks of otherwise useful and commonly used radiometric (Fry et al., 1992) and iodine-complexation-based methods (Sulová et al., 1995) to measure xyloglucan transglycosylation include a need to synthesize and handle radioactive XGO acceptor substrates and a lack of correlation of kinetic data with specific turnover rates, respectively.

As an alternative, an HPAEC-PAD-based method has been devised (Baumann et al., 2007), which allows simultaneous quantitation of hydrolysis and transglycosylation products from enzymatic action on XGO_{Glc8}. Under initial-rate conditions (substrate conversion < 1–10%), an individual hydrolytic event yields two molecules of XGO_{Glc4}, while a transglycosylation event yields one molecule of XGO_{Glc4} and one of XGO_{Glc12} (due to interception of the XGO_{Glc4} glycosyl enzyme by a second molecule of XGO_{Glc8} substrate). Quantitation of XGO "monomer" versus XGO "trimer" production thus allows v_0 versus [S] plots for both pathways to be generated, which explicitly show the extent of transglycosylation across a range of donor/acceptor substrate concentrations (Fig. 6.4). Apparent Michaelis–Menten kinetic parameters can also be extracted, although their interpretation is complex (Baumann et al., 2007).

For convenience, we routinely use the variably galactosylated mixture of XGO_{Glc8} resulting from the controlled digestion of tamarind xyloglucan (Section 3.4), with the additional motivation that the side chain polydispersity mirrors that of the natural substrate. However, the pure tetradecasaccharide XXXGXXXG can also be used in the method below for exacting kinetic studies.

1. Prepare standard samples of at least six (6) points from 1 to 100 µM of XGO_{Glc4} and XGO_{Glc12}.(the two can be combined into a single solution to reduce the number of HPLC runs).
2. Add 10 µL 1 M NaOH to 25 µL of each standard sample and load into a cooled (4 °C) autosampler.
3. Prepare substrate solutions of XGO_{Glc8} (20–5000 µM) in buffer.
4. At fixed time intervals, add 5 µL of enzyme to 20 µL of each substrate solution (pre-equilibrated at the desired assay temperature) and incubate for a defined assay length (10–30 min). Note: Trial and error range-finding experiments will be required to identify the optimal assay time and enzyme load to ensure less than 10% conversion at each substrate concentration.
5. Stop each reaction by addition of 10 µL 1 M NaOH and place directly on ice.

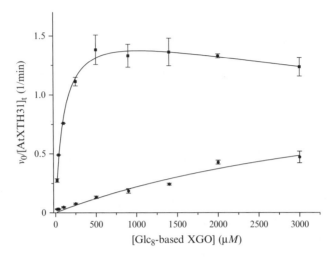

Figure 6.4 Plot of initial rates ($v_0/[E_t]$) versus substrate concentration ([S]) for the hydrolysis (filled squares) and transglycosylation (filled circles) of XGO_{Glc8} by XTH31 from *Arabidopsis thaliana*, measured simultaneously by HPAEC-PAD (data from Kaewthai, 2011).

6. Control samples, to which buffer but no enzyme is added, should be similarly prepared to quantify any trace amounts of XGO_{Glc4} and XGO_{Glc12} in the XGO_{Glc8} preparation.
7. Transfer each solution to an autosampler vial and load into a cooled (4 °C) autosampler.
8. Analyze the samples by injecting 10 μL on a HPAEC-PAD system (Gradient 2, Section 2.2.1).
9. Quantify the amount XGO_{Glc4} and XGO_{Glc12} versus the standard curves.

The rate of transglycosylation is calculated directly from the molar concentration of XGO_{Glc12} produced and the assay time. The hydrolysis rate is calculated from the concentration of XGO_{Glc4}, following subtraction of the XGO_{Glc12} concentration (to compensate for the stoichiometric amount of XGO_{Glc4} generated during transglycosylation); this adjusted XGO_{Glc4} molar concentration is subsequently divided by two, as each hydrolytic event generates two molecules of XGO_{Glc4} (see above).

4.5. Use of aryl β-glycosides of xyloglucan oligosaccharides to determine the contribution of individual enzyme subsites to catalysis

A further advanced technique to characterize xyloglucanase versus glucanase specificity builds on classic cellulase biochemistry, in which aryl β-glycosides (and especially substituted phenyl β-glycosides) of cello-oligosaccharides are used as chromogenic substrates for convenient spectrophotometric assays (see Chapter 1). By analogy, phenyl β-glycosides XGO$_{Glc4}$ have been synthesized using a combined enzymatic and chemical synthesis-based approach. These allow the effect of extended branching of the cello-tetraosyl core to be determined by comparative kinetic analysis (Ariza *et al.*, 2011; Ibatullin *et al.*, 2008). Although requiring some specialist skill, we would argue that the resynthesis of such substrates (which are currently not commercially available), in-house, or via collaboration, is well justified in terms of the additional molecular insight they bring (analogously, cello-oligo aryl β-glycosides have a rich history in the study of cellulases—see Claeyssens and Aerts, 1992; Claeyssens and Henrissat, 1992; MacLeod *et al.*, 1996 and citing references—despite that the larger molecules, with a degree of polymerization >2, have not been marketed until rather recently).

Michaelis–Menten parameters for aglycone cleavage are measured under initial-rate conditions for a series of congeners of increasing complexity, for example, GGGG-β-Ar (cello-tetraosyl), XXXG-β-Ar (heptasaccharide), and XLLG-β-Ar (nonasaccharide). Quantitation of the contribution of xylosylation and further galactosylation to substrate binding and catalysis can then be performed by the direct comparison of the k_{cat}/K_m values. Free-energy contributions to catalysis can be ascertained using Eq. (6.1), where $(k_{cat}/K_m)_a$ is the value for a reference substrate (e.g., GGGG-β-Ar) and $(k_{cat}/K_m)_b$ is the value for a more substituted congener. A caveat is that such analysis only provides information on carbohydrate interactions in the negative enzyme subsites (see Davies *et al.*, 1997 for subsite nomenclature); longer, more complex oligosaccharides are required to probe positive subsite interactions (see, e.g., Saura-Valls *et al.*, 2008).

$$\Delta\Delta G^{\ddagger} = \Delta G_b - \Delta G_a = -RT \ln\left(\frac{(k_{cat}/K_m)_b}{(k_{cat}/K_m)_a}\right) \quad (6.1)$$

While *para*-nitrophenyl (formally 4-nitrophenyl) β-glycosides of XGOs are convenient substrates for endpoint (stopped) assays, we prefer the 2-chloro-4-nitrophenyl aglycone developed and widely popularized for cellulase substrates by Prof. Marc Claeyssens and collaborators (Claeyssens and Aerts, 1992; Claeyssens and Henrissat, 1992). Adventitiously, 2-chloro-4-nitrophenol has a lower pK_a value (5.4) than 4-nitrophenol (7.2), which allows continuous spectrophotometric assays at pH values suited to plant and microbial *endo*-(xylo)glucanases (Ibatullin *et al.*, 2008; Fig. 6.5). Similarly, resorufin XGO glycosides (leaving group pK_a 5.8) are useful for the real-time visualization of xyloglucanase activity both *in vitro* and *in vivo* (Ibatullin *et al.*, 2009). In lieu of a detailed protocol, the reader is referred to the primary literature for adaptable assay methods using these substrates.

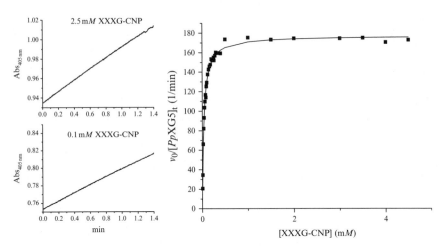

Figure 6.5 Initial-rate kinetics of a *Paenibacillus pabuli* GH5 xyloglucanase on XXXG-β-CNP (data from Gloster *et al.*, 2007). Using 2-chloro-4-nitrophenyl (CNP) substrates, aglycone release can be measured in real time (left panels), which allows direct inspection of the linearity of the reaction across the investigated substrate concentration range. Michaelis–Menten parameters can be subsequently extracted by curve-fitting data in $v_0/[E_t]$ versus [S] plots (right panel).

5. Coda: Xyloglucan Hydrolysis by β-Glucanases — Obvious but Often Overlooked

In conclusion, we hope that a growing appreciation of the presence and importance of xyloglucans in plant cell wall structure might stimulate a greater scrutiny of the xyloglucan specificity of putative β-glucanases, facilitated by the straightforward analytical methods outlined above. We thus advocate the routine inclusion of one or more xyloglucans in the panel of substrates against which putative β-glucanases are tested, highlighting that the ready availability (commercially or by extraction) of the archetypal xyloglucan from tamarind seed should further eliminate any hesitation. Together with assays against traditional low-substituted "celluloses," such as HEC, barley β-glucan, CMC, and PASC, xyloglucanase assays will contribute to a holistic understanding of the enzymes involved in biomass deconstruction, which continues to be topical from both fundamental and applied perspectives (Chanliaud *et al.*, 2004; Chundawat *et al.*, 2011; Himmel *et al.*, 2007; Irwin *et al.*, 2003; Pauly *et al.*, 1999).

ACKNOWLEDGMENTS

The authors wish to acknowledge the contributions of all past members of the Brumer group, especially Dr. Martin J. Baumann and Mr. Farid Ibatullin, who have contributed to the original publications from which these methods have been abstracted. The authors also wish to express their gratitude to Prof. Gideon Davies and colleagues at the University of York, UK, for continuing fruitful collaborations on diverse xyloglucan-active enzymes. The Swedish Research Council (*Vetenskapsrådet*) and The Swedish Research Council Formas (via CarboMat—The KTH Advanced Carbohydrate Materials Consortium, A Formas Strong Research Environment) are acknowledged for project grant funding, and H. B. acknowledges faculty support from the University of British Columbia.

REFERENCES

Anthon, G. E., and Barrett, D. M. (2002). Determination of reducing sugars with 3-methyl-2-benzothiazolinonehydrazone. *Anal. Biochem.* **305**, 287–289.
Ariza, A., Eklöf, J. M., Spadiut, O., Offen, W. A., Roberts, S. M., Besenmatter, W., Friis, E. P., Skøjt, M., Wilson, K. S., Brumer, H., and Davies, G. (2011). Structure and activity of *Paenibacillus polymyxa* xyloglucanase from glycoside hydrolase family 44. *J. Biol. Chem.* **286**, 33890–33900.
Baumann, M. J., Eklöf, J. M., Michel, G., Kallas, Å.M., Teeri, T. T., Czjzek, M., and Brumer, H., III (2007). Structural evidence for the evolution of xyloglucanase activity from xyloglucan endo-transglycosylases: Biological implications for cell wall metabolism. *Plant Cell* **19**, 1947–1963.
Benkö, Z., Siika-aho, M., Viikari, L., and Reczey, K. (2008). Evaluation of the role of xyloglucanase in the enzymatic hydrolysis of lignocellulosic substrates. *Enzyme Microb. Technol.* **43**, 109–114.
Buckeridge, M. S., dos Santos, H. P., and Tine, M. A. S. (2000). Mobilisation of storage cell wall polysaccharides in seeds. *Plant Physiol. Biochem.* **38**, 141–156.
Chanliaud, E., De Silva, J., Strongitharm, B., Jeronimidis, G., and Gidley, M. J. (2004). Mechanical effects of plant cell wall enzymes on cellulose/xyloglucan composites. *Plant J.* **38**, 27–37.
Chundawat, S. P. S., Beckham, G. T., Himmel, M. E., and Dale, B. E. (2011). Deconstruction of lignocellulosic biomass to fuels and chemicals. *In* "Annual Review of Chemical and Biomolecular Engineering," (J. M. Prausnitz, ed.)Vol. 2, pp. 121–145.
Claeyssens, M., and Aerts, G. (1992). Characterization of cellulolytic activities in commercial *Trichoderma reesei* preparations—An approach using small, chromogenic substrates. *Bioresour. Technol.* **39**, 143–146.
Claeyssens, M., and Henrissat, B. (1992). Specificity mapping of cellulolytic enzymes—Classification into families of structurally related proteins confirmed by biochemical analysis. *Protein Sci.* **1**, 1293–1297.
Claeyssens, M., Vantilbeurgh, H., Kamerling, J. P., Berg, J., Vrsanska, M., and Biely, P. (1990). Studies of the cellulolytic system of the filamentous fungus *Trichoderma reesei* QM-9414—Substrate-specificity and transfer activity of endoglucanase-I. *Biochem. J.* **270**, 251–256.
Cosgrove, D. J. (2005). Growth of the plant cell wall. *Nat. Rev. Mol. Cell Biol.* **6**, 850–861.
Davies, G. J., Wilson, K. S., and Henrissat, B. (1997). Nomenclature for sugar-binding subsites in glycosyl hydrolases. *Biochem. J.* **321**, 557–559.
Desmet, T., Cantaert, T., Gualfetti, P., Nerinckx, W., Gross, L., Mitchinson, C., and Piens, K. (2007). An investigation of the substrate specificity of the xyloglucanase Cel74A from *Hypocrea jecorina*. *FEBS J.* **274**, 356–363.
Eklöf, J. M., and Brumer, H. (2010). The XTH gene family: An update on enzyme structure, function, and phylogeny in xyloglucan remodeling. *Plant Physiol.* **153**, 456–466.
Faure, R., Saura-Valls, M., Brumer, H., Planas, A., Cottaz, S., and Driguez, H. (2006). Synthesis of a library of xylogluco-oligosaccharides for active-site mapping of xyloglucan endo-transglycosylase. *J. Org. Chem.* **71**, 5151–5161.

Fry, S. C. (1986). In vivo formation of xyloglucan nonasaccharide—A possible biologically-active cell-wall fragment. *Planta* **169,** 443–453.
Fry, S. C., Smith, R. C., Renwick, K. F., Martin, D. J., Hodge, S. K., and Matthews, K. J. (1992). Xyloglucan endotransglycosylase, a new wall-loosening enzyme-activity from plants. *Biochem. J.* **282,** 821–828.
Fry, S. C., York, W. S., Albersheim, P., Darvill, A., Hayashi, T., Joseleau, J. P., Kato, Y., Lorences, E. P., Maclachlan, G. A., Mcneil, M., Mort, A. J., Reid, J. S. G., et al. (1993). An unambiguous nomenclature for xyloglucan-derived oligosaccharides. *Physiol. Plant.* **89,** 1–3.
Gilbert, H. J., Stalbrand, H., and Brumer, H. (2008). How the walls come crumbling down: Recent structural biochemistry of plant polysaccharide degradation. *Curr. Opin. Plant Biol.* **11,** 338–348.
Gloster, T. M., Ibatullin, F. M., Macauley, K., Eklöf, J. M., Roberts, S., Turkenburg, J. P., Bjørnvad, M. E., Jorgensen, P. L., Danielsen, S., Johansen, K. S., Borchert, T. V., Wilson, K. S., et al. (2007). Characterization and three-dimensional structures of two distinct bacterial xyloglucanases from families GH5 and GH12. *J. Biol. Chem.* **282,** 19177–19189.
Greffe, L., Bessueille, L., Bulone, V., and Brumer, H. (2005). Synthesis, preliminary characterization, and application of novel surfactants from highly branched xyloglucan oligosaccharides. *Glycobiology* **15,** 437–445.
Grishutin, S. G., Gusakov, A. V., Markov, A. V., Ustinov, B. B., Semenova, M. V., and Sinitsyn, A. P. (2004). Specific xyloglucanases as a new class of polysaccharide-degrading enzymes. *Biochim. Biophys. Acta Gen. Subj.* **1674,** 268–281.
Harvey, D. J. (2011). Analysis of carbohydrates and glycoconjugates by matrix-assisted laser desorption/ionization mass spectrometry: An update for the period 2005–2006. *Mass Spectrom. Rev.* **30,** 1–100.
Henrissat, B. (1997). A new cellulase family. *Mol. Microbiol.* **23,** 848–849.
Hilz, H., de Jong, L. E., Kabel, M. A., Verhoef, R., Schols, H. A., and Voragen, A. G. J. (2007). Bilberry xyloglucan—Novel building blocks containing beta-xylose within a complex structure. *Carbohydr. Res.* **342,** 170–181.
Himmel, M. E., Ding, S. Y., Johnson, D. K., Adney, W. S., Nimlos, M. R., Brady, J. W., and Foust, T. D. (2007). Biomass recalcitrance: Engineering plants and enzymes for biofuels production. *Science* **315,** 804–807.
Hoffman, M., Jia, Z. H., Pena, M. J., Cash, M., Harper, A., Blackburn, A. R., Darvill, A., and York, W. S. (2005). Structural analysis of xyloglucans in the primary cell walls of plants in the subclass Asteridae. *Carbohydr. Res.* **340,** 1826–1840.
Hsieh, Y. S. Y., and Harris, P. J. (2009). Xyloglucans of monocotyledons have diverse structures. *Mol. Plant* **2,** 943–965.
Ibatullin, F. M., Baumann, M. J., Greffe, L., and Brumer, H. (2008). Kinetic analyses of retaining endo-(xylo)glucanases from plant and microbial sources using new chromogenic xylogluco-oligosaccharide aryl glycosides. *Biochemistry* **47,** 7762–7769.
Ibatullin, F. M., Banasiak, A., Baumann, M. J., Greffe, L., Takahashi, J., Mellerowicz, E. J., and Brumer, H. (2009). A real-time fluorogenic assay for the visualization of glycoside hydrolase activity in planta. *Plant Physiol.* **151,** 1741–1750.
Irwin, D. C., Cheng, M., Xiang, B. S., Rose, J. K. C., and Wilson, D. B. (2003). Cloning, expression and characterization of a family-74 xyloglucanase from Thermobifida fusca. *Eur. J. Biochem.* **270,** 3083–3091.
Johansson, P., Brumer, H., Baumann, M. J., Kallas, Å.M., Henriksson, H., Denman, S. E., Teeri, T. T., and Jones, T. A. (2004). Crystal structures of a poplar xyloglucan endotransglycosylase reveal details of transglycosylation acceptor binding. *Plant Cell* **16,** 874–886.
Kaewthai, N. (2011). In vitro and in vivo approaches in the characterization of XTH gene products. Ph.D. Thesis. Royal Institute of Technology, Stockholm, p. 48 (plus appendices). Available at, http://urn.kb.se/resolve?urn=urn:nbn:se:kth:diva-28222.

Larsbrink, J., Izumi, A., Ibatullin, F. M., Nakhai, A., Gilbert, H. J., Davies, G. J., and Brumer, H. (2011). Structural and enzymatic characterization of a glycoside hydrolase family 31 alpha-xylosidase from *Cellvibrio japonicus* involved in xyloglucan saccharification. *Biochem. J.* **436,** 567–580.

Lever, M. (1972). New reaction for colorimetric determination of carbohydrates. *Anal. Biochem.* **47,** 273–279.

MacLeod, A. M., Tull, D., Rupitz, K., Warren, R. A. J., and Withers, S. G. (1996). Mechanistic consequences of mutation of active site carboxylates in a retaining beta-1,4-glycanase from *Cellulomonas fimi*. *Biochemistry* **35,** 13165–13172.

Mark, P., Baumann, M. J., Eklöf, J. M., Gullfot, F., Michel, G., Kallas, Å.M., Teeri, T. T., Brumer, H., and Czjzek, M. (2009). Analysis of nasturtium TmNXG1 complexes by crystallography and molecular dynamics provides detailed insight into substrate recognition by family GH16 xyloglucan *endo*-transglycosylases and *endo*-hydrolases. *Proteins.* **75,** 820–836.

Marry, M., Cavalier, D. M., Schnurr, J. K., Netland, J., Yang, Z. Y., Pezeshk, V., York, W. S., Pauly, M., and White, A. R. (2003). Structural characterization of chemically and enzymatically derived standard oligosaccharides isolated from partially purified tamarind xyloglucan. *Carbohydr. Polym.* **51,** 347–356.

Martinez-Fleites, C., Guerreiro, C. I., Baumann, M. J., Taylor, E. J., Prates, J. A., Ferreira, L. M., Fontes, C. M., Brumer, H., and Davies, G. J. (2006). Crystal structures of Clostridium thermocellum xyloglucanase, XGH74A, reveal the structural basis for xyloglucan recognition and degradation. *J. Biol. Chem.* **281,** 24922–24933.

Master, E. R., Zheng, Y., Storms, R., Tsang, A., and Powlowski, J. (2008). A xyloglucan-specific family 12 glycosyl hydrolase from *Aspergillus niger*: Recombinant expression, purification and characterization. *Biochem. J.* **411,** 161–170.

McDougall, G. J., and Fry, S. C. (1991). Purification and analysis of growth-regulating xyloglucan derived oligosaccharides by high-pressure liquid-chromatography. *Carbohydr. Res.* **219,** 123–132.

McFeeters, R. F. (1980). A manual method for reducing sugar determinations with 2,2'-bicinchoninate reagent. *Anal. Biochem.* **103,** 302–306.

Mishra, A., and Malhotra, A. V. (2009). Tamarind xyloglucan: A polysaccharide with versatile application potential. *J. Mater. Chem.* **19,** 8528–8536.

Nelson, N. (1944). A photometric adaptation of the Somogyi method for the determination of glucose. *J. Biol. Chem.* **153,** 375–380.

Nishikubo, N., Awano, T., Banasiak, A., Bourquin, V., Ibatullin, F., Funada, R., Brumer, H., Teeri, T. T., Hayashi, T., Sundberg, B., and Mellerowicz, E. J. (2007). Xyloglucan endo-transglycosylase (XET) functions in gelatinous layers of tension wood fibers in poplar—A glimpse into the mechanism of the balancing act of trees. *Plant Cell Physiol.* **48,** 843–855.

Nishikubo, N., Takahashi, J., Roos, A. A., Derba-Maceluch, M., Piens, K., Brumer, H., Teeri, T. T., Stalbrand, H., and Mellerowicz, E. J. (2011). Xyloglucan endo-transglycosylase-mediated xyloglucan rearrangements in developing wood of hybrid aspen. *Plant Physiol.* **155,** 399–413.

Noelting, G., and Bernfeld, P. (1948). Sur les enzymes amylolytiques III. La beta-amylase: Dosage d'activité et contrôle de l'absence d'alpha-amylase. *Helv. Chim. Acta* **31,** 286–290.

Olson, B. J. S. C., and Markwell, J. (2007). Assays for determination of protein concentration. *Current Protocols in Pharmacology, Appendix 3A*. John Wiley & Sons, Inc., New York pp. 1–29.

Pauly, M., Albersheim, P., Darvill, A., and York, W. S. (1999). Molecular domains of the cellulose/xyloglucan network in the cell walls of higher plants. *Plant J.* **20,** 629–639.

Popper, Z. A., Michel, G., Herve, C., Domozych, D. S., Willats, W. G. T., Tuohy, M. G., Kloareg, B., and Stengel, D. B. (2011). Evolution and diversity of plant cell walls: From algae to flowering plants. *In* "Annual Review of Plant Biology," (S. S. Merchant, W. R. Briggs, and D. Ort, eds.), Vol. 62, pp. 567–588.

Powlowski, J., Mahajan, S., Schapira, M., and Master, E. R. (2009). Substrate recognition and hydrolysis by a fungal xyloglucan-specific family 12 hydrolase. *Carbohydr. Res.* **344,** 1175–1179.

Saura-Valls, M., Faure, R., Brumer, H., Teeri, T. T., Cottaz, S., Driguez, H., and Planas, A. (2008). Active-site mapping of a *Populus* xyloglucan endo-transglycosylase with a library of xylogluco-oligosaccharides. *J. Biol. Chem.* **283,** 21853–21863.

Smith, P. K., Krohn, R. I., Hermanson, G. T., Mallia, A. K., Gartner, F. H., Provenzano, M. D., Fujimoto, E. K., Goeke, N. M., Olson, B. J., and Klenk, D. C. (1985). Measurement of protein using bicinchoninic acid. *Anal. Biochem.* **150,** 76–85.

Sulová, Z., Lednicka, M., and Farkaš, V. (1995). A colorimetric assay for xyloglucan-endotransglycosylase from germinating seeds. *Anal. Biochem.* **229,** 80–85.

Vincken, J. P., Beldman, G., and Voragen, A. (1994). The effect of xyloglucans on the degradation of cell-wall-embedded cellulose by the combined action of cellobiohydrolase and endoglucanases from *Trichoderma viride*. *Plant Physiol.* **104,** 99–107.

Vincken, J. P., Beldman, G., and Voragen, A. G. J. (1997a). Substrate specificity of endoglucanases: What determines xyloglucanase activity? *Carbohydr. Res.* **298,** 299–310.

Vincken, J. P., York, W. S., Beldman, G., and Voragen, A. G. J. (1997b). Two general branching patterns of xyloglucan, XXXG and XXGG. *Plant Physiol.* **114,** 9–13.

Vlasenko, E., Schulein, M., Cherry, J., and Xu, F. (2010). Substrate specificity of family 5, 6, 7, 9, 12, and 45 endoglucanases. *Bioresour. Technol.* **101,** 2405–2411.

Vogel, J. (2008). Unique aspects of the grass cell wall. *Curr. Opin. Plant Biol.* **11,** 301–307.

Wong, D., Chan, V. J., McCormack, A. A., and Batt, S. B. (2010). A novel xyloglucan-specific endo-beta-1,4-glucanase: Biochemical properties and inhibition studies. *Appl. Microbiol. Biotechnol.* **86,** 1463–1471.

Yaoi, K., Kondo, H., Hiyoshi, A., Noro, N., Sugimoto, H., Tsuda, S., Mitsuishi, Y., and Miyazaki, K. (2007). The structural basis for the exo-mode of action in GH74 oligoxyloglucan reducing end-specific cellobiohydrolase. *J. Mol. Biol.* **370,** 53–62.

York, W. S., and Hawkins, R. (2000). Preparation of oligomeric beta-glycosides from cellulose and hemicellulosic polysaccharides via the glycosyl transferase activity of a Trichoderma reesei cellulase. *Glycobiology* **10,** 193–201.

York, W. S., Vanhalbeek, H., Darvill, A. G., and Albersheim, P. (1990). The structure of plant cell walls. 29. Structural analysis of xyloglucan oligosaccharides by H-1-NMR spectroscopy and fast atom bombardment mass spectrometry. *Carbohydr. Res.* **200,** 9–31.

York, W. S., Harvey, L. K., Guillen, R., Albersheim, P., and Darvill, A. G. (1993). The structure of plant cell walls. 36. Structural analysis of tamarind seed xyloglucan oligosaccharides using beta-galactosidase digestion and spectroscopic methods. *Carbohydr. Res.* **248,** 285–301.

CHAPTER SEVEN

Methods for Structural Characterization of the Products of Cellulose- and Xyloglucan-Hydrolyzing Enzymes

Maria J. Peña,* Sami T. Tuomivaara,*,† Breeanna R. Urbanowicz,* Malcolm A. O'Neill,* and William S. York*,†

Contents

1. Introduction	122
2. Preparation of Substrates	124
3. Purification of the Oligosaccharide Products	124
3.1. Isolation of soluble products generated by enzymatic hydrolysis of insoluble substrates	124
3.2. Isolation of soluble products generated by enzymatic hydrolysis of soluble polymeric substrates	125
3.3. Purification of oligosaccharides by liquid chromatography	125
4. Chemical and Structural Analysis of the Reaction Products	127
4.1. Glycosyl residue composition analysis by gas chromatography with mass spectrometric and flame ionization detection	127
4.2. Converting oligosaccharides to their corresponding oligoglycosyl alditols	129
4.3. Glycosyl-linkage composition analysis	129
4.4. Analysis of oligosaccharides by matrix-assisted laser desorption/ionization time-of-flight mass spectrometry	131
4.5. Analysis of the reaction products by NMR spectroscopy	131
4.6. Electrospray ionization mass spectrometry	135
Acknowledgments	137
References	138

Abstract

Structural characterization of oligosaccharide products generated by enzymatic hydrolysis of plant cell wall polysaccharides provides valuable

* Complex Carbohydrate Research Center, University of Georgia, Athens, Georgia, USA
† Department of Biochemistry and Molecular Biology, University of Georgia, Athens, Georgia, USA

information about the enzyme's activity and substrate specificity. In this chapter, we describe some of the chemical, chromatographic, and spectroscopic methods that we routinely use to isolate and characterize oligosaccharides formed by enzymatic fragmentation of cellulose and xyloglucan. These include techniques to determine glycosyl residue and glycosyl linkage compositions by gas chromatography and mass spectrometry. We also illustrate the use of electrospray ionization with multistage mass spectrometry, matrix-assisted laser desorption/ionization time-of-flight mass spectrometry, and nuclear magnetic resonance spectroscopy to perform detailed structural analysis of these oligosaccharides.

1. INTRODUCTION

The polysaccharide-rich plant cell wall provides structural support to cells, tissues, and organs and is a recalcitrant barrier that wards off diverse biotic and abiotic challenges. All land plant cell walls consist of several structurally and functionally independent networks that together form recalcitrant composite structure that is nevertheless biologically responsive and pliable upon developmental and environmental cues. The networks consist of cellulose microfibrils tethered by cross-linking glycans which are embedded in a pectin and hemicellulosic matrix and may be further fortified by structural proteins and/or polyphenolic substances together with small amounts of inorganic compounds and hydrophobic molecules (Carpita and Gibeaut, 1993; O'Neill and York, 2003). Cell wall composition and architecture vary among plant species and also between tissues and developmental stages within the same plant. With the growing desire to use plant biomass as a renewable carbon-rich substrate for industrial and biomedical applications, there is an impetus for structural characterization of cell wall polysaccharides. Linkage-specific glycosyl hydrolases (GHs) have become a powerful tool for such studies. GHs are also the major players required to depolymerize and ultimately convert the plant cell wall into sugars for downstream processes. For a recent review describing the enzymes involved in deconstruction of plant cell wall polysaccharides, see Gilbert (2010).

One feature that is consistent among all land plant cell walls is the presence of cellulose, which is the most abundant component of these walls. Cellulose is composed of 1,4-linked β-D-glucosyl residues that form linear chains which are interconnected by hydrogen bonds and van der Waals forces to form rigid and insoluble cellulose microfibrils (Harris *et al.*, 2010; O'Neill and York, 2003). Effective and complete cellulose hydrolysis requires the combined action of several types of enzymes including endo-β-1,4-glucanases (EC 3.2.1.4), exo-β-1,4-glucanases or cellobiohydrolases

(E.C. 3.2.1.91), and β-glucosidases (E.C. 3.2.1.21) from a variety of GH families (Cantarel et al., 2009; Gilbert, 2010; http://www.cazy.org/). Recent evidence suggests that some cellulose-degrading fungi also utilize an oxidoreductive cellulose-depolymerizing system that functions synergistically with the canonical hydrolytic cellulase system and utilizes the combined activities of cellobiose dehydrogenase (CDH, EC1.199.18) and GH61 copper-dependent oxidases (Langston et al., 2011; Quinlan et al., 2011).

Xyloglucans (XGs) are branched hemicellulosic polysaccharides that have a linear backbone of 1,4-linked β-D-glucosyl residues. Stretches containing two to five of these residues are consecutively substituted at O-6 with α-D-xylosyl residues. The xylosyl residues themselves may be further extended with additional sugars leading to side chain structures that can vary significantly among plant species (Hoffman et al., 2005; Peña et al., 2008; York et al., 1990, 1996). Due to the structural complexity of XGs, a nomenclature system has been devised to describe different side chain substitutions (Fry et al., 1993). Briefly, the letter G represents an unsubstituted Glc$p$ residue. When the Glc$p$ is substituted at O-6 with α-D-Xyl$p$, this is termed the X side chain, which may be substituted with a β-D-Gal$p$ (L side chain), which itself may be substituted with an α-L-Fuc$p$ (F side chain). Many endo-β-1,4-glucanases that hydrolyze cellulose are also able to cleave the XG backbone. However, several XG-specific endo-β-1,4-glucanases (XG endohydrolases, XEH, EC 3.2.1.151), which are members of the CAZy glycoside hydrolase families GH5, GH7, GH12, GH16, GH44, and GH74, have been identified (Ariza et al., 2011; Cantarel et al., 2009; Eklöf and Brumer, 2010; Gilbert, 2010; Gloster et al., 2007; Irwin et al., 2003; Pauly et al., 1999). Some members of family GH74 have been shown to be oligoxyloglucan reducing end-specific cellobiohydrolases (EC 3.2.1.150) (Bauer et al., 2005; Yaoi and Mitsuishi, 2002). Together these enzymes have become valuable tools for determining the fine structures of XGs and XG oligosaccharides.

Complete structural analysis of enzyme-generated oligosaccharides requires the determination of many features including monosaccharide composition and interglycosidic linkages, molecular mass, anomeric configurations, glycosidic sequence, and the presence and position of side chains. In this chapter, we describe some of the methods that are used to isolate and characterize the oligosaccharides formed by enzymatic fragmentation of cellulose and XG polysaccharides. The procedures described here can be used on a wide variety of polysaccharides from various sources. These methods are also applicable with minor modification for the characterization of oxidized cellodextrin products of GH61s and related enzymes. We describe how different techniques, including mass spectrometry (MS) and nuclear magnetic resonance (NMR) spectroscopy, are used to determine the fine structure of enzymatically generated oligosaccharides.

2. Preparation of Substrates

Microcrystalline cellulose (Avicel) and cellulose derivatives (e.g., carboxymethyl cellulose) are substrates typically used to determine cellulase activity. These and several other plant cell wall polysaccharides are available from different chemical companies. Megazyme (http://secure.megazyme.com/Homepage.aspx), in particular, offers plant cell wall polysaccharides (e.g., XGs, xylans, and mannans), as well as purified cello- and other oligosaccharides.

Methods to isolate cell walls from plant tissues and the protocols to extract and purify individual wall polysaccharides have been previously described (Hoffman et al., 2005; York et al., 1986).

Oligosaccharide substrates can be prepared by chemical or enzymatic hydrolysis of the respective polysaccharides and purified using methods described in Section 3. In-house purification of oligosaccharides may be time-consuming, but typically yields compounds that are more homogenous than commercially available products.

3. Purification of the Oligosaccharide Products

This review is focused on structural characterization of enzymatically produced oligosaccharides, and not enzyme assay procedures. Heterogeneity of plant cell wall components and the GHs involved in their metabolism and breakdown ultimately translate to a variety of experimental details in the assay conditions. The procedures described here are applicable to a wide variety of polysaccharides from various sources, including insoluble cellulose and soluble XG. Many of these procedures rely on the differential solubility of the substrate and oligosaccharide products.

3.1. Isolation of soluble products generated by enzymatic hydrolysis of insoluble substrates

This protocol is used for the isolation of soluble products from an insoluble substrate. For example, the isolation of low molecular weight cello-oligosaccharides from Avicel cellulose.

1. Separate the soluble products from the insoluble material by centrifugation for 15 min at $3600 \times g$. Collect the supernatant and transfer to a clean tube. To maximize the product yield, wash the solid residue with a small amount of water or buffer, centrifuge again, and pool the supernatants.

2. Inactivate the enzyme by heating the supernatant for 10 min in a boiling water bath.
3. Lyophilize the solution.

Further cleanup of the lyophilized reaction products can be largely circumvented if volatile buffers are used. We routinely use ammonium formate or ammonium acetate buffers (in the range of 10–50 mM) in our enzyme assays because of their volatility as well as their optimum buffering ranges (pH 4–6), which coincides with the pH optimum of many plant cell wall-degrading GHs. Alternatively, ammonium bicarbonate buffer can be used for enzymes requiring neutral to slightly alkaline pH (pH 6–9).

3.2. Isolation of soluble products generated by enzymatic hydrolysis of soluble polymeric substrates

This protocol applies if both the substrate and the product are water soluble, but the substrate precipitates in high concentrations of ethanol. This is the case for XG and several other polysaccharides. This protocol is very simple and can often substitute for labor-intensive chromatographic techniques and is frequently used as a preliminary fractionation step even if further purification is eventually required.

1. Add ethanol to the enzymatic reaction mixture to a final concentration of 75% and incubate at 4 °C for several hours to precipitate any undigested polymeric materials. Separate the soluble products from the precipitated material by centrifugation for 15 min at $3600 \times g$. XG polysaccharides and oligosaccharides with three or more repeating structural units can be separated from smaller XG oligosaccharides with 75% ethanol. Lower final ethanol concentration yields higher-molecular weight oligosaccharides in the soluble fraction.
2. Collect the supernatant and remove the ethanol by rotary evaporation or centrifugal evaporation. Dissolve the material in water and lyophilize the solution.

Note that the temperature and time requirement as well as the final ethanol concentration for efficient precipitation vary according to the characteristics of both the substrate and the product. Empirical testing is suggested with new substances.

3.3. Purification of oligosaccharides by liquid chromatography

Liquid chromatographic techniques are used extensively to isolate and purify oligosaccharides. The choice of the chromatographic mode typically depends on the sample characteristics. We have found that size-exclusion and reversed-phase chromatographies are versatile and are usually sufficient to obtain homogeneous oligosaccharides. Moreover, these methods employ volatile buffers and/or organic solvents that are easily removed and thus do not interfere with subsequent structural or functional analyses.

3.3.1. Size-exclusion liquid chromatography

Size-exclusion chromatography (SEC) is used to separate molecules according to their hydrodynamic size. However, if the columns are eluted with water rather than with buffer, anionic oligosaccharides typically elute in the column void volume (Peña et al., 2007). The protocol below is effectively used to separate enzymatically generated oligosaccharides from undigested high molecular weight substrate.

1. Dissolve the lyophilized enzymatic digestion products in 50 mM ammonium formate pH 5 ($\sim$5 mg/ml).
2. Inject aliquots (200 μl) of the solution to a Supedex-75 HR10/300 column (GE Healthcare) previously equilibrated with the same buffer. These columns typically have a capacity of $\sim$5 mg.
3. Elute the oligosaccharides with the same buffer at a flow rate of 0.5 ml/min. The eluate can be monitored with a refractive index or an evaporative light scattering (ELS) detector.
4. Collect peak fractions and combine those that contain the oligosaccharides. Lyophilize the solution.
5. Dissolve the lyophilized material in water and lyophilize again to remove the volatile ammonium salt. Repeat this step up to three times to ensure complete removal of the ammonium formate. Alternatively, the oligosaccharides can be desalted by using LC-18 cartridges (following the instructions of the manufacturer) (York et al., 1996) or by using a Bio-Gel P2 or Sephadex G-10 desalting column.

Mixtures of cello- and XG oligosaccharides can be separated into individual components using a Bio-Gel P2 (400 mesh, extra fine; Bio-Rad) column (1 × 100 cm) eluted with water using gravity flow ($\sim$2 m hydrostatic head). Fractions are collected with a fraction collector and assayed colorimetrically for the presence of carbohydrate using the phenol–sulfuric acid method (Masuko et al., 2005).

3.3.2. Reversed-phase high-performance liquid chromatography

Reversed-phase high-performance liquid chromatography (RP-HPLC) is used to separate individual components from oligosaccharide mixtures based on their hydrophobic characteristics. Usually the oligosaccharides have been partially purified by SEC prior to RP-HPLC.

1. Dissolve the oligosaccharides or oligoglycosyl alditols (see Section 4.2) in water (∼1 mg/ml) and filter the solution using a centrifugal filter device (0.2 μm).
2. Inject aliquots (200 μl) of the solution onto a Zorbax reversed-phase C-18 column (250 × 4.6 mm; Agilent Technologies Inc., Santa Clara, CA). Other manufacturers' C 18 columns are also suitable.
3. Elute the oligosaccharides at a flow rate of 1 ml/min with a linear gradient from 6% to 15% of aqueous methanol over 35 min followed by a linear gradient from 15% to 35% over 15 min. The column can be reconditioned by washing with aqueous 50% methanol for 10 min followed by washing with 6% aqueous methanol for an additional 10 min. In our laboratory, we detect the oligosaccharides using a Sedex 55 ELS detector (Sedere, Alfortville, France).
4. Fractions containing the oligosaccharides are collected and lyophilized.

4. Chemical and Structural Analysis of the Reaction Products

Several methods can be used, individually or in combination, to characterize the purified oligosaccharides. The selection of the methods depends on the scope of the experiments and the purity and previous knowledge of enzymes and substrates.

4.1. Glycosyl residue composition analysis by gas chromatography with mass spectrometric and flame ionization detection

Determination of the glycosyl residue composition of the soluble products of a reaction is usually a preliminary step in the analysis. When working with a GH of unknown specificity, composition analysis of the crude hydrolysate or purified components can provide insight into the enzyme properties, including its specificity. Neutral sugar analysis is a good starting point of analysis and is a powerful technique that gives information on both the types and amounts of monosaccharides present in the sample. The information obtained is used to guide both the analytical and purification strategies and requirements. This method to convert neutral sugars in their per-O-acetylated alditol (AA) derivatives is based on the protocols developed by Blakeney *et al.* (1983) and Carpita and Shea (1989). The analysis of AA was performed by gas chromatography (GC) using mass spectrometric detection (MSD) and flame ionization detection (FID). One advantage of using a MSD is that per-O-acetylated alditols give diagnostic mass spectra and thus can be readily distinguished from contaminating noncarbohydrate peaks including phthalates. However, quantification using FID is preferred for determining the amounts of monosaccharides in the samples since it gives more reproducible results than MSD. Quantification by ion monitoring presents specific problems, including differences due to changes in mass tuning and ion suppression. Response factors are calculated by analysis of known standards and determination of the peak areas for each sugar derivative relative to the internal standard, myo-inositol.

1. Prepare a 20 mM solution of standard monosaccharides in water corresponding to the monosaccharides present in the sample (e.g., for plant cell wall materials use L-Rha, L-Fuc, L-Ara, D-Xyl, D-Man, D-Gal, D-Glc). Transfer 10–100 µl aliquots of the solution to Teflon-lined screw-cap tubes to make a series of standards with increasing concentrations of the monosaccharides and then lyophilize the samples.
2. All subsequent steps are performed in a fume hood. Transfer 200–500 µg of the oligosaccharides to Teflon-lined screw-cap tubes. Add 1 ml of 2 M trifluoroacetic acid (TFA) and 50 µl of 20 mM myo-inositol to both the samples and standards. Cap the tubes tightly and heat for 90 min at 120 °C in a temperature-controlled heating block.
3. Cool the tubes to room temperature and then evaporate the TFA at 40–45 °C using a stream of filtered air or nitrogen. Trace amounts of TFA are then removed by adding 1 ml of volatile organic solvent such as methanol or acetone and evaporated as above. The next step requires alkaline conditions, and unintentional acidification can result in incomplete reduction. Dry samples can be stored overnight if necessary.
4. Prepare a 100 mg/ml solution of sodium borohydride (or borodeuteride if isotopic labeling is required) in 1 M ammonium hydroxide and then dilute with 5 volumes of dimethyl sulfoxide (DMSO). Add 600 µl of the diluted solution to each sample and incubate for 90 min at 40–45 °C. Vortex every 30 min.
5. Destroy the unreacted $NaBH_4$ by adding 100 µl of glacial acetic acid and mixing thoroughly. Add 100 µl of 1-methylimidazole and 75 µl of anhydrous acetic anhydride. Mix and keep for 30 min at 40–45 °C.
6. To terminate the reaction, add 2 ml of deionized water and shake vigorously. Allow the mixture to cool to room temperature and then add 1 ml dichloromethane and shake the mixture for 1 min. Centrifuge for 5 min at $3000 \times g$ and discard the upper aqueous layer.
7. Add 2 ml of deionized water to the organic layer and shake the mixture for 1 min. Centrifuge for 5 min at $3000 \times g$ and again discard the aqueous layer. Repeat this step four more times to ensure complete removal of water-soluble salts and DMSO, which can interfere with the downstream analysis.
8. Concentrate the organic layer to dryness by a stream of air or nitrogen and dissolve the residue in 100–200 µl of dichloromethane or acetone.
9. The alditol acetate derivatives are analyzed using Hewlett Packard 5890A gas chromatograph coupled to a Hewlett Packard 5970 mass

selective detector (MSD) or an Agilent 7890A gas chromatograph with a flame ionization detector. In both cases, derivatized monosaccharides are separated on a SP 2330 column (30 m × 0.25 mm, 0.25 μm film thickness, Supelco) with helium as the carrier gas and the following oven temperature gradient: 80 °C held for 2 min, 80–170 °C at 30 °C/min, 170–240 °C at 4 °C/min, 240 °C held for 20 min.

4.2. Converting oligosaccharides to their corresponding oligoglycosyl alditols

For some applications, oligosaccharides are converted to their corresponding oligoglycosyl alditols prior to analysis. There are several advantages to working with alditols rather than native carbohydrates. First, unlike native reducing carbohydrates, oligoglycosyl alditols do not have reducing end α and β anomeric protons and thus have less complex NMR spectra. In some modes of chromatography, the anomeric forms can separate, and the chromatograms become unnecessarily complex. Also, oligoglycosyl alditols are less likely to undergo base-catalyzed β-elimination during per-O-methylation reactions (Waeghe et al., 1983).

1. All steps are performed in a fume hood. Transfer 250–500 μg of the oligosaccharides to a Teflon-lined screw-cap tube. Add 250 μl of a solution of 1 M sodium borohydride (or borodeuteride if isotopic label is required) in 1 M ammonium hydroxide and keep for at least 1 h at room temperature.
2. Terminate the reaction by dropwise addition of glacial acetic acid until the release of hydrogen gas stops. Then, add 0.5 ml of 10% (v/v) acetic acid in methanol and evaporate the solvents using a stream of air or nitrogen gas to remove borate as its volatile trimethyl ester. Repeat the acetic acid/methanol treatment three times. Add 0.5 ml of anhydrous methanol and concentrate to dryness to remove traces of acetic acid; repeat three times.

4.3. Glycosyl-linkage composition analysis

A fundamental step in understanding the fine structure of an oligosaccharide is to determine how the individual sugars are linked to one another. One well-established method for this is to methylate all the free (unsubstituted) hydroxyl groups of the intact oligosaccharides prior to acid hydrolysis. The hydroxyl groups generated during the hydrolysis to monosaccharides and reduction to the alditols are then O-acetylated to form the partially-O-methylated alditol acetate (PMAA) derivatives. The fragmentation pattern of these derivatives during electron ionization (EI) MS provides information about the type of monosaccharide (i.e., hexose, pentose, deoxyhexose) as well as the position of the O-methyl and O-acetyl groups, which can then be used to deduce the positions that the original sugars were linked (Carpita and Shea, 1989).

4.3.1. Preparation of the solid NaOH/DMSO reagent

Methylation of oligosaccharides using solid NaOH in DMSO is a faster and less hazardous approach than some other methylation procedures. This method is based on the protocol developed by Ciucanu and Kerek (1984).

1. All steps are performed in a fume hood. Combine 100 µl of aqueous 50% (w/v) NaOH and 200 µl of anhydrous methanol in a screw-cap tube and vortex to give a clear solution. Add 1 ml of anhydrous DMSO, vortex and then centrifuge for 5 min at $3600 \times g$.
2. Remove and discard the supernatant. Add 1 ml of anhydrous DMSO, vortex and centrifuge as before.
3. Repeat Step 2 five times to prepare an opalescent NaOH pellet. Suspend the pellet in 200–300 µl of anhydrous DMSO and use immediately for the subsequent methylation reaction.

4.3.2. Methylation of oligosaccharides

1. All steps are performed in a fume hood. Suspend 100–200 µg of the lyophilized oligosaccharide or oligoglycosyl alditol in 200–500 µl of anhydrous DMSO in a screw-cap vial. Sonicate or heat gently (60 °C) to increase solubility if necessary.
2. Add ~200 µl of freshly prepared NaOH in DMSO (from Section 4.3.1) and 10 µl of water, which is added to limit oxidative degradation (Ciucanu and Costello, 2003). Stir the mixture for 15 min at room temperature.
3. Add 300 µl of methyl iodide (caution, this reagent is toxic) and keep the mixture for 15 min at room temperature, then cool the reaction mixture in ice, add 1 ml of water and vortex. Remove excess methyl iodide by carefully bubbling air (or nitrogen) through the reaction mixture.
4. Add 1 ml of chloroform, vortex, centrifuge for 5 min at $3000 \times g$, and then transfer the lower organic layer to a clean tube.
5. Wash the combined organic phase three times with 1 ml of deionized water. Evaporate the chloroform and dissolve the residue containing the methylated materials in methanol.
6. The per-O-methylated glycans are converted to their PMAA derivatives and analyzed by GC-MS as described in Section 4.1, using sodium borodeuteride. The symmetry of some alditols (e.g., xylitol and galactitol) makes it impossible to unambiguously determine the location of the O-acetyl and O-methyl groups of some PMAA derivatives (e.g., 1,3,4,5-tetra-O-acetyl-2-O-methyl xylitol and 1,2,3,5-tetra-O-acetyl-4-O-methyl xylitol) unless the C1 is labeled by reduction with

deuterium. The positions of the O-methyl and O-acetyl substituents in per-O-methylated alditol acetates are readily determined from the primary fragment ions present in their EI mass spectra. A collection of spectra of partially methylated alditol acetate derivatives generated from diverse glycosidically linked sugars is available at http://www.ccrc.uga.edu/specdb/ms/pmaa/pframe.html or in the literature (Carpita and Shea, 1989).

4.4. Analysis of oligosaccharides by matrix-assisted laser desorption/ionization time-of-flight mass spectrometry

Matrix-assisted laser desorption/ionization time-of-flight mass spectrometry (MALDI-TOF MS) analysis is used to determine the mass distribution of the different species present in a mixture of oligosaccharides and is routinely used to monitor enzymatic reactions. The technique requires small amounts of sample (0.1–0.5 µg) to obtain good spectra and is rapid and relatively tolerant to the presence of low concentrations of salts and other contaminants. Another advantage of this technique is that the oligosaccharides can be analyzed without derivatization.

1. Mix 1 µl of 0.5–1 mg/ml oligosaccharide solution in water with an equal volume of matrix solution (20 mg/ml 2,5-dihydroxybenzoic acid in aqueous 50% methanol) on the MALDI target plate and then dry with a stream of warm air (hair dryer).
2. Record a positive ion MALDI-TOF mass spectrum (e.g., using a Bruker Microflex LT mass spectrometer and workstation). Spectra generated by 10–200 laser shots are typically collected and averaged.

In the MALDI-TOF mass spectra shown in Fig. 7.1A, XG isolated from cell walls prepared from Arabidopsis inflorescence stems was treated with a GH12 xyloglucan-specific endoglucanase (XEH; EC 3.2.1.151) from *Aspergillus aculeatus* (Novozymes, Bagsvaerd, Denmark) purified as described (Pauly *et al.*, 1999). The XG oligosaccharides formed were then treated (Fig. 7.1B) with an GH74 oligoxyloglucan reducing end-specific xyloglucanobiohydrolase (OREX; EC 3.2.1.150) overexpressed in *Pichia pastoris* and purified as described (Bauer *et al.*, 2005, 2006). Both reactions were performed for 16 h in 50 mM ammonium formate pH 5, at 25 °C.

4.5. Analysis of the reaction products by NMR spectroscopy

NMR spectroscopy is a powerful tool for studying the enzymatic hydrolysis of polysaccharides and oligosaccharides as it provides information on the catalytic mechanism and substrate specificity (Rudsander *et al.*, 2008). NMR analysis is especially useful when the enzyme activity is unknown

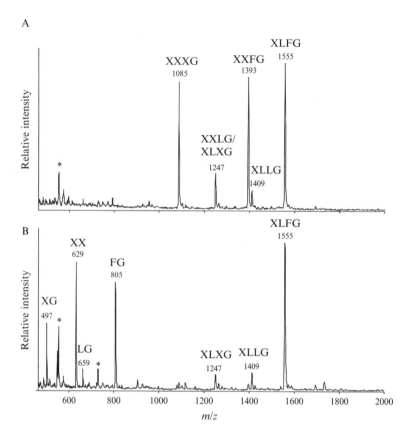

Figure 7.1 MALDI-TOF mass spectra of oligosaccharides generated from Arabidopsis xyloglucan by (A) xyloglucan-specific endoglucanase (XEH) and (B) both XEH and oligoxyloglucan reducing end-specific cellobiohydrolase (OREX). XEH hydrolyzes the glycosidic bond at the reducing end side of an unbranched glucosyl residue generating subunits with an XXXG-type branching pattern, namely XXXG, XXFG, XLFG, and the isobaric pair XXLG/XLXG. OREX further degrades xyloglucan oligosaccharides by hydrolyzing the glycosidic bond between the second and third glucosyl units from the reducing end, if the xylose at the third glucose from the reducing end is not further substituted. Thus, XXXG, XXLG, and XXFG are degraded to the common fragment XX, as well as XG, LG, and FG, respectively. XLXG and XLFG are not substrates for OREX. For detailed nomenclature of the xyloglucan side chains, see Section 1. The peaks marked with an asterisk are matrix signals.

or when the substrate to be analyzed is complex (e.g., plant cell wall). The enzyme-generated oligosaccharides, purified as described in Section 3, can be structurally characterized by one- and two-dimensional (1D and 2D, respectively) NMR. Analysis of the 2D spectra can be used to identify isolated spin systems and to obtain chemical shifts and coupling constants for each type of residue in the oligosaccharide. These data are then used to determine the identity, the ring form, and the anomeric configuration of each residue. Experiments including ^{1}H—^{13}C HMBC provide data to deduce glycosidic linkages between the residues in the oligosaccharide.

In the case of water-soluble polysaccharides (e.g., XG) or short oligosaccharides (e.g., cello-oligosaccharides), their enzymatic hydrolysis can be monitored by NMR. One advantage of this approach is that the composition and structure of both substrates and products can be determined. Most of these substrates and products have been characterized previously, and the chemical shifts and coupling constants diagnostic for their most common structural features are available in the literature (Harjunpää et al., 1996; Hoffman et al., 2005; York et al., 1990). In particular, a database consisting of a searchable table of ^{1}H NMR chemical shift of XG oligoglycosyl alditols is available at http://ccrc.uga.edu/db/nmr. Structural features of the residues in the oligoglycosyl alditols can be used to retrieve their chemical shifts from the database. Also, chemical shift data can be used as a query to obtain information about the residue identity and chemical environment.

4.5.1. Determination of the hydrolysis stereochemistry by real-time ^{1}H NMR

Real-time ^{1}H NMR can be used to determine the anomeric stereochemistry of enzymatic hydrolysis products, and thus distinguish between the retaining and inverting mechanisms of a GH (Rudsander et al., 2008). In this technique, an enzyme is added to a solution of the oligosaccharide in an NMR tube, which is then immediately placed in the instrument. 1D ^{1}H-NMR spectra are acquired at fixed time points over the course of the enzyme reaction. The cleavage of the glycosidic bond results in the formation of a new oligosaccharide with α or β reducing glycose depending on the mechanism that the enzyme uses to hydrolyze the substrate as well as the original anomericity. If the rate of enzymatic hydrolysis is faster than the rate of anomeric interconversion, the α or β anomeric proton signal that appears in the 1D spectra will indicate the cleavage mechanism (see Fig. 7.2). The anomeric conversion of glucose, for example, at or near neutral pH in D$_2$O occurs at time constant of approximately 1 h (Harjunpää et al., 1996). Thus, the amounts of substrate and enzyme, as well as temperature and pH, have to be suitable in order to get a workable ratio of the rates of anomeric conversion and product (new anomeric signal) generation. The presence of α and β anomeric signals from the reducing end of the initial substrate itself may complicate interpretation of such results. Thus, oligosaccharide substrates are typically converted to their corresponding oligoglycosyl alditols prior to analysis (Section 4.2) so that their NMR spectra contain no reducing end anomeric resonances.

1. Polysaccharides, oligosaccharides, or oligoglycosyl alditols (1–2 mg) are dissolved in a suitable buffer for NMR analysis at a concentration and pH/pD appropriate for the enzymatic reaction. Nonvolatile inorganic buffers such as sodium phosphate, that lack nonexchangeable protons, are preferred, so the buffering capacity is retained during lyophilization, and the NMR spectrum is not "contaminated" with additional signals that can overlap with the signals from the sample.
2. Lyophilize the solution and dissolve the dry material in 1 ml of deuterium oxide (99.9%; Cambridge Isotope Laboratories). Repeat this step to remove all exchangeable protons from the carbohydrate and buffer molecules.
3. Dissolve the dry material in 200 µl of deuterium oxide and transfer the solution to a 3 mm NMR tube. Place the tube with the oligosaccharide in the instrument (e.g., in a Varian Inova NMR spectrometer operating

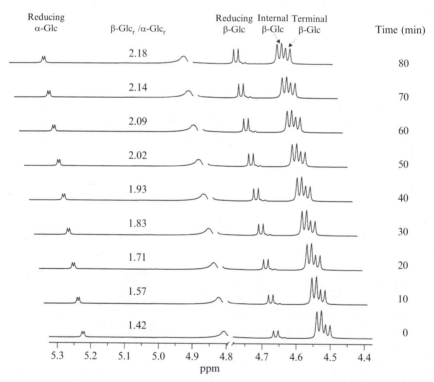

Figure 7.2 Real-time ^{1}H NMR spectra of cellopentaose hydrolysis with a GH12 β-(1 → 4)-endoglucanase from *Aspergillus niger* (Megazyme) at 25 °C. Ratio of the integrals of anomeric signals of reducing β-Glc$p$ (β-Glc$_r$) and reducing α-Glc$p$ (α-Glc$_r$) residues indicates that the hydrolysis product has a β-Glc$p$ residue at the reducing end, demonstrating that the enzyme cleaves the glycosidic linkage by a retaining mechanism.

at 600 MHz and equipped with a 3 mm cold probe) and set the desired sample temperature for the experiment (this will be the temperature for the enzymatic reaction).
4. Allow sufficient time for the sample temperature and the substrate reducing end α/β anomeric ratio to reach equilibrium. Tune the probe and then lock and shim the instrument. Set the conditions for an array experiment including the time interval between collection of spectra. It is recommendable to use an experiment with a water suppression technique (e.g., presaturation) to record the 1D spectra.
5. Remove the sample from the instrument and add the enzyme to the oligosaccharide solution. The enzyme should be in the same buffer as the sample in order to maintain the initial anomeric ratio (α/β) at the reducing end of the oligosaccharide. Immediately return the tube to the instrument, check and adjust the shimming if necessary and start the array experiment. A series of ^{1}H NMR spectra summing 16 scans each are recorded at defined time intervals during the enzymatic reaction.
6. Data are processed using commercially available software (e.g., MNova, Universidad de Santiago de Compostela, Spain). Chemical shifts are calibrated relative to a diagnostic resonance previously identified in the spectrum of the sample.

In the example in Fig. 7.3, the cellopentaose substrate was generated from Avicel by phosphoric acid hydrolysis (Liebert *et al.*, 2008) and purified using a Bio-Gel P2 as described in Section 3.3.1.

4.6. Electrospray ionization mass spectrometry

Electrospray ionization with multiple-stage MS (ESI-MS^n) is a useful technique to determine the glycosyl sequences and branching pattern of oligosaccharides. These structural features may be difficult to determine by other analytic techniques (e.g., NMR) due to the redundancy of the characteristic signals as well as higher amounts of sample required. One advantage of ESI-MS^n is its ability to selectively fragment a molecular ion to generate fragment ions that can be further fragmented. This process can be repeated many times, with every generation giving structural information about the parent ion (see Fig. 7.3). To resolve ambiguities during the identification of the fragments, the analysis is usually performed on per-O-methylated oligosaccharides or, preferably, per-O-methylated oligoglycosyl alditols. Treatment with a reducing agent such as NaBH$_4$ prior to methylation converts the reducing end residue to an alditol, allowing the so-called A, B, and C fragments, which contain the nonreducing end to be distinguished from X, Y, and Z fragments, which contain the alditol (Domon and Costello, 1988; Mazumder and York, 2010). Methylation of the free hydroxyl groups increases sensitivity and, more

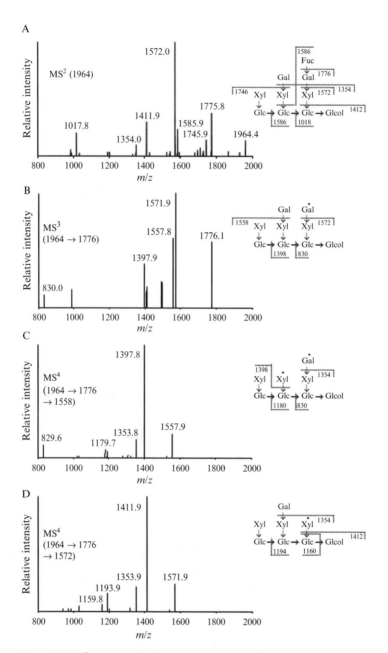

Figure 7.3 ESI-MS^n spectra of a per-O-methylated xyloglucan oligoglycosyl alditol (XLFGol) generated from Arabidopsis xyloglucan. All labeled peaks, except the ion at m/z 1160 in panel D, correspond to sodiated Y-type ions. The pseudomolecular ion at m/z 1964 was selected from the full scan spectrum (not shown) for collision-induced dissociation (CID) fragmentation. The resulting MS^2 mass spectrum (A) shows extensive

importantly, helps to identify the hydroxyl groups involved in glycosic linkages, since fragmentation at the glycosidic bond exposes nonmethylated hydroxyl group analogously to acid hydrolysis of methylated oligosaccharides.

1. Dissolve the per-O-methylated oligosaccharides or oligoglycosyl alditols in methanol and mix with 50% aqueous acetonitrile containing 0.1% TFA to give a final concentration of ~ 1 ng/μl.
2. Introduce the sample into the MS source through a fused silica capillary (150 μm i.d. × 60 cm; Thermo Finnigan, USA) at flow rate of 3 μl/min using a syringe pump.
3. Record ESI mass spectra (e.g., with a Thermo Scientific LTQ XL ion trap mass spectrometer). The electrospray source is operated at a voltage of 5.0 kV and a capillary temperature of 275 °C. To obtain MS^n spectra, the collision energy is adjusted to obtain optimal fragmentation.

ACKNOWLEDGMENTS

This work was supported in part by US Department of Energy Grant DE-FG02-96ER20220, US Department of Energy-funded Center for Plant and Microbial Complex Carbohydrates (DE-FG02-93ER20097), and National Science Foundation Grant DBI-0421683. This research was also supported by the BioEnergy Science Center. The BioEnergy Science Center is a US Department of Energy Bioenergy Research Center supported by the Office of Biological and Environmental Research in the DOE Office of Science.

The *P. pastoris* clone (AN1542.2) carrying the gene encoding oligoxyloglucan reducing end-specific xyloglucanobiohydrolase was obtained from the Fungal Genetics Stock Center (Kansas City, Missouri, USA; http://www.fgsc.net/).

---

fragmentation, to generate a sodiated Y-ion (XLLG, m/z 1776) with a single hydroxyl group exposed by cleavage of the glycosidic bond that linked the fucosyl residue to the galactosyl residue. Such "scars," which are indicated as dots (●), provide sequence information not available when native oligosaccharides are analyzed. The diagnostic ion at m/z 1018 (FG with one scar, panel A) establishes the location of the fucosylated side chain. The ion at m/z 1586 can arise from two distinct fragmentation pathways, forming either LFG with one scar or XGFG with one scar. The ion at m/z 1776 was then selected from the MS^2 spectrum and fragmented to generate the MS^3 spectrum (B), which included signals corresponding to fragment ions at m/z 1558 (XXLG with two scars) and m/z 1572 (XLXG with one scar). These two ions were individually selected for further fragmentation. The resulting MS^4 (C and D) included additional diagnostic signals. The ion at m/z 1160 (D) arises from two MS^4 fragmentation events, which resulted in the loss of one side chain (Y-type fragmentation) and the alditol residue (B-type fragmentation). Together, these data unambiguously show that the fucosyl residue is attached to the galactosyl residue closest to the alditol end.

REFERENCES

Ariza, A., Eklöf, J. M., Spadiut, O., Offen, W. A., Roberts, S. M., Besenmatter, W., Friis, E. P., Skjot, M., Wilson, K. S., Brumer, H., and Davies, G. (2011). Structure and activity of *Paenibacillus polymyxa* xyloglucanase from glycoside hydrolase family 44. *J. Biol. Chem.* **286,** 33890–33900.
Bauer, S., Vasu, P., Mort, A. J., and Somerville, C. R. (2005). Cloning, expression, and characterization of an oligoxyloglucan reducing end-specific xyloglucanobiohydrolase from *Aspergillus nidulans*. *Carbohydr. Res.* **340,** 2590–2597.
Bauer, S., Vasu, P., Persson, S., Mort, A. J., and Somerville, C. R. (2006). Development and application of a suite of polysaccharide-degrading enzymes for analyzing plant cell walls. *Proc. Natl. Acad. Sci. USA* **103,** 11417–11422.
Blakeney, A. B., Harris, P. J., Henry, R. J., and Stone, B. A. (1983). A simple and rapid preparation of alditol acetates for monosaccharide analysis. *Carbohydr. Res.* **113,** 291–299.
Cantarel, B. L., Coutinho, P. M., Rancurel, C., Bernard, T., Lombard, V., and Henrissat, B. (2009). The Carbohydrate-Active EnZymes database (CAZy): An expert resource for glycogenomics. *Nucleic Acids Res.* **37,** D233–D238.
Carpita, N. C., and Gibeaut, D. M. (1993). Structural models of primary cell walls in flowering plants—Consistency of molecular structure with the physical properties of the walls during growth. *Plant J.* **3,** 1–30.
Carpita, N. C., and Shea, E. M. (1989). Linkage structure of carbohydrates by gas chromatography-mass spectrometry (GC-MS) of partially methylated alditol acetates. *In* "Analysis of Carbohydrates by GLC and MS," (C. J. Biermann and G. D. McGinnis, eds.), pp. 157–215. CRC Press, Baton Rouge, FL.
Ciucanu, I., and Costello, C. E. (2003). Elimination of oxidative degradation during the per-O-methylation of carbohydrates. *J. Am. Chem. Soc.* **125,** 16213–16219.
Ciucanu, I., and Kerek, F. (1984). A simple and rapid method for the permethylation of carbohydrates. *Carbohydr. Res.* **131,** 209–217.
Domon, B., and Costello, C. E. (1988). A systematic nomenclature for carbohydrate fragmentations in FAB-MS/MS spectra of glycoconjugates. *Glycoconj. J.* **5,** 397–409.
Eklöf, J. M., and Brumer, H. (2010). The XTH gene family: An update on enzyme structure, function, and phylogeny in xyloglucan remodelling. *Plant Physiol.* **153,** 456–466.
Fry, S. C., York, W. S., Albersheim, P., Darvill, A., Hayashi, T., Joseleau, J. P., Kato, Y., Lorences, E. P., Maclachlan, G. A., McNeil, M., Mort, A. J., Reid, J. S. G., et al. (1993). An unambiguous nomenclature for xyloglucan-derived oligosaccharides. *Physiol. Plant.* **89,** 1–3.
Gilbert, H. J. (2010). The biochemistry and structural biology of plant cell wall deconstruction. *Plant Physiol.* **153,** 444–455.
Gloster, T. M., Ibatullin, F. M., Macauley, K., Eklof, J. M., Roberts, S., Turkenburg, J. P., Bjornvad, M. E., Jorgensen, P. L., Danielsen, S., Johansen, K. S., Borchert, T. V., Wilson, K. S., et al. (2007). Characterization and three-dimensional structures of two distinct bacterial xyloglucanases from families GH5 and GH12. *J. Biol. Chem.* **282,** 19177–19189.
Harjunpää, V., Teleman, A., Koivula, A., Ruohonen, L., Teeri, T. T., Teleman, O., and Drakenberg, T. (1996). Cello-oligosaccharide hydrolysis by cellobiohydrolase II from *Trichoderma reesei*. Association and rate constants derived from an analysis of progress curves. *Eur. J. Biochem.* **240,** 584–591.
Harris, D., Bulone, V., Ding, S.-Y., and DeBolt, S. (2010). Tools for cellulose analysis in plant cell walls. *Plant Physiol.* **153,** 420–426.
Hoffman, M., Jia, Z., Peña, M. J., Cash, M., Harper, A., Blackburn, A. R., Darvill, A. G., and York, W. S. (2005). Structural analysis of xyloglucan in the primary cell walls of plants in the subclass Asteridae. *Carbohydr. Res.* **340,** 1826–1840.

Irwin, D. C., Cheng, M., Xiang, B. S., Rose, J. K. C., and Wilson, D. B. (2003). Cloning, expression and characterization of a family-74 xyloglucanase from *Thermobifida fusca*. *Eur. J. Biochem.* **270,** 3083–3091.

Langston, J. A., Shaghasi, T., Abbate, E., Xu, F., Vlasenko, E., and Sweeney, M. D. (2011). Oxidoreductive cellulose depolymerization by the enzymes cellobiose dehydrogenase and glycoside hydrolase 61. *Appl. Environ. Microbiol.* **77,** 7007–7015.

Liebert, T., Seifert, M., and Heinze, T. (2008). Efficient method for the preparation of pure, water-soluble cellodextrines. *Macromol. Symp.* **262,** 140–149.

Masuko, T., Minami, A., Iwasaki, N., Majima, T., Nishimura, S.-I., and Lee, Y. C. (2005). Carbohydrate analysis by a phenol–sulfuric acid method in microplate format. *Anal. Biochem.* **339,** 69–72.

Mazumder, K., and York, W. S. (2010). Structural analysis of arabinoxylans isolated from ball-milled switchgrass biomass. *Carbohydr. Res.* **345,** 2183–2193.

O'Neill, M. A., and York, W. S. (2003). The composition and structure of plant primary cell walls. *In* "The Plant Cell Wall," (J. K. C. Rose, ed.), pp. 1–54. Blackwell, Oxford.

Pauly, M., Andersen, L. N., Kauppinen, S., Kofod, L. V., York, W. S., Albersheim, P., and Darvill, A. (1999). A xyloglucan-specific endo-beta-1,4-glucanase from *Aspergillus aculeatus*: Expression cloning in yeast, purification and characterization of the recombinant enzyme. *Glycobiology* **9,** 93–100.

Peña, M. J., Zhong, R., Zhou, G.-K., Richardson, E. A., O'Neill, M. A., Darvill, A. G., York, W. S., and Ye, Z.-H. (2007). *Arabidopsis irregular xylem8* and *irregular xylem9*: Implication for the complexity of glucuronoxylan biosynthesis. *Plant Cell* **19,** 549–563.

Peña, M. J., Darvill, A. G., Eberhard, S., York, W. S., and O'Neill, M. A. (2008). Moss and liverwort xyloglucans contain galacturonic acid and are structurally distinct from the xyloglucans synthesized by hornworts and vascular plants. *Glycobiology* **18,** 891–904.

Quinlan, R. J., Sweeney, M. D., Lo Leggio, L., Otten, H., Poulsen, J.-C. N., Johansen, K. S., Krogh, K. B. R. M., Jorgensen, C. I., Tovborg, M., Anthonsen, A., Tryfona, T., Walter, C. P., *et al.* (2011). Insights into the oxidative degradation of cellulose by a copper metalloenzyme that exploits biomass components. *Proc. Natl. Acad. Sci. USA* **108,** 15079–15084.

Rudsander, U. J., Sandstrom, C., Piens, K., Master, E. R., Wilson, D. B., Brumer, H., Kenne, L., and Teeri, T. T. (2008). Comparative NMR analysis of cellooligosaccharide hydrolysis by GH9 bacterial and plant endo-1,4-ß-glucanases. *Biochemistry* **47,** 5235–5241.

Waeghe, T., Darvill, A., McNeil, M., and Albersheim, P. (1983). Determination, by methylation analysis, of the glycosyl-linkage compositions of microgram quantities of complex carbohydrates. *Carbohydr. Res.* **123,** 281–304.

Yaoi, K., and Mitsuishi, Y. (2002). Purification, characterization, cloning, and expression of a novel xyloglucan-specific glycosidase, oligoxyloglucan reducing end-specific cellobiohydrolase. *J. Biol. Chem.* **277,** 48276–48281.

York, W. S., Darvill, A. G., McNeil, M., Stevenson, Y. T., and Albersheim, P. (1986). Isolation and characterization of plant cell walls and cell wall components. *Meth. Enzymol.* **118,** 3–40.

York, W. S., van Halbeek, H., Darvill, A. G., and Albersheim, P. (1990). Structural analysis of xyloglucan oligosaccharides by ^{1}H-n.m.r. Spectroscopy and fast-atom bombardment mass spectrometry. *Carbohydr. Res.* **200,** 9–31.

York, W. S., Kumar Kolli, V. S., Orlando, R., Albersheim, P., and Darvill, A. G. (1996). The structures of arabinoxyloglucans produced by solanaceous plants. *Carbohydr. Res.* **285,** 99–128.

CHAPTER EIGHT

The Crystallization and Structural Analysis of Cellulases (and Other Glycoside Hydrolases): Strategies and Tactics

Shirley M. Roberts *and* Gideon J. Davies

Contents

1. General Crystallization Considerations	144
1.1. Before you start: Protein quality control	144
1.2. Commercial and bespoke crystallization screens	146
2. Optimization and Additive Screens	149
2.1. Seeding strategies	150
3. Cellulase- and Glycosidase-Specific Complex Tactics	152
3.1. Product/pseudo-product complexes	152
3.2. Enzyme (inactive) variants and longer oligosaccharides	153
3.3. Enzyme inhibitors	154
3.4. Inhibitor soaking strategies	158
3.5. Prospecting: Panning ligand mixtures for the best binders	158
4. Buffer and Cryo-protection Problems	159
4.1. Buffers	159
4.2. Cryo-protectants	161
4.3. Problems of inexperience	161
5. Toward a Cellulase-Specific Approach?	162
Acknowledgments	162
References	163

Abstract

The three-dimensional (3-D) structures of cellulases, and other glycoside hydrolases, are a central feature of research in carbohydrate chemistry and biochemistry. 3-D structure is used to inform protein engineering campaigns, both academic and industrial, which are typically used to improve the stability or activity of an enzyme. Examples of classical protein engineering goals

York Structural Biology Laboratory, Department of Chemistry, The University of York, Heslington, York, YO10 5DD, United Kingdom

Methods in Enzymology, Volume 510
ISSN 0076-6879, DOI: 10.1016/B978-0-12-415931-0.00008-2

© 2012 Elsevier Inc.
All rights reserved.

include higher thermal stability, reduced metal-ion dependency, detergent and protease resistance, decreased product inhibition, and altered specificity. 3-D structure may also be used to interpret the behavior of enzyme variants that are derived from screening or random mutagenesis approaches, with a view to establishing an iterative design process. In other areas, 3-D structure is used as one of the many tools to probe enzymatic catalysis, typically dovetailing with physical organic chemistry approaches to provide complete reaction mechanisms for enzymes by visualizing catalytic site interactions at different stages of the reaction. Such mechanistic insight is not only fundamentally important, impacting on inhibitor and drug design approaches with ramifications way beyond cellulose hydrolysis, but also provides the framework for the design of enzyme variants to use as biocatalysts for the synthesis of bespoke oligosaccharides. Here we review some of the strategies and tactics that may be applied to the X-ray structure solution of cellulases (and other carbohydrate-active enzymes). The general approach is first to decide why you are doing the work, then to establish correct domain boundaries for truncated constructs (typically the catalytic domain only), and finally to pursue crystallization of pure, homogeneous, and monodisperse protein with appropriate ligand and additive combinations. Cellulase-specific strategies are important for the delineation of domain boundaries, while glycoside hydrolases generally also present challenges and opportunities for the selection and optimization of ligands to both aid crystallization, and also provide structural and mechanistic insight. As the many roles for plant cell wall degrading enzymes increase, so does the need for rapid high-quality structure determination to provide a sound structural foundation for understanding mechanism and specificity, and for future protein engineering strategies.

Why? That is the first question the budding structural enzymologist working on cellulases should ask himself or herself. Why do I want to know the three-dimensional (3-D) structure of the enzyme? This question is not as stupid as it may seem, for we all have different motivations and needs. For many, the 3-D structure may simply be a route to aid the design, and subsequent interpretation, of protein variants designed to act in (industrial) applications, or to assist the establishment of intellectual property pertaining to such inventions. For others, the goals may be chemical and mechanistic for which cellulases may be an ideal model system. Others may see enzymes as biocatalysts for the production of bespoke oligosaccharides. Many scientists in the applied areas will be interested to see how the macroscopic properties of proteins on polymeric substrates are reflected in their surface topographies. These diverse constituencies all have different needs in terms of protein structure, crystallographic resolution, complexes, etc. So the first question one needs to address before embarking on structure solution is why am I doing this? What do you hope to achieve?

The question of *why* is of the utmost importance since it dictates both initial strategy and subsequent tactics. If the goal is a simple *apo* structure, perhaps informed by complexes on homologous enzymes (and there are many complex structures in most cellulase CAZy families; www.cazy.org (Cantarel *et al.*, 2009)), then a much simpler strategy involving far fewer costs can be envisaged. In this case, 3-D structure at any credible resolution is likely satisfactory. If the goal is to dissect chemistry, unlock mechanistic secrets using specifically conceived and synthesized inhibitors, then it is best to consider the work, and the resource demands, from the outset. In this latter case, the work required to fine-tune constructs, optimize crystals and data collection strategies, and to obtain informative ligand complexes is considerable, but may be worthwhile if the enzyme is sufficiently novel, and if the weight of homologous structures fails to provide sufficient insight.

Having decided why you are interested in 3-D structure, the most important issue to address is the genetic (or other) construct that you will be expressing for protein production. The first question, linked completely to the *"why?"* issue, is whether you are tied to a single protein, or are willing to clone a spectrum of close homologs to improve the chances of crystallization. Once the decision on target enzyme(s) has been reached, the crucial issue in cellulase (and related) enzyme research is the, often complex, modularity of the protein. Crystallization of multimodular glycosidases (Henrissat and Davies, 2000) is rarely successful, especially when the domains are linked by flexible linker domains (harder still if they are glycosylated), although there are some precedents (e.g., Bott *et al.*, 2008; Fujimoto *et al.*, 2000, 2010; Pell *et al.*, 2004). Thus, it is likely that your journey will start with crucial decisions about the construction of truncated enzymes; typically the catalytic domain only. There are two obvious strategies here. The first is the historical method of tweaking your protein into stable soluble domains through the use of limited proteolysis after which the individual domains can be purified and crystallized. The next stage would be to analyze the proteolytic fragments by mass spectrometry and sequencing in order to make the equivalent, stable, construct using genetic methods. Although the limited proteolysis technique is still valid, it had its heyday in structural research on cellulases in the mid 1990s where, especially with the pioneering work of van Tilbeurgh (Tomme *et al.*, 1988; Van Tilbeurgh *et al.*, 1986) coupled to the strategies employed by industry, it was the favored approach. These days, it is more likely that a bioinformatics-based strategy will be used with the CAZy resource (www.cazy.org; Cantarel *et al.*, 2009), backed up with sequence alignments central to the assignment of reliable domain boundaries. For most proteins, it will simply be a matter of delineating the catalytic domain from associated ancillary domains such as noncatalytic carbohydrate-binding modules (reviewed in Boraston *et al.*, 2004), and expressing the required domain in isolation. While many boundaries are clear, others may be less so and thus comparing many different sequences may be informative. Furthermore, there are occasions where extra domains need to be incorporated in the construct as the protein may be unstable without certain additional modules (see, e.g., Adams *et al.*, 2006).

The curators of the CAZy resource may also be able to provide domain boundary information as such knowledge underpins the annotation of domains in CAZy.

Although more relevant after the initial structure solution—and again this is a question which results from your purpose in doing the work—it can often be useful to go back and reengineer a genetic construct that reflects what you see ordered in the crystal lattice. For example, should the last 10 residues be disordered in the crystal structure, it may improve crystal quality if a truncated construct is made that lacks the disordered amino acids. We have used such approaches many times in our own research (e.g., recently on a glycosyltransferase; Flint et al., 2005).

1. General Crystallization Considerations

Once you have obtained your pure protein domain, the bottom line with strategies for the crystallization of cellulases is that they are actually little different from any protein structure crystallization. It is therefore worth investing a few words outlining the general strategies and tactics that apply to all protein crystallizations.

1.1. Before you start: Protein quality control

Protein crystallization can be a real hold up in the progress toward achieving structural information. A few experiments determining protein quality, and hence "crystallizability," can be done prior to, or even alongside, the initial crystallization screening. Polyacrylamide SDS gels, part of the protein purification process, should show a single band at the expected molecular weight, but given that the protein has been denatured as part of the technique, additional information on the behavior of the true mass of the protein and its aggregation state in solution is required. MALDI-MS (matrix-assisted laser desorption-ionization mass spectroscopy) is convenient for mass determination, as it can deal with low levels of buffers and salts needed for protein solubility, but ultimately electrospray ionization mass spectroscopy (ESI-MS) will give more precise information. The protein needs to be exchanged into a volatile buffer such as 25 mM ammonium acetate, which may prove problematic. High-throughput approaches to the preparation of samples for ESI-MS have recently been described and offer a more facile route to efficient and robust sample preparation (Sundqvist et al., 2007).

The solution behavior of the protein is central to its potential for crystallization. Ideally, for crystallization, the protein should be >95% pure and homogeneous both in content and size (i.e., monodisperse in solution). A polydispersity of <25% gives an optimistic outlook for crystallization although certain protein modifications such as glycosylation may confuse this prediction. Certainly, the presence of aggregated protein in the solution needs to be minimized as random aggregation of protein is a poison to crystallization (Bergfors, 2009; Zulauf and D'Arcy, 1992).

Native gels can be used to assess the extent of aggregation but dynamic light scattering (DLS) has the advantage of testing the levels of protein mono/polydispersity in the presence of different buffers and additives. Varying the pH, the type of buffer and the addition of NaCl or different cations/anions in the protein solution can make a big difference to the level of polydispersity (Fig. 8.1A–C shows an example of changing the buffer

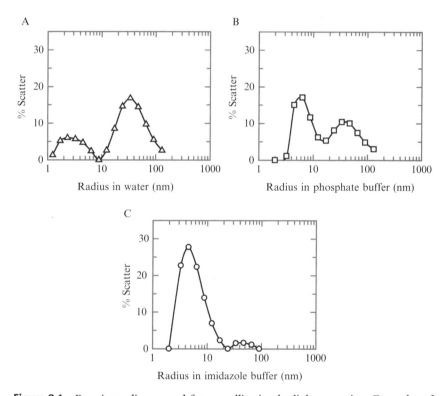

Figure 8.1 Protein quality control for crystallization by light scattering. Examples of the effect of buffer choice on the dispersity (and crystallizability) of a glycosidase; here the *Thermatoga maritima* β-glucosidase TmGH1 (Zechel *et al.*, 2003). (A) Protein, in water, after the Ni-affinity column is highly polydisperse with the predominant scatter from a large, but soluble aggregate; (B) the protein in 50 mM phosphate buffer (pH 7.0) is also polydisperse; (C) only in 40 mM imidazole/40 mM NaCl (pH 7.6) buffer did the sample become monodisperse with a radius corresponding to the dimer eventually seen in the crystal structure. We have observed this imidazole phenomenon on other systems and believe it to reflect competition for extraneous Ni (which otherwise causes aggregation around the His-tags). What makes this situation complex is that it is also preparation dependent, likely reflecting the age of "charging" of the Ni column.

which had marked impact on protein monodispersity and crystallization of a β-glucosidase).

A rather newer technique, the fluorescence-based thermal shift assay (often termed "Thermofluor®," after one of the commercial approaches), is becoming popular in screening conditions that aid protein stability and therefore probably crystallization (Ericsson et al., 2006; Niesen et al., 2007). This fluorescence-based assay uses quantitative polymerase chain reaction (qPCR) equipment and was originally developed for rapid screening of fragment libraries in fragment-based drug discovery (Pantoliano et al., 2001). The protein is mixed with a screen consisting of individual buffers at various pHs and a broad range of additives, and subjected to an incremental increase in temperature in the presence of a dye, typically SYPRO® orange. The dye has little fluorescence in aqueous solution but as the protein starts to unfold and melt the dye binds to exposed hydrophobic parts of the protein and a significant increase in fluorescence emission occurs. A comparison of the thermally induced melting points will reveal stabilizing buffers and additives that can be included in subsequent protein crystallization screening. Some proteins do not show a clear melting curve with SYPRO® orange, however, alternate dyes are available; in the YSBL we have also been using Deep Purple™ in such assays. If protein crystallization proves to be a real problem, even after investigation using these methods, then additional physical biochemistry techniques such as NMR, CD, and N-terminal sequencing may prove useful to further clarify protein characteristics.

A final point to consider is the presence or absence of protein purification affinity tags (such as His-tags). An affinity tag-labeled protein may behave differently in crystallization trials in the presence or absence of that tag. If the protein has an enzymatically cleavable tag, it may be worth digesting away the tag and repeating screening, in case the tag itself is preventing crystallization, perhaps due to interfering with crystal packing, or posttranslational modifications to the tag introducing heterogeneity (such as spontaneous alpha-N-6-gluconoylation at the amino group of His-tagged proteins; Kim et al., 2001). Sometimes it is beneficial to reclone a gene with a tag at the other terminus of the protein, or with a different affinity tag altogether. Sadly, there are no hard and fast rules and all options may need to be considered.

1.2. Commercial and bespoke crystallization screens

The process of obtaining protein crystal growth is largely empirical in nature, and demands patience, perseverance, and intuition (McPherson, 2004). The experiments begin with a rather trial-and-error approach until some success is obtained with the growth of microcrystals or a microcrystalline precipitate. However, recognizing a "promising" precipitate can be tricky; a good binocular microscope with dark field optics is an essential laboratory equipment; a polarizer is also useful although both protein and salt crystals can polarize light. The color of the protein precipitate is a good indicator, brown is usually bad, a rice-like glistening granular precipitate is considered more promising. Salt crystals will inevitably occur if phosphate or sulfate anions in the well solution meet divalent cations included in the protein solution or *vice versa*. Dyes are available, often based around methylene blue, which can help identify protein rather than salt crystals (e.g., Izit Crystal Dye, available from Hampton Research). UV attachments to microscopes can also be used, but the deciding factor will be to actually mount the crystal and test it in the X-ray beam and to check if the diffraction is from protein (large unit cell dimensions) or salt (small unit cell dimensions).

It should be emphasized that well-diffracting crystals, which give good quality data, are required and this may take several rounds of optimization to achieve, if it is possible at all. Although the physics of protein crystallization has been studied extensively, usually using lysozyme as a test case, predicting the conditions for crystallizing a new protein target is not possible at the present time, although tools for predicting the crystallizability of a target have been published (Rupp and Wang, 2004). Some propose a connection between the calculated pI and the pH linked to crystal growth (Kantardjieff and Rupp, 2004; Kantardjieff *et al.*, 2004). In the recent past, biophysical methods to determine pI, such as iso-electric focusing, were routine in order to inform subsequent purification procedures, but with the advent of affinity-tagged proteins and their subsequent ease of purification, pI determinations are less common and any link to pH of crystallization is rarely considered.

The process of crystal growth has two main stages, the formation of nuclei and the steady growth of nuclei into a crystal. A supersaturated state of protein in solution is required for the critical stage of nucleation and this is attempted in the initial round of crystallization screening by mixing concentrated protein with different precipitants, the most popular being various molecular weight polyethylene glycol (PEG) polymers, organic compounds, or concentrated salt solutions. There are many references explaining the crystallization process, covering both theoretical and practical aspects that go into far more detail than allowed in the present paper (Benvenuti and Mangani, 2007; Bergfors, 2009; McPherson, 1999, 2004). They are recommended reading for anyone starting in this research field. The list of useful web pages is extensive; technical support pages from companies selling protein crystallization consumables and screening kits contain many tips and photos (e.g., Hampton Research, Molecular Dimensions).

In practice, purified protein (usually the final stage being the middle of the peak off a size-exclusion column) is concentrated between 5 and 20 mg/ml in the buffer used for the gel filtration (GF). The final concentration of protein and the buffer used is rather dependent on protein solubility and is one of the many variables of protein crystallization experiments. The peaks off the column can be analyzed immediately by DLS and, of course, SDS-PAGE is used to compare the protein before GF with the subsequent separated fractions. There is a specialized equipment available enabling size-exclusion chromatography linked to multi-angle light scattering (SEC-MALS) that gives accurate sizing and information on dispersity as the protein comes off the column. This multidetector system measures refractive index, absorbance, and light scattering in real time as the column is running. If the protein is from a new project, it is recommended that a few trays are set up as soon as possible after purification. The presence of a crystallization robot in the laboratory (e.g., Mosquito (TTPLabtech, Herts.) or Phoenix (ARI, Sunnyvale, CA)) means that two or three 96-well screens can be set up using less than a milligram of protein; the use of these robots in a particular laboratory has recently been reviewed (Newman et al., 2008). Inspection of the results, notably the amount of precipitation, yields considerable preliminary information. If more than a third to a half of the tray has precipitated, then the protein concentration may need to be lowered. There are simple screens available which will provide information on whether the protein gives more promising results in PEGs or salts and at which pH (Stura et al., 1992). A solubility screen in which pH and buffers are varied can be used to discover optimal conditions where the protein remains soluble (Jancarik et al., 2004). This screen may be followed with DLS and determines the best starting buffers for storage and setup of crystallization experiments. Screens can be mixed using existing laboratory chemicals (which makes the reproducibility of the crystals more straightforward) or can be purchased "ready-mixed" (e.g., Hampton Research, Molecular Dimensions, Jena Biosciences, Qiagen, Emerald BioSciences). Twenty-five years ago the procedure was to set up manually many grid screens in 24-well plates, methodically going through increasing concentrations of popular precipitants such as PEG 4000 (PEG 4K; a PEG polymer with a mean molecular weight of ~4000), PEG 8K and concentrated salts such as ammonium sulfate and sodium/potassium phosphate at different pHs. Since then, groups have taken crystallization data from many published structures and devised screens that sample the most successful conditions (Carter, 1979; Jancarik and Kim, 1991). These are termed sparse matrix screens as they better sample the available crystallization "space," unlike grid screening which samples a small area more exhaustively; in either case there is no possibility of sampling all available conditions and mixes. With the advent of crystallization robotics and structural genomics consortia and thus the huge increase in available data, analysis of successful "hits" has led to more sophisticated screens such as Index^{TM} (Hampton Research), PACT, and JSCG+ (Newman et al., 2005), which are all commercially available to buy from the companies listed earlier, as well as screens using unusual polymers (e.g., PGA^{TM}, Molecular Dimensions Hu et al., 2008). There is quite a lot of overlap between available screens and the new student should avoid the rather mindless "overscreening" of a protein, especially if this is made easy (albeit expensive) by the presence of robotics and ready made-up screens; a useful web tool is available to look at this overlap (Newman et al., 2010). If no leads have resulted after the use of ~6–

8 screens, then it might be worth revisiting protein characterization and considering using a different construct or protein from a different organism. The presence and position of the affinity tag, discussed above, becomes additionally relevant at this point. Long-term trays should be checked as crystals can appear after days, weeks, months, and even a year or so. Sometimes this is due to a small amount of *in situ* proteolysis to give a crystallizable fragment (Dong *et al.*, 2007). If the crystallization cannot be repeated, running the crystals on a gel or in a mass spectrometer may be possible to check the molecular weight of what has actually crystallized. Temperature is a crystallization variable and ideally protein should be screened identically at two different temperatures, often one that is "cold" (between ~ 4 and $\sim 10\ ^\circ C$) and one at "room temperature" (between ~ 18 and $\sim 22\ ^\circ C$). Temperature-controlled rooms or incubators ($\pm 2\ ^\circ C$) are therefore beneficial for protein crystallization experiments.

2. Optimization and Additive Screens

Once a crystal hit is obtained, the process of crystal optimization begins. At this point, it is important to emphasize how crucial it is to make good notes of what is in the tray, what protein concentration, temperature, and what buffer/salt is present and the ratio of protein: well solution. Unfortunately, repeating crystallization hits is not always straightforward. A first stage of optimization is usually to try and repeat the crystals, or promising precipitate conditions, and also to test around the concentration of precipitant and vary the pH slightly either way. Different buffers but the same pH as the hit can be tried as well as slightly different PEG sizes to the original if the hit was in PEG. Higher molecular weight PEGs may need to be used at slightly lower precipitant concentrations. If the hit was in a PEG/salt mixture, a few different salts can be tried that have not already been tested. Varying the protein to well solution ratio can make a big difference to the rate of crystal growth and should always be tried. Although initial rounds of optimization can be carried out using the 96-well nanoliter trays and crystallization robots, for crystals to be soaked with ligand it is usually necessary to move into larger drops set up manually in 24- or 48-well trays. The most popular option is 24-well hanging drops but sometimes the switch from the sitting drop (used by the robots) is not straightforward. There are now 48-well sitting drop trays available that can be set up using a robot or by hand and take larger drops. It is worth trying both methods as, although both are described as vapor diffusion where water leaves the protein/precipitant drop until equilibrium is achieved with the well solution, results can differ widely, perhaps due to the different rates of equilibration. Some laboratories use micro-batch methods where the protein and precipitant are mixed and put under oil (Chayen *et al.*, 1990; D'Arcy *et al.*,

2004). Oil can be used in vapor-diffusion experiments too (Chayen, 1999); the rate of evaporation can be varied by the type of oil used (Chayen, 1997; D'Arcy et al., 1996; D'Arcy et al., 2003). While useful for screening, especially if the crystallization robot owned by the laboratory preferably sets up this method, obtaining crystals for soaking experiments may have to be done by one of the vapor-diffusion methods. Counter-diffusion methods in gels (Ng et al., 2003; Otalora et al., 2009) and microfluidics (Zheng et al., 2003) are two more specialist techniques used for optimizing protein crystal growth. *In situ* collection of diffraction data from crystals in microfluidic chips or 96-well plates is becoming a reality.

One way of improving crystal growth is testing the effect of various additives; they are added in small amounts to the "best" well solution obtained in the initial optimization process. A 96-component additive screen is available from Hampton Research as are various detergent screens (Cudney et al., 1994). Various cations, anions, organic compounds, detergents, sugars, and protein stabilizers can be tested. Of course, these additives may sometimes improve crystal growth by binding in a ligand-binding site, a result not always wanted if it cannot be displaced by specific ligands of the research project (see below). Some newer screens are now available, for example, Morpheus^{TM} (Gorrec, 2009) that include popular additives or indeed hundreds of small molecules such as Silver Bullets^{TM} (Larson et al., 2008; McPherson and Cudney, 2006) that may facilitate crystal contacts and enhance crystallization. Ionic liquids have more recently been used to improve crystallization results (Judge et al., 2009; Kennedy et al., 2011; Pusey et al., 2007).

2.1. Seeding strategies

If crystal growth cannot be improved by the above methods, then seeding is always worth a try although knowing where to start in this technique can prove rather overwhelming as there are so many variables to consider. The rationale is to introduce small crystal "seeds" to remove the need for the nucleation step in crystal growth. It has been noted that the conditions for crystal nucleation are different to those required for crystal growth (Stura and Wilson, 1990), so separating the two stages gives the opportunity to study far more conditions that may well support crystal growth but would not be capable of initiating a protein nucleation site. There are micro, macro, and matrix seeding methods and seeding tools may be natural materials such as human or horse hair, cat or chinchilla whiskers, or man-made such as commercially available seed beads (Hampton Research), or hypodermic needles. For micro-seeding, the seeds can be prepared by crushing clusters of crystals and used either neat or by first diluting in the crystal growth solution (mother liquor). Seeds can be microcrystals or even a promising precipitate. They can be introduced into new drops of protein/well solution using a whisker (seed several drops before reloading) or a long-stemmed crystal harvesting rayon loop (an example of successful seeding is shown in Fig. 8.2). The drops can be freshly mixed or equilibrated for an hour, several hours, or overnight before seeding; it is good to try several time points and compare the results. The protein solution may be diluted slightly or the well solution can be at a slightly lower precipitant concentration (although not low enough to dissolve the seeds completely). The aim is to use drops that would not grow crystals without seeding intervention, although it is worth trying a range of protein and precipitant concentrations. There are papers outlining the method (Bergfors, 2003; Stura and Wilson, 1991). A more recent seeding method uses a seed stock introduced into a new round of crystal screening rather than the hit conditions (D'Arcy et al., 2007; Ireton and Stoddard, 2003). In our experience, sometimes it is just the small amount of mother

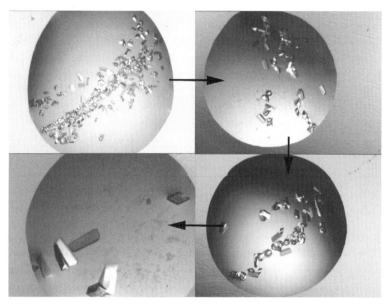

Figure 8.2 Seeding crystals. Examples of streak-seeding of protein crystals in order to aid nucleation of tricky proteins. The example shown is that of the SeMet version of CBM36 which was recalcitrant to nucleation and this was overcome though successive streak-seeding with native CBM36 microcrystals (Jamal et al., 2004). Crushed crystals of CBM36, suspended in approximately 10 μl of mother liquor, were brushed with a human hair and then the hair successively streaked, without recharging, through a series of crystallizing droplets. Many microcrystals act as seeds initially, but as the concentration of deposited seeds decreases so the crystals grow larger. In this case, data to >1.0 Å were achieved. Such streaking techniques are often helpful for overcoming problems with TRIS bound in the active site as TRIS-grown crystals will often nucleate crystallization in other buffer conditions thus avoiding the problems of active-center buffer problems. (For the color version of this figure, the reader is referred to the Web version of this chapter.)

liquor added with the seeds that gives new crystallization conditions so it is worth doing this control in the successful new conditions to check. A possible "universal nucleant" is under development (Saridakis and Chayen, 2009).

3. Cellulase- and Glycosidase-Specific Complex Tactics

For many, a simple uncomplexed 3-D "*apo*-enzyme" structure may be entirely sufficient for the purposes of identifying regions for mutagenesis and engineering, or to aid in the interpretation of preexisting enzyme variants. For others however, people whose "why?" was because they were interested in substrate binding and catalysis, or because pragmatically, protein complexes may deliver a more interesting and respected publication or perhaps a stronger patent, then an insightful ligand complex, better still complexes, is the next goal on the structural itinerary. That said, there are now many cellulase 3-D structures in most if not all the CAZy families (www.cazy.org) and it is quite likely that considerable insight into ligand binding will be derived through comparison with the many known complex structures of cellulases. A thorough review of related structures (which may even be in related GH families if the enzyme is part of a higher-order "clan"; Henrissat and Davies, 1997; Henrissat *et al.*, 1995) may well provide all the ligand-binding information that is required. If not, then strategies to obtain ligand complexes need to be considered.

3.1. Product/pseudo-product complexes

Undoubtedly the easiest, and certainly the cheapest, of all ligand complexes of cellulases are those obtained with short di- and trisaccharides: cellobiose and cellotriose (Fig. 8.3). The disaccharide cellobiose is likely to be the first port of call both as the tight-binding product of cellobiohydrolases and also as a generic cellulose-binding site reporter. We would advocate soaking or cocrystallization using 1–50 mM (we typically use 10 mM) cellobiose or cellotriose. These two ligands are generally (unless crystal packing or buffer/cryo-protectant issues intervene; see below) informative and may well provide most of the insight required. It is worth noting that many commercial supplies of cellobiose and longer cellooligosaccharides can be impure—with the main impurities cellooligosaccharides of different degrees of polymerization. So, occasionally one can be surprised and see longer species than added; most likely these represent impurities but they may also be derived by catalytic transglycosylation events. Complexes with shorter cellooligosaccharides should certainly be considered as routine following the initial structure solution and they take little time. Beyond these approaches, however, complex

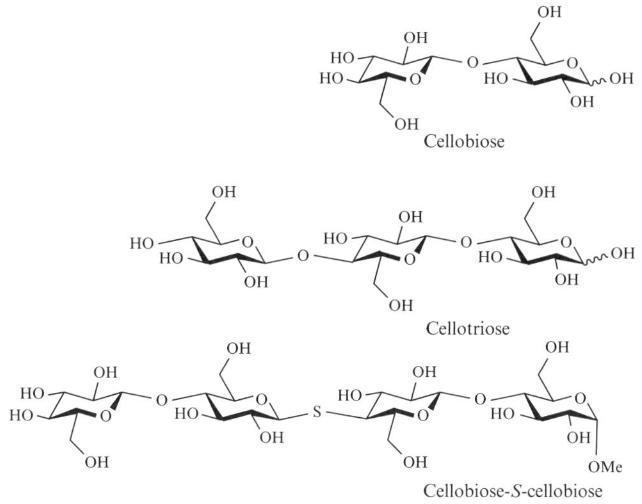

Figure 8.3 Cellooligosaccharides and their nonhydrolyzable analogs. The first port of call for cellulase complexes has to be the product species cellobiose and cellotriose. Longer, perhaps active-center spanning complexes require the use either of inactive enzyme variants or specifically synthesized nonhydrolyzable analogs such as thiooligosaccharides (exemplified here by cellobiose-S-cellobiose).

structures become much more demanding on resources, both staff time and cost and reagents. Any of the approaches outlined below are likely an order of magnitude more work than those discussed above; more still if you require custom-synthesized ligands.

3.2. Enzyme (inactive) variants and longer oligosaccharides

In order to obtain complexes with longer cellooligosaccharides, it is necessary to render the enzyme inactive to prevent hydrolysis of the substrate. Typically, this involves mutation of the genes encoding the catalytic acid/base or nucleophile of a retaining enzyme and either the catalytic acid or the catalytic base of an inverting enzyme (enzyme catalytic mechanisms are beyond the scope of this *Methods* article, the reader is directed to Vocadlo and Davies, 2008; Zechel and Withers, 2000). An important, but often forgotten, caveat here is that mutation of the codon encoding a given amino acid should always involve mutation of at least two of the nucleotide bases. Single base changes result in a translational misincorporation of approximately 0.1% of the wild-type protein (i.e., if you only make a single base change in the codon, 0.1% of your protein will still contain the wild-type amino acid at that position; Schimmel, 1989; Toth *et al.*, 1988); enough to mislead all kinetics and to hydrolyze substrates in the crystallization experiment! Similarly, misleading kinetic and structural work will arise from inappropriate recycling of purification media which may be contaminated with wild-type proteins.

Although many authors on many different systems have used mutagenesis approaches, no enzyme variant guarantees the holy grail of an active-site spanning complex that provides conformational insight (reviewed in Davies *et al.*, 2003, 2012). Indeed, it is common to obtain complexes with a ligand either side of the point of cleavage rather than spanning across, and occasionally spanning complexes simply bypass the catalytic site in a manner that is either artifact or perhaps represent an initial encounter complex between enzyme and ligand (e.g., Hrmova *et al.*, 2001; Juers *et al.*, 2001; Varrot *et al.*, 2001). Furthermore, the "longer oligosaccharide complex" is also the one most open to misinterpretation either accidental or mischievous. Optimistic overinterpretation of complexes, as spanning the active site (when often they are simply pseudo-product complexes, often with a glycerol molecule (see below) in the -1 subsite (subsite nomenclature in Davies *et al.*, 1997)), is common in the field. Crystallographic techniques such as unbiased electron density maps can help here and should always be available for supervisor and referee alike. Indeed, with these potentially more challenging interpretations, it is essential to only build and refine the ligand at the very *end* of the structure refinement process in order to minimize the possibility of bias in the structure refinement and to have the best possible phases for map calculation.

An alternative approach we have used, albeit with slowly turned over 2-fluoro glycosides, is to reduce catalytic activity using a pH jump in the crystals. In the case of endocellulase Cel5A, this involved dropping the pH to ~ 5 affording the Michaelis complex with the 2,4-dinitrophenyl "2-F cellobioside" (Davies *et al.*, 1998); most likely through protonation to the nucleophile to render the enzyme inactive.

3.3. Enzyme inhibitors

Perhaps the most insightful complexes of cellulases, and related enzymes, come from complexes with specifically conceived inhibitors. Such complexes, however, come at a price for the many subsites of a cellulase demand that ligands are di-, tri-, or oligosaccharides and the synthesis of oligosaccharide enzyme inhibitors is both long and costly, and there is only a handful of academic laboratories worldwide working in the area. Furthermore, academic synthetic chemists are reluctant to resynthesize known compounds to aid structural interpretation as this lacks the kudos of a novel synthesis. In the history of cellulase structural research, different "classes" of cellulase inhibitor have been used with great success; some of these are discussed below.

3.3.1. Nonhydrolyzable substrate mimics

The goal of an unhydrolyzed oligosaccharide spanning the enzyme active site had been a goal for glycosidase chemists dating right back to the early work on lysozyme. David Phillips, the pioneering structural biologist who solved the structure of the first enzyme (hen egg white lysozyme), himself had pondered trying to obtain a longer oligosaccharide complex using a nonhydrolyzable ligand but this "...*never got beyond the wondering stage*" (personal correspondence to G.J.D.). It was not until the 1990s that Hugues Driguez pioneered the use of thiooligosaccharides as cellulase inhibitors (reviewed in Driguez, 1995, 2001). In these compounds, one or more of the glycosidic oxygens are replaced by sulfur rending that bond less (effectively un-) hydrolyzable by cellulases. Such compounds had immediate impact allowing access to active-center spanning pseudo-Michaelis complexes on families GH6 (Varrot *et al.*, 2002, 2005; Zou *et al.*, 1999) and GH7 (Sulzenbacher *et al.*, 1996) and thus providing key insight into the unusual substrate distortions in these families (conformational aspects of glycosidase action are reviewed in Davies *et al.*, 2003, 2012). Although not always conformationally insightful, thiooligosaccharides have also provided extensive descriptions of substrate binding (and movement) in cellulases from families GH5 (Varrot *et al.*, 2001), GH6 (Varrot *et al.*, 2003a), and GH48 (Parsiegla *et al.*, 1998, 2000, 2008). Perhaps one area that has not been greatly exploited is the longer oligosaccharides with alternating *S*- and *O*-glycosidic bonds. These "hemi-thio" cellooligosaccharides (Parsiegla *et al.*, 2008) are extremely soluble, much more so than regular cellooligosaccharides with high degrees of polymerization (d.p.) accessible. Although long (d.p. 12–20) hemi-thiooligosaccharides have not received widespread use, it is certainly possible to conceive of new applications for such extended species such as in mapping the link from catalytic domain to CBM.

3.3.2. Covalent inactivators: Withers 2-fluoro compounds and other trapping reagents

Many covalent glycosidase inhibitors have been devised (for a review, see Rempel and Withers, 2008). In the field of retaining glycosidase research, the most powerful have undoubtedly been the 2-deoxy-2-fluoro glycosides, pioneered by Steve Withers (Withers and Aebersold, 1995; Withers *et al.*, 1987). These are classical mechanism-based inhibitors, which trap catalytically competent covalent glycosyl-enzyme intermediates on retaining enzymes. The inductive effect of the fluorine substituent adjacent to the reactive anomeric center renders the formation and breakdown of the covalent intermediate slow but inclusion of a good chemical leaving group (such as 2,4-dinitrophenolate or fluoride) mitigates the rate-retardation effect on the first chemical step and renders the intermediate accessible (Fig. 8.4A). 2-Fluoro di- and trisaccharides have found extensive use on

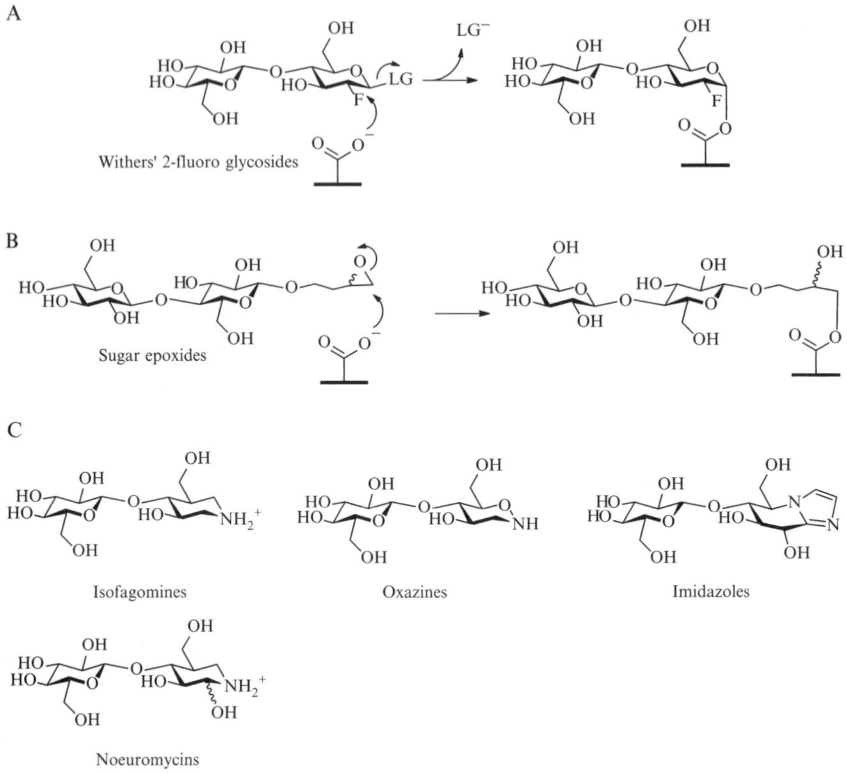

Figure 8.4 Cellulase inhibitors used in structural and mechanistic analyses. (A) Trapping of catalytically competent covalent intermediates using the "Withers" 2-fluoro glycosides in conjunction with a good leaving group (LG). (B) Use of alkyl sugar epoxides to trap the nucleophile of retaining enzymes. Epoxide trapping is notoriously promiscuous and, correctly, out of vogue but the development of more specific epoxides, notably by Overkleeft, may stimulate resurgence in their use as activity-based probes for cellulase prospecting in environmental samples. (C) Examples (illustrated here in their disaccharide incarnations) of putative transition-state mimics that have been used to obtain cellulase complexes: sugar imidazoles, isofagomines, oxazines, and noeuromycins.

many retaining cellulase families both to identify the covalent intermediate and also to map the ligand binding in the −1 to −3 subsites; examples include families GH5 (Davies et al., 1998), GH12 (Sulzenbacher et al., 1999), GH26 (Money et al., 2008) with the first example being the GH10 endoxylanase CEX (White et al., 1996). Occasionally (although not yet seen for any β-D-glucosidases, including cellulases), the trapped 2-fluoro covalent intermediate is still turned over rapidly and in these cases an acid/base variant may be required to further slow the deglycosylation step (see, e.g., Ducros et al., 2002; Vocadlo et al., 2001).

One class of compounds that had initial use, but are no longer considered useful for polysaccharidases, are the sugar epoxides. Such compounds typically have a disaccharide core, an alkyl linker, and the epoxide warhead (Fig. 8.4B). The basis for their action is that the enzymatic nucleophile of a retaining enzyme ring opens the epoxide by nucleophilic attack with protonic assistance from the catalytic acid; thus trapping the nucleophile irreversibly. While this can sometimes work (e.g., family GH7; Sulzenbacher et al., 1997) the inactivation can be slow, but worse still it can be highly promiscuous and label the incorrect residues proving actively misleading (a good example in the field of xylanase research is Havukainen et al., 1996). Withers 2-fluoro compounds have entirely superseded the use of oligosaccharide epoxides for cellulase structural research.

One reason for even discussing epoxides in this *Methods* article is that, despite the caveats above, monosaccharide epoxides (especially the recently described, more specific, compounds; Gloster et al., 2007b) and related aziridines are still extremely useful on exoglycosidases; indeed, they have found new use as the basis of activity-based protein probes in the emerging area of chemical biology. One can certainly conceive of new applications, in prospecting for new cellulases in environmental mixtures, using similar approaches to those pioneered by Overkleeft using epoxides (Witte et al., 2010) and also Vocadlo and Bertozzi (2004) using 2-fluoro glycosides.

3.3.3. Transition-state mimics and related compounds

The tightest binding (and therefore likely most useful) enzyme inhibitors of cellulases are the diverse compounds that might usefully be termed (putative) "transition-state mimics" (recently reviewed in Gloster and Davies, 2010). Again, there are serious synthetic considerations here for while monosaccharide inhibitors are commercially available, the requirement for oligosaccharide inhibitors for cellulase research renders these ligands difficult to synthesize and hence costly or challenging to obtain.

The basic design of all the putative transition-state mimics is that they feature both a tight-binding moiety in combination with extended oligosaccharide chains, with typically di-, tri-, and tetrasaccharide versions used. Putative transition-state mimics, including sugar imidazoles (Varrot et al., 1999), isofagomines and noeuromycins (Meloncelli et al., 2007), and (uncharged) oxazines (Gloster et al., 2004) (illustrated for their disaccharide versions in Fig. 8.4C), have all found use in cellulase research. Complexes have been observed (example references are given) for enzymes in families GH5 (Varrot et al., 2003c), GH6 (Varrot et al., 2003b), GH10 (Notenboom et al., 2000), GH26 (Meloncelli et al., 2007), and GH44 (Ariza et al., 2011). Although the charged inhibitors such as the noeuromycins and isofagomines are classically believed to bind through strong interactions with the enzymatic nucleophile of retaining enzymes, these compounds can also be used with inverting enzymes where they often show the putative catalytic water and, in the case of Cel6 enzymes, they can provide insight into pyranoside ring distortion and thus on the conformational itinerary of catalysis (Varrot et al., 2003b, 2005). Although binding of the oligosaccharide inhibitors increases with chain length, in a manner that reflects the number of thermodynamically significant enzyme subsites, hydrolysis of the longer oligosaccharides is observed both structurally and often reflected in unusual inhibition kinetics for the longer species.

3.4. Inhibitor soaking strategies

With bespoke ligands either purchased at enormous cost, or acquired from academic synthetic groups, the challenge is to obtain complexes with very little ligand; typically <2 mg. Often such small quantities prevent extensive cocrystallization approaches that would simply use up too much ligand (although robotic crystallization in nano-droplets does allow access to more conditions than using manual droplets on the microliter scale). Where possible, inhibitors are normally added using soaking techniques (although conformational changes if they occur will likely destroy the crystals; in which case some limited screening will be required). In most areas of crystallography, soaking would involve making up mother liquor with the compound dissolved at an appropriate concentration and soaking for different lengths of time prior to crystal freezing. Where tiny quantities of inhibitor are available, this approach is not feasible and the approach used by the authors is to introduce tiny amounts of powdered inhibitor, adjacent to a single crystal, using a fine needle and then to observe the crystal as the inhibitor dissolves and slowly migrates (visible in the Schlieren optics). We favor continual observance of the crystal(s) followed by mounting crystals soaked for a few seconds/minutes/hours/overnight. It is hard to judge which length of soaking will produce the most informative complex and, if data collection time is not rate limiting then collecting data on a series of inhibitor soaks, as outlined above should yield the best results.

3.5. Prospecting: Panning ligand mixtures for the best binders

Although strictly not relevant to cellulase research, where the soluble substrate is unbranched and uniform, there are occasions when cocrystallization in a mixture of ligands can provide useful structural information. Such approaches, more typically used on glycosidases with branched or heterogeneous substrates, work on the basis that the enzyme itself will "pan out" (the analogy being with panning for gold) the best binding ligand from a complex mixture. This can be helpful where the exact ligand is either not known or unavailable and instead only impure mixtures of compounds are accessible. This is an approach we have used successfully, several times, on xyloglucanases, for example (Ariza et al., 2011; Gloster et al., 2007a). While this approach can work, and is powerful, it does place an increased burden on the correct and careful interpretation of electron density.

4. Buffer and Cryo-protection Problems

4.1. Buffers

The issue of buffer choice for the crystallization of cellulases (indeed all glycoside hydrolases) is a thorny one. When setting up crystal screens, one is either saddled with the buffer in a commercial screen or one chooses one's own buffer(s) to cover appropriate pH ranges; just how many buffers and how much pH space one can cover is a matter for sample availability and enzyme stability. TRIS-based buffers (TRIS, bis-TRIS, and bis-TRIS propane; Fig. 8.5) are many people's "favorite" buffers and they are also highly prevalent in commercial screens. Yet, TRIS-based buffers are all extremely potent glycosidase inhibitors, often in the low micromolar level. The reason for such tight binding is that the positively charged amino-center partially mimics the oxocarbenium-ion like transition state of the reaction while the various flexible alkyl hydroxyl moieties mimic the oxygen atoms of the sugar substrates. TRIS-based buffers almost inevitably bind in the active

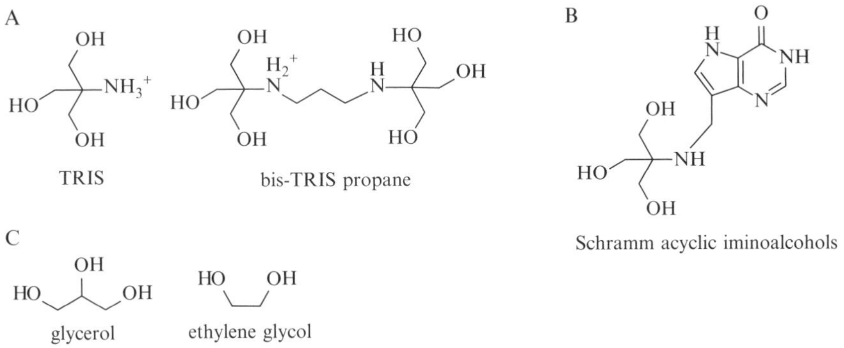

Figure 8.5 Buffers and cryo-protectant problems. (A) TRIS-based buffers, exemplified here by TRIS and bis-TRIS-propane, are classical glycosidase inhibitors. They bind tightly to the active center (reflecting a degree of transition-state mimicry) which no doubt improves the chances of crystallization but this renders it more difficult to obtain complexes subsequently. (B) Schramm has actually pioneered the incorporation of TRIS-like species into potent enzyme inhibitors in the (related) area of N-glycosidase inhibition. (C) Glycerol, and to a lesser extent ethylene glycol (ethane-1,2-diol), used as cryo-protectants also bind in place of sugars in the active center and can also be problematic for glycosidase research (see text). Should this be the case then low molecular weight PEG species make suitable, noninvasive, cryo-protectants.

site and prove very difficult to remove. Indeed, Schramm has harnessed the tight binding of TRIS and TRIS-like species to design potent enzyme inhibitors (Taylor et al., 2007) (Fig. 8.5).

While this is a review on cellulase structural methods, it is worth noting that TRIS is still used as a buffer in many reports, for cellulase and other glycosidase kinetics. We would advise that TRIS-based buffers are never used for kinetic analyses of any glycoside hydrolase; indeed, we have even seen active-center blocking TRIS groups on glycosyltransferases too (Offen et al., 2006).

Harking back to the initial discussion of strategies, if *any* 3-D structure is all that is required then TRIS-based buffers are actually a sound choice; their tight binding almost certainly stabilizes the 3-D structure and *improves* one's chance of crystallization (the basis for many of the additive screens discussed previously). But, if your goal is to obtain mechanistically insightful complexes then prepare to be disappointed and potentially await lots of work trying to find new crystallization conditions. The basis for partial mimicry of the positive transition state by TRIS—the tight binding of a positively charged species at the active center—also means that any metal ion present may also bind in the active center and act as cellulase inhibitors; a recent Cel6A structure reveals that even Li^+ can bind at the site of the positively charged anomeric carbon of the transition state of the enzyme (Thompson and Davies, unpublished observations).

What tactics can one employ to overcome the presence of TRIS? The most simple is clearly to try crystallizing in similar conditions of pH but with a panel of different buffers. Should the simple fail, and no crystals are formed in the new buffer conditions, then one solution is to break up the TRIS-derived crystals and use them as "seeds" in the crystallization process with the alternative buffers (see previous seeding section and Fig. 8.2). Often such a simple seeding process is sufficient to allow nucleation in the new conditions and the growth of TRIS-free crystals. Another option is to complex your enzyme with the appropriate ligand prior to setting up the crystal screens. This approach often works especially with tight-binding transition-state mimics and covalent inactivators; it is, however, far less successful with weak-binding ligands including Michaelis complexes.

When desperate, both in the search for crystal forms allowing complexes but also crystallization of novel proteins *de novo* then we have occasionally resorted to making crystallization variants such as *via* surface entropy reduction strategies (recently reviewed in Derewenda, 2011). Such approaches target regions of the protein surface with consecutive runs of flexible residues, which are then typically mutated, to alanine. Two or three sets of such crystallization variants often allow access to new packing arrangements and crystal forms that may diffract better, or have more easily accessed active centers for complexes. While such approaches may be mocked, there have been several examples of extremely sought-after structures only yielding to surface variant approaches; a classical example being the trypanosomal *trans*-sialidase from the Alzari group (Buschiazzo et al., 2002). In the Alzari case, the approach taken was not a surface entropy reduction one, but a more rational one based upon trying to engineer the surface to look like those of more crystallizable sialidases but a key point here is that it delivered a 3-D structure—but not one that crystallized the way it was being engineered to do! This does suggest that even random surface engineering may have a role to play in the crystallization of recalcitrant proteins.

4.2. Cryo-protectants

Another potential problem facing the crystallographer is which cryo-protectant to choose for the freezing of protein crystals prior to data collection. The molecule of choice throughout the world of protein crystallography is glycerol (Fig. 8.5) followed by ethylene glycol (ethane-1,2-diol). But glycerol especially, and to a lesser extent ethylene glycol, have the disadvantage for the glycosidase structural biologist that they are sugar mimics and are often found in crystal structures in sugar-binding sites. While this may be considered an advantage, perhaps giving one insight into the likely location of carbohydrates, in practice, it is both a hindrance to complex formation and also potentially highly misleading. Glycerol is often used at 5–30% (v/v) (Garman and Mitchell, 1996) and at these concentrations it can easily displace weak-binding ligand complexes, or prevent ligands being soaked into crystals at all. If one is forced to use glycerol, then it is important to ensure that soaked ligands are also included in the cryo-protectant and that crystal exposure to the cryo-liquor is kept to a bare minimum. Worse still, glycerol has, very many times, been incorrectly interpreted in glycoside hydrolase and glycosyltransferase crystal structures as a partially ordered sugar. This can be disastrous when it comes to functional and mechanistic interpretation. If one wants a noninvasive cryo-protectant that should not interfere with ligand binding, then small molecular weight PEGs, and their derivatives, or oils offer good alternatives. Frequently, a simple shift to ethylene glycol can also work as this would appear to bind far less tightly than glycerol. Sometimes the precipitant and salt concentrations in the crystal mother liquor reach a sufficient concentration wherby crystals can be removed straight from the drop without any cryoprotectant.

4.3. Problems of inexperience

It is increasingly common for structure determination to be performed by nonexperts, as commented upon by others (Wlodawer et al., 2008). The improvement in automated data collection and in powerful easy-to-use crystallographic software often driven by user-friendly interfaces makes it comparatively easy for a newcomer to enter the structural world. But such ease of access does bring its own problems, as crystallography is a technical discipline and it can be easy to make serious errors in space group assignment, structure solution, and model building. Mindful of these problems many laboratories and software developers (including both CCP4 and

PHENIX) offer training courses for the nonspecialist and these are to be strongly recommended, as are some of the textbooks appropriate for these audiences (Rhodes, 2006; Rupp, 2009).

Many of the cellulase-specific pitfall considerations have been discussed previously. In our experience, the inexperienced will often misinterpret glycerol density as a mobile sugar, often believe things to be spanning the active center that might better be modeled as two distinct species either side of the catalytic residues and can also misinterpret poor electron density (reflecting mobile or mixed occupancy species) as "distortion." As discussed in the introduction, if insightful and unambiguous complexes are the goal then additional effort has to be made in construct and crystal optimization, crystal soaking strategies, and in data collection.

5. Toward a Cellulase-Specific Approach?

As discussed at the start of the review, cellulase structural biology is essentially like any other and most of the considerations are the same. But if one were to come up with a typical "decision chart" for cellulase structure solution, it would be summarized thus: Firstly, decide why you are doing the work and identify target protein(s). Then delineate domain boundaries using appropriate bioinformatic or experimental (proteolysis) approaches. Make the appropriate domain construct(s) and express the recombinant gene in a suitable host. Probe the protein quality through gel electrophoresis, mass spectrometry, and light scattering; the latter in different buffers. Consider employing deglycosylation strategies for eukaryotically expressed proteins. For crystallization, the first approach will be to tackle the *apo*-enzyme structure and the facile complexes with commercially available di- and trisaccharides. More advanced work might then involve inactive variants with cellooligosaccharides and perhaps the use of mechanism and structure-based inhibitors. As the known roles for plant cell wall degrading enzymes increase, so does the desire for 3-D structural insight into 3-D fold, catalytic action, mechanism, and protein stability. 3-D structure is useful for both directing protein engineering campaigns and for retrospective interpretation of past data in order to inform, better, future enzyme variants for the applied milieu. With a little prior thought, successful crystallization and structure solution strategies may be introduced and cellulase-specific tactics for ligand complexes established.

ACKNOWLEDGMENTS

Research on plant cell wall degrading enzymes in the Davies laboratory is funded by the Biotechnology and Biological Sciences Research Council (BBSRC) through grant BB/I014802. G.J.D. is a Royal Society/Wolfson Research Merit award recipient.

REFERENCES

Adams, J. J., Pal, G., Jia, Z. C., and Smith, S. P. (2006). Mechanism of bacterial cell-surface attachment revealed by the structure of cellulosomal type II cohesin–dockerin complex. *Proc. Natl. Acad. Sci. USA* **103,** 305–310.
Ariza, A., Eklöf, J. M., Spadiut, O., Offen, W. A., Roberts, S. M., Besenmatter, W., Friis, E. P., Skjøt, M., Wilson, K. S., Brumer, H., and Davies, G. (2011). Structure and activity of a *Paenibacillus polymyxa* xyloglucanase from glycoside hydrolase family 44. *J. Biol. Chem.* **286,** 33890–33900.
Benvenuti, M., and Mangani, S. (2007). Crystallization of soluble proteins in vapor diffusion for X-ray crystallography. *Nat. Protoc.* **2,** 1633–1651.
Bergfors, T. (2003). Seeds to crystals. *J. Struct. Biol.* **142,** 66–76.
Bergfors, T. M. (2009). Protein Crystallization: Strategies, Techniques, and Tips. International University Line, La Jolla, CA.
Boraston, A. B., Bolam, D. N., Gilbert, H. J., and Davies, G. J. (2004). Carbohydrate-binding modules: Fine-tuning polysaccharide recognition. *Biochem. J.* **382,** 769–781.
Bott, R., Saldajeno, M., Cuevas, W., Ward, D., Scheffers, M., Aehle, W., Karkehabadi, S., Sandgren, M., and Hansson, H. (2008). Three-dimensional structure of an intact glycoside hydrolase family 15 glucoamylase from *Hypocrea jecorina*. *Biochemistry* **47,** 5746–5754.
Buschiazzo, A., Amaya, M. F., Cremona, M. L., Frasch, A. C., and Alzari, P. M. (2002). The crystal structure and mode of action of trans-sialidase, a key enzyme in *Trypanosoma cruzi* pathogenesis. *Mol. Cell* **10,** 757–768.
Cantarel, B. L., Coutinho, P. M., Rancurel, C., Bernard, T., Lombard, V., and Henrissat, B. (2009). The Carbohydrate-Active EnZymes database (CAZy): An expert resource for glycogenomics. *Nucleic Acids Res.* **37,** D233–D238.
Carter, C. W. (1979). Protein crystallization using incomplete factorial experiments. *J. Biol. Chem.* **254,** 2219–2223.
Chayen, N. E. (1997). The role of oil in macromolecular crystallization. *Structure* **5,** 1269–1274.
Chayen, N. E. (1999). Crystallization with oils: A new dimension in macromolecular crystal growth. *J. Crystal Growth* **196,** 434–441.
Chayen, N. E., Stewart, P. D. S., Maeder, D. L., and Blow, D. M. (1990). An automated system for microbatch protein crystallization and screening. *J. Appl. Cryst.* **23,** 297–302.
Cudney, R., Patel, S., Weisgraber, K., Newhouse, Y., and McPherson, A. (1994). Screening and optimization strategies for macromolecular crystal growth. *Acta Crystallogr. D Biol. Crystallogr.* **50,** 414–423.
D'Arcy, A., Mac Sweeney, A., Stihle, M., and Haber, A. (2003). The advantages of using a modified microbatch method for rapid screening of protein crystallization conditions. *Acta Crystallogr. D Biol. Crystallogr.* **59,** 396–399.
D'Arcy, A., MacSweeney, A., and Haber, A. (2004). Practical aspects of using the microbatch method in screening conditions for protein crystallization. *Methods* **34,** 323–328.
D'Arcy, A., Villard, F., and Marsh, M. (2007). An automated microseed matrix-screening method for protein crystallization. *Acta Crystallogr. D Biol. Crystallogr.* **63,** 550–554.
D'Arcy, A., Elmore, C., Stihle, M., and Johnston, J. E. (1996). A novel approach to crystallising proteins under oil. *J. Cryst. Growth* **168,** 175–180.
Davies, G. J., Wilson, K. S., and Henrissat, B. (1997). Nomenclature for sugar-binding subsites in glycosyl hydrolases. *Biochem. J.* **321,** 557–559.
Davies, G. J., Mackenzie, L., Varrot, A., Dauter, M., Brzozowski, A. M., Schulein, M., and Withers, S. G. (1998). Snapshots along an enzymatic reaction coordinate: Analysis of a retaining beta-glycoside hydrolase. *Biochemistry* **37,** 11707–11713.
Davies, G. J., Ducros, V. M.-A., Varrot, A., and Zechel, D. L. (2003). Mapping the conformational itinerary of β-glycosidases by X-ray crystallography. *Biochem. Soc. Trans.* **31,** 523–527.

Davies, G. J., Planas, A., and Rovira, C. (2012). Conformational analyses of the reaction coordinate of glycosidases. *Accounts Chem. Res.* **45**, 308–316.
Derewenda, Z. S. (2011). It's all in the crystals. *Acta Crystallogr. Sect. D—Biol. Crystallogr.* **67**, 243–248.
Dong, A., Xu, X., Edward, A. M., and Mcsg, Sgc (2007). In situ proteolysis for protein crystallization and structure determination. *Nat. Methods* **4**, 1019–1021.
Driguez, H. (1995). Thiooligosaccharides: Toys or tools for the studies of glycanases. In "Carbohydrate Bioengineering," (S. B. Petersen, B. Svennson, and S. Pedersen, eds.), pp. 113–124. Elsevier, Amsterdam.
Driguez, H. (2001). Thiooligosaccharides as tools for structural biology. *Chembiochem* **2**, 311–318.
Ducros, V., Zechel, D. L., Murshudov, G. N., Gilbert, H. J., Szabo, L., Stoll, D., Withers, S. G., and Davies, G. J. (2002). Substrate distortion by a β-mannanase: Snapshots of the Michaelis and covalent intermediate complexes suggest a $B_{2,5}$ conformation for the transition state. *Angew. Chem. Int. Ed.* **41**, 2824–2827.
Ericsson, U. B., Hallberg, B. M., DeTitta, G. T., Dekker, N., and Nordlund, P. (2006). Thermofluor-based high-throughput stability optimization of proteins for structural studies. *Anal. Biochem.* **357**, 289–298.
Flint, J., Taylor, E., Yang, M., Bolam, D. N., Tailford, L. E., Martinez-Flietes, C., Dodson, E. J., Davis, B. G., Gilbert, H. J., and Davies, G. J. (2005). Structural dissection and high-throughput screening of mannosylglycerate synthase. *Nat. Struct. Mol. Biol.* **12**, 608–614.
Fujimoto, Z., Kuno, A., Kaneko, S., Yoshida, S., Kobayashi, H., Kusakabe, I., and Mizuno, H. (2000). Crystal structure of Streptomyces olivaceoviridis E-86 beta-xylanase containing xylan-binding domain. *J. Mol. Biol.* **300**, 575–585.
Fujimoto, Z., Ichinose, H., Maehara, T., Honda, M., Kitaoka, M., and Kaneko, S. (2010). Crystal structure of an Exo-1,5-{alpha}-L-arabinofuranosidase from Streptomyces avermitilis provides insights into the mechanism of substrate discrimination between exo- and endo-type enzymes in glycoside hydrolase family 43. *J. Biol. Chem.* **285**, 34134–34143.
Garman, E. F., and Mitchell, E. P. (1996). Glycerol concentrations required for cryoprotection of 50 typical protein crystallization solutions. *J. Appl. Cryst.* **29**, 584–587.
Gloster, T. M., and Davies, G. J. (2010). Glycosidase inhibition: Assessing mimicry of the transition state. *Org. Biomol. Chem.* **8**, 305–320.
Gloster, T. M., Macdonald, J. M., Tarling, C. A., Stick, R. V., Withers, S. G., and Davies, G. J. (2004). Structural, thermodynamic and kinetic analysis of tetrahydrooxazine-derived inhibitors bound to β-glucosidases. *J. Biol. Chem.* **279**, 49236–49242.
Gloster, T. M., Ibatullin, F. M., Macauley, K., Eklöf, J. M., Roberts, S., Turkenburg, J. P., Bjørnvad, M. E., Jørgensen, P. L., Danielsen, S., Johansen, K. S., Borchert, T. V., Wilson, K. S., et al. (2007a). Characterisation and 3-D structures of two distinct bacterial xyloglucanases from families GH5 and GH12. *J. Biol. Chem.* **282**, 19177–19186.
Gloster, T. M., Madsen, R., and Davies, G. J. (2007b). Structural basis for cyclophellitol inhibition of a β-glucosidase. *Org. Biomol. Chem.* **5**, 444–446.
Gorrec, F. (2009). The MORPHEUS protein crystallization screen. *J. Appl. Biol.* **42**, 1035–1042.
Havukainen, R., Torronen, A., Laitinen, T., and Rouvinen, J. (1996). Covalent binding of three epoxyalkyl xylosides to the active site of endo-1,4-xylanase II from Trichoderma reesei. *Biochemistry* **35**, 9617–9624.
Henrissat, B., and Davies, G. (1997). Structural and sequence-based classification of glycoside hydrolases. *Curr. Opin. Struct. Biol.* **7**, 637–644.
Henrissat, B., and Davies, G. J. (2000). Glycoside hydrolases and glycosyltransferases. Families, modules, and implications for genomics. *Plant Physiol.* **124**, 1515–1519.
Henrissat, B., Callebaut, I., Fabrega, S., Lehn, P., Mornon, J. P., and Davies, G. (1995). Conserved catalytic machinery and the prediction of a common fold for several families of glycosyl hydrolases. *Proc. Natl. Acad. Sci. USA* **92**, 7090–7094.

Hrmova, M., Varghese, J. N., De Gori, R., Smith, B. J., Driguez, H., and Fincher, G. B. (2001). Catalytic mechanisms and reaction intermediates along the hydrolytic pathway of a plant beta-D-glucan glucohydrolase. *Structure* **9**, 1005–1016.

Hu, T. C., Korczynska, J., Smith, D. K., and Brzozowski, A. M. (2008). *Acta Crystallogr. D Biol. Crystallogr.* **64**, 957–963.

Ireton, G. C., and Stoddard, B. L. (2003). Microseed matrix screening to improve crystals of yeast cytosine deaminase. *Acta Crystallogr. D Biol. Crystallogr.* **60**, 601–605.

Jamal, S., Boraston, A. B., Turkenburg, J. P., Tarbouriech, N., Ducros, V. M.-A., and Davies, G. J. (2004). Ab initio structure determination and functional characterization of CBM36: A new family of calcium dependent carbohydrate-binding modules. *Structure* **12**, 1177–1187.

Jancarik, J., and Kim, S. H. (1991). Sparse-matrix sampling—A screening method for crystallization of proteins. *J. Appl. Cryst.* **24**, 409–411.

Jancarik, J., Pufan, R., Hong, C., Kim, S. H., and Kim, R. (2004). Optimum solubility (OS) screening: An efficient method to optimize buffer conditions for homogeneity and crystallization of proteins. *Acta Crystallogr. D Biol. Crystallogr.* **60**, 1670–1673.

Judge, R. A., Takahashi, S., Longenecker, K. L., Fry, E. H., Abad-Zapatero, C., and Chiu, M. L. (2009). The effect of ionic liquids on protein crystallization and X-ray diffraction resolution. *Cryst. Growth Des.* **9**, 3463–3469.

Juers, D. H., Heightman, T. D., Vasella, A., McCarter, J. D., Mackenzie, L., Withers, S. G., and Matthews, B. W. (2001). A structural view of the action of Escherichia coli (lacZ) beta-galactosidase. *Biochemistry* **40**, 14781–14794.

Kantardjieff, K. A., and Rupp, B. (2004). Protein isoelectric point as a predictor for increased crystallization screening efficiency. *Bioinformatics* **20**, 2162–2168.

Kantardjieff, K. A., Jamshidian, M., and Rupp, B. (2004). Distributions of pI versus pH provide prior information for the design of crystallization screening experiments: Response to comment on 'Protein isoelectric point as a predictor for increased crystallization screening efficiency'. *Bioinformatics* **20**, 2171–2174.

Kennedy, D. F., Drummond, C. J., Peat, T. S., and Newman, J. (2011). Evaluating protic ionic liquids as protein crystallization additives. *Crystal Growth & Design* **11**, 1777–1785.

Kim, K. M., Yi, E. C., Baker, D., and Zhang, K. Y. J. (2001). Post-translational modification of the N-terminal His tag interferes with the crystallization of the wild-type and mutant SH3 domains from chicken src tyrosine kinase. *Acta Crystallogr. D Biol. Crystallogr.* **57**, 759–762.

Larson, S. B., Day, J. S., Nguyen, C., Cudney, R., and McPherson, A. (2008). Progress in the development of an alternative approach to macromolecular crystallization. *Crystal Growth & Design* **8**, 3038–3052.

McPherson, A. (1999). *Crystallization of Biological Macromolecules*. Cold Spring Harbor Laboratory Press, Cold Spring Harbor, NY.

McPherson, A. (2004). Introduction to protein crystallization. *Methods* **34**, 254–265.

McPherson, A., and Cudney, B. (2006). Searching for silver bullets: An alternative strategy for crystallizing macromolecules. *J. Struct. Biol.* **156**, 387–406.

Meloncelli, P. J., Gloster, T. M., Money, V. A., Tarling, C. A., Davies, G. J., Withers, S. G., and Stick, R. V. (2007). D-glucosylated derivatives of Isofagomine and Noeuromycin and their potential as inhibitors of β-glycoside hydrolases. *Aust. J. Chem.* **60**, 549–565.

Money, V., Cartmell, A., Guerreiro, C. I. P. D., Ducros, V. M.-A., Fontes, C. M. G. A., Gilbert, H. J., and Davies, G. J. (2008). Probing the β-1,3:1,4 glucanase, *Ct*Lic26A, with a thio-oligosaccharide and enzyme variants. *Org. Biomol. Chem.* **6**, 851–853.

Newman, J., Egan, D., Walter, T. S., Meged, R., Berry, I., Ben Jelloul, M., Sussman, J. L., Stuart, D. I., and Perrakis, A. (2005). Towards rationalization of crystallization screening for small- to medium-sized academic laboratories: The PACT/JCSG plus strategy. *Acta Crystallogr. D Biol. Crystallogr.* **61**, 1426–1431.

Newman, J., Pham, T. M., and Peat, T. S. (2008). Phoenito experiments: Combining the strengths of commercial crystallization automation. *Acta Crystallogr. Sect. F Struct. Biol. Cryst. Commun.* **64**, 991–996.
Newman, J., Fazio, V. J., Lawson, B., and Peat, T. S. (2010). The C6 Web Tool: A resource for the rational selection of crystallization conditions. *Crystal Growth & Design* **10**, 2785–2792.
Ng, J. D., Gavira, J. A., and Garcia-Ruiz, J. M. (2003). Protein crystallization by capillary counterdiffusion for applied crystallographic structure determination. *J. Struct. Biol.* **142**, 218–231.
Niesen, F. H., Berglund, H., and Vedadi, M. (2007). The use of differential scanning fluorimetry to detect ligand interactions that promote protein stability. *Nat. Protoc.* **2**, 2212–2221.
Notenboom, V., Williams, S. J., Hoos, R., Withers, S. G., and Rose, D. R. (2000). Detailed structural analysis of glycosidase/inhibitor interactions: Complexes of Cex from Cellulomonas fimi with xylobiose-derived aza-sugars. *Biochemistry* **39**, 11553–11563.
Offen, W., Martinez-Fleites, C., Yang, M., Kiat-Lim, E., Davis, B. G., Tarling, C. A., Ford, C. M., Bowles, D. J., and Davies, G. J. (2006). Structure of a flavonoid glucosyltransferase reveals the basis for plant natural product modification. *EMBO J.* **25**, 1396–1405.
Otalora, F., Antonio Gavira, J., Ng, J. D., and Manuel Garcia-Ruiz, J. (2009). Counterdiffusion methods applied to protein crystallization. *Prog. Biophys. Mol. Biol.* **101**, 26–37.
Pantoliano, M. W., Petrella, E. C., Kwasnoski, J. D., Lobanov, V. S., Myslik, J., Graf, E., Carver, T., Asel, E., Springer, B. A., Lane, P., and Salemme, F. R. (2001). High-density miniaturized thermal shift assays as a general strategy for drug discovery. *J. Biomol. Screen.* **6**, 429–440.
Parsiegla, G., Juy, M., Reverbel-Leroy, C., Tardif, C., Belaich, J. P., Driguez, H., and Haser, R. (1998). The crystal structure of the processive endocellulase CelF of Clostridium cellulolyticum in complex with a thiooligosaccharide inhibitor at 2.0 angstrom resolution. *EMBO J.* **17**, 5551–5562.
Parsiegla, G., Reverbel-Leroy, C., Tardif, C., Belaich, J. P., Driguez, H., and Haser, R. (2000). Crystal structures of the cellulase Ce148F in complex with inhibitors and substrates give insights into its processive action. *Biochemistry* **39**, 11238–11246.
Parsiegla, G., Reverbel, C., Tardif, C., Driguez, H., and Haser, R. (2008). Structures of mutants of cellulase Ce148F of Clostridium cellulolyticum in complex with long hemithiocellooligosaccharides give rise to a new view of the substrate pathway during processive action. *J. Mol. Biol.* **375**, 499–510.
Pell, G., Szabo, L., Charnock, S. J., Xie, H., Gloster, T. M., Davies, G. J., and Gilbert, H. J. (2004). Structural and biochemical analysis of Cellvibrio japonicus xylanase 10C: How variation in substrate-binding cleft influences the catalytic profile of family GH-10 xylanases. *J. Biol. Chem.* **279**, 11777–11788.
Pusey, M. L., Paley, M. S., Turner, M. B., and Rogers, R. D. (2007). Protein crystallization using room temperature ionic liquids. *Crystal Growth & Design* **7**, 787–793.
Rempel, B. P., and Withers, S. G. (2008). Covalent inhibitors of glycosidases and their applications in biochemistry and biology. *Glycobiology* **18**, 570–586.
Rhodes, G. (2006). Crystallography Made Crystal Clear: A Guide for Users of Macromolecular Models. 3rd edn. Academic Press, New York, NY.
Rupp, B. (2009). Biomolecular Crystallography: Principles, Practice, and Application to Structural Biology Garland Science. Garland Science, New York, NY.
Rupp, B., and Wang, J. W. (2004). Predictive models for protein crystallization. *Methods* **34**, 390–407.
Saridakis, E., and Chayen, N. E. (2009). Towards a 'universal' nucleant for protein crystallization. *Trends Biotechnol.* **27**, 99–106.
Schimmel, P. (1989). Hazards of deducing enzyme structure–activity relationships on the basis of chemical applications of molecular biology. *Acc. Chem. Res.* **22**, 232–233.

Stura, E. A., and Wilson, I. A. (1990). Analytical and production seeding techniques. *Methods* **1**, 38–49.

Stura, E. A., and Wilson, I. A. (1991). Applications of the streak seeding technique in protein crystallization. *J. Cryst. Growth* **110**, 270–282.

Stura, E. A., Nemerow, G. R., and Wilson, I. A. (1992). Strategies in the crystallization of glycoproteins and protein complexes. *J. Cryst. Growth* **122**, 273–285.

Sulzenbacher, G., Driguez, H., Henrissat, B., Schülein, M., and Davies, G. J. (1996). Structure of the *Fusarium oxysporum* endoglucanase I with a non-hydrolysable substrate analogue: Substrate distortion gives rise to a pseudo-axial orientation for the leaving group. *Biochemistry* **35**, 15280–15287.

Sulzenbacher, G., Schulein, M., and Davies, G. J. (1997). Structure of the endoglucanase I from Fusarium oxysporum: Native, cellobiose, and 3,4-epoxybutyl beta-D-cellobioside-inhibited forms, at 2.3 angstrom resolution. *Biochemistry* **36**, 5902–5911.

Sulzenbacher, G., Mackenzie, L. F., Wilson, K. S., Withers, S. G., Dupont, C., and Davies, G. J. (1999). The crystal structure of a 2-fluorocellotriosyl complex of the Streptomyces lividans endoglucanase CelB2 at 1.2 angstrom resolution. *Biochemistry* **38**, 4826–4833.

Sundqvist, G., Stenvall, M., Berglund, H., Ottosson, J., and Brumer, H. (2007). A general, robust method for the quality control of intact proteins using LC-ESI-MS. *J. Chromatogr. B* **852**, 188–194.

Taylor, E. A., Clinch, K., Kelly, P. M., Li, L., Evans, G. B., Tyler, P. C., and Schramm, V. L. (2007). Acyclic ribooxacarbenium ion mimics as transition state analogues of human and malarial purine nucleoside phosphorylases. *J. Am. Chem. Soc.* **129**, 6984–6985.

Tomme, P., Vantilbeurgh, H., Pettersson, G., Vandamme, J., Vandekerckhove, J., Knowles, J., Teeri, T., and Claeyssens, M. (1988). Studies of the cellulolytic system of Trichoderma reesei qm-9414—Analysis of domain function in 2 cellobiohydrolases by limited proteolysis. *Eur. J. Biochem.* **170**, 575–581.

Toth, M. J., Murgola, E. J., and Schimmel, P. (1988). Evidence for a unique first position codon–anticodon mismatch in vivo. *J. Mol. Biol.* **201**, 451–454.

Van Tilbeurgh, H., Tomme, P., Claeyssens, M., Bhikhabhai, R., and Pettersson, G. (1986). Limited proteolysis of the cellobiohydrolase I from Trichoderma reesei separation of functional domains. *FEBS Lett.* **204**, 223–227.

Varrot, A., Schulein, M., Pipelier, M., Vasella, A., and Davies, G. J. (1999). Lateral protonation of a glycosidase inhibitor. Structure of the Bacillus agaradhaerens Cel5A in complex with a cellobiose-derived imidazole at 0.97 angstrom resolution. *J. Am. Chem. Soc.* **121**, 2621–2622.

Varrot, A., Schulein, M., Fruchard, S., Driguez, H., and Davies, G. J. (2001). Atomic resolution structure of endoglucanase Cel5A in complex with methyl 4,4(II),4(III),4(IV) tetrathio-alpha-cellopentoside highlights the alternative binding modes targeted by substrate mimics. *Acta Crystallogr. D Biol. Crystallogr.* **57**, 1739–1742.

Varrot, A., Frandsen, T., Driguez, H., and Davies, G. J. (2002). Structure of the Humicola insolens cellobiohydrolase, Cel6A, D416A mutant in complex with a non-hydrolysable substrate analogue, methyl-cellobiosyl-4-deoxy-4-thio-b-cellobioside at 1.9 Å. *Acta Crystallogr. D Biol. Crystallogr.* **58**, 2201–2204.

Varrot, A., Frandsen, T. P., von Ossowski, I., Boyer, V., Driguez, H., Schülein, M., and Davies, G. J. (2003a). Structural basis for ligand binding and processivity in cellobiohydrolase Cel6A from Humicola insolens. *Structure* **11**, 855–864.

Varrot, A., Macdonald, J., Stick, R. V., Pell, G., Gilbert, H. J., and Davies, G. J. (2003b). Distortion of a cellobio-derived isofagomine highlights the potential conformational itinerary of inverting β-glucosidases. *Chem. Commun.* **8**, 946–947.

Varrot, A., Tarling, C. A., Macdonald, J., Stick, R. V., Zechel, D., Withers, S. G., and Davies, G. J. (2003c). Direct observation of the protonation of an imino-sugar glycosidase inhibitor upon binding revealed by the 1Å structure of Cel5A with a cellobio-derived isofagomine. *J. Am. Chem. Soc.* **125,** 7496–7497.

Varrot, A., Leydier, S., Pell, G., Macdonald, J. M., Stick, R. V., Gilbert, H. J., and Davies, G. J. (2005). *Mycobacterium tuberculosis* strains possess functional cellulases. *J. Biol. Chem.* **280,** 20181–20184.

Vocadlo, D. J., and Bertozzi, C. R. (2004). A strategy for functional proteomic analysis of glycosidase activity from cell lysates. *Angew. Chem. Int. Ed. Engl.* **43,** 5338–5342.

Vocadlo, D., and Davies, G. J. (2008). Mechanistic insights into glycosidase chemistry. *Curr. Opin. Chem. Biol.* **12,** 539–555.

Vocadlo, D. J., Davies, G. J., Laine, R., and Withers, S. G. (2001). Catalysis by hen egg-white lysozyme proceeds via a covalent intermediate. *Nature* **412,** 835–838.

White, A., Tull, D., Johns, K., Withers, S. G., and Rose, D. R. (1996). Crystallographic observation of a covalent catalytic intermediate in a β-glycosidase. *Nat. Struct. Biol.* **3,** 149–154.

Withers, S. G., and Aebersold, R. (1995). Approaches to labeling and identification of active site residues in glycosidases. *Protein Sci.* **4,** 361–372.

Withers, S. G., Street, I. P., Bird, P., and Dolphin, D. H. (1987). 2-Deoxy-2-fluoroglucosides: A novel class of mechanism-based glucosidase inhibitors. *J. Am. Chem. Soc.* **109,** 7530–7531.

Witte, M. D., Kallemeijn, W. W., Aten, J., Li, K.-Y., Strijland, A., Donker-Koopman, W. E., van den Nieuwendijk, A. M. C. H., Bleijlevens, B., Kramer, G., Florea, B. I., Hooibrink, B., Hollak, C. E. M., *et al.* (2010). Ultrasensitive in situ visualization of active glucocerebrosidase molecules. *Nat. Chem. Biol.* **6,** 907–913.

Wlodawer, A., Minor, W., Dauter, Z., and Jaskolski, M. (2008). Protein crystallography for non-crystallographers, or how to get the best (but not more) from published macromolecular structures. *FEBS J.* **275,** 1–21.

Zechel, D. L., and Withers, S. G. (2000). Glycosidase mechanisms: Anatomy of a finely tuned catalyst. *Acc. Chem. Res.* **33,** 11–18.

Zechel, D. L., Boraston, A. B., Gloster, T. M., Boraston, C. M., Macdonald, J. M., Tilbrook, M. G., Stick, R. V., and Davies, G. J. (2003). Iminosugar glycosidase inhibitors: Structural and thermodynamic dissection of the binding of isofagomine and 1-deoxynojirimycin to two β-glucosidases. *J. Am. Chem. Soc.* **125,** 14313–14323.

Zheng, B., Roach, L. S., and Ismagilov, R. F. (2003). Screening of protein crystallization conditions on a microfluidic chip using nanoliter-size droplets. *J. Am. Chem. Soc.* **125,** 11170–11171.

Zou, J.-Y., Kleywegt, G. J., Ståhlberg, J., Driguez, H., Nerinckx, W., Claeyssens, M., Koivula, A., Teeri, T. T., and Jones, T. A. (1999). Crystallographic evidence for substrate ring distortion and protein conformational changes during catalysis in cellobiohydrolase Cel6A from *Trichoderma reesei*. *Structure* **7,** 1035–1045.

Zulauf, M., and D'Arcy, A. (1992). Light-scattering of proteins as a criterion for crystallization. *J. Cryst. Growth* **122,** 102–106.

CHAPTER NINE

Visualization of Cellobiohydrolase I from *Trichoderma reesei* Moving on Crystalline Cellulose Using High-Speed Atomic Force Microscopy

Kiyohiko Igarashi,[*] Takayuki Uchihashi,[†] Anu Koivula,[‡]
Masahisa Wada,[*,§] Satoshi Kimura,[*,§] Merja Penttilä,[‡]
Toshio Ando,[†] and Masahiro Samejima[*]

Contents

1. Introduction	170
2. Sample Preparation for AFM Observations	171
2.1. Preparation of crystalline cellulose	171
2.2. Purification of cellulase	172
2.3. Highly oriented pyrolytic graphite disk	173
3. Observation of Cellulase Molecules on Crystalline Cellulose	174
3.1. Immobilization of crystalline cellulose on graphite surface	174
3.2. Observation of crystalline cellulose by HS-AFM	175
3.3. Observation of cellulase	176
4. Image Analysis	177
5. Conclusion	179
Acknowledgments	180
References	180

Abstract

Cellulases hydrolyze β-1,4-glucosidic linkages of insoluble cellulose at the solid/liquid interface, generating soluble cellooligosaccharides. We describe here our method for real-time observation of the behavior of cellulase molecules on the substrate, using high-speed atomic force microscopy (HS-AFM). When glycoside hydrolase family 7 cellobiohydrolase from *Trichoderma reesei* (*Tr*Cel7A) was incubated with crystalline cellulose, many enzyme molecules

---

[*] Department of Biomaterial Sciences, Graduate School of Agricultural and Life Sciences, The University of Tokyo, Bunkyo-ku, Tokyo, Japan
[†] Department of Physics, Kanazawa University, Kanazawa, Japan
[‡] VTT Technical Research Centre of Finland, P.O. Box 1000, VTT, Finland
[§] College of Life Sciences, Kyung Hee University, Gyeonggi-do, Yongin-si, Republic of Korea

were observed to move unidirectionally on the surface of the substrate by HS-AFM. The velocity of the moving molecules of TrCel7A on cellulose I crystals was estimated by means of image analysis.

1. INTRODUCTION

Cellulose is a linear polymer of β-1,4-linked glucose residues and is the major component of plant cell walls (Hon, 1994). The degree of polymerization in native celluloses ranges from thousands to tens of thousands of glucose units. In nature, cellulose chains are packed into ordered arrays to form insoluble microfibrils (Nishiyama *et al.*, 2002, 2003; Wolfenden and Yuan, 2008). Microfibrils generally consist of a mixture of disordered amorphous cellulose and cellulose I, which forms highly ordered crystalline regions stabilized by intra- and intermolecular hydrogen bonds. To degrade cellulose, many organisms produce cellulases, which hydrolyze β-1,4-glycosidic linkages of the polymer. Cellulase is a generic term for enzymes hydrolyzing these linkages. However, if we consider the structure of microfibrils, cellulases can be subdivided into two categories, since all cellulases can hydrolyze amorphous cellulose, whereas only a limited number can hydrolyze crystalline cellulose (Teeri, 1997). The enzymes that hydrolyze crystalline cellulose are called cellobiohydrolases (CBHs) because the major product of the reaction is cellobiose, a soluble β-1,4-linked dimer (Teeri *et al.*, 1998). Many CBHs share a two-domain structure, having a catalytic domain (CD) and a cellulose-binding domain (CBD) (Abuja *et al.*, 1988a,b, 1989). As the initial step of the reaction, CBHs are adsorbed on the surface of crystalline cellulose via the CBD, and then glycosidic linkages are hydrolyzed by the CD (Johansson *et al.*, 1989; Ståhlberg *et al.*, 1991). Since the reaction produces mainly soluble cellobiose from insoluble substrates, the hydrolysis of crystalline cellulose occurs at a solid/liquid interface.

CBH belonging to glycoside hydrolase family 7 (Cel7) is the major secreted protein of many cellulolytic fungi, and Cel7A CBH from the industrially important cellulolytic ascomycete fungus *Trichoderma reesei* (*Tr*Cel7A) is one of the best-studied enzymes hydrolyzing crystalline cellulose to cellobiose (Tomme *et al.*, 1988; http://www.cazy.org/). This enzyme has a two-domain structure: a 50 kDa CD and a small (3 kDa) CBD connected by a highly O-glycosylated linker region (Abuja *et al.*, 1988b, 1989). Loss of the CBD causes a significant decrease of crystalline cellulose decomposition but has less effect on the hydrolysis of soluble or amorphous cellulose, suggesting that adsorption of the enzyme on the surface via the CBD is essential for the effective hydrolysis of crystalline cellulose (Ståhlberg *et al.*, 1991). Presumably the two domains work cooperatively during the hydrolysis of solid substrates.

*Tr*Cel7A is thought to hydrolyze crystalline cellulose chain in a processive manner, making consecutive cuts without releasing the chain (Kurasin and Väljamäe, 2011; von Ossowski et al., 2003). This is attributed to the long, tunnel-shaped active site topology of the CBH (Divne et al., 1994, 1998).

Although the kinetics of crystalline cellulose hydrolysis by cellulases has been investigated intensively, the mechanism of crystalline cellulose degradation by CBHs is still not well understood. The main reason for the difficulty in elucidating the mechanism is the absence of analytical methods to monitor enzymatic reaction at a solid/liquid interface. However, in the last decade, methodology to visualize biomolecules using biological atomic force microscopy (AFM) has been developed (Ando et al., 2001, 2007, 2008a,b; Yamamoto et al., 2010), and dynamic protein behaviors have been observed using high-speed AFM (HS-AFM) (Kodera et al., 2010; Shibata et al., 2010; Uchihashi et al., 2011; Yokokawa et al., 2006). We have applied this methodology to track the movement of cellulase molecules on crystalline cellulose in order to better understand the mechanism of enzymatic degradation of cellulose (Igarashi et al., 2009, 2011).

2. SAMPLE PREPARATION FOR AFM OBSERVATIONS

2.1. Preparation of crystalline cellulose

In order to avoid the effect of heterogeneity due to amorphous regions, we prepared highly crystalline cellulose consisting mainly of cellulose I_α (>95% crystallinity) from the cell wall of green algae as a substrate for HS-AFM study (Araki et al., 1998; Igarashi et al., 2006, 2007). The green alga *Cladophora* sp. was harvested from the sea of Chikura, Chiba, Japan and the crystalline cellulose was purified by the following method.

1. Whole algae were immersed overnight in 5% KOH at room temperature. They were thoroughly washed in water and purified by bleaching in 0.3% $NaClO_2$ at 70 °C for 3 h. These treatments were repeated several times until the sample became perfectly white.
2. The purified cell wall was homogenized into small fragments using a double-cylinder-type homogenizer (US-150; Microtec Co., Ltd).
3. The sample was treated with 4 N HCl at 80 °C for 6 h with continuous stirring. The samples were then washed with deionized water by successive dilution and centrifugation at 3200 g for 5 min until the supernatant became turbid.

The obtained crystalline cellulose samples were examined by means of transmission electron microscopy (TEM) and FT-IR and X-ray diffraction analysis, as shown in Fig. 9.1.

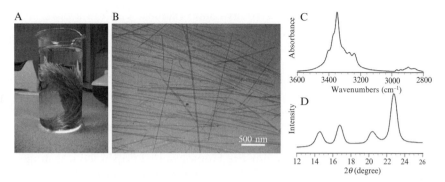

Figure 9.1 (A) Picture of green algae *Cladophora* spp. used in the experiments. Transmission electron microscopic picture (B), FT-IR spectrum (C), and X-ray diffraction (D) of crystalline cellulose purified from *Cladophora* and used in the present studies. (For color version of this figure, the reader is referred to the Web version of this chapter.)

2.2. Purification of cellulase

Cellulases typically interact with other cellulolytic enzymes in nature, showing a synergistic effect in cellulose hydrolysis. In the present experiment, highly purified *Tr*Cel7A was prepared, in order to avoid the effect of contaminating enzyme(s), by means of affinity column chromatography using a substrate/product analogue, 4-aminophenyl-1-thio-β-D-cellobioside, as described previously (Tilbeurgh *et al.*, 1984). Since we are observing single molecules in the present experiment, purity of the protein preparation is of particular importance. Therefore, the experiments were carried out carefully using different protein batches, which should have different levels of contamination, to check reproducibility. Cel7A from *T. reesei* was purified from a commercial cellulase mixture, Celluclast® 1.5L (Novozyme, available from Sigma-Aldrich) as follows (Igarashi *et al.*, 2006; Imai *et al.*, 1998; Samejima *et al.*, 1997):

1. Celluclast® (2.5 ml) was applied to a PD-10 column (GE Healthcare) equilibrated with 20 mM potassium phosphate buffer, pH 7.0, for desalting and buffer exchange.
2. After four repetitions of Step 1, crude enzyme (~ 15 ml) was applied to a DEAE-Toyopearl 650S column (22 × 400 mm, column volume of 150 ml) equilibrated with 20 mM potassium phosphate buffer, pH 7.0. Cel7A was eluted with a linear gradient of KCl concentration from 0 to 0.5 M.
3. The fractions containing Cel7A from the anion-exchange column in Step 2 were pooled and further purified on a Phenyl-Toyopearl 650S (Tosoh) column (22 × 260 mm, column volume of 100 ml) equilibrated

with 20 mM potassium phosphate buffer, pH 7.0, with a linear gradient of ammonium sulfate from 1.0 to 0 M.
4. The fractions from the Phenyl-Toyopearl column were dialyzed against 50 mM sodium acetate buffer, pH 5.0, and purified on an affinity column (Affigel-10; Biorad) with 4-aminophenyl-1-thio-β-D-cellobioside, as described previously. Purity of the enzyme was confirmed by SDS-PAGE. No β-glucosidase or hydroxyethylcellulose-degrading activity was detectable in the Cel7A preparations.

2.3. Highly oriented pyrolytic graphite disk

A bare mica surface is often used for AFM observations because one can easily obtain an atomically flat surface by cleavage of the top layer. However, crystalline celluloses are not immobilized on a mica surface under aqueous conditions because hydrophobic cellulose is poorly adsorbed on the hydrophilic mica surface in liquid environments. Only when a sample droplet containing cellulose is placed on the mica surface, and then naturally dried, can the polysaccharide be physically immobilized on the matrix. However, the AFM observations would then have to be carried out in air, not in a liquid environment, because the cellulose becomes detached from the mica when it is immersed in a liquid. Since further drying would affect the surface condition of cellulose, it is important to choose an immobilization method that does not involve drying the cellulose sample. In the present study, we used highly oriented pyrolytic graphite (HOPG) as a substrate to immobilize crystalline cellulose in a liquid environment because cellulose with its hydrophobic surface is readily adsorbed on the HOPG hydrophobic surface. Thus, when we use hydrophobic HOPG as a substrate, cellulose crystals are expected to be oriented with their hydrophobic surfaces in contact with the HOPG surface, and because of their symmetry, a hydrophobic surface is exposed at the top face as well. Cellulases typically bind on the hydrophobic surface of crystalline cellulose via the CBD. Therefore, we considered that HOPG would be preferable to mica as a support for observing cellulase molecules on cellulose.

The HOPG substrate should be fixed on the z-scanner for HS-AFM observations. The size and flatness of the substrate are important for fast scanning because a large substrate decreases the z-piezo resonant frequency, and a rough surface destabilizes HS-AFM observations. HOPG disks are prepared from a HOPG plate with a sharp punch having a diameter of 1.5 mm. A HOPG disk thus obtained was fixed with glue to the AFM stage, and the top layer of the HOPG surface was cleaved to expose a clean surface. Splinters on the surface, which are often formed by partial cleavage, should be avoided because hydrodynamic turbulent flow around the rapidly scanning AFM tip would induce significant vibration of the substrate through partially cleaved layers. The following handling protocol is recommended:

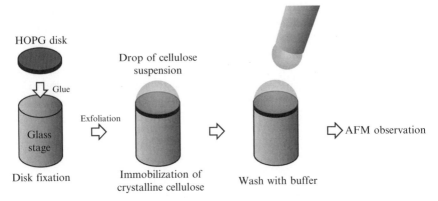

Figure 9.2 Schematic representation of fixation of crystalline cellulose on highly oriented pyrolytic graphite surface.

1. Cut out a HOPG plate with a thickness of ~0.1 mm from a HOPG block (10 × 10 × 2 mm; SPI Supplies) with a cutter blade.
2. Punch out a HOPG disk (ø1.5 mm) with a punch.
3. Glue the HOPG disk onto a columnar glass stage (2 mm in height, 1.5 mm in diameter) with epoxy glue. After gluing, leave the sample for more than 1 h for the glue to set (see Fig. 9.2).
4. Fix the glass stage on the z-piezo with either nail polish or instant glue. Wait until the bond is tight, usually at least 10 min.
5. Press Scotch tape onto the HOPG surface. To obtain a flat area of the HOPG, ensure the tape is evenly stuck to the HOPG surface by wiping with a cotton bud. Then, peel off the top layer of the HOPG surface by removing the tape.
6. The flatness of the surface should be carefully inspected because a rough surface with many large terraces causes scattering of the laser light, resulting in serious interference between the incident and reflected lights.

3. Observation of Cellulase Molecules on Crystalline Cellulose

3.1. Immobilization of crystalline cellulose on graphite surface

As shown in Fig. 9.2, 2 μl of crystalline cellulose suspension in water (0.1–0.5%) was dropped on the freshly cleaved surface of a HOPG disk and incubated for 5–10 min. The disk was then rinsed three to five times with 18 μl of 20 mM sodium acetate buffer, pH 5.0. In the present experiment, we typically used 0.2% cellulose suspension, 5-min incubation, and rinsing three times by the same buffer. Changing these parameters greatly affects the number of crystals available for AFM observation. A higher concentration of cellulose suspension, a longer incubation period, and fewer washings increase the number of visible crystals. Since it is impossible to estimate how many cellulose crystals are fixed on the HOPG surface, only qualitative, but not quantitative, evaluation of the reaction products is possible in this experiment. The product analysis using HPLC has been described elsewhere (Igarashi et al., 2008).

3.2. Observation of crystalline cellulose by HS-AFM

The washing procedure does not completely remove crystalline cellulose weakly adsorbed on the HOPG surface, and consequently, the amplitude of the cantilever often starts to reduce at a point where the tip still seems to be far from the substrate, as the tip is moved toward the surface. A large cantilever amplitude (>10 nm) should be used for the approach because then the AFM tip can punch through the weakly bound cellulose layer. After the tip approaches the surface, we sometimes clean the surface by scanning it with the tip at a small set-point amplitude over a wide area (normally 1 μm × 1 μm) to remove weakly bound cellulose. This cellulose sometimes sticks to the tip and/or the cantilever, causing fluctuation of the amplitude, preventing AFM images. In this case, the tip is withdrawn from the substrate and a new approach should be tried. After removal of the weakly bound cellulose, we set the cantilever amplitude at less than 1–3 nm and start imaging cellulose firmly immobilized on the substrate, as shown in Fig. 9.3A–C. Use of the small amplitude is essential not only to avoid

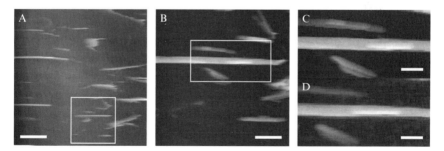

Figure 9.3 (A) AFM pictures of crystalline cellulose using our modified HS-AFM apparatus, 1500 × 1500 nm in 200 × 200 pixels. Bar indicates 300 nm. (B) Close-up view of the white square in A, 500 × 500 nm, 200 × 200 pixels. Bar indicates 100 nm. (C) Close-up view of the white rectangle in B, 300 × 150 nm, 200 × 100 pixels. Bar indicates 50 nm. (D) The same crystalline cellulose observed in C after addition of 2.0 μM TrCel7A. Bar indicates 50 nm. (For color version of this figure, the reader is referred to the Web version of this chapter.)

removal of crystalline cellulose from the substrate but also to prevent disturbance of cellulase binding to the cellulose.

3.3. Observation of cellulase

Crystalline cellulose on the HOPG surface was first visualized as described in Section 3.2. For fast- and less-invasive imaging of cellulases on the crystalline cellulose surface, it is desirable that the fast-scanning direction of the AFM tip along the x direction is parallel to the long axis of the cellulose crystal. After identifying crystalline cellulose tightly immobilized in an appropriate orientation on the HOPG surface, without interference from other crystals, images of the crystal were monitored for 10–20 sec. Typically, crystalline cellulose that is not tightly bound to the surface is easily lost into solution during scanning. On the other hand, crystals that survive the initial observation can be observed again for at least 30 min after addition of enzyme. Highly purified TrCel7A, prepared as described above, was added from the scanner inlet to give a final concentration of 0.2–20 μM (Figs. 9.3D and 9.4A); 2 μM TrCel7A (final concentration) was used routinely. After injection of the cellulase into the observation solution, the solution was slowly and carefully stirred by pipetting. In general, floating

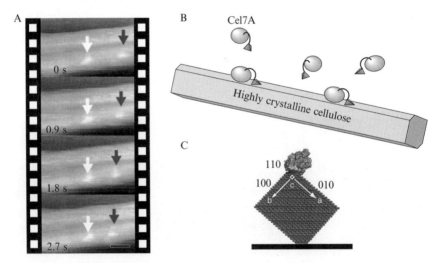

Figure 9.4 (A) Time-lapse images of TrCel7A molecules on crystalline cellulose observed by HS-AFM. Red and white arrows indicate individual mobile and immobile molecules, respectively. Bar indicates 50 nm. Schematic representation of the interaction of TrCel7A and the hydrophobic surface of crystalline cellulose (B) and slice image of crystalline cellulose with TrCel7A adsorbed at the 110 face of the crystal (C). (For interpretation of the references to color in this figure legend, the reader is referred to the Web version of this chapter.)

molecules in the buffer solution during scanning significantly reduce the AFM imaging quality. This is because the sensitivity of the optical beam deflection is altered when the refractive index of the solution is changed. To minimize such disturbance of imaging quality, we used automatic drift compensation, in which the free oscillation amplitude is kept constant by using a higher harmonic oscillation amplitude with slow feedback control. During AFM observations, it is difficult to identify the features in the images because AFM just shows the shape of the molecule. We therefore compared the AFM images with TEM images.

4. Image Analysis

For statistical analysis of the velocities of individual cellulase molecules linearly moving on crystalline cellulose, we developed analysis software in IGOR Pro (WaveMetrics Inc.), which semiautomatically tracks the mass-center position of molecules on the cellulose in successive AFM frames. Since AFM images sometimes contain large noise spikes, probably due to adsorption of cellulase molecules on the cellulose, it is difficult to achieve completely automatic tracking, so we used a semiautomatic tracking method. The calculation is carried out using the following procedures.

1. To reduce noise spikes in the AFM images, all raw images (Fig. 9.5A) were processed with an averaging filter of 2×2 pixels (Fig. 9.5B) before the analysis.
2. Choose a single cellulase molecule on the first frame and set the region of interest (ROI) by enclosing the molecule within a rectangle (Fig. 9.5C).
3. Within the ROI, the slope of the image is compensated by first-order plane fitting and the mass-center position is calculated (Fig. 9.5C).

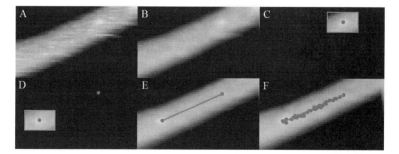

Figure 9.5 Image analysis of HS-AFM pictures using IGOR Pro (version 6.1.2). Pictures A–F are representatives of the steps described in Section 4. (For color version of this figure, the reader is referred to the Web version of this chapter.)

4. At the last frame of the successive images, the same molecule is indicated by enclosing the molecule within a rectangle and the mass-center position is calculated by slope compensation as before (Fig. 9.5D). Then, the positions of the target molecule in the first and last frames are determined. The expected trajectory is calculated as a linear function of the first and the last positions (Fig. 9.5E).
5. At the second frame, the rectangular ROI is moved to the second position along the expected trajectory. The two-dimensional correlation coefficient r for the ROI of the second frame is calculated, using the ROI of the first frame as the reference image. The two-dimensional correlation coefficient r is defined as follows;

$$r = \frac{\sum_m \sum_n (I_{mn} - \bar{I})(R_{mn} - \bar{R})}{\sqrt{\left(\sum_m \sum_n (I_{mn} - \bar{I})^2\right)\left(\sum_m \sum_n (R_{mn} - \bar{R})^2\right)}}$$

Here, I_{mn} and R_{mn} are pixel intensities at a point (m, n) in the analyzing and the reference images, respectively. $\bar{I}$ and $\bar{R}$ are mean values of the intensity matrices I and R, respectively. In the present case, the analyzing and the reference images correspond to the second and the first images, respectively. The ROI of the second frame is moved around the expected position with a tolerance of 5×5 pixels. The position of the ROI in the second frame is fixed to give the maximum value. Then, the mass-center position is calculated within the ROI; this represents the position of the molecule in the second frame.

6. The procedures described above are repeated up to the last frame to track the molecular position for all frames (Fig. 9.5E).
7. After the positions of the molecule in all frames have been calculated, we check that the calculated positions are consistent with the images because coincidental large noise signals sometimes produce errors. If that appears to be the case, we manually choose the ROI in the relevant frames and the mass-center position is recalculated in the corrected ROI.

The movement from the initial position and the off-axis movement of the molecule (distance between the actual center of molecule and the line in Fig. 9.5E) analyzed in Fig. 9.5 are plotted in Fig. 9.6. Since CBH molecules move linearly on a crystalline surface, as mentioned above, off-axis movement was quite small (± 0.83 nm), suggesting that this analysis can provide a reasonable estimate of the velocity. The distance from the initial position clearly increased with time, indicating that this image analysis technique can track enzyme molecules on the substrate.

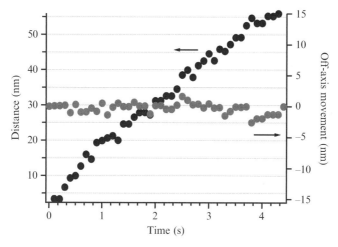

Figure 9.6 Time courses of mobility (black filled circles) and off-axis movement (red filled circles) obtained from the image analysis, as described in Section 4 and Fig. 9.5. (For interpretation of the references to color in this figure legend, the reader is referred to the Web version of this chapter.)

5. CONCLUSION

We have successfully visualized CBH (*Tr*Cel7A) molecules moving on crystalline cellulose, using HS-AFM. The key features of the technique are the use of fine crystalline substrates, a highly purified enzyme preparation, and a flat hydrophobic surface suitable for adhesion of the substrates, in addition to the development of the image analysis technique and customization of the HS-AFM itself. We have shown in our previous reports (Igarashi *et al.*, 2009, 2011) that the observed movement reflects the catalytic reaction and processivity of the enzyme. Therefore, detailed analysis of the movement should throw light on the mechanism of enzymatic decomposition of crystalline cellulose. This technique is expected to be applicable not only to cellulases but also to other types of glycoside hydrolases, that is, hemicellulases, chitinases, and amylases, and should lead to a deeper mechanistic understanding of how these enzymes act at a solid/liquid interface and how biomass degradation occurs in nature. Moreover, this type of enzymatic reaction should be involved in the degradation of many insoluble materials, such as biodegradable plastics, so HS-AFM is expected to be a powerful tool for analysis of the reactions of various biomacromolecules on surfaces.

ACKNOWLEDGMENTS

This research was supported by Grants-in-Aid for Scientific Research to K. I. (19688016 and 21688023), T. U. (21023010 and 21681017), and T. A. (20221006) from the Japanese Ministry of Education, Culture, Sports, and Technology; by a grant of the Knowledge Cluster Initiative to T. A.; by a Grant for Development of Technology for High Efficiency Bioenergy Conversion Project to M. S. (07003004-0) from the New Energy and Industrial Technology Development Organization; and by an Advanced Low Carbon Technology Research and Development Program from the Japan Science and Technology Agency to K. I. and T. U.

REFERENCES

Abuja, P. M., Pilz, I., Claeyssens, M., and Tomme, P. (1988a). Domain structure of cellobiohydrolase II as studied by small angle X-ray scattering: Close resemblance to cellobiohydrolase I. *Biochem. Biophys. Res. Commun.* **156,** 180–185.

Abuja, P. M., Schmuck, M., Pilz, I., Tomme, P., Claeyssens, M., and Esterbauer, H. (1988b). Structural and functional domains of cellobiohydrolase I from *Trichoderma reesei*. A small angle X-ray scattering study of the intact enzyme and its core. *Eur. Biophys. J.* **15,** 339–342.

Abuja, P. M., Pilz, I., Tomme, P., and Claeyssens, M. (1989). Structural changes in cellobiohydrolase I upon binding of a macromolecular ligand as evident by SAXS investigations. *Biochem. Biophys. Res. Commun.* **165,** 615–623.

Ando, T., Kodera, N., Takai, E., Maruyama, D., Saito, K., and Toda, A. (2001). A high-speed atomic force microscope for studying biological macromolecules. *Proc. Natl. Acad. Sci. USA* **98,** 12468–12472.

Ando, T., Uchihashi, T., Kodera, N., Yamamoto, D., Taniguchi, M., Miyagi, A., and Yamashita, H. (2007). High-speed atomic force microscopy for observing dynamic biomolecular processes. *J. Mol. Recognit.* **20,** 448–458.

Ando, T., Uchihashi, T., and Fukuma, T. (2008a). High-speed atomic force microscopy for nano-visualization of dynamic biomolecular processes. *Prog. Surf. Sci.* **83,** 337–437.

Ando, T., Uchihashi, T., Kodera, N., Yamamoto, D., Miyagi, A., Taniguchi, M., and Yamashita, H. (2008b). High-speed AFM and nano-visualization of biomolecular processes. *Pflugers Arch.* **456,** 211–225.

Araki, J., Wada, M., Kuga, S., and Okano, T. (1998). Flow properties of microcrystalline cellulose suspension prepared by acid treatment of native cellulose. *Colloids Surf. A* **142,** 75–82.

Divne, C., Ståhlberg, J., Reinikainen, T., Ruohonen, L., Pettersson, G., Knowles, J. K., Teeri, T. T., and Jones, T. A. (1994). The three-dimensional crystal structure of the catalytic core of cellobiohydrolase I from *Trichoderma reesei*. *Science* **265,** 524–528.

Divne, C., Ståhlberg, J., Teeri, T. T., and Jones, T. A. (1998). High-resolution crystal structures reveal how a cellulose chain is bound in the 50 Å long tunnel of cellobiohydrolase I from *Trichoderma reesei*. *J. Mol. Biol.* **275,** 309–325.

Hon, D. N.-S. (1994). Cellulose: A random walk along its historical path. *Cellulose* **1,** 1–25.

Igarashi, K., Wada, M., Hori, R., and Samejima, M. (2006). Surface density of cellobiohydrolase on crystalline celluloses. A critical parameter to evaluate enzymatic kinetics at a solid-liquid interface. *FEBS J.* **273,** 2869–2878.

Igarashi, K., Wada, M., and Samejima, M. (2007). Activation of crystalline cellulose to cellulose III(I) results in efficient hydrolysis by cellobiohydrolase. *FEBS J.* **274,** 1785–1792.

Igarashi, K., Ishida, T., Hori, C., and Samejima, M. (2008). Characterization of an endoglucanase belonging to a new subfamily of glycoside hydrolase family 45 of the basidiomycete *Phanerochaete chrysosporium*. *Appl. Environ. Microbiol.* **74,** 5628–5634.

Igarashi, K., Koivula, A., Wada, M., Kimura, S., Penttilä, M., and Samejima, M. (2009). High speed atomic force microscopy visualizes processive movement of *Trichoderma reesei* cellobiohydrolase I on crystalline cellulose. *J. Biol. Chem.* **284,** 36186–36190.

Igarashi, K., Uchihashi, T., Koivula, A., Wada, M., Kimura, S., Okamoto, T., Penttilä, M., Ando, T., and Samejima, M. (2011). Traffic jams reduce hydrolytic efficiency of cellulase on cellulose surface. *Science* **333,** 1279–1282.

Imai, T., Boisset, C., Samejima, M., Igarashi, K., and Sugiyama, J. (1998). Unidirectional processive action of cellobiohydrolase Cel7A on *Valonia* cellulose microcrystals. *FEBS Lett.* **432,** 113–116.

Johansson, G., Ståhlberg, J., Lindeberg, G., Engstrom, A., and Pettersson, G. (1989). Isolated fungal cellulase terminal domains and a synthetic minimum analog bind to cellulose. *FEBS Lett.* **243,** 389–393.

Kodera, N., Yamamoto, D., Ishikawa, R., and Ando, T. (2010). Video imaging of walking myosin V by high-speed atomic force microscopy. *Nature* **468,** 72–76.

Kurasin, M., and Väljamäe, P. (2011). Processivity of cellobiohydrolases is limited by the substrate. *J. Biol. Chem.* **286,** 169–177.

Nishiyama, Y., Langan, P., and Chanzy, H. (2002). Crystal structure and hydrogen-bonding system in cellulose Iβ from synchrotron X-ray and neutron fiber diffraction. *J. Am. Chem. Soc.* **124,** 9074–9082.

Nishiyama, Y., Sugiyama, J., Chanzy, H., and Langan, P. (2003). Crystal structure and hydrogen bonding system in cellulose Iα from synchrotron X-ray and neutron fiber diffraction. *J. Am. Chem. Soc.* **125,** 14300–14306.

Samejima, M., Sugiyama, J., Igarashi, K., and Eriksson, K.-E. L. (1997). Enzymatic hydrolysis of bacterial cellulose. *Carbohydr. Res.* **305,** 281–288.

Shibata, M., Yamashita, H., Uchihashi, T., Kandori, H., and Ando, T. (2010). High-speed atomic force microscopy shows dynamic molecular processes in photoactivated bacteriorhodopsin. *Nat. Nanotechnol* **5,** 208–212.

Ståhlberg, J., Johansson, G., and Pettersson, G. (1991). A new model for enzymatic hydrolysis of cellulose based on the two-domain structure of cellobiohydrolase I. *BioTechnology* **9,** 286–290.

Teeri, T. T. (1997). Crystalline cellulose degradation: New insight into the function of cellobiohydrolases. *Trends Biotechnol.* **15,** 160–167.

Teeri, T. T., Koivula, A., Linder, M., Wohlfahrt, G., Divne, C., and Jones, T. A. (1998). *Trichoderma reesei* cellobiohydrolases: Why so efficient on crystalline cellulose? *Biochem. Soc. Trans.* **26,** 173–178.

Tilbeurgh, H., Bhikhabhai, R., Pettersson, G., and Claeyssens, M. (1984). Separation of endo- and exo-type cellulases using a new affinity chromatography method. *FEBS Lett.* **169,** 215–218.

Tomme, P., Van Tilbeurgh, H., Pettersson, G., Van Damme, J., Vandekerckhove, J., Knowles, J., Teeri, T., and Claeyssens, M. (1988). Studies of the cellulolytic system of *Trichoderma reesei* QM 9414. Analysis of domain function in two cellobiohydrolases by limited proteolysis. *Eur. J. Biochem.* **170,** 575–581.

Uchihashi, T., Iino, R., Ando, T., and Noji, H. (2011). High-speed atomic force microscopy reveals rotary catalysis of rotorless F-ATPase. *Science* **333,** 755–758.

von Ossowski, I., Ståhlberg, J., Koivula, A., Piens, K., Becker, D., Boer, H., Harle, R., Harris, M., Divne, C., Mahdi, S., Zhao, Y., Driguez, H., *et al.* (2003). Engineering the exo-loop of *Trichoderma reesei* cellobiohydrolase, Cel7A. A comparison with *Phanerochaete chrysosporium* Cel7D. *J. Mol. Biol.* **333,** 817–829.

Wolfenden, R., and Yuan, Y. (2008). Rates of spontaneous cleavage of glucose, fructose, sucrose, and trehalose in water, and the catalytic proficiencies of invertase and trehalas. *J. Am. Chem. Soc.* **130,** 7548–7549.

Yamamoto, D., Uchihashi, T., Kodera, N., Yamashita, H., Nishikori, S., Ogura, T., Shibata, M., and Ando, T. (2010). High-speed atomic force microscopy techniques for observing dynamic biomolecular processes. *Methods Enzymol.* **475,** 541–564.

Yokokawa, M., Wada, C., Ando, T., Sakai, N., Yagi, A., Yoshimura, S. H., and Takeyasu, K. (2006). Fast-scanning atomic force microscopy reveals the ATP/ADP-dependent conformational changes of GroEL. *EMBO J.* **25,** 4567–4576.

CHAPTER TEN

Small-Angle X-Ray Scattering and Crystallography: A Winning Combination for Exploring the Multimodular Organization of Cellulolytic Macromolecular Complexes

Mirjam Czjzek,[*,†] Henri-Pierre Fierobe,[‡] and Véronique Receveur-Bréchot[§]

Contents

1. Introduction	184
2. Measuring and Analyzing SAXS Data	186
2.1. Sample preparation	186
2.2. Experimental requirements for SAXS	187
2.3. The experimental data and quality assessment	187
2.4. Data analysis and determination of structural values obtained with SAXS	188
2.5. Toward 3D models	190
3. Combining SAXS and Crystallography to Analyze Multimodular Organization of Cellulolytic Enzymes and Complexes	191
3.1. Cellulases appended by one CBM and two modular enzymes: The importance of the linker	192
3.2. The dissect and build approach: Analysis of more complex, multimodular cellulolytic assemblies	194
3.3. A concrete example: Constructing the 3D organization of the multimodular enzyme XynZ	196
3.4. Combination with computational modeling to construct the "missing parts": Dealing with flexibility	200

[*] Université Pierre et Marie Curie, Paris 6, France
[†] Centre National de la Recherche Scientifique, Marine Plants and Biomolecules, UMR 7139, Station Biologique de Roscoff, Roscoff, France
[‡] Laboratoire de Chimie Bactérienne, CNRS-UPR 9043, Marseille, France
[§] CRCM, UMR7258 CNRS, INSERM, Aix-Marseille University, IPC, Marseille cedex 9, France

4. Conclusions and Outlook	203
Acknowledgments	205
References	205

Abstract

Small-angle X-ray scattering (SAXS) is an increasingly popular method to obtain low-resolution structures of complex macromolecules and their complexes in solution, in part due to recent technical and computational advances that make this method more and more accessible. However, to obtain unambiguous molecular interpretation from SAXS envelopes, the efficient use of and combination with additional structural methods are crucial. The multimodular character of cellulases and their assemblage in the cellulosome are ideally analyzed by such a combination of structural methods. Here, we describe how information from different sources can be combined with SAXS to determine the molecular organization and we depict the recent advancements and trends that are leading to a more comprehensive picture of the molecular architecture of these multimodular enzymes and their organization in macro-assemblages such as cellulosomes.

1. INTRODUCTION

Cellulose, the natural substrate of microbial cellulases, is of particular recalcitrant character. Moreover, the crystallinity and occurrence within the solid and compact cell wall of plants, interconnected with various other hemicelluloses, necessitate that the catalytic reaction must take place within the heterogeneous phase. Despite these difficulties and due to the abundance of the substrate, many microbes have specialized to utilize this extraordinary source of carbon and energy, thus contributing a central component of the global carbon cycle. Early on, Reese et al. (1950) already assumed that several components were required to be able to produce soluble sugars from the recalcitrant, crystalline cellulose fibers present in the plant cell wall. But the fact that the enzymes responsible for the breakdown of this particular substrate are conceived as multimodular components, consisting of core catalytic modules assembled with additional modules capable of binding to the solid substrate, has only been discovered almost 40 years later. Indeed, the modular architecture of plant cell wall active enzymes was first discovered for the cellulases from *Hypocrea jecorina* (previously named *Trichoderma reesei*) by controlled proteolysis of the full-length enzymes (Tomme et al., 1988; van Tilbeurgh et al., 1986). Rapidly, and understanding that the instability and flexibility of the full-length cellulases would hinder crystallization, the first small-angle X-ray measurements were performed and confirmed their particular overall architecture

(Abuja et al., 1988a,b, 1989). Such a modular constitution has later been found to occur also in xylanases and other plant cell-wall-degrading enzymes (Gilbert et al., 1992; Gilkes et al., 1991). Actually, the flood of gene information coming from "cazome" assessment of microbial genomes reveals that the modular architecture of polysaccharide-degrading enzymes (Fig. 10.1) is a very frequently occurring character of these enzyme families (http://www.cazy.org/cazy). In parallel, the discovery of the macromolecular assembly, the cellulolytic machinery secreted by anaerobic bacteria called cellulosome (Bayer et al., 2004; Lamed et al., 1983), highlighted an extremely complex multimodular arrangement. In this context, it is believed that the anaerobic environment induces a greater selective pressure for the evolution of highly effective machineries for the extracellular degradation of polymeric substrates. These macromolecular assemblies not only challenged biochemists to understand the synergistic interplay of all components, but also defied structural biologists to get a three-dimensional (3D) picture of how the architectural arrangement of this large number of modules and

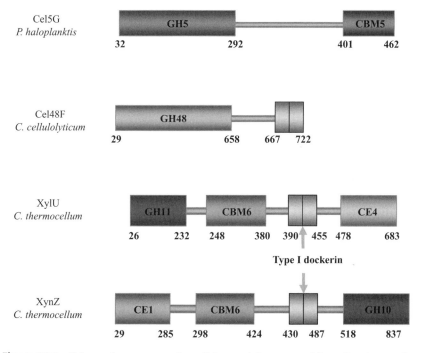

Figure 10.1 Schematic representation of the modular composition of various carbohydrate-degrading enzymes. The names and the organisms of origin are given to the left of each polypeptide representation. The numbers underneath each protein correspond to the residue numbers delimiting the different modules, as determined by multiple sequence alignments and/or experimental evidence. (See Color Insert.)

proteins allows the extraordinary efficiency of the ensemble. While 3D structures by crystallography and NMR (Carvalho et al., 2003; Davies et al., 1993; Dominguez et al., 1995; Ducros et al., 1995; Johnson et al., 1996; Kleywegt et al., 1997; Mosbah et al., 2000; Sakon et al., 1996; Shimon et al., 1997; Tormo et al., 1996) of constructs of the isolated modules and enzymes gave access to local information of specific interactions and modes of recognition, 3D structures of full-length multimodular (in fact multi-domain) enzymes or complexes stayed exceptional (Adams et al., 2010; Carvalho et al., 2003; Sakon et al., 1997).

Recently, there have been significant developments in automated bioinformatics, and technical advances in structural methods based on X-ray scattering that include extensive use of synchrotron radiation, anomalous dispersion methods in protein crystallography, and the adaptation of beamlines to perform small-angle X-ray scattering (SAXS) of proteins in solution. These advances allow more sophisticated experimental design, which has enabled us to tackle the challenging and complex cellulolytic assemblies. This chapter summarizes some of the major achievements and trends of recent lines of research combining SAXS with other structural methods to dissect and analyze the multimodular organization of cellulolytic macromolecular enzymes and complexes.

2. Measuring and Analyzing SAXS Data

2.1. Sample preparation

Purity of the sample to be measured is a prerequisite for the performance of high-quality small-angle scattering measurements. Therefore, special care has to be taken when purifying the proteins to be analyzed; particularly higher molecular weight (MW) impurities, which will hamper the collection of the correct scattering signals. Thus, a final step of size exclusion chromatography for protein polishing should systematically be included. The scattering signal is proportional to the square of the MW, and, therefore, already minor amounts of large contaminants or aggregates will disproportionately contribute to the scattering and bias the scattering curve toward larger particle sizes. Furthermore, in contrast to other structural methods such as X-ray crystallography, small-angle scattering patterns can *always* be obtained from samples of any quality, and, as such, verification that the scattering particles are monodisperse and identical is essential before data analysis and construction of a structural model inferred from the data. In general, SDS-PAGE analyses cannot detect aggregation, and sample quality should be also checked using dynamic light scattering, to reveal the putative presence of multiple oligomeric states.

2.2. Experimental requirements for SAXS

Although the development of small-angle scattering instrumentation poses several technical challenges for data acquisition, schematically the experimental setup is rather simple. A highly collimated, single wavelength, X-ray beam is used to illuminate the sample, a protein or macromolecular complex in solution. Similar to all biophysical methods, SAXS requires large quantities of very pure (>99%) samples. Consequently, many efforts of experimental setups have been concentrated on the automatic pipetting and the use of smaller volumes (<100 μL) to measure the scattering profiles of the protein samples (typically >1 mg mL^{-1} in 30–100 μL for X-ray scattering). The scattered radiation is recorded on a detector, while the direct beam is usually absorbed by a beam stop, the size, position, and distance from the sample of which are key factors determining the minimum angle measured in an experiment. Another advantage over X-ray crystallography resides in the fact that the 3D structure information, although at low resolution (typically 15 Å), is obtained from proteins in solution. In contrast to X-ray crystallography, SAXS is inherently a contrast method and depends on the difference of the average electron density of the scattering solute molecule (for proteins ~ 0.44 e$^-$/Å^3) compared to that of the bulk solvent (for pure water ~ 0.33 e$^-$/Å^3); the larger the difference the stronger the scattering signal and the more accurate are the data on dilute solutions. In practice, data are collected from a series of at least four concentrations of sample solutions and alternated with measurements on a blank buffer, in order to control how protein concentration and intermolecular interactions may affect the measured scattered intensity. Subsequent subtraction of the buffer signal from the observed sample scattering yields the signal due to the macromolecule present in solution. The subtraction of the buffer signal is a crucial step in data treatment and must be done as precisely as possible to accurately measure differences over the entire scattering curve. Only gel filtration buffer, or dialysis buffer, allows a proper acquisition of the scattering from the bulk solvent surrounding the macromolecules/complexes of interest. Today, measurements using *in situ* gel filtration, such as those enabled by the experimental setup developed on beamline SWING, Soleil, France (David and Perez, 2009), also available on the BL-10C station at the Photon factory, Tsukuba, Japan (Watanabe and Inoko, 2011) and the SAXS/WAXS beamline of the Australian synchrotron, Melbourne, Australia (Gunn et al., 2011), allow a perfect subtraction of the buffer signal.

2.3. The experimental data and quality assessment

The intensity of the scattering curve resulting from the subtraction of the buffer from the sample, $I(q)$, is radially isotropic due to the randomly oriented distribution of particles in solution (Fig. 10.2). $I(q)$ is a function

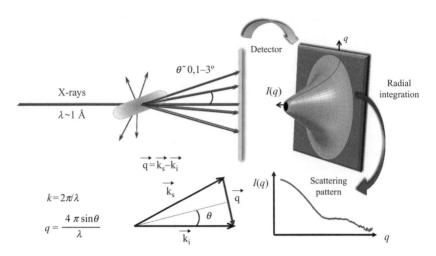

Figure 10.2 Schematic representation of the setup and analyses of a small-angle scattering experiment. The scattering vector q is defined as the difference between the scattered wavevector k_s and the incident wavevector k_i. (See Color Insert.)

of the momentum transfer $q = (4\pi \sin\theta)/\lambda$, where 2θ is the scattering angle and λ is the wavelength of the incident X-ray beam. The units of the momentum transfer q are the inverse of the wavelength expressed in Å^{-1} or nm^{-1}, in analogy to Bragg's law for crystallography, where $q = 2\pi/d$ and $1/d$ is the reciprocal resolution. In contrast to crystallography, where diffraction provides a direct measure of data quality, it is more difficult to assess the quality of a SAXS curve, although some empirical guidelines do exist. As small-angle scattering yields data that are inherently one-dimensional (1D) and the user invariably is looking to support a 3D model, overinterpretation of the data will always be a risk if not given careful consideration. Broadly accepted standards for quality assessments have still to be developed and accepted. A recent, excellent review summarizes this issue (Jacques and Trewhella, 2010), and thus this aspect will not be further described here. Briefly, the coherence of data measured at various concentrations and the calibration of the scattered intensity using a standard protein or even better, using water, are indispensable to have independent indicators of data quality before analyzing (and overinterpreting) any further.

2.4. Data analysis and determination of structural values obtained with SAXS

Three experimental values describing the protein can be directly extracted from the subtracted sample scattering curve. These are the radius of gyration (R_g), the forward scattering intensity $I(0)$, and the maximal distance (D_{max}), which provide different parts of information about the size of the protein.

The radius of gyration is defined as the root mean square distance of the protein atoms from the centroid, weighted by their scattering length density (electron density for SAXS). The radius of gyration, thus, reflects the average dimension of the protein (see representation in Fig. 10.3A). The forward scattering intensity $I(0)$ is directly proportional to the molecular mass of the macromolecule. Determining the value of $I(0)$ from the experimental scattering pattern thus allows to directly infer the oligomeric state of a protein or to check the molecular mass of a multicomponent complex.

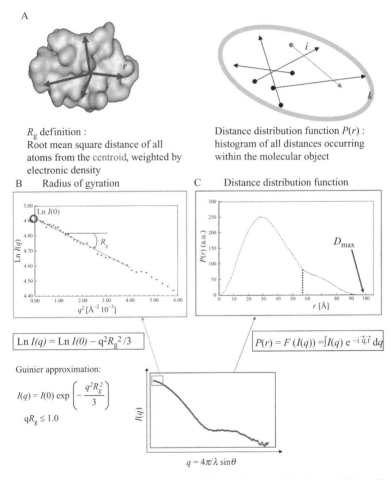

Figure 10.3 (A) Graphical representations illustrating the definitions of the radius of gyration (R_g) and the distance distribution function ($P(r)$). (B, C) Graphical representations of how the values R_g, $P(r)$, and D_{max} are extracted from the experimental data, respectively. (See Color Insert.)

The radius of gyration, together with the intensity $I(0)$, is usually inferred from the scattering curve through the Guinier law: $I(q) = I(0) \exp(-q^2 R_g^2/3)$. Plotting the logarithm of the intensity versus q^2 allows to infer the R_g and the $I(0)$ from the slope and the intercept, respectively, of the regression line of the experimental curve (Fig. 10.3B). The Guinier law is valid over a q range limited by $q < 1/R_g$.

Transforming the scattering intensity from the reciprocal space to the real space provides the distance distribution function $P(r)$, which is the histogram of all the intramolecular distances between atoms of the molecule (Fig. 10.3A). Among them, the maximum dimension D_{max} of the particle is very informative on the anisotropy of the macromolecule and on its degree of extension. The $P(r)$ function is yielded by the Fourier transform of the scattering intensity using the program GNOM (Svergun, 1992) or GIFT (Bergmann et al., 2000), for example (Fig. 10.3C).

All these first data analyses can be conveniently performed, for example, with the widely distributed program suite ATSAS, starting with the program PRIMUS, but other programs for data treatment do also exist.

2.5. Toward 3D models

As shown above, SAXS measurements give access to the experimental determination of the MW, R_g, and D_{max} values and to the Fourier transform of the experimental curve, $P(r)$ the distance distribution function, which is correlated to the overall shape of the individual molecules in solution. But in general, a scientist desires to visually describe the objects in solution, and therefore creates a 3D envelope by fitting the experimental SAXS curve using various algorithms. During the past decade, the development of new such algorithms and programs in concert with advances in data processing techniques allowed to infer much more structural information from a given scattering pattern. The application of *ab initio* shape restoration programs using bead models, such as DAMMIN, DAMMIF (Svergun, 1999), GAS-BOR (Svergun et al., 2001), Saxs3D (Walther et al., 2000), and DAILA_GA (Chacon et al., 1998), has thus been developed and constitutes the first step toward the determination of the 3D arrangement of proteins and complexes in solution. The shapes obtained already represent a level of interpretation of the data, since in most cases the shapes that can be generated by the data are not unique, mainly because SAXS is an underdetermined technique that infers 3D models from 1D data. In the case of flexible systems, where multiple conformers coexist in solution, the retrieved shape reflects the average envelope of the protein in solution (Bernardo, 2010).

Data analysis becomes even more powerful when crystal structures of the individual modules exist. Given that crystal structures (or NMR structures) of the macromolecules are available, the 3D shapes generated from the experimental SAXS data allow to position the crystal structure within the envelope and therefore, by combination, gives access to quaternary organization of multimers, for example, or the relative position of modules and domains of multimodular proteins. A similar approach has been developed for cryo-electron microscopy (CryoEM) studies, with which crystal or atomic structures of smaller subdomains can be docked into the CryoEM envelope of macromolecular complexes (Forster and Villa, 2010). A significant advantage of SAXS relies on the possibility to calculate the theoretic scattering curve of a given atomic model, and therefore to be able to directly validate the model by comparing it to the experimental scattering pattern (Svergun et al., 1995). This often allows assessment of flexibility of proteins or parts of the proteins in solution with respect to their conformationally trapped crystal structure. This approach also allows not only to build models of macromolecular complexes, where the crystal or NMR structures of the various partners or subdomains are known, but also to decipher the conformation of missing regions within the complex, such as linkers tethering these subdomains or modules together (Petoukhov and Svergun, 2005; Petoukhov et al., 2002). The most recent advances in SAXS analyses also allow retrieving a distribution of conformations adopted by a flexible domain such as linkers (Receveur-Brechot and Durand, 2012).

3. COMBINING SAXS AND CRYSTALLOGRAPHY TO ANALYZE MULTIMODULAR ORGANIZATION OF CELLULOLYTIC ENZYMES AND COMPLEXES

As mentioned before, the bimodularity of cellulases, consisting of generally a large catalytic module (200–700 aa) tethered to a smaller CBM module (30–200 aa) by a Thr/Ser-rich linker of variable length, not only challenged the experimental biochemist, but also resisted to all trials of crystallization. Constructs of isolated modules and domains needed to be produced to have at least access to the partial structural information of the isolated modules. Being too large for NMR studies and too small for EM analyses, these molecular objects were ideal for analyses using solution scattering. The questions concerned their 3D arrangement in solution: Are the individual modules in close contact or extended? What is the role of the linker? Is the nature and length of importance and do these factors vary the spatial arrangement of the enzyme molecules in solution?

The recent advances of SAXS data collection and analyses described above were timely and helped to perform new structural studies combining mainly SAXS and crystallography that relied on all this considerable spadework. This chapter, therefore, deals with the specific application of this "combining methods strategy" to have further insights to the global architecture of cellulolytic complexes.

3.1. Cellulases appended by one CBM and two modular enzymes: The importance of the linker

The first SAXS studies that were performed on CBHI and CBHII from *H. jecorina* described the resulting form as tadpole, with a globular "head" part corresponding to the catalytic module and a "tail" that included the linker and the short (45 aa) CBM (Abuja *et al.*, 1988a).

This description partially hinted toward the answers to the questions that were analyzed in much more detail 14 years later using a more systematic approach (Receveur *et al.*, 2002). The clue to this study was to have access to different constructs of the bimodular cellulase Cel45 from *Humicola insolens* (provided by the late M. Schülein from Novo-Nordisk) that was analyzed in different forms, as full-length, catalytic module only and catalytic module plus linker, as well as several mutants with varying length and nature of the connecting linker (Table 10.1). The interesting conclusion of this study was that the early tadpole description by Abuja *et al.* (1988a) is indeed confirmed but, furthermore, the results showed that the linker was in all cases essentially extended (roughly 0.7–1.3 residues per/Å as compared to 4.7 residues per/Å for a globular, compact catalytic module). Notably, the linker conformation was not modified in presence or in absence of the cellulose-binding domain (Receveur *et al.*, 2002). Moreover, and rather expectedly, the results indicated that the nature of the amino acid composition of the linker region was crucial for the extendedness of the polypeptide. Interesting in this respect was the fact that heavy glycosylation (*H. insolens* Cel45; Receveur *et al.*, 2002), the presence of glycines (*H. insolens* Cel6A; von Ossowski *et al.*, 2005), or the predominance of negatively charged residues combined with the presence of short disulfide-bridged loops, such as for Cel5G (*Pseudoalteromonas haloplanktis*; Violot *et al.*, 2005) or Cel7B (*H. insolens*; V. Receveur-Bréchot, unpublished data), represented alternative compositions, all leading to the extended forms of linker regions (Table 10.1).

At the same time, the comparison of the experimental molecular dimensions, R_g and D_{max}, of Cel45 from *H. insolens* with a wild-type (WT) linker and in a mutant form, where two residues were mutated into prolines (leading to a stretch of five prolines in a row), also gave a first glimpse on the flexibility of the linker region (Receveur *et al.*, 2002). Indeed, the radius of gyration reflects the average dimension of all conformers in solution, while D_{max} is due to the most extended conformation attained by the protein in solution, significantly populated to be detected by SAXS. The two proteins displayed similar D_{max}, but WTCel45 exhibited a smaller R_g than modified Cel45. This indicated that WTCel45 adopts both compact and extended conformations, while the extended conformations are more populated in modified Cel45, thereby revealing that the mutant with the polyproline stretch was more rigid than the WT protein. Further aspects of

Table 10.1 Properties of different cellulase linkers as deduced from SAXS measurements

Cellulase	P. haloplanktis Cel5G	H. insolens Cel45WT	H. insolens Cel45 Δ S219-T235	H. insolens Cel45 PP	T. terrestris Cel45	H. insolens Cel6A	H. insolens Cel7B
No aa in linker	110	36	19	36	30	38	17
R_g (Å) (of entire enzyme)	53.2±1.3	33.5±0.5	30.0±0.4	35.5±0.6	29.0±1.5	26.5±0.8	25.6±0.3
D_{max} (Å) (of linker)	148	49	19	46	55	30	22
Density residues/Å	0.7	0.7	1.0	0.8	0.6	1.3	0.8
Charged aa	−23	0	0	0	0	0	+5/−2
S-T (glycosylation)	0	61%	50%	59%	37%	15%	0
Glycines (%)	12	0	0	0	33	32	12
Prolines (%)	5	19	18	25	17	18	24

analyzing the linker region by SAXS with respect to its flexibility are described in Section 3.4.

3.2. The dissect and build approach: Analysis of more complex, multimodular cellulolytic assemblies

The functioning of the large cellulolytic assemblies called cellulosomes is even more intriguing than that of free enzyme systems. In most cases, the cellulosomal cellulases are devoid of genuine cellulose-specific CBMs (Blouzard et al., 2010) but are robustly attached to a scaffolding protein which, in the case of model mesophilic and thermophilic *Clostridia*, hosts a powerful family-3a CBM (Pages et al., 1996; Poole et al., 1993), thus anchoring the whole complex at the surface of cellulose fibers. The scaffolding protein contains copies of a receptor module, called cohesin, which strongly interacts with a complementary module dockerin, borne by the catalytic subunits (Gerngross et al., 1993; Pages et al., 1999; Shoseyov and Doi, 1990). Thus, a simple cellulosomal enzyme is composed of a catalytic module tethered to a small (~50 residues) dockerin through a linker (Fig. 10.1). The picture of enzyme efficiency and specific protein–protein interaction in the cohesin–dockerin couple became clearer through numerous structures of isolated catalytic modules or scaffolding modules that were established by X-ray diffraction or NMR studies, followed by solution structures of a dockerin module (Lytle et al., 2001) and crystal structures of cohesin/dockerin complexes (Carvalho et al., 2003; Pinheiro et al., 2008). The latter structures revealed, for two different bacterial cellulosomes, a dual binding mode of the dockerin to its cohesin partner. As for bimodular cellulases appended with a CBM, related questions arose about the spatial arrangement of catalytic modules with respect to their dockerin or the dockerin/cohesin interface. Some intrinsic flexibility induced by the linker was suspected since trials to crystallize entire cellulosomal enzymes failed systematically. The overall plasticity of cellulosomes was also suggested rapidly by electron microscopy observations on the complexes produced by *Clostridium papyrosolvens* (Pohlschroder et al., 1995), but the molecular mechanisms generating the conformational flexibility remained obscure. In particular, are the enzymatic linkers or the intercohesin sequences of the scaffoldins, or both, involved in flexibility of the complex?

Based on the available structural data on cellulosomal modules and enzymes, a "step-by-step" or "dissect and build" strategy was developed to conduct a complementary SAXS study that turned out to provide some useful new information on this basic interrogation. In addition, these experiments were possible only because the groups of E. Bayer and H. P. Fierobe developed an *in vitro* recombinant approach to tailor "minicellulosomes" by exploiting the species specificity of the respective scaffoldin cohesins (Fierobe et al., 2002, 2001). The strategy measured SAXS data of the isolated modules, allowing comparison with the crystal structures, and then measuring multimodular constructs containing several of the individually measured modules. These experiments led to the analysis of "mini-cellulosomes" that comprised two interlinked but independent scaffoldin-modules in complex with two different full-length enzymes, made up of a catalytic module and their respective dockerins.

A WT *Clostridium cellulolyticum* family GH48 cellulase, one of the most prominent cellulases in bacterial cellulosomes (Gal *et al*., 1997; Reverbel-Leroy *et al*., 1997) and for which the crystal structure had been determined (Parsiegla *et al*., 1998), was examined by SAXS together with engineered forms lacking its native C-terminal dockerin, or bearing a dockerin derived from *Clostridium thermocellum* (Fierobe *et al*., 2001). SAXS analysis showed that in the free state, the enzyme linker connecting the catalytic module to the dockerin adopts multiple extended conformations, thus confirming a noticeable intrinsic flexibility (Hammel *et al*., 2004). Nevertheless, similar analyses performed on the GH48 enzyme bound to either a *C. cellulolyticum* or a *C. thermocellum* cohesin revealed that the complex formation triggers a pleating of the cellulase linker and induces a diminution of the conformational flexibility, although a limited mobility persists (Hammel *et al*., 2004). The observed compaction of the enzyme linker was so drastic (Fig. 10.4) that the maximal particle dimension of the cellulase/cohesin complex ($D_{max} = 122 \pm 2$ Å) was smaller than that determined for the free cellulase ($D_{max} = 142 \pm 6$ Å).

This dissect and build strategy was further extended to the simplest available hybrid "mini-cellulosome." The latter was based on a chimeric "mini"-scaffoldin containing two divergent cohesins connected by a 48-residue linker (Fierobe *et al*., 2002). SAXS studies of the free form revealed extended and flexible conformations of the intercohesins linker (Hammel *et al*., 2005). In contrast to enzyme linkers, the flexibility of the scaffoldin linker persists when one or both cohesins interact with the corresponding dockerins hosted by medium- (50 kDa) or large- (80 kDa) size cellulosomal cellulases. These observations strongly supported a pivotal role of the scaffoldin's intercohesin linkers in cellulosome plasticity and their adaptation to the local topology of the substrate (Hammel *et al*., 2005). Additional information about the conformational freedom of cohesin arrangements within the scaffoldin was obtained by Noach *et al*. (2009). Indeed, they were able to trap different conformations of intercohesin linkers when crystallizing the two adjacent cohesins with their respective linkers in different crystal forms. In these structures, different linker conformations were observed in the individual molecules within the asymmetric unit of each structure. The structural information therefore allowed reconstructing possible scaffoldin structures composed of two consecutive cohesins, as derived from the crystal structure of each cohesin appended by the same linker. This conformational diversity implies that the linkers may adopt

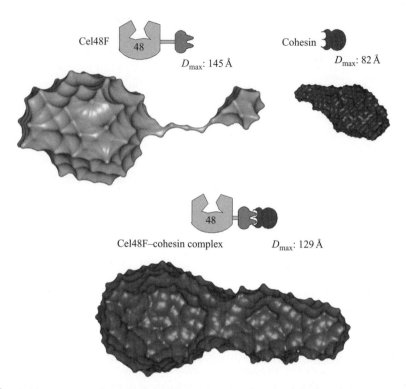

Figure 10.4 Schematic icons and calculated envelopes using GASBOR corresponding to full-length Cel48F (top left), the cognate cohesin (top right), and the complex of both (bottom), as obtained by SAXS experiments. The experimental D_{max} values obtained for each molecule in solution are indicated. (See Color Insert.)

alternative conformations in their natural environment, consistent with varying environmental conditions (Noach et al., 2009).

3.3. A concrete example: Constructing the 3D organization of the multimodular enzyme XynZ

The multimodular xylanase Z (XynZ) from *C. thermocellum* has extensively been studied for its interesting activity (Fontes et al., 1995a,b; Grepinet et al., 1988a,b) and its outstanding multimodular organization (Fig. 10.1), consisting of a N-terminal feruloyl-esterase module (family CE1), a C-terminal xylanase module belonging to family GH10, interconnected through a CBM from family 6, and a type I dockerin module. The four modules of this complex enzyme are connected through 6–30 residue-containing linkers. Trials to crystallize the full-length enzyme failed, but 3D crystal structures of all modules (CE1 (Schubot et al., 2001) pdbcode 1JJF, xylanase

GH10 (Dominguez et al., 1995; Souchon et al., 1994) pdbcode 1XYZ) or homologs of the modules (CBM6 (Czjzek et al., 2001) pdbcode 1GMM and cohesin/dockerin complex type I (Carvalho et al., 2003) pdbcode 1OHZ) are available. The full-length enzyme is produced recombinantly and appears to be stable in time, and the protein in complex with its natural cohesin can also be purified to homogeneity (H. P. Fierobe, unpublished data). As such, this full-length enzyme and its molecular complex with a cohesin represent an ideal model to study the spatial arrangement using the SAXS combined methods strategy.

SAXS experiments were performed at the European Synchrotron Radiation Facility (Grenoble, France) on beamline ID02. The wavelength λ was 0.99 Å. The sample-to-detector distances were set at 4.0 and 1.0 m, resulting in scattering vectors, q ranging from 0.010 to 0.15 Å^{-1} and 0.03 to 0.46 Å^{-1}, respectively. The data acquired at both sample-to-detector distances of 4 and 1 m were merged for the calculations using the entire scattering spectrum. All experiments were performed at 20 °C on a series of concentrations ranging from 1.2 to 10 mg/mL for both the full-length enzyme in the unbound form (XynZfree) and in complex with its natural cohesin module (XynZcoh). The experimental SAXS data for all samples were linear in a Guinier plot of the low q region, indicating that the proteins did not undergo any aggregation. The radius of gyration (R_g) was derived by the Guinier approximation as described in Section 2.4. The program GNOM (Svergun, 1992) was used to compute the pair-distance distribution function $P(r)$ from which the maximum distance (D_{max}) values of the macromolecules were also determined, and used to subsequently restore the overall shapes from the experimental data using the program GASBOR (Svergun et al., 2001) (Fig. 10.5C and F). While the D_{max} value was identical (150 Å) for both XynZfree and XynZcoh, the R_g values were calculated to be 42 ± 2 Å for the former and 45 ± 2 Å for the latter. The atomic structures of the individual modules were then positioned into the low-resolution envelopes using the program SUPCOMB (Kozin and Svergun, 2000). The missing linker residues were added with the graphical interface TURBO (Roussel and Cambillau, 1991).

At this stage, it was evident that supplemental information was needed to identify the best arrangement of the modules to fit the experimental SAXS curve. Therefore, molecular modeling was used to determine the best-fitting relative spatial position of these modules within the envelopes, obtained by GASBOR. The constructed atomic models of the full-length enzymes described above were used as templates for the rigid body modeling. The program CHARMM, version 28.b1 (Brooks et al., 1983), with the all-atom CHARMM force field version 22.0 (MacKerell et al., 1998), was used for molecular dynamics (MD) simulation. In all cases, only the atoms of the linker regions were allowed to move relatively to each other, while the modules of each assembly were treated as rigid bodies, with no internal

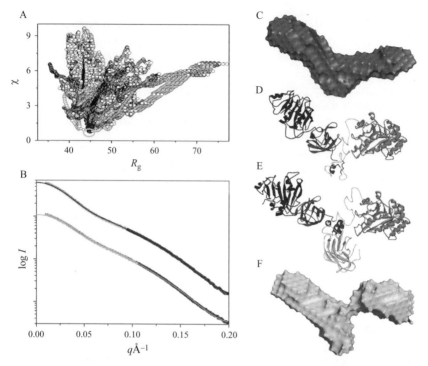

Figure 10.5 SAXS analyses of XynZfree and XynZcoh. (A) The graph represents the discrepancy (ξ) values for 15,000 models as a function of their R_g values, calculated for XynZfree (XynZcoh, data not shown). The best-fitting model is indicated by a yellow circle. (B) Experimental SAXS curves and scattering profiles computed from the models of XynZfree (top) and XynZcoh (bottom). Black circles: experimental data; dark blue line (top): theoretical scattering curve from best-fit model obtained by rigid body modeling of XynZfree using the program CRYSOL; light blue line (bottom): theoretical scattering curve from best-fit model obtained by rigid body modeling of XynZcoh using the program CRYSOL. (C) The molecular envelope shape in surface representation of XynZfree calculated with GASBOR and (D) the corresponding best-fit model obtained from the molecular dynamics simulation in ribbon representation. The modules are colored as follows: CE1 in dark red, CBM6 in blue, dockerin in red, and the GH10 xylanase in green. (E) The best-fit model of XynZcoh obtained from the molecular dynamics simulation in ribbon representation. The modules are colored as above and the cohesin in beige and (F) the corresponding molecular envelope shape in surface representation calculated with GASBOR. (See Color Insert.)

motion. The simulations were performed in implicit solvent represented by setting the relative dielectric constant to 80.0. A time step of 1 fs was employed for MD simulation. All simulations were performed with the following protocol. The system was subjected to energy minimization with harmonic constraints on the protein atoms, followed by 30 ps MD heating up to 1500 K, keeping the protein atoms fixed. During the subsequent "production phase" (7.5 ns), the resulting structures of the protein were

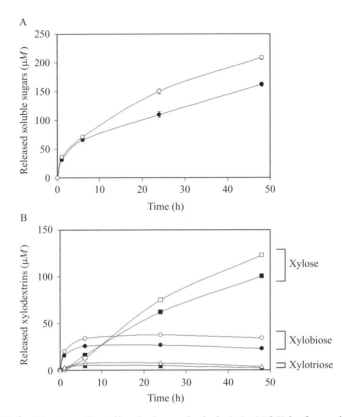

Figure 10.6 Kinetic studies of hatched straw hydrolysis (at 37 °C) by free and cohesin-bound XynZ. (A) Soluble reducing sugars released. Curves are labeled as follows: XynZfree, black circles; XynZcoh, open circles. The data show the mean of three independent experiments and the bars indicate the standard deviation. (B) Analysis of the oligosaccharides released by free and complexed XynZ on hatched straw. The most representative kinetic (panel A) was selected in the case of XynZfree and XynZcoh: curves are labeled as follows: xylose, squares; xylobiose, circles; and xylotriose, triangles. XynZfree, black symbols; XynZcoh, open symbols. Note that trace amounts (<1 μM) of xylotretraose were detected in the samples extracted at 1 and 6 h for both XynZfree and XynZcoh complex (not shown in the graph), but xylopentaose was not detected in any sample. (For the interpretation of the references to color in this figure legend, the reader is referred to the Web version of this chapter.)

recorded every 0.5 ps in a trajectory file. For each registered conformation, the theoretical SAXS profile, the R_g, and the corresponding fit to the experimental data were calculated using the program CRYSOL (Svergun et al., 1995) (Fig. 10.5A). Coherently, the best fits to the experimental data (represented by lowest χ^2 values; Fig. 10.5B) leading to χ^2-values of 0.76 for XynZfree and 0.89 for XynZcoh were obtained for 3D models that had R_g values that corresponded to those determined from the experimental SAXS data. Close inspection of the 10 best models indicated that they were all

Table 10.2 Specific activity of free and cohesin-bound XynZ on xylan (5 g/L) at 37 °C and 60 °C

XynZ state	37 °C	60 °C
XynZfree (IU/μmol)	$657^a \pm 18^b$	6129 ± 64
XynZcoh (IU/μmol)	681 ± 48	6362 ± 43

[a] Average of two independent experiments.
[b] Standard deviation.

consistent with one relative spatial arrangement of the different modules as represented in Fig. 10.5D for XynZfree and Fig. 10.5E for XynZcoh. The common feature in these best-fitting spatial arrangements (Fig. 10.5D and E) is that the relative position and conformation of the CBM6 and the dockerin changed in the free and bound form of the full-length enzyme. The relative exposure of the modules was consistent with the CBM6 being more exposed to solvent or substrate when the enzyme is bound into the cellulosome complex and less accessible when the enzyme was in the free state. This makes sense in that the enzyme is "protected" from binding to the solid surface of substrate through the CBM6, while it is not part of the cellulosome, risking to be "lost" for efficient substrate uptake by the bacteria, and the substrate binding module CBM6 is more exposed, thus "ready" to bind to the substrate when the enzyme is integrated into the cellulosomal machinery. Our immediate reflex was to verify the bound XynZcoh form for more efficient xylan degradation, especially in the bulk. Indeed, XynZcoh was slightly more efficient (Fig. 10.6), but the experimental data were not as clear cut as expected (Table 10.2), again showing that our understanding of the mechanisms is not yet complete.

3.4. Combination with computational modeling to construct the "missing parts": Dealing with flexibility

While the linkers are at the heart of the cooperativity between the different modules in cellulases and in cellulolytic assemblies, the way they fulfill their synergistic function remained enigmatic and had been poorly documented from the molecular and structural basis. Several of the examples described in the preceding chapters hinted toward the crucial role of the flexibility displayed by the linkers (Hammel et al., 2005; Violot et al., 2005; von Ossowski et al., 2005). A molecular description of the linker was therefore required to understand the molecular basis of their specific role. As SAXS conveys information on the structure adopted by every conformer in solution, the structural properties and the flexibility of disordered regions such as linkers can be investigated using SAXS, especially when combined with atomistic information arising from crystallography, NMR, or molecular modeling, together with MD simulations (Bernardo, 2010; Putnam et al., 2007).

The flexible nature of the linker was already strongly suspected because it remained virtually impossible to obtain any crystal of full-length cellulases, unlike for the individual modules. While SAXS established that the linker was extended, a first glimpse on the flexibility of the linker was caught by analyzing the SAXS data of Cel45 and mutant forms from *H. insolens*. The only scientific interpretation of the differences, observed in the R_g values while D_{max} was essentially the same, was given by more flexibility of the WT linker with respect to the modified, proline containing one (Receveur et al., 2002).

Advances in computational methods analyzing SAXS data then allowed the 3D modeling of a missing loop, typically such as a linker, in the atomic structure of a full-length protein (Petoukhov and Svergun, 2005; Petoukhov et al., 2002). The disordered nature of a 109-residue-long linker in the psychrophilic cellulase Cel5A from *P. haloplanktis* was ascertained for the first time when the structural mechanisms of cold adaptation of this bacterial cellulase were investigated by SAXS (Violot et al., 2005). Running the program several times yielded different possible models that all fitted equally well with the data. Interestingly, an average of the calculated scattering profiles of all these possible conformations was consistent with the experimental SAXS curve (Fig. 10.7), indicating that modeling the behavior of the cellulase with a population of different conformations better accounts for the observed scattering curve than individual models. On the other hand, a close examination of the structures calculated for the linker revealed that the program was able to retrieve the transversal loops formed through disulfide linkages between cysteines in the linker (Violot et al., 2005). These loops were proposed to add steric constraints to stabilize the most extended conformations.

A thorough study was further conducted on a chimeric double cellulase in order to quantitatively assess this flexibility (von Ossowski et al., 2005). The designed double cellulase contained an 88-residue-long linker. In this case, no individual model could be obtained accounting for the experimental SAXS data. Therefore, a large number of models with varying linker lengths between the two catalytic domains were generated using molecular modeling approaches. A weighting scheme was then applied to establish the distribution of conformations adopted by the double cellulase and describing properly the observed SAXS data. This computational analysis showed that the linker follows a nonrandom distribution with a preference for the more compact conformations. Notably, it also revealed that the linker can unfold from compact to extended conformation at a low energetic cost, which strongly supports the idea that the linker behaves as a molecular spring between the different modules, thereby triggering their cooperativity.

In parallel, studies on the flexibility of linkers within cellulosomes also provided clues on their role in these multienzymatic assemblies. As explained above, a pleating of the linker tethering the cellulase module to

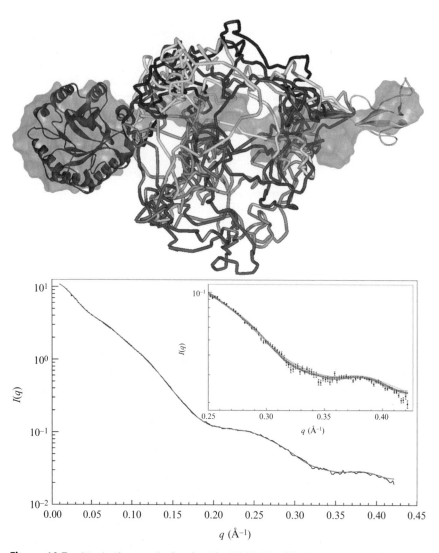

Figure 10.7 (Top) Shape calculated with GASBOR (blue) superimposed with 10 different models of Cel5G provided by GLOOPY represented by secondary structure element type. (Bottom) Fit on the experimental scattering curve obtained with the average form factor of the different models provided by GLOOPY. (Inset) Fits on the experimental scattering curve obtained with individual form factors of the different models provided by GLOOPY. (See Color Insert.)

the dockerin module was observed upon binding of the dockerin onto a cohesin, as revealed by the dimensions and 3D models of the complexes inferred from SAXS data (Hammel *et al.*, 2004). Normal mode analysis on the full-length cellulase with the linker and the dockerin, in presence and absence of the cohesin indeed confirmed the drastic decrease in flexibility of the linker in the complex with respect to the free protein. This provided a very original example of a "remote" induced folding, where the disordered region undergoes an induced folding upon complex formation, similarly as many intrinsically disordered proteins involved in molecular recognition (Dyson and Wright, 2002; Wright and Dyson, 2009), except that the flexible region here is not directly involved in the binding interface. Further studies on larger assemblies corresponding to simplified cellulosomes revealed that the intercohesin linkers remain much more flexible even after complex formation (Hammel et al., 2005). The amplitude of the conformations attained by the intercohesin linkers was later thoroughly explored for several linkers varying in length and composition by establishing their distribution of conformation using SAXS (Molinier et al., 2011). This study showed that modifications did indeed influence the flexibility of the scaffolding protein, but unexpectedly had no impact on the activity or synergy on crystalline cellulose of studied cellulase pairs bound onto these modified hybrid scaffoldins. The various degrees of the linkers' flexibilities in the complex, deciphered by SAXS, allowed proposing a model where the cellulosome conformation is finely tuned by adjusting the linker length, therefore being able to perfectly fit to the topology of the substrate, leading to a dramatic gain in enzymatic efficiency (Fig. 10.8).

All these studies investigating the structural and dynamic properties of the linkers in cellulolytic complexes not only shed new light on the functional role of these linkers as molecular springs, but also contributed to classify them in the growing family of intrinsically disordered proteins, and thus to further underline the importance of such disordered but essential domains.

4. Conclusions and Outlook

Determining the 3D arrangements of multimodular enzymes and complexes to understand their associated function *in vivo* is a considerable challenge. The combination of biophysical and structural methods with the determination of solution structures by SAXS of these objects has greatly advanced our understanding on the importance of linkers and intermodular dynamics and represents a powerful tool to go even further beyond the actual state of art.

An interesting result obtained by the combination of techniques resides in the fact that despite variation of linker compositions, such as glycosylations and/or distribution of glycines and negatively charged residues, all these features have an important effect on the extendedness of the intermodule linkers. With this respect, a recent NMR study assessed the local flexibility of

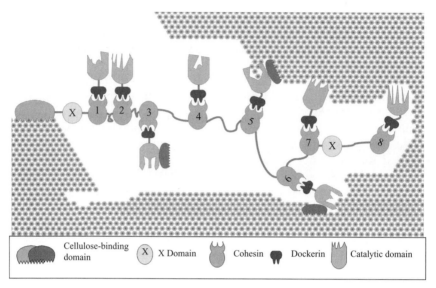

Figure 10.8 Schematic representation depicting a functional model of a cellulosome and the interaction of its component parts with the cellulose substrate. The scaffoldin subunit (based on the scaffoldin from *C. cellulolyticum*) and its complement of enzymes are bound to the cellulose component of the plant cell wall by virtue of the potent family-3a CBM. In the presence of cellulose, the intercohesin modules of the scaffoldin undergo large-scale motion to adjust the respective positions of the complexed catalytic subunits according to the topography of the substrate. In this context, some cellulosomal enzyme subunits include a CBM that mediates a relatively weak interaction with the substrate. The names of the cellulosomal modules are given in the legend. This figure was originally published in Hammel *et al.* (2005). © The American Society for Biochemistry and Molecular Biology. (For the color version of this figure, the reader is referred to the Web version of this chapter.)

the linker in *Cellulomonas fimi* xylanase Cex and showed that O-glycosylation partially dampens its flexibility (Poon *et al.*, 2007). A SAXS study could be extremely complementary as it would provide information on the long-range distances, in particular on the effect of glycosylation on the separation distance between the two globular domains. This would also provide an experimental validation of a recent molecular dynamic simulation on *H. jecorina* cellulase Cel7A (Beckham *et al.*, 2010) that suggests that glycosylation does not alter the stiffness of the linker, but rather prevents collapse of the linker, and stabilizes conformations with long separation distances between the two globular modules.

The next step to understanding the mechanisms by which multimodularity and large macromolecular complexes enhance the efficiency of the cellulolytic enzymes consist in experimental design allowing to study the discovered flexibility and 3D arrangement directly in interaction with its natural substrate. In absence of experimental data, models provided by computer simulations (Beckham *et al.*, 2011; Zhong *et al.*, 2009) show that distances between the catalytic domain and the CBM fluctuate over the simulation, but the authors admit that the time span of the simulations is too short to infer definitive conclusions. The development of a mechanochemical model to describe the kinetics of a multimodular cellulase (Ting *et al.*, 2009) showed that the enzyme may be considered as "random walker" on the substrate, and that the linker length and stiffness play a critical role in the cooperative action of the CD and CBM domains. Small-angle scattering might again play a crucial role in getting experimental evidence to confirm these interesting theoretical models. Indeed, the use of neutron scattering by setting up contrast variation experiments (Jacques and Trewhella, 2010; Putnam *et al.*, 2007) of complexes consisting of macromolecules, which have different scattering densities, would open the possibility to "see" and compare the 3D arrangement of the enzymes on their natural substrates. One of the largest differences in neutron scattering power is that between naturally abundant hydrogen (^{1}H) and its stable isotope deuterium (^{2}H or D). It would therefore be possible, in a small-angle neutron scattering experiment, to have different contrasts for two interacting molecules if one molecule is deuterated. Nevertheless, two major bottlenecks exist to perform these experiments: (i) deuteration of proteins or carbohydrates is not trivial and (ii) cellulose chains with a polymerization degree above six units are insoluble. Confidently, progress in sample preparation and experimental techniques will make these experiments possible in the near future.

ACKNOWLEDGMENTS

The authors are indebted to beamline staff for precious help on the different SAXS beamlines X33, DESY, Hamburg (D. Svergun, M. Roessle, A. Kikhney); ID02, ESRF, Grenoble (T. Narayanan, P. Panine, S. Finet); and LURE, Paris (P. Vachette). The authors thank M.E. Hammel for the molecular dynamics calculations of XynZ, and also wish to thank Eric Marcon for his help in preparing Fig. 10.2.

REFERENCES

Abuja, P. M., Pilz, I., Claeyssens, M., and Tomme, P. (1988a). Domain structure of cellobiohydrolase II as studied by small angle X- ray scattering: Close resemblance to cellobiohydrolase I. *Biochem. Biophys. Res. Commun.* **156,** 180–185.

Abuja, P. M., Schmuck, M., Pilz, I., Tomme, P., Claeyssens, M., and Esterbauer, H. (1988b). Structural and functional domains of cellobiohydrolase I from *Trichoderma reesei*. A small angle X-ray scattering study of the intact enzyme and its core. *Eur. Biophys. J.* **15,** 339–342.

Abuja, P. M., Pilz, I., Tomme, P., and Claeyssens, M. (1989). Structural changes in cellobiohydrolase I upon binding of a macromolecular ligand as evident by SAXS investigations. *Biochem. Biophys. Res. Commun.* **165,** 615–623.

Adams, J. J., Currie, M. A., Ali, S., Bayer, E. A., Jia, Z., and Smith, S. P. (2010). Insights into higher-order organization of the cellulosome revealed by a dissect-and-build approach: Crystal structure of interacting *Clostridium thermocellum* multimodular components. *J. Mol. Biol.* **396,** 833–839.

Bayer, E. A., Belaich, J. P., Shoham, Y., and Lamed, R. (2004). The cellulosomes: Multienzyme machines for degradation of plant cell wall polysaccharides. *Annu. Rev. Microbiol.* **58,** 521–554.

Beckham, G. T., Bomble, Y. J., Matthews, J. F., Taylor, C. B., Resch, M. G., Yarbrough, J. M., Decker, S. R., Bu, L., Zhao, X., McCabe, C., Wohlert, J., Bergenstrahle, M., et al. (2010). The O-glycosylated linker from the *Trichoderma reesei* family 7 cellulase is a flexible, disordered protein. *Biophys. J.* **99,** 3773–3781.

Beckham, G. T., Bomble, Y. J., Bayer, E. A., Himmel, M. E., and Crowley, M. F. (2011). Applications of computational science for understanding enzymatic deconstruction of cellulose. *Curr. Opin. Biotechnol.* **22,** 231–238.

Bergmann, A., Fritz, G., and Glatter, O. (2000). Solving the generalized indirect Fourier transformation (GIFT) by Boltzmann simplex simulated annealing (BSSA). *J. Appl. Crystallogr.* **33,** 1212–1216.

Bernardo, P. (2010). Effect of interdomain dynamics on the structure determination of modular proteins by small-angle scattering. *Eur. Biophys. J.* **39,** 769–780.

Blouzard, J. C., Coutinho, P. M., Fierobe, H. P., Henrissat, B., Lignon, S., Tardif, C., Pages, S., and de Philip, P. (2010). Modulation of cellulosome composition in *Clostridium cellulolyticum*: Adaptation to the polysaccharide environment revealed by proteomic and carbohydrate-active enzyme analyses. *Proteomics* **10,** 541–554.

Brooks, B. R., Bruccoleri, R. E., Olafson, B. D., States, D. J., Swaminathan, S., and Karplus, M. (1983). CHARMM: A program for macromolecular energy, minimization, and dynamics calculations. *J. Comp. Chem.* **4,** 187–217.

Carvalho, A. L., Dias, F. M., Prates, J. A., Nagy, T., Gilbert, H. J., Davies, G. J., Ferreira, L. M., Romao, M. J., and Fontes, C. M. (2003). Cellulosome assembly revealed by the crystal structure of the cohesin-dockerin complex. *Proc. Natl. Acad. Sci. USA* **100,** 13809–13814.

Chacon, P., Moran, F., Diaz, J. F., Pantos, E., and Andreu, J. M. (1998). Low-resolution structures of proteins in solution retrieved from X-ray scattering with a genetic algorithm. *Biophys. J.* **74,** 2760–2775.

Czjzek, M., Bolam, D. N., Mosbah, A., Allouch, J., Fontes, C. M., Ferreira, L. M., Bornet, O., Zamboni, V., Darbon, H., Smith, N. L., Black, G. W., Henrissat, B., et al. (2001). The location of the ligand-binding site of carbohydrate-binding modules that have evolved from a common sequence is not conserved. *J. Biol. Chem.* **276,** 48580–48587.

David, G., and Perez, J. (2009). Combined sampler robot and high-performance liquid chromatography: A fully automated system for biological small-angle X-ray scattering experiments at the Synchrotron SOLEIL SWING beamline. *J. Appl. Crystallogr.* **42,** 892–900.

Davies, G. J., Dodson, G. G., Hubbard, R. E., Tolley, S. P., Dauter, Z., Wilson, K. S., Hjort, C., Mikkelsen, J. M., Rasmussen, G., and Schülein, M. (1993). Structure and function of endoglucanase V. *Nature* **365,** 362–364.

Dominguez, R., Souchon, H., Spinelli, S., Dauter, Z., Wilson, K. S., Chauvaux, S., Beguin, P., and Alzari, P. M. (1995). A common protein fold and similar active site in two distinct families of beta-glycanases. *Nat. Struct. Biol.* **2,** 569–576.

Ducros, V., Czjzek, M., Belaich, A., Gaudin, C., Fierobe, H. P., Belaich, J. P., Davies, G. J., and Haser, R. (1995). Crystal structure of the catalytic domain of a bacterial cellulase belonging to family 5. *Structure* **3,** 939–949.

Dyson, H. J., and Wright, P. E. (2002). Coupling of folding and binding for unstructured proteins. *Curr. Opin. Struct. Biol.* **12,** 54–60.

Fierobe, H. P., Mechaly, A., Tardif, C., Belaich, A., Lamed, R., Shoham, Y., Belaich, J. P., and Bayer, E. A. (2001). Design and production of active cellulosome chimeras. Selective incorporation of dockerin-containing enzymes into defined functional complexes. *J. Biol. Chem.* **276**, 21257–21261.

Fierobe, H. P., Bayer, E. A., Tardif, C., Czjzek, M., Mechaly, A., Belaich, A., Lamed, R., Shoham, Y., and Belaich, J. P. (2002). Degradation of cellulose substrates by cellulosome chimeras. Substrate targeting versus proximity of enzyme components. *J. Biol. Chem.* **277**, 49621–49630.

Fontes, C. M., Hall, J., Hirst, B. H., Hazlewood, G. P., and Gilbert, H. J. (1995a). The resistance of cellulases and xylanases to proteolytic inactivation. *Appl. Microbiol. Biotechnol.* **43**, 52–57.

Fontes, C. M., Hazlewood, G. P., Morag, E., Hall, J., Hirst, B. H., and Gilbert, H. J. (1995b). Evidence for a general role for non-catalytic thermostabilizing domains in xylanases from thermophilic bacteria. *Biochem. J.* **307**, 151–158.

Forster, F., and Villa, E. (2010). Integration of cryo-EM with atomic and protein-protein interaction data. *Methods Enzymol.* **483**, 47–72.

Gal, L., Pages, S., Gaudin, C., Belaich, A., Reverbel-Leroy, C., Tardif, C., and Belaich, J. P. (1997). Characterization of the cellulolytic complex (cellulosome) produced by *Clostridium cellulolyticum*. *Appl. Environ. Microbiol.* **63**, 903–909.

Gerngross, U. T., Romaniec, M. P., Kobayashi, T., Huskisson, N. S., and Demain, A. L. (1993). Sequencing of a *Clostridium thermocellum* gene (cipA) encoding the cellulosomal SL-protein reveals an unusual degree of internal homology. *Mol. Microbiol.* **8**, 325–334.

Gilbert, H. J., Hazlewood, G. P., Laurie, J. I., Orpin, C. G., and Xue, G. P. (1992). Homologous catalytic domains in a rumen fungal xylanase: Evidence for gene duplication and prokaryotic origin. *Mol. Microbiol.* **6**, 2065–2072.

Gilkes, N. R., Henrissat, B., Kilburn, D. G., Miller, R. C., Jr., and Warren, R. A. (1991). Domains in microbial beta-1, 4-glycanases: Sequence conservation, function, and enzyme families. *Microbiol. Rev.* **55**, 303–315.

Grepinet, O., Chebrou, M. C., and Beguin, P. (1988a). Nucleotide sequence and deletion analysis of the xylanase gene (xynZ) of *Clostridium thermocellum*. *J. Bacteriol.* **170**, 4582–4588.

Grepinet, O., Chebrou, M. C., and Beguin, P. (1988b). Purification of *Clostridium thermocellum* xylanase Z expressed in *Escherichia coli* and identification of the corresponding product in the culture medium of *C. thermocellum*. *J. Bacteriol.* **170**, 4576–4581.

Gunn, N. J., Gorman, M. A., Dobson, R. C., Parker, M. W., and Mulhern, T. D. (2011). Purification, crystallization, small-angle X-ray scattering and preliminary X-ray diffraction analysis of the SH2 domain of the Csk-homologous kinase. *Acta Crystallogr.* **F67**, 336–339.

Hammel, M., Fierobe, H. P., Czjzek, M., Finet, S., and Receveur-Brechot, V. (2004). Structural insights into the mechanism of formation of cellulosomes probed by small angle X-ray scattering. *J. Biol. Chem.* **279**, 55985–55994.

Hammel, M., Fierobe, H. P., Czjzek, M., Kurkal, V., Smith, J. C., Bayer, E. A., Finet, S., and Receveur-Brechot, V. (2005). Structural basis of cellulosome efficiency explored by small angle X-ray scattering. *J. Biol. Chem.* **280**, 38562–38568.

Jacques, D. A., and Trewhella, J. (2010). Small-angle scattering for structural biology—Expanding the frontier while avoiding the pitfalls. *Protein Sci.* **19**, 642–657.

Johnson, P. E., Joshi, M. D., Tomme, P., Kilburn, D. G., and McIntosh, L. P. (1996). Structure of the N-terminal cellulose-binding domain of *Cellulomonas fimi* CenC determined by nuclear magnetic resonance spectroscopy. *Biochemistry* **35**, 14381–14394.

Kleywegt, G. J., Zou, J. Y., Divne, C., Davies, G. J., Sinning, I., Stahlberg, J., Reinikainen, T., Srisodsuk, M., Teeri, T. T., and Jones, T. A. (1997). The crystal structure of the catalytic core domain of endoglucanase I from *Trichoderma reesei* at 3.6 Å resolution, and a comparison with related enzymes. *J. Mol. Biol.* **272**, 383–397.

Kozin, M. B., and Svergun, D. I. (2000). Automated matching of high- and low-resolution structural models. *J. Appl. Crystallogr.* **34**, 33–41.

Lamed, R., Setter, E., and Bayer, E. A. (1983). Characterization of a cellulose-binding, cellulase-containing complex in *Clostridium thermocellum*. *J. Bacteriol.* **156**, 828–836.

Lytle, B. L., Volkman, B. F., Westler, W. M., Heckman, M. P., and Wu, J. H. (2001). Solution structure of a type I dockerin domain, a novel prokaryotic, extracellular calcium-binding domain. *J. Mol. Biol.* **307**, 745–753.

MacKerell, J. A. D., Bashford, D., Bellott, M., Dunbrack, R. L. J., Evenseck, J. D., Field, M. J., Ficher, S., Gao, J., Guo, H., Ha, S., Joseph-McCarthy, D., Kuchnir, L., *et al.* (1998). All-atom empirical potential for molecular modeling and dynamics studies of proteins. *J. Phys. Chem. B* **102**, 3586–3616.

Molinier, A. L., Nouailler, M., Valette, O., Tardif, C., Receveur-Brechot, V., and Fierobe, H. P. (2011). Synergy, structure and conformational flexibility of hybrid cellulosomes displaying various inter-cohesins linkers. *J. Mol. Biol.* **405**, 143–157.

Mosbah, A., Belaich, A., Bornet, O., Belaich, J. P., Henrissat, B., and Darbon, H. (2000). Solution structure of the module X2 1 of unknown function of the cellulosomal scaffolding protein CipC of *Clostridium cellulolyticum*. *J. Mol. Biol.* **304**, 201–217.

Noach, I., Frolow, F., Alber, O., Lamed, R., Shimon, L. J., and Bayer, E. A. (2009). Intermodular linker flexibility revealed from crystal structures of adjacent cellulosomal cohesins of *Acetivibrio cellulolyticus*. *J. Mol. Biol.* **391**, 86–97.

Pages, S., Belaich, A., Tardif, C., Reverbel-Leroy, C., Gaudin, C., and Belaich, J. P. (1996). Interaction between the endoglucanase CelA and the scaffolding protein CipC of the *Clostridium cellulolyticum* cellulosome. *J. Bacteriol.* **178**, 2279–2286.

Pages, S., Belaich, A., Fierobe, H. P., Tardif, C., Gaudin, C., and Belaich, J. P. (1999). Sequence analysis of scaffolding protein CipC and ORFXp, a new cohesin-containing protein in *Clostridium cellulolyticum*: Comparison of various cohesin domains and subcellular localization of ORFXp. *J. Bacteriol.* **181**, 1801–1810.

Parsiegla, G., Juy, M., Reverbel-Leroy, C., Tardif, C., Belaich, J. P., Driguez, H., and Haser, R. (1998). The crystal structure of the processive endocellulase CelF of *Clostridium cellulolyticum* in complex with a thiooligosaccharide inhibitor at 2.0 A resolution. *EMBO J.* **17**, 5551–5562.

Petoukhov, M. V., and Svergun, D. I. (2005). Global rigid body modeling of macromolecular complexes against small-angle scattering data. *Biophys. J.* **89**, 1237–1250.

Petoukhov, M. V., Eady, N. A., Brown, K. A., and Svergun, D. I. (2002). Addition of missing loops and domains to protein models by x-ray solution scattering. *Biophys. J.* **83**, 3113–3125.

Pinheiro, B. A., Proctor, M. R., Martinez-Fleites, C., Prates, J. A., Money, V. A., Davies, G. J., Bayer, E. A., Fontesm, C. M., Fierobe, H. P., and Gilbert, H. J. (2008). The *Clostridium cellulolyticum* dockerin displays a dual binding mode for its cohesin partner. *J. Biol. Chem.* **283**, 18422–18430.

Pohlschroder, M., Canale-Parola, E., and Leschine, S. B. (1995). Ultrastructural diversity of the cellulase complexes of *Clostridium papyrosolvens* C7. *J. Bacteriol.* **177**, 6625–6629.

Poole, D. M., Hazlewood, G. P., Huskisson, N. S., Virden, R., and Gilbert, H. J. (1993). The role of conserved tryptophan residues in the interaction of a bacterial cellulose binding domain with its ligand. *FEMS Microbiol. Lett.* **106**, 77–83.

Poon, D. K., Withers, S. G., and McIntosh, L. P. (2007). Direct demonstration of the flexibility of the glycosylated proline-threonine linker in the *Cellulomonas fimi* Xylanase Cex through NMR spectroscopic analysis. *J. Biol. Chem.* **282**, 2091–2100.

Putnam, C. D., Hammel, M., Hura, G. L., and Tainer, J. A. (2007). X-ray solution scattering (SAXS) combined with crystallography and computation: Defining accurate macromolecular structures, conformations and assemblies in solution. *Q. Rev. Biophys.* **40**, 191–285.

Receveur, V., Czjzek, M., Schulein, M., Panine, P., and Henrissat, B. (2002). Dimension, shape, and conformational flexibility of a two domain fungal cellulase in solution probed by small angle X-ray scattering. *J. Biol. Chem.* **277**, 40887–40892.

Receveur-Brechot, V., and Durand, D. (2012). How random are intrinsically disordered proteins? A small angle scattering perspective. *Curr. Prot. Pept. Sci.* **13**(1), 55–75.

Reese, E. T., Siu, R. G., and Levinson, H. S. (1950). The biological degradation of soluble cellulose derivatives and its relationship to the mechanism of cellulose hydrolysis. *J. Bacteriol.* **59**, 485–497.

Reverbel-Leroy, C., Pages, S., Belaich, A., Belaich, J. P., and Tardif, C. (1997). The processive endocellulase CelF, a major component of the *Clostridium cellulolyticum* cellulosome: Purification and characterization of the recombinant form. *J. Bacteriol.* **179**, 46–52.

Roussel, A., and Cambillau, C. (1991). TURBO-FRODO. *In* SIlicon Graphics Geometry Partners Directory pp. 86–87. Silicon Graphics, Mountain View, CA.

Sakon, J., Adney, W. S., Himmel, M. E., Thomas, S. R., and Karplus, P. A. (1996). Crystal structure of thermostable family 5 endocellulase E1 from *Acidothermus cellulolyticus* in complex with cellotetraose. *Biochemistry* **35**, 10648–10660.

Sakon, J., Irwin, D., Wilson, D. B., and Karplus, P. A. (1997). Structure and mechanism of endo/exocellulase E4 from *Thermomonospora fusca*. *Nat. Struct. Biol.* **4**, 810–818.

Schubot, F. D., Kataeva, I. A., Blum, D. L., Shah, A. K., Ljungdahl, L. G., Rose, J. P., and Wang, B. C. (2001). Structural basis for the substrate specificity of the feruloyl esterase domain of the cellulosomal xylanase Z from *Clostridium thermocellum*. *Biochemistry* **40**, 12524–12532.

Shimon, L. J., Bayer, E. A., Morag, E., Lamed, R., Yaron, S., Shoham, Y., and Frolow, F. (1997). A cohesin domain from *Clostridium thermocellum*: The crystal structure provides new insights into cellulosome assembly. *Structure* **5**, 381–390.

Shoseyov, O., and Doi, R. H. (1990). Essential 170-kDa subunit for degradation of crystalline cellulose by *Clostridium cellulovorans* cellulase. *Proc. Natl. Acad. Sci. USA* **87**, 2192–2195.

Souchon, H., Spinelli, S., Beguin, P., and Alzari, P. M. (1994). Crystallization and preliminary diffraction analysis of the catalytic domain of xylanase Z from *Clostridium thermocellum*. *J. Mol. Biol.* **235**, 1348–1350.

Svergun, D. (1992). Determination of the regularization parameter in indirect-transform methods using perceptual criteria. *J. Appl. Crystallogr.* **25**, 495–503.

Svergun, D. I. (1999). Restoring low resolution structure of biological macromolecules from solution scattering using simulated annealing. *Biophys. J.* **76**, 2879–2886.

Svergun, D., Baraberato, C., and Koch, M. H. (1995). CRYSOL—A program to evaluate X-ray solution scattering of biological macromolecules from atomic coordinates. *J. Appl. Crystallogr.* **28**, 768–773.

Svergun, D. I., Petoukhov, M. V., and Koch, M. H. (2001). Determination of domain structure of proteins from X-ray solution scattering. *Biophys. J.* **80**, 2946–2953.

Ting, C. L., Makarov, D. E., and Wang, Z. G. (2009). A kinetic model for the enzymatic action of cellulase. *J. Phys. Chem. B* **113**, 4970–4977.

Tomme, P., Van Tilbeurgh, H., Pettersson, G., Van Damme, J., Vandekerckhove, J., Knowles, J., Teeri, T., and Claeyssens, M. (1988). Studies of the cellulolytic system of *Trichoderma reesei* QM 9414. Analysis of domain function in two cellobiohydrolases by limited proteolysis. *Eur. J. Biochem.* **170**, 575–581.

Tormo, J., Lamed, R., Chirino, A. J., Morag, E., Bayer, E. A., Shoham, Y., and Steitz, T. A. (1996). Crystal structure of a bacterial family-III cellulose-binding domain: A general mechanism for attachment to cellulose. *EMBO J.* **15**, 5739–5751.

van Tilbeurgh, H., Tomme, P., Claeyssens, M., Bhikhabhai, R., and Pettersson, G. (1986). Limited proteolysis of the cellobiohydrolase I from *Trichoderma reesei*. Separation of the functional domains. *FEBS Lett.* **204,** 223–227.

Violot, S., Aghajari, N., Czjzek, M., Feller, G., Sonan, G. K., Gouet, P., Gerday, C., Haser, R., and Receveur-Brechot, V. (2005). Structure of a full length psychrophilic cellulase from *Pseudoalteromonas haloplanktis* revealed by X-ray diffraction and small angle X-ray scattering. *J. Mol. Biol.* **348,** 1211–1224.

von Ossowski, I., Eaton, J. T., Czjzek, M., Perkins, S. J., Frandsen, T. B., Schülein, M., Panine, P., Henrissat, B., and Receveur-Brechot, V. (2005). Protein disorder: Conformational distribution of the flexible linker in a chimeric double cellulase. *Biophys. J.* **88,** 2823–2832.

Walther, D., Cohen, F. E., and Doniach, S. (2000). Reconstruction of low-resolution three-dimensional density maps from one-dimensional small-angle X-ray solution scattering data for biomolecules. *J. Appl. Crystallogr.* **33,** 350–363.

Watanabe, Y., and Inoko, Y. (2011). Further application of size-exclusion chromatography combined with small-angle X-ray scattering optics for characterization of biological macromolecules. *Anal. Bioanal. Chem.* **399,** 1449–1453.

Wright, P. E., and Dyson, H. J. (2009). Linking folding and binding. *Curr. Opin. Struct. Biol.* **19,** 31–38.

Zhong, L., Matthews, J. F., Hansen, P. I., Crowley, M. F., Cleary, J. M., Walker, R. C., Nimlos, M. R., Brooks, C. L. I., Adney, W. S., Himmel, M. E., and Brady, J. W. (2009). Computational simulations of the *Trichoderma reesei* cellobiohydrolase I acting on microcrystalline cellulose Ib: The enzyme–substrate complex. *Carbohydr. Res.* **344,** 1984–1992.

CHAPTER ELEVEN

Quantitative Approaches to The Analysis of Carbohydrate-Binding Module Function

D. Wade Abbott[*] and Alisdair B. Boraston[†]

Contents

1. Introduction	212
2. Experimental Approaches to Investigating CBM Function	215
2.1. General derivation of binding constants	215
2.2. Solid state depletion assay	215
2.3. Affinity gel electrophoresis	217
2.4. UV difference	220
2.5. Isothermal titration calorimetry	223
3. Summary and New Directions for CBM Research	228
References	229

Abstract

Carbohydrate-binding modules (CBMs) are important components of carbohydrate-active enzymes. Their primary functions are to assist in substrate turnover by targeting appended catalytic modules to substrate and concentrating appended catalytic modules on the surface of substrate. Presented here are four well-established methodologies for investigating and quantifying the CBM–polysaccharide binding relationship. These methods include: (1) the solid state depletion assay, (2) affinity gel electrophoresis, (3) UV difference and fluorescence spectroscopy, and (4) isothermal titration calorimetry. In addition, entropy-driven CBM–crystalline cellulose binding events and differential approaches to calculating stoichiometry with polyvalent polysaccharide ligands are also discussed.

---

[*] Lethbridge Research Station, Agriculture and Agri-Food Canada, Lethbridge, Alberta, Canada
[†] Biochemistry & Microbiology, University of Victoria, P.O. Box 3055 STN CSC, Victoria, British Columbia, Canada

1. Introduction

A carbohydrate-binding module (CBM) is defined as a contiguous, independently folding sequence of amino acids found within the primary structure of a carbohydrate-active enzyme that interacts with a carbohydrate, but does not chemically modify its structure. There are two classical roles for CBMs that have been described in the literature (Boraston et al., 2004): (i) targeting an appended catalytic fragment to specific carbohydrate, and (ii) concentrating or retaining catalytic fragments in proximity of a substrate. These functions, although distinct in mechanism, operate to increase the overall rate of substrate attack, a process that is significant for insoluble polysaccharides such as cellulose that are difficult to degrade, and for carbohydrate-active biocatalysts that are secreted into environments with strong diffusionary forces (e.g., soil and oceans). Often CBMs display a binding specificity that reflects the catalytic activity of its parent enzyme, an observation which can help to inform the biochemical characterization of new CBMs. This is not an exclusive partnership in nature, however, as several examples have been described in which a coupled enzyme and CBM display unique specificities (Gregg et al., 2008; Hagglund et al., 2003; Montanier et al., 2009; Newstead et al., 2005). Also, in the case of highly modular enzymes, which contain multiple CBMs that can be unrelated in sequence (Abbott et al., 2008; Boraston et al., 2007), the process of experimentally elucidating CBM function can be further confounded.

Out of what originated as a relatively small number of "cellulose-binding domains" (CBDs), a transformed CBM paradigm has emerged. Binding specificities have been defined not for just cellulose but for nearly every major class of plant cell wall structural polysaccharide, α-glucan storage polysaccharide, and several of the complex carbohydrates decorating the surfaces of animal cells (see Table 11.1), thus resulting in the redefining of CBDs as CBMs. Subsequently, large-scale genome and metagenome platforms have inundated the field with sequences belonging to known CBM families, and illuminated a multitude of new potential families still awaiting characterization. Overall, the CBM "sequence space" has vastly expanded and in this quickly evolving field, which has implications for carbon cycling within the biosphere, climate change, bioprocessing, biofuels, and human health and disease; techniques to investigate and quantify the mechanism of CBM–carbohydrate interactions have become of interest to a much broader range of scientists. Therefore, to help in the growing survey of CBM biology and its applications, this review will discuss the methodology of four key experimental approaches for quantifying CBM–polysaccharide interactions: (1) the solid state depletion assay, (2) affinity gel electrophoresis (AGE), (3) UV difference spectroscopy (with mention of fluorescence spectroscopy), and (4) isothermal titration calorimetry (ITC).

Table 11.1 Structural and functional summary of known CBM families[a]

Family	Protein fold	Demonstrated binding specificities
1	Cysteine Knot	Cellulose (chitin one case)
2	β-Sandwich	Cellulose, chitin, xylan
3	β-Sandwich	Cellulose and chitin
4	β-Sandwich	Xylan, β-1,3-glucan, β-1,3-1,4-glucan, β-1,6-glucan, and amorphous cellulose
5	Unique	Chitin
6	β-Sandwich	Amorphous cellulose, β-1,4-xylan, β-1,3-glucan, β-1,3-1,4-glucan, and β-1,4-glucan
7	Deleted	
8	Unknown	Cellulose
9	β-Sandwich	Cellulose
10	OB fold	Cellulose
11	β-Sandwich	β-1,4-glucan and β-1,3-1,4-mixed-linked glucans
12	Unique	Chitin
13	β-Trefoil	Mannose, xylan, N-acetylgalactosamine
14	Unique	Chitin
15	β-Sandwich	Xylan and xylooligosaccharides
16	β-Sandwich	Cellulose and glucomannan
17	β-Sandwich	Amorphous cellulose, cellooligosaccharides, and derivatized cellulose
18	Hevein fold	Chitin
19	Unknown	Chitin
20	β-Sandwich	Granular starch, cyclodextrins
21	β-Sandwich	Starch
22	β-sandwich	Xylan, β-1,3/β-1,4-glucans
23	Unknown	Mannan
24	Unknown	α-1,3-glucan
25	β-Sandwich	Starch
26	β-Sandwich	Starch
27	β-Sandwich	Mannan
28	β-Sandwich	Noncrystalline cellulose, cellooligosaccharides, and β-(1,3)(1,4)-glucans
29	β-Sandwich	Mannan and glucomannan
30	β-Sandwich	Cellulose
31	β-Sandwich	β-1,3-xylan
32	β-Sandwich	Galactose, lactose, polygalacturonic acid, β-D-galactosyl-1,4-β-D-N-acetylglucosamine
33	β-Sandwich	Chitin and chitosan
34	β-Sandwich	Granular starch

(*Continued*)

Table 11.1 (*Continued*)

Family	Protein fold	Demonstrated binding specificities
35	β-Sandwich	4,5-deoxygalaturonic acid, glucuronic acid, xylan, β-galactan
36	β-Sandwich	Xylan and xylooligosaccharides
37	Unknown	Xylan, chitin, microcrystalline and phosphoric acid-swollen cellulose, alfalfa cell walls, banana stem, and wheat straw
38	Unknown	Inulin
39	β-Sandwich	β-1,3-glucan, lipopolysaccharide, and lipoteichoic acid
40	β-Sandwich	Sialic acid
41	β-Sandwich	Amylose, amylopectin, pullulan, and α-glucan oligosaccharide fragments
42	β-Trefoil	Arabinofuranose
43	CtD-Ole e 9	β-1,3-glucan
44	β-Sandwich	Cellulose and xyloglucan
45	Unknown	Starch
46	Unknown	Cellulose
47	β-Sandwich	Fucose
48	β-Sandwich	Glycogen
49	Unknown	Cellulose
50	LysM-domain	Chitopentaose
51	β-Sandwich	Galactose and to blood group A/B-antigens
52	Unknown	β-1,3-glucan
53	Unknown	Starch
54	Unknown	Xylan, yeast cell wall glucan and chitin
55	Unknown	Chitin
56	Unknown	β-1,3-glucan
57	β-Sandwich	Glucose oligomers
58	β-Sandwich	Maltoheptaose
59	β-Sandwich	Mannan, xylan, and cellulose
60	β-Sandwich	Xylan
61	β-Sandwich	β-1,4-galactan
62	β-Sandwich	Galactose moieties found on xyloglucan, arabinogalactan, and galactomannan
63	Expansin-like	Cellulose
64	Unknown	Cellulose

[a] Adapted from the CAZy database: http://www.cazy.org/Carbohydrate-Binding-Modules.html; Cantarel et al. (2009).

2. Experimental Approaches to Investigating CBM Function

2.1. General derivation of binding constants

Each of the different quantitative methodologies presented below are based upon variations of the general binding reaction:

$$H + G \leftrightarrow HG,$$

where H represents Host, G represents Guest, and HG represents their complex. At equilibrium the association constant is given by:

$$K_a = [GH]/[G][H]. \quad (11.1)$$

This general nomenclature is used as it provides the flexibility to represent the various types of CBM–carbohydrate interactions described in the sections below. Defining $[H_{total}] = [H_{free}] + [H]$ and rearranging the equation we can derive a general Langmuir model as

$$[GH]/[H_{total}] = K_a[G]/(1 + K_a[G]), \quad (11.2)$$

where $[GH]/[H_{total}]$ is often considered as the fraction bound or υ.

2.2. Solid state depletion assay

The solid state depletion (also referred to as the adsorption) assay is the simplest method for qualitative and quantitative assessment of CBM interactions with insoluble polysaccharides. This approach monitors the depletion of CBM from solution following adsorption to sorbent (i.e., cellulose; Gilkes et al., 1992; Fig. 11.1). Although considered to operate in a binding equilibrium, some CBM adsorption may be irreversible (Greenwood et al., 1994; Ong et al., 1991). Inexpensive affinity purification technologies have been developed based upon CBM–cellulose binding, which has opened up numerous potential recombinant enzyme biotechnology applications (Levy and Shoseyov, 2002; Shoseyov et al., 2006).

2.2.1. Experimental setup

Insoluble polysaccharide, for example, crystalline cellulose or chitin, is prepared by weighing dry powder and suspending it in buffer. Dialysis can also be performed to ensure complete buffer exchange. To keep the

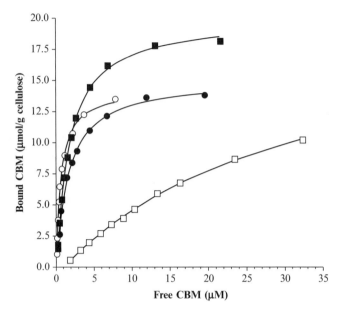

Figure 11.1 Adsorption of various family 6 CBM tandem constructs from *Clostridium stercorarium* onto PASA (phosphoric acid-swollen Avicel): CBM6-3 (□), CBM6-1.2 (●), CBM6-2.3 (■), and CBM6-1.2.3 (○). Isotherms represent differences in strength of binding affinity for various constructs. Figure reproduced and modified with permission from Boraston et al. (2002).

polysaccharide suspended adsorption assays are performed under constant tumbling. Reaction series are set up with identical cellulose concentrations (here this is considered as *H*) and variable CBM concentrations (*G*), which extend between ~1/10 of the dissociation constant (K_d) and a concentration in ~5- to 10-fold excess of the K_d to ensure that saturation of the polysaccharide is reached. In addition, protein samples without cellulose are prepared in an identical volume and used as a baseline to compensate for tube adsorption or protein precipitation. Following equilibration of the polysaccharide with titrated quantities of CBM, reactions are centrifuged (20,000 × *g* for 5 min) to separate solid and liquid phase. The supernatant is then removed and can be analyzed qualitatively by SDS-PAGE and quantitatively by UV absorbance. In the latter case, the protein concentration in the liquid can be determined at 280 nm using a calculated extinction coefficient for the CBM (Gasteiger et al., 2005). All reactions are typically performed in triplicate for statistical significance. More sophisticated solid state competition experiments derived from this method have been described (Boraston et al., 2006; McLean et al., 2000; the reader is referred to these sources for methodology on such techniques).

2.2.2. Quantification of binding constants
As defined above

$$[GH]/[H_{total}] = K_a[G]/(1+K_a[G]). \quad (11.3)$$

In this case $[G]$ is the free CBM concentration measured in the supernatant of the samples containing polysaccharide, which we redefine as $[P]$. $[GH]$ is the concentration of CBM bound to the polysaccharide and is redefined more accurately as $[P_{bound}]$. $[P_{bound}]$ is calculated as $[P_{total}] - [P]$, where $[P_{total}]$ is the total CBM concentration calculated from the samples that did not have polysaccharide added. $[H_{total}]$ is then the total concentration of the available binding sites on the cellulose, which is unknown, and is typically defined as N_o or the binding capacity. CBM adsorption onto cellulose can be mathematically described as

$$[P_{bound}]/N_o = K_a[P_{free}]/(1+K_a[P_{free}]), \quad (11.4)$$

where nonoverlapping independent binding sites on the cellulose are assumed and a single binding site on the CBM is present. Since N_o is unknown it can be rearranged to yield the modified Langmuir-type equation:

$$[P_{bound}] = N_o K_a[P_{free}]/(1+K_a[P_{free}]), \quad (11.5)$$

where N_o with the units of moles of CBM per g cellulose and the association constant, K_a, with units of M^{-1}, are derived from nonlinear regression using this model. Under conditions where two noninteracting, nonidentical classes of binding site with significantly difference binding affinities are suspected on the polysaccharide, the following two-site binding model has been used (Creagh et al., 1996)

$$[P_{bound}] = N_{o,1} K_{a,1}[P_{free}]/(1+K_{a,1}[P_{free}]) + N_{o,2} K_{a,2}[P_{free}]/(1+K_{a,2}[P_{free}]), \quad (11.6)$$

where $N_{o,1}$ and $K_{a,1}$ represent the binding parameters for one class of binding site and $N_{o,2}$ and $K_{a,2}$ the binding parameters for the other.

2.3. Affinity gel electrophoresis

AGE is a technique originally pioneered in the lectin field to assay the interactions between lectins and soluble polysaccharides (Takeo, 1984). AGE was first used to study a CBM in 2000 (Tomme et al., 2000), and this technique has now been used to study CBM interactions with an abundance of polysaccharides including α-glucan, β-glucan, mannan, xylan, and various pectins (e.g., see: Abbott et al., 2007; Abou Hachem et al., 2000; Boraston et al., 2000; Freelove et al., 2001; Lammerts van Bueren et al., 2004; Sunna et al., 2001). Affinity gels are prepared by embedding a soluble polysaccharide into a polyacrylamide matrix. The electrophoretic mobility of a CBM in the polysaccharide-infused gel is compared with its mobility in a native polyacrylamide gel. The interaction of the CBM with the polysaccharide results in a complex of larger size and thus reduced mobility in the polysaccharide containing gel, providing a rapid and convenient readout for binding (Fig. 11.2A). Bovine serum albumin, which has no affinity for carbohydrates, is often used as a negative control. Most often, AGE is used as a qualitative method to efficiently screen several potential polysaccharide ligands and/or rapidly determine the effects of mutagenesis on binding. In contrast, quantitative AGE is currently underutilized as an experimental technique.

2.3.1. Experimental setup

Affinity gels are typically comprised of common nondenaturing polyacrylamide gels, with a low percentage stacking gel (~4% acrylamide) and a separating gel (8.0–12.5% acrylamide) all at pH 8.8. In series, the separating gel is supplemented with polysaccharide ranging in final concentrations from 0% to 0.5% (w/v, polysaccharide are dissolved in water). The concentration of acrylamide can be optimized to provide greater resolution of the CBMs if required. All gels are prepared synchronously to limit preparative error between different batches. Typically ~5–20 µg of protein in native sample buffer is loaded per lane and electrophoresed. Gels can be stained with Commassie Blue or other common gel staining methods to detect protein bands. The mobility of CBM and the noninteracting control (e.g., BSA) are measured as the distance traveled from the top of the separating gel. The relative mobility of the sample, R, is calculated as the ratio of the CBM mobility to the noninteracting control mobility. A decreased R in the presence of polysaccharide indicates an interaction.

2.3.2. Quantification of binding constants

In AGE the concentrations of free and bound CBM, in this case $[G]$ and $[GH]$, respectively, are unknown. However, the relative change in mobility of the CBM gives an indirect quantitative indication of CBM–carbohydrate complex formation. Thus, $R_o - R$, where R_o is the relative mobility (to BSA) in the absence of polysaccharide and R in the presence of polysaccharide, is proportional to $[GH]$ at a given polysaccharide concentration (Fig. 11.2B). $R_o - R_{max}$, where R_{max} is the maximum change in relative mobility when the CBM is fully bound, is then proportional to the maximum amount of complex that could be formed, which is equivalent to $[H_{total}]$. It therefore follows that $(R_o - R)/(R_o - R_{max})$ is equal to the

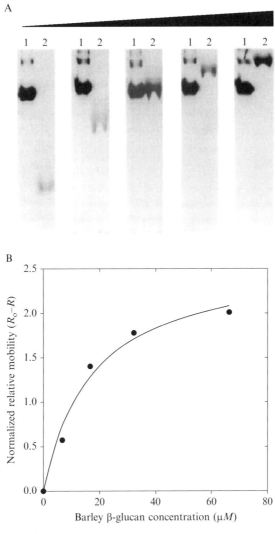

Figure 11.2 (A) AGE of the CBM4-1 from *Cellumonas fimi* with an increasing gradient of barley β-glucan (0–0.1%). Lane 1 contains a BSA control and Lane 2 contains the CBM. (B) Isotherm displaying CBM4-1 relative mobilities as a function of barley β-glucan concentration. The fit was obtained using Eq. (11.6). The units of barley β-glucan, represented as μM binding sites is based upon microcalorimetry performed previously. Figures reproduced and modified with permission from Tomme et al. (2000).

fraction of complex formed, υ, or $[GH]/[H_{total}]$. The binding isotherm can then be written as

$$(R_o - R)/(R_o - R_{max}) = K_a[G]/(1 + K_a[G]), \quad (11.7)$$

where [G] is the concentration of polysaccharide used. It is assumed that the polysaccharide concentration is in excess and that the amount present in the CBM–polysaccharide complex, GH, is negligible. Thus, the free polysaccharide concentration [G] is approximated by $[G_{total}]$, or the concentration of polysaccharide incorporated into the gel. To better isolate the dependent and independent variables the equation can be rewritten as

$$R = R_o - \{(R_o - R_{max})K_a[G]/(1+K_a[G])\}. \tag{11.8}$$

R_o is a measured value and is set as a constant, while R_{max} and K_a are determined from nonlinear regression analysis. However, it is often convenient to precalculate $R_o - R$ values allowing the use of the following equation:

$$R_n = R_{max,n}K_a[G]/(1+K_a[G]), \tag{11.9}$$

where R_n are $R_o - R$ normalized values and $R_{max,n}$ is again a regressed variable representing the maximum $R_o - R$ normalized value when the CBM is fully complexed.

It has been estimated that the association constants ranging between 10^2 and $10^5 \, M^{-1}$ can be most accurately determined with this method (Tomme et al., 2000). Above this range the assumption of the total polysaccharide concentration approximating to the free polysaccharide concentration is invalid. Ultimately, this technique is limited by the sensitivity of protein detection and more sophisticated visualization approaches may help to extend the experimental range (Tomme et al., 2000).

2.4. UV difference

Carbohydrate recognition very often involves aromatic amino acid side chains in the binding site of the protein. Sugar rings can pack against these residues through CH-π bonds (Fig. 11.3A; Brandl et al., 2001). The spectroscopic properties of such aromatic amino acid side chains, mainly tryptophan and to a lesser extent tyrosine, are sensitive to their microenvironment. Interaction with sugars, therefore, usually shields the amino acid side chains from solvent (i.e., water) resulting in small perturbations in their spectroscopic profiles, much in the manner that perturbing the polarity of N-acetyltryptophan and N-acetyltyrosine solutions with DMSO changes their UV absorbance profiles (Fig. 11.3B). These changes in spectroscopic properties are easily visualized for CBMs binding to soluble carbohydrates by UV difference spectroscopy, which measures the change in the absorbance of UV light by the CBM induced by the binding of ligand. The patterns of change detected in the UV difference spectra are influenced by

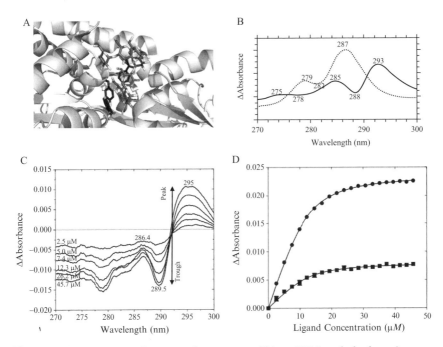

Figure 11.3 Spectroscopic approaches to quantifying CBM–carbohydrate interactions. (A) The structure of the binding site of MalX from *Strepococcus pneumoniae* in complex with maltoheptaose (dark gray sticks, PDB code: 2XD3; Abbott et al., 2010). Relevant aromatic amino acids in the active site that are shielded from solvent upon sugar binding are shown as black sticks. (B) Perturbations of *N*-acetyltryptophan (solid line) and *N*-acetyltyrosine (dotted line) with 20% DMSO demonstrating the changes in UV absorbance induced in these compounds when moving from a polar to an apolar environment. (C) The UV difference spectra of MalX collected in the presence of six different concentrations of maltotetraose demonstrating the relationship between UV difference signal and complex formation. Relevant peaks and troughs are labeled with their respective wavelengths. An example of a peak-to-trough distance used to quantify the signal is indicated by the dashed arrow. (D) Isotherms of MalX binding to maltotetraose which incorporates the partial data set shown in (C). The circles (●) indicate the data for the 295–289.5 peak–trough height and the squares (■) indicate the data for the 286.4–289.5 peak–trough height. Figure B is reproduced and modified with permission from Boraston et al. (2000).

the type of aromatic side chain involved, often providing a small degree of structural insight into the interaction (Fig. 11.3C). This method can also be used to quantify interactions (Fig. 11.3D).

2.4.1. Experimental setup

Individual UV difference spectra provide a qualitative indication of carbohydrate binding. Spectra are taken for CBM solutions with an absorbance at 280 nm in the range of 0.5–1.0 in the presence and absence of carbohydrate over a suitable wavelength range (~270–310 nm). Assuming that the CBM concentrations in the samples are identical (this is often achieved by adding solid carbohydrate to the CBM-only sample after taking the baseline spectrum, which avoids significant changes in volume) and that the carbohydrate does not absorb in the selected range of wavelengths, the CBM-only spectrum can be used as a baseline and subtracted from the CBM sample with carbohydrate. If a CBM–carbohydrate complex forms with the participation of a tryptophan or tyrosine, a distinctive pattern of peaks and troughs is observed in the UV difference spectrum. This can often be used in a medium-throughput mode to screen binding to soluble sugars. The magnitude of the signals, measured as peak–trough heights in the UV difference spectrum, is proportional to the amount of complex that is formed, allowing the method to be used in a quantitative manner (Fig. 11.3C and D). In this case, titrations are typically performed by analyzing CBM samples in the presence of increasing amounts of carbohydrate and examining one, two, or three prominent peak-to-trough pairs. Runs are performed in triplicate for statistical significance. Reactions can be executed manually (Boraston et al., 2000) or in an automated fashion (Boraston et al., 2001b). Controlled reaction temperature, thorough and continuous mixing, and precise titrating are essential to generating high quality data.

2.4.2. Quantification of binding constants
As defined above

$$[GH]/[H_{total}] = K_a[G]/(1 + K_a[G]). \qquad (11.10)$$

UV difference cannot directly measure the reactant concentrations at equilibrium; however, the UV difference signal, S, defined as a peak-to-trough value ($\Delta A_{peak} - \Delta A_{trough}$) is proportional to the amount of complex formed, $[GH]$, where the CBM is considered the Host, H, and the carbohydrate the Guest, G. The maximum possible signal, S_{max}, which represents the point at which the CBM is fully saturated with sugar and no more complex can be formed, is proportional to the maximum amount of GH formed and, therefore, $[H_{total}]$. Thus, similar to what was described for AGE analysis, S/S_{max} is equal to the fraction of complex formed, υ, or $[GH]/[H_{total}]$. Thus, the binding equation can be written as

$$S/S_{max} = K_a[G]/(1 + K_a[G]). \qquad (11.11)$$

Analogous to the binding capacity of cellulose (N_o) described above, S_{max} is unknown; therefore, the equation can be rearranged to yield

$$S = S_{max}K_a[G]/(1 + K_a[G]). \qquad (11.12)$$

Under conditions where $[H_{total}]$ (i.e., the total concentration of CBM used in the experiment) $\ll 1/K_a$, $[G]$, or the free carbohydrate concentration, is approximated by $[G_{total}]$, or the total concentration of added carbohydrate. Nonlinear regression is used to determine S_{max} and the association constant, K_a. When $[G]$ cannot be approximated by $[G_{total}]$, and ligand depletion must be considered, the following relationship can be substituted into Eq. (11.9) to express it in terms of $[G_{total}]$

$$[G] = 0.5\left[([G_{total}] - 1/K_a - [H_{total}]) + \sqrt{(\{[H_{total}] - [G_{total}] + 1/K_a\}^2 + 4[G_{total}]/K_a)}\right],$$

(11.13)

$[H_{total}]$ is the known CBM concentration and is set as a constant. This assumes a 1:1 binding stoichiometry. Under circumstances where $[H_{total}] \gg 1/K_a$, $[H_{total}]$ in Eq. (11.13) can be substituted with $n[H_{total}]$, where n represents the stoichiometry of the binding reaction and can be determined by the nonlinear regression analysis.

2.4.3. Fluorescence spectroscopy

Similar principles apply to the use of fluorescence spectroscopy to monitor CBM–carbohydrate interactions except for rather than measuring changes in tryptophan or tyrosine absorbance the changes in intrinsic fluorescence properties of these residues in response to carbohydrate binding is monitored (Boraston et al., 2000; Eftink, 1997).

2.5. Isothermal titration calorimetry

Calorimetry, the science of measuring heat changes during a chemical reaction, is one of the oldest techniques known to experimental science, dating back to the mid-1750s (Breslauer et al., 1992). Within the past 30 years substantial improvements to microcalorimeters have revolutionized molecular glycobiology (Dam and Brewer, 2002). In particular, ITC, which measures the increases (exothermic) and decreases (endothermic) in temperature within an isothermally regulated solution following the titration of ligand (i.e., carbohydrate, G) into acceptor (i.e., protein, H) has emerged as the gold standard in the quantification of CBM–carbohydrate interactions. A single ITC experiment can determine Gibb's free energy (ΔG), enthalpy (ΔH), entropy (ΔS), and stoichiometry (n) of a designated interaction. In addition, reaction series at varying temperature can be performed to determine the heat capacity of the system (ΔC_p). Dissecting these parameters has helped to define fundamental mechanisms of how CBMs interact with insoluble polysaccharides (Type A CBMs), soluble poly- and oligosaccharide chains (Type B), and small oligosaccharides and monosaccharides (Type C) (Boraston et al., 2004).

2.5.1. Experimental setup

Preparation of stable and soluble protein for an ITC run is often the limiting step. Typically, an acceptor concentration is within the 10–100 μM range, which can be lowered in the case of high affinity interactions or increased in the case of lower affinity interactions. The development of small volume microcalorimeters (Baranauskiene et al., 2009) has recently changed the ITC landscape as significantly less protein and ligand (approximately five-fold) are consumed per reaction. This increased sensitivity, particularly useful for difficult to solublize protein and expensive complex carbohydrate reagents, comes at the cost of lower accuracy. For each approach, the experimental ligand concentration, [G], is dependent upon the affinity and typically requires some preliminary testing to produce ideal heats and saturation profiles. Initially, concentration is approximated from the following equation (Wiseman et al., 1989)

$$C_{\text{value}} = nK_{\text{a}}[H_{\text{total}}], \tag{11.14}$$

where n is the stoichiometry of reaction (for a 1-site bimolecular interaction, it is set to 1), K_{a} is the affinity constant, and $[H_{\text{total}}]$ is the total Host or acceptor concentration in the cell. For accurate ITC reactions the resulting C_{value}, a dimensionless constant, should range between 1 and 1000 (Wiseman et al., 1989). If possible, the experimental C_{value} can be increased by increasing the $[H_{\text{total}}]$ in the system.

To limit the background heat caused by dilution effects, both reactant solvents should be identical. In the case of large or insoluble polysaccharides it may be possible to first dialyze both reagents in the same container as the acceptor protein. In the case of smaller soluble poly- and oligosaccharides, it is best to dissolve the dry carbohydrate in the dialysate or filtrate of the protein. Small background heats of dilution can be corrected by titrating buffer into acceptor protein and subtracting this dataset from the experimental during analysis.

When setting up crystalline polysaccharide–ITC experiments, careful preparation of the carbohydrate is recommended. Previously a method for regenerated Avicel was reported (Boraston et al., 2001a). Briefly, Avicel is completely dissolved in 85% phosphoric acid (1 g:200 mL) followed by precipitation with the addition of five volumes of cold acetone. The precipitated cellulose is extensively washed with distilled water using a vacuum filtration system and glass fiber filter, resuspended in distilled water, and homogenized in a blender. The concentration of suspended cellulose is determined by using the quantified mass of cellulose present in dried samples.

2.5.2. Determination of thermodynamic parameters

Modern microcalorimetry systems have made the calculation of thermodynamic parameters a relatively simple process. Units are equipped with step-by-step software that enables quick and user-friendly analysis of raw data. The exothermic and endothermic events observed during each titration are reported as energetic values (μcal/s). These values, represented as data spikes accompanying each titration, indicate whether the calorimeter has to increase (endothermic) or lower (exothermic) energy to the reaction system to maintain a constant temperature (Fig. 11.4A). Once G has bound to available H following each titration, the system will return to baseline. Similarly, when the acceptor sites reach saturation there will be no significant changes in heat generated after a titration. Nonlinear least-squares analysis of the data give rise to the characteristic sigmoidal isotherm represented as kcal/mol injectant against the molar ratio, enabling the calculation of the association constant (K_a) (Fig. 11.4A).

The natural log of the determined K_a can be used to calculate the change in Gibb's free energy (ΔG) by the following equation:

$$\Delta G = -RT \ln K_a, \tag{11.15}$$

where R is the universal gas constant (8.34 J K^{-1} mol^{-1}) and T is the absolute reaction temperature (Kelvin). With quantified enthalpies and a calculated ΔG (kJ mol^{-1}), the entropy of binding ΔS (kJ mol^{-1} K^{-1}) can be determined using the Gibb's free energy equation

$$\Delta G = \Delta H - T\Delta S. \tag{11.16}$$

A positive $T\Delta S$ (kJ mol^{-1}) value reflects a thermodynamically favorable event, as it equates to an overall decrease in the ΔG. Entropic contributions to CBM–polysaccharide binding may include desolvation of binding surfaces and release of order waters back to solvent, a significant factor in the binding of Type A CBMs to crystalline polysaccharides. In other CBM systems (Type B and Type C), entopic penalties tend to predominant the $T\Delta S$ contributions due to restricted rotational, translational, and conformational freedoms of the CBM and ligand, which overshadow any favorable desolvation effects (see Section 2.5.4) (Chervenak and Toone, 1995; Tomme et al., 1996).

To explore the role of enthalpy and entropy at various experimental temperatures, the change in heat capacity (ΔC_p) can be calculated. ΔC_p is defined as $\delta H/\delta T$ and can be determined by measuring the change in ΔH as a function of experimental temperature (Fig. 11.4B). Calculating the ΔC_p for a binding interaction is helpful for defining the polar and apolar solvent accessible binding surface area (Edgcomb and Murphy, 2000). Large negative ΔC_p effects indicate an accumulation of weak interactions, such as burying apolar groups away from solvent and release of solvating waters

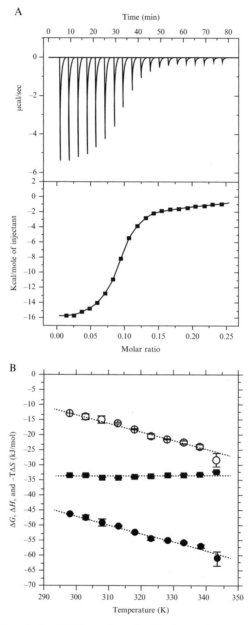

Figure 11.4 ITC analysis of CBM–polysaccharide interactions. (A) Raw (top panel) and integrated (lower panel) heats produced by titrating a family 61 CBM from *Thermotoga maritima* (Guest) into pectic potato galactan (Host). (B) Relationship between ΔG (■), which remains relatively constant, ΔH (●) which decreases, and ΔS (○) which decreases, as a function of temperature for the recognition of cellobiose by the family 9 CBM from *Thermotoga maritima*. Figures (A) and (B) reproduced and modified with permission from Cid et al. (2010) and Boraston et al. (2001a), respectively.

during complex formation, which can make a significant contribution to the binding energy (Brandts et al., 1974; Cooper, 2005; Dill, 1990). This is often accompanied by an increased ΔS due to increased rotational and translational freedom of released water molecules (Creagh et al., 1996; Lammerts van Bueren and Boraston, 2004). A small ΔC_p value suggests that polar interactions are balanced and the hydrophobic effect is not a substantial driving force across the reaction landscape (Tomme et al., 1996).

2.5.3. A note on stoichiometric calculations for CBM–polysaccharide interactions by ITC

CBM–polysaccharide interactions rarely occur in a 1:1 stoichiometry (*n*) as linear, branched, and crystalline polysaccharides may present many internal binding sites for a CBM along the chain. This complicates the experimental determination of *n* and consequently affects the determination of all other binding parameters. For example, a single polysaccharide chain comprising 100 monosaccharide units may present up to 10–20 CBM binding sites; however, this relationship may be unknown at the start of the experiment. This problem is addressed by determining the stoichiometries during data fitting. Within the literature there are two different experimental approaches described to accomplish this. One method treats the polysaccharide as Guest (*G*) and the protein as Host (*H*), where *G* is titrated into *H*. During the fitting analysis the concentration of *G* is manually altered until the fitting results in an *n* of 1. This, however, is a somewhat biased approach and places the responsibility of iteratively "minimizing" the value of the polysaccharide concentration in the hands of the user. Also this approach tends to mask the possible presence of different classes of binding sites on the polysaccharide, which is not uncommon. It is possible to back-calculate values such as the number of monosaccharide units that comprise a CBM binding site. For example, if the known concentration of the polysaccharide is equivalent to 1 mM of monosaccharide units but the concentration during the fitting that results in an *n* of 1 is 0.2 mM then, on average, a binding site is found every 5 monosaccharide units. A more robust approach reverses the titrant and acceptor, treating the polysaccharide as *H* and protein as *G* (i.e., CBM is titrated into polysaccharide). [*H*] is estimated by the mass of polysaccharide (w/v) converted to an equivalent molar equivalent concentration of monosaccharide (or other defined oligosaccharide structure), as in the example above. The CBM, or *G*, which is typically monovalent, is titrated into the polysaccharide. The data are fitted allowing *n* to be minimized with no user intervention; the resulting *n* then represents the number of CBM binding sites recognizing a monosaccharide equivalent (*n* is always less than 1). The "footprint" of the CBM is calculated by taking the inverse of *n*, which gives the number of monosaccharide equivalents that, on average, contain a single CBM binding site. Recent examples of this include a family 61 CBM from *Thermotoga maritima* with galactose for β1,4-galactan (Cid et al., 2010) and a family 32 CBM from *Yersinia enterocolitica* with trigalacturonic acid for α1,4-polygalacturonic acid (Abbott et al., 2007). An important feature of this approach is that it readily reveals the presence of nonequivalent classes of binding sites on the polysaccharide or other complicated features of binding.

2.5.4. Entropy, CBMs, and crystalline cellulose

Generally, carbohydrate–CBM interactions are enthalpically driven with a small offsetting entropic penalty. The etiology of this thermodynamic signature arises from accumulative contributions of cognate protein–carbohydrate interactions ($-\Delta H$), the release of ordered waters (favorable, $+\Delta S$) and the loss of translational, rotational, and conformational freedoms of the binding partners (unfavorable, $-\Delta S$). In addition to small sugars and soluble polysaccharides, similar thermodynamic profiles have been reported for CBMs binding cellooligosaccharides (Boraston et al., 2001a; Tomme et al., 1996), amorphous cellulose (Tomme et al., 1996), and insoluble noncrystalline cellulose (Boraston, 2005). Contrasting this maturing paradigm is a seminal microcalorimetric study that characterized the interaction between the CBD from the *Cellulomonas fimi* β1,4-xylanase, Cex (named CBD_{Cex}), and suspended insoluble bacterial microcrystalline cellulose (BMCC) fragments (Creagh et al., 1996). BMCC is an array of parallel cellulose chains packed into crystalline microfibrils and displays an estimated 76% crystallinity (Gilkes et al., 1992; Tomme et al., 1996), the heterogeneity of which likely creates alternate binding environments to the crystalline surface. Two binding models (one site and two site) were presented, both of which indicate the free energy is dominated by entropic forces. In protein–carbohydrate interactions involving soluble carbohydrates and regions of flexible β1,4-glucan the net loss of conformational freedom in the ligand is unfavorable and outweighs any entropically favorable events such as desolvation. In the case of crystalline cellulose–CBD_{Cex} interactions, however, the penalizing offset would be absent due to the inherent rigidity of ligand, and coupled to the relatively small enthalpic contribution observed in the binding mechanism (i.e., few polar interactions), the cumulative thermodynamics would result in an free energy dominated by the entropy contributed from desolvation. Our understanding of this binding model would greatly benefit from further ITC studies using other crystalline polysaccharides.

3. Summary and New Directions for CBM Research

This review has discussed four quantitative approaches useful for determining the binding constant, changes in enthalpy, changes in entropy, changes in heat capacity, and stoichiometry of CBM–polysaccharide interactions. Depending upon the physiochemical nature of the ligand in question, a researcher may need to be selective in the chosen experimental method. For example, the solid state depletion assay is a robust technique to study CBM interactions with crystalline polysaccharides; whereas AGE, UV difference/fluorescence, or ITC are well-established approaches with soluble polysaccharides. As our knowledge of CBM biology continues to expand, exceptions to the general targeting and concentrating roles of CBMs have begun to surface. These include the retention of ligand within cellular compartments by an independent CBM (Abbott et al., 2007), and CBM oligomerization in a ligand-dependent (Freelove et al., 2001) and calcium-dependent (Montanier et al., 2011) manner. Such growing functional diversity for these abundant molecules suggests that the current experimental approaches described above may need to be revisited in the coming years.

REFERENCES

Abbott, D. W., Hrynuik, S., and Boraston, A. B. (2007). Identification and characterization of a novel periplasmic polygalacturonic acid binding protein from Yersinia enterolitica. *J. Mol. Biol.* **367**, 1023–1033.

Abbott, D. W., Eirin-Lopez, J. M., and Boraston, A. B. (2008). Insight into ligand diversity and novel biological roles for family 32 carbohydrate-binding modules. *Mol. Biol. Evol.* **25**, 155–167.

Abbott, D. W., Higgins, M. A., Hyrnuik, S., Pluvinage, B., Lammerts van Bueren, A., and Boraston, A. B. (2010). The molecular basis of glycogen breakdown and transport in Streptococcus pneumoniae. *Mol. Microbiol.* **77**, 183–199.

Abou Hachem, M., Nordberg Karlsson, E., Bartonek-Roxa, E., Raghothama, S., Simpson, P. J., Gilbert, H. J., Williamson, M. P., and Holst, O. (2000). Carbohydrate-binding modules from a thermostable Rhodothermus marinus xylanase: Cloning, expression and binding studies. *Biochem. J.* **345**(Pt 1), 53–60.

Baranauskiene, L., Petrikaite, V., Matuliene, J., and Matulis, D. (2009). Titration calorimetry standards and the precision of isothermal titration calorimetry data. *Int. J. Mol. Sci.* **10**, 2752–2762.

Boraston, A. B. (2005). The interaction of carbohydrate-binding modules with insoluble non-crystalline cellulose is enthalpically driven. *Biochem. J.* **385**, 479–484.

Boraston, A. B., Chiu, P., Warren, R. A., and Kilburn, D. G. (2000). Specificity and affinity of substrate binding by a family 17 carbohydrate-binding module from Clostridium cellulovorans cellulase 5A. *Biochemistry* **39**, 11129–11136.

Boraston, A. B., Creagh, A. L., Alam, M. M., Kormos, J. M., Tomme, P., Haynes, C. A., Warren, R. A., and Kilburn, D. G. (2001a). Binding specificity and thermodynamics of a family 9 carbohydrate-binding module from *Thermotoga maritima* xylanase 10A. *Biochemistry* **40**, 6240–6247.

Boraston, A. B., Warren, R. A., and Kilburn, D. G. (2001b). beta-1,3-Glucan binding by a thermostable carbohydrate-binding module from *Thermotoga maritima*. *Biochemistry* **40**, 14679–14685.

Boraston, A. B., McLean, B. W., Chen, G., Li, A., Warren, R. A., and Kilburn, D. G. (2002). Co-operative binding of triplicate carbohydrate-binding modules from a thermophilic xylanase. *Mol. Microbiol.* **43**, 187–194.

Boraston, A. B., Bolam, D. N., Gilbert, H. J., and Davies, G. J. (2004). Carbohydrate-binding modules: Fine-tuning polysaccharide recognition. *Biochem. J.* **382,** 769–781.

Boraston, A. B., Healey, M., Klassen, J., Ficko-Blean, E., Lammerts van Bueren, A., and Law, V. (2006). A structural and functional analysis of alpha-glucan recognition by family 25 and 26 carbohydrate-binding modules reveals a conserved mode of starch recognition. *J. Biol. Chem.* **281,** 587–598.

Boraston, A. B., Ficko-Blean, E., and Healey, M. (2007). Carbohydrate recognition by a large sialidase toxin from Clostridium perfringens. *Biochemistry* **46,** 11352–11360.

Brandl, M., Weiss, M. S., Jabs, A., Suhnel, J., and Hilgenfeld, R. (2001). C-H...pi-interactions in proteins. *J. Mol. Biol.* **307,** 357–377.

Brandts, J. F., Jackson, W. M., and Ting, T. Y. (1974). A calorimetric study of the thermal transitions of three specific transfer ribonucleic acids. *Biochemistry* **13,** 3595–3600.

Breslauer, K. J., Freire, E., and Straume, M. (1992). Calorimetry: A tool for DNA and ligand-DNA studies. *Methods Enzymol.* **211,** 533–567.

Cantarel, B. L., Coutinho, P. M., Rancurel, C., Bernard, T., Lombard, V., and Henrissat, B. (2009). The Carbohydrate-Active EnZymes database (CAZy): An expert resource for Glycogenomics. *Nucleic Acids Res.* **37,** D233–D238.

Chervenak, M. C., and Toone, E. J. (1995). Calorimetric analysis of the binding of lectins with overlapping carbohydrate-binding ligand specificities. *Biochemistry* **34,** 5685–5695.

Cid, M., Pedersen, H. L., Kaneko, S., Coutinho, P. M., Henrissat, B., Willats, W. G., and Boraston, A. B. (2010). Recognition of the helical structure of beta-1,4-galactan by a new family of carbohydrate-binding modules. *J. Biol. Chem.* **285,** 35999–36009.

Cooper, A. (2005). Heat capacity effects in protein folding and ligand binding: A re-evaluation of the role of water in biomolecular thermodynamics. *Biophys. Chem.* **115,** 89–97.

Creagh, A. L., Ong, E., Jervis, E., Kilburn, D. G., and Haynes, C. A. (1996). Binding of the cellulose-binding domain of exoglucanase Cex from *Cellulomonas fimi* to insoluble microcrystalline cellulose is entropically driven. *Proc. Natl. Acad. Sci. USA* **93,** 12229–12234.

Dam, T. K., and Brewer, C. F. (2002). Thermodynamic studies of lectin-carbohydrate interactions by isothermal titration calorimetry. *Chem. Rev.* **102,** 387–429.

Dill, K. A. (1990). Dominant forces in protein folding. *Biochemistry* **29,** 7133–7155.

Edgcomb, S. P., and Murphy, K. P. (2000). Structural energetics of protein folding and binding. *Curr. Opin. Biotechnol.* **11,** 62–66.

Eftink, M. R. (1997). Fluorescence methods for studying equilibrium macromolecule-ligand interactions. *Methods Enzymol.* **278,** 221–257.

Freelove, A. C., Bolam, D. N., White, P., Hazlewood, G. P., and Gilbert, H. J. (2001). A novel carbohydrate-binding protein is a component of the plant cell wall-degrading complex of Piromyces equi. *J. Biol. Chem.* **276,** 43010–43017.

Gasteiger, E., Hoogland, C., Gattiker, A., Duvaud, S., Wilkins, M. R., Appel, R. D., and Bairoch, A. (2005). Protein identification and analysis tools on the ExPASy server. *In* "The Proteomics Protocols Handbook," (J. M. Walker, ed.), pp. 571–607. Humana Press.

Gilkes, N. R., Jervis, E., Henrissat, B., Tekant, B., Miller, R. C., Jr., Warren, R. A., and Kilburn, D. G. (1992). The adsorption of a bacterial cellulase and its two isolated domains to crystalline cellulose. *J. Biol. Chem.* **267,** 6743–6749.

Greenwood, J. M., Gilkes, N. R., Miller, R. C., Jr., Kilburn, D. G., and Warren, R. A. (1994). Purification and processing of cellulose-binding domain-alkaline phosphatase fusion proteins. *Biotechnol. Bioeng.* **44,** 1295–1305.

Gregg, K. J., Finn, R., Abbott, D. W., and Boraston, A. B. (2008). Divergent modes of glycan recognition by a new family of carbohydrate-binding modules. *J. Biol. Chem.* **283,** 12604–12613.

Hagglund, P., Eriksson, T., Collen, A., Nerinckx, W., Claeyssens, M., and Stalbrand, H. (2003). A cellulose-binding module of the Trichoderma reesei beta-mannanase Man5A increases the mannan-hydrolysis of complex substrates. *J. Biotechnol.* **101**, 37–48.

Lammerts van Bueren, A., and Boraston, A. B. (2004). Binding sub-site dissection of a carbohydrate-binding module reveals the contribution of entropy to oligosaccharide recognition at "non-primary" binding subsites. *J. Mol. Biol.* **340**, 869–879.

Lammerts van Bueren, A., Finn, R., Ausio, J., and Boraston, A. B. (2004). Alpha-glucan recognition by a new family of carbohydrate-binding modules found primarily in bacterial pathogens. *Biochemistry* **43**, 15633–15642.

Levy, I., and Shoseyov, O. (2002). Cellulose-binding domains: Biotechnological applications. *Biotechnol. Adv.* **20**, 191–213.

McLean, B. W., Bray, M. R., Boraston, A. B., Gilkes, N. R., Haynes, C. A., and Kilburn, D. G. (2000). Analysis of binding of the family 2a carbohydrate-binding module from *Cellulomonas fimi* xylanase 10A to cellulose: Specificity and identification of functionally important amino acid residues. *Protein Eng.* **13**, 801–809.

Montanier, C., van Bueren, A. L., Dumon, C., Flint, J. E., Correia, M. A., Prates, J. A., Firbank, S. J., Lewis, R. J., Grondin, G. G., Ghinet, M. G., Gloster, T. M., Herve, C., et al. (2009). Evidence that family 35 carbohydrate binding modules display conserved specificity but divergent function. *Proc. Natl. Acad. Sci. USA* **106**, 3065–3070.

Montanier, C. Y., Correia, M. A., Flint, J. E., Zhu, Y., Basle, A., McKee, L. S., Prates, J. A., Polizzi, S. J., Coutinho, P. M., Lewis, R. J., Henrissat, B., Fontes, C. M., et al. (2011). A novel, noncatalytic carbohydrate-binding module displays specificity for galactose-containing polysaccharides through calcium-mediated oligomerization. *J. Biol. Chem.* **286**, 22499–22509.

Newstead, S. L., Watson, J. N., Bennet, A. J., and Taylor, G. (2005). Galactose recognition by the carbohydrate-binding module of a bacterial sialidase. *Acta Crystallogr. D Biol. Crystallogr.* **61**, 1483–1491.

Ong, E., Gilkes, N. R., Miller, R. C., Jr., Warren, A. J., and Kilburn, D. G. (1991). Enzyme immobilization using a cellulose-binding domain: Properties of a beta-glucosidase fusion protein. *Enzyme Microb. Technol.* **13**, 59–65.

Shoseyov, O., Shani, Z., and Levy, I. (2006). Carbohydrate binding modules: Biochemical properties and novel applications. *Microbiol. Mol. Biol. Rev.* **70**, 283–295.

Sunna, A., Gibbs, M. D., and Bergquist, P. L. (2001). Identification of novel beta-mannan- and beta-glucan-binding modules: Evidence for a superfamily of carbohydrate-binding modules. *Biochem. J.* **356**, 791–798.

Takeo, K. (1984). Affinity electrophoresis: Principles and applications. *Electrophoresis* **5**, 187–195.

Tomme, P., Creagh, A. L., Kilburn, D. G., and Haynes, C. A. (1996). Interaction of polysaccharides with the N-terminal cellulose-binding domain of *Cellulomonas fimi* CenC. 1. Binding specificity and calorimetric analysis. *Biochemistry* **35**, 13885–13894.

Tomme, P., Boraston, A., Kormos, J. M., Warren, R. A., and Kilburn, D. G. (2000). Affinity electrophoresis for the identification and characterization of soluble sugar binding by carbohydrate-binding modules. *Enzyme Microb. Technol.* **27**, 453–458.

Wiseman, T., Williston, S., Brandts, J. F., and Lin, L. N. (1989). Rapid measurement of binding constants and heats of binding using a new titration calorimeter. *Anal. Biochem.* **179**, 131–137.

CHAPTER TWELVE

In Situ Detection of Cellulose with Carbohydrate-Binding Modules

J. Paul Knox

Contents

1. Introduction	234
1.1. Detection methods using CBM molecular probes	235
2. Preparation of Plant Materials	236
2.1. Fixation of plant materials for cellulose detection with CBMs	236
2.2. Embedding and sectioning of plant and cellulose-based materials	238
3. CBM-Labeling Procedures	240
3.1. Section pretreatments prior to cellulose detection—Enzymatic pretreatments (pectic HG removal)	241
3.2. *In situ* cellulose detection with directly coupled CBMs	241
3.3. Indirect fluorescence imaging methods for *in situ* cellulose detection with CBMs	242
3.4. Post-CBM-labeling treatments prior to fluorescence imaging	243
4. Summary	244
Acknowledgment	244
References	244

Abstract

Cellulose is generally found in the context of complex plant cell wall materials and mostly in association with other glycans. Cellulose-directed carbohydrate-binding modules (CBMs) can be readily adapted to a range of methods for the *in situ* imaging of cellulose structures within plant cell walls or other cellulose-based materials. Protocols for the preparation and selection of plant materials, their fixation and processing for preparation of sections for CBM labeling, and fluorescence imaging procedures are described. Approaches to direct methods in which CBMs are directly coupled to fluorophores and indirect methods in which staged incubations with secondary reagents are used for the fluorescence imaging of CBM binding to materials are discussed and presented.

Faculty of Biological Sciences, Centre for Plant Sciences, University of Leeds, Leeds, United Kingdom

1. Introduction

Cellulose is a widespread and abundant polysaccharide found in the cell walls of land plants and many algae, and it is also produced by tunicates and some bacteria. Cellulose molecules can form a range of macromolecular configurations that contain both crystalline and amorphous regions and these fibrillar configurations can be important structural components of biomaterials (Stone, 2005). In addition to biological roles, cellulose-based substrates have significant industrial applications.

The bulk of the cellulose on Earth occurs in plant cell walls and that will be the focus here. In structural terms, plant cell walls are composites of glycans, and cellulose comprises in the region of 30% (or more) of these glycans. In some cases, such as cotton fibers, cellulose content may reach well over 90%. Techniques to detect or image cellulose within materials have been considered for many years. Initial approaches involved the use of stains although these are rarely specific to cellulose. The fluorescent brightener Calcofluor White (Sigma-Aldrich) is used as a counter stain as it binds widely to β-glycans, including cellulose, and fluoresces under UV excitation and therefore can indicate the location of cell walls in sections and is useful for orientation in relation to organ and tissue anatomy. Recently, Pontamine Fast Scarlet 4B has been identified as showing greater specificity toward cellulose and this has been utilized in studies of cellulose microfibril orientation (Anderson *et al.*, 2010). In recent years, interest in the *in situ* detection of cellulose has developed alongside interest in the detection of other major plant cell wall components including hemicellulosic and pectic polysaccharides (Knox, 2008). The first approaches used cellulose-directed enzymes inactivated or otherwise attached to marker molecules such as gold particles (Berg *et al.*, 1988; Bonfante-Fasolo *et al.*, 1990; Seibert *et al.*, 1978). Although antibodies to pectins and hemicellulose became available, none were available to cellulose due to the difficulty of the generation and isolation of antibodies to insoluble and crystalline structures. However, the identification of naturally occurring noncatalytic cellulose-binding domains (CBDs) of carbohydrate-active enzymes led to techniques that used these protein domains, with specific cellulose recognition capabilities, as molecular probes for cellulose. CBDs became incorporated into a wider grouping of carbohydrate-binding modules (CBMs) that are defined as noncatalytic modules of carbohydrate-active enzymes (Boraston *et al.*, 2004). Recent work has shown that these can act *in situ* to potentiate the action of the associated catalytic modules against intact cell walls (Hervé *et al.*, 2010). It is increasingly apparent that CBMs are widely present in microbial carbohydrate-active enzymes and display a range of amino acid sequences, protein folding motifs, and also recognition specificities toward polysaccharides that include xylan and mannan, in addition to cellulose

(Boraston et al., 2004). The methods and principles discussed here for the use of CBMs for the in situ detection of cellulose are also of direct applicability to CBMs that bind to noncellulosic glycans.

1.1. Detection methods using CBM molecular probes

An early demonstration of the potential of cellulose-specific CBMs to act as molecular probes made use of a fluorescently labeled CBM to directly detect cellulose in the context of wood structures (Hildén et al., 2003). Subsequently, it was demonstrated that CBMs in the form of polyhistidine(His_6)-tagged proteins could be detected with secondary reagents and readily adapted to antibody-based protocols and used in the equivalent of immunochemistry assays and immunofluorescence procedures (Hervé et al., 2011; McCartney et al., 2004). A striking feature of these approaches was that CBMs directed to xylans (used at ~ 10 μg/ml in the primary stage) generated similar fluorescence signals to xylan-directed monoclonal antibodies (McCartney et al., 2006). These approaches opened up versatile and flexible opportunities to exploit the wide range of CBMs found in microbial enzymes. This includes cellulose-directed CBMs that can discriminate crystalline cellulose and amorphous cellulose. Such sets of CBM probes were found to bind to distinct regions of cell walls and thus revealed in situ distinctions in the fine structure of cellulose within a range of cell wall materials (Blake et al., 2006).

The focus of the discussion and protocols here will be fluorescence in situ detection. There are three broad approaches to the generation of CBM-induced fluorescence imaging to identify the location of specific cellulose ligands within plant materials: two direct methods and an indirect method. In the first direct method, a recombinant CBM is chemically coupled to a fluorophore such as FITC/Alexa Fluor using a coupling procedure provided by the manufacturers. A second direct method involves a genetic approach to generating a fusion protein in which the CBM is linked to green fluorescent protein (GFP). The CBM-fluorophore molecules are then used in straight-forward, single-step incubations with plant materials. The indirect method (reflecting widely used antibody techniques) involves the use of the His-tag of a CBM as the epitope for a series of secondary reagents such as anti-his-IgG coupled to a fluorophore. These methods generate similar results and each has particular advantages and potential disadvantages (Hervé et al., 2011). Directly coupled CBMs are small and their application is quick and, due to the stability and binding mechanism of CBMs, may not require secure buffering such as the use of phosphate-buffered saline (PBS), which is essential for effective antibody recognition. The potential disadvantages are that chemically coupled or GFP fusion CBMs may not always have the appropriate binding capacity/affinity due to fluorophore interference of the ligand binding site, and this should be tested on known samples in binding assays with a range of isolated celluloses and other appropriate glycans.

The indirect methodology allows for great flexibility in CBM use; it provides a direct comparison of a range of CBMs under the same conditions of detection and allows the use of the same batch of CBMs in range of assays with diverse secondary markers such as enzymes, fluorescence tags, or gold particles. The use of secondary or indeed tertiary reagents (see below) may also allow amplification of signal intensities. The disadvantages of the indirect methodologies include stricter requirements for buffers for secondary antibody reagents and longer time requirements for protocols—although this is not extensive.

The focus here will be on fluorescence methods as they are the most flexible and the most widely used. The principles, approaches, and procedures can be directly applied to gold detection of CBM binding using electron microscopy or enzyme-based systems that generate insoluble or soluble colored substrates.

2. Preparation of Plant Materials

There is potential to detect cellulose in all plant materials, and there are a wide variety of ways cell walls can be presented for CBM binding analysis. The most widely method used is the assessment of CBM binding to thin sections taken through an organ and adhered to glass slides. This approach by cutting across cell walls exposes all cell wall layers from outer cell walls and truly extracellular material to the plasma membrane—exposing all domains to be probed with CBMs as shown in Fig. 12.1A and B. However, approaches can be readily extended beyond this, and with careful manipulation and consideration, probes can be used to challenge the intact surfaces of plant organs such as roots (where no waxy cuticle prevents access to cell wall polysaccharides), the outer face of cell walls as seen on the surface of tip-growing cells or the surface of intact cotton fibers (Kljun et al., 2011), as shown in Fig. 12.1C–F. Cellulose in the context of cell wall architectures at the outer surface of intact cells released from coherent tissues by chemical or enzymatic means or, after protoplast removal, on the inner surface of cell walls (formerly adjacent to plasma membranes) can also be explored in this way. In principle, any plant material can be studied, and in some cases, whole mount procedures are directly applicable and represent very powerful approaches.

2.1. Fixation of plant materials for cellulose detection with CBMs

Although it is by no means essential, it is generally the case that plant material, when prepared for *in situ* analysis of glycans, is initially excised and fixed to stop all endogenous reactions. In principle, however, it is entirely possible to image probe binding to living cells such as the surface of tip-growing cells or *Arabidopsis* roots.

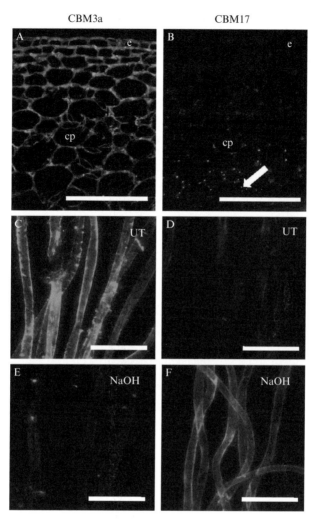

Figure 12.1 Examples of fluorescence micrographs showing CBM imaging of cellulose in plant materials using crystalline cellulose-directed CBM3a and amorphous cellulose-directed CBM17. (A, B) Transverse sections of tobacco stem showing epidermis (e) and cortical parenchyma (cp). CBM3a binds to all cell walls and CBM17 binds most abundantly to cell corners of intercellular spaces (arrow). See Blake et al. (2006) for details. (C–F) CBM binding to the surface of intact cotton fibers showing a comparison of binding to untreated fibers (UT) with fibers pretreated with 7 M NaOH. CBM3a binding declines in response to alkali treatment whereas that of CBM17 increases. See Kljun et al. (2011) for details. Bars = 100 μm.

Aldehyde fixatives are widely used to kill and preserve materials. It has been common practice that formaldehyde is used prior to fluorescence imaging procedures as glutaraldehyde (a more effective fixative and used for electron microscopy imaging) may induce cell wall autofluorescence.

However, the post-CBM labeling of sections with Toluidine Blue O (see Section 3.4.2 below) can quench cell wall fluorescence while maintaining CBM/antibody-induced fluorescence. This, however, would need to be checked in each case. Aldehyde fixatives directly cross-link proteins but not polysaccharides, and therefore some may remain soluble and can be lost from sections/plant materials during incubation steps.

1. Prepare fixative solutions. 4% Formaldehyde in PEM buffer (50 mM PIPES (piperazine-N,N'-bis(2-ethanesulfonic acid)), 5 mM EGTA, 5 mM MgSO$_4$; adjusted to pH 7.0 with KOH). Make a 12% or 16% (w/v) stock solution of paraformaldehyde in water by heating up to 70 °C and adding 1 M NaOH dropwise until the cloudy solution turns clear. Cool to room temperature (RT). A good alternative is 16% formaldehyde solution (Agar Scientific, Stansted, UK). Aliquots can be stored at -20 °C for up to 6 months. For glutaraldehyde, use a 2.5% (w/v) solution in PEM buffer by a 10-fold dilution of 25% (w/v) glutaraldehyde in H$_2$O (Grade 1; Sigma-Aldrich).
2. Excise small regions of plant material of interest and place immediately into an appropriate fixative solution at RT. For wax- and resin-embedding procedures, pieces of material (generally no thicker than 5 mm) are excised from plant organs and placed in fixative solution. Small samples such as *Arabidopsis* seeds or seedlings can be plunged directly into fixative.
3. Maintain in fixative for at least 2 h and for no more than overnight. Placing material under vacuum (to expel air) can help with infiltration of the fixative solutions.
4. Transfer to PEM buffer or PBS and store at 4 °C until use.

2.2. Embedding and sectioning of plant and cellulose-based materials

For best results, plant materials should be embedded in some material for support during the sectioning process. In some cases, dependent upon the properties of the material, hand-cut sections can be prepared with a razor blade and can be cut from a fresh stem directly into fixative solution or into water if the material is prefixed. We routinely use one or two methods involving intact materials, hand-cut sections, wax-embedded and hard resin-embedded materials—the choice being dependent upon the nature of the plant material being studied. This sequence from hand-cut sections (without embedding), through wax-embedding with wax removal to resin-embedded materials, will generally reflect decreasing maintenance of glycan recognition (i.e., antigenicity for antibodies) although increasing preservation and hence image quality.

2.2.1. Embedding protocol for low-melting point polyester wax

An effective wax for embedding plant materials for CBM labeling is a polyester wax known as Steedman's wax (Steedman, 1957). This is a low-melting point (35–37 °C) polyester wax with good sectioning properties. It is soluble in ethanol allowing easy removal prior to labeling with CBMs, resulting in good maintenance and presentation of ligands and substrates.

1. Wash fixed plant material twice for 10 min each time in PEM buffer or PBS. (Prepare $10\times$ stock with 1.37 M NaCl, 27 mM KCl, 100 mM Na_2HPO_4, and 18 mM KH_2PO_4 (pH 7.4) and autoclave before storage at RT. Use a 10-fold dilution of the stock. Alternatively, use prepared $10\times$ PBS (Severn Biotech, Kidderminster, UK.))
2. Dehydrate by incubation in an ascending ethanol series (30%, 50%, 70%, 90%, and 100% (v/v)) with 30-min incubation for each change at 4 °C.
3. Move samples to 37 °C for following steps.
4. Prepare wax. The low-melting point polyester wax is prepared from a mixture of polyethylene glycol 400 distearate and 1-hexadecanol (cetyl alcohol) (Sigma-Aldrich, Gillingham, UK). Melt 900 g of polyethylene glycol 400 distearate and 100 g 1-hexadecanol in a large beaker in an incubator at 65 °C. When melted, stir wax very thoroughly using a stirring bar. Pour the wax into a tray lined with aluminum foil (or 50-ml plastic conical tubes) and leave at RT to harden. Prepared wax can be stored at RT indefinitely.
5. Melt an appropriate amount of prepared wax at 37 °C and if using a water bath ensure that container is closed to keep out moisture.
6. Incubate in molten wax and ethanol (1:1, overnight) and then 100% wax (2 times 1 h).
7. Keep wax molten using a 37 °C oven.
8. Fill disposable base mold (15 × 15 × 5 mm; Electron Microscopy Sciences, Hatfield, USA) with molten wax and place sample in the wax. Take care to orientate the sample for optimal sectioning. Fill up with molten wax and when almost set, apply embedding cassette (Simport, Beloeil, Canada).
9. Leave at RT overnight to solidify. Can be used 12–24 h after embedding or can be stored in a cool, dry place for several months.

2.2.1.1. Sectioning wax-embedded material This procedure relates to the use of a HM325 rotary microtome (Microm, Bicester, UK), but it can be readily adapted to other microtomes.

1. Cut sections to a thickness of $\sim$10–12 μm to produce ribbons which are transferred to paper.
2. Sections are selected and placed on polysine-coated microscope slides (VWR, Lutterworth, UK) over a drop of water to promote spreading.

3. Allow slides to air dry.
4. Dewax and rehydrate sections by incubation of slides with 100% ethanol (3 times 10 min, 90% ethanol/water 10 min, 50% ethanol/water 10 min, water 10 min, water 90 min).
5. Slides are then air-dried and can be stored at RT indefinitely.

2.2.2. Embedding protocol for LR White resin

1. Wash plant material in buffer minus fixative for three times 10 min (or overnight at 4 °C).
2. Dehydrate using an ascending ethanol series (30%, 50%, 70%, 90%, and 100% (v/v)) with 30 min each change.
3. Infiltrate with LR White resin (hard grade, containing 0.5% of the catalyst benzoin methyl ether; Agar Scientific) at 4 °C by increasing from 10% resin in ethanol 1 h, to 20% 1 h, 30% 1 h, 50% 1 h, 70% 1 h, 90% 1 h, 100% resin overnight, then 8 h, then overnight.
4. Transfer to gelatin capsules (Agar Scientific) and ensure appropriate orientation of plant material. Fill to the top with resin and seal to exclude air.
5. Allow polymerization of resin either at 37 °C for 5 days or by action of UV light at −20°C for 24 h.

2.2.2.1. Sectioning of resin-embedded materials
These instructions relate to the use of a Reichert-Jung Ultracut Ultramicrotome.

1. Prepare glass knives or use a diamond knife.
2. For light microscopy, cut sections to a thickness of 1–2 µm into water.
3. Transfer sections to a drop of water on Vectabond-coated slides (Vector Laboratories, Peterborough, UK) or coated multitest eight-well glass slides (MP Biomedicals, Solon, USA) and allow them to dry on the slide in air.

3. CBM-LABELING PROCEDURES

This section will deal with CBM-labeling procedures for prepared sections of plant materials. These are essentially identical (unless otherwise stated) for both the sections prepared by wax-embedding and those prepared and maintained in hard resin.

The core protocol of this chapter is the use of the CBM to detect ligands in the complex plant materials. In some cases, a pretreatment may be useful to optimize recognition and detection and some posttreatment processes are useful for subsequent fluorescence imaging.

3.1. Section pretreatments prior to cellulose detection—Enzymatic pretreatments (pectic HG removal)

It is now known that the presence of one polysaccharide in a cell wall composite can block the access of binding proteins to other polysaccharides in the composite. To date, the capacity of pectic homogalacturonan (HG) to reduce or block access to cellulose or hemicellulose has been most extensively studied (Blake et al., 2006; Marcus et al., 2008; Hervé et al., 2009). Pectate lyase or polygalacturonase can effectively remove pectic HG and as both of these enzymes act on de-esterified pectic HG, a high pH pretreatment to remove pectic HG methylesters may optimize subsequent enzyme action and HG removal. Application of pectin-degrading enzymes to material not fixed to a glass slide is likely to result in separation of cells and may cause degradation of samples. Section pretreatments can also be extended for the enzymatic removal of other cell wall polysaccharides, and the enzymes, buffers, and conditions required will need to be determined accordingly.

1. Incubate section with a solution of 0.1 M sodium carbonate (pH 11.4) for 2 h. This high pH solution will result in pectic HG de-esterification.
2. Wash two times for 10 min with PBS.
3. Incubate with pectate lyase (e.g., from *Cellvibrio japonicus*; Megazyme, Bray, Ireland) at 10 μg/ml in CAPS (*N*-cyclohexyl-3-aminopropanesulfonic acid) buffer (50 mM CAPS, 2 mM CaCl$_2$, pH 10) for 2 h. Effectiveness of a particular enzyme can be determined using an antipectic HG probe.
4. Wash with three changes of PBS.
5. Sections are ready for CBM labeling as detailed in Sections 3.2 and 3.3.

3.2. *In situ* cellulose detection with directly coupled CBMs

In the case of a GFP tag, care must be taken to assess the binding ability of the fused CBM. Indeed, depending on the recombinant target, the folding of this bulky tag may reduce the recognition ability of the appended CBM by covering its binding site. It is also important to check that CBMs directly coupled to fluorophores such as FITC/Alexa Fluor retain the appropriate recognition capabilities after coupling. Always ensure that there is a no-CBM-control to assess the extent of cell wall autofluorescence present in the material.

1. Use a hydrophobic pen (Super PAP hydrophobic pen; Agar Scientific) for marking buffer incubation regions on glass slides around the sections to be labeled. The regions will contain the incubation solutions. This is not required when using eight-well glass slides.
2. Block the section to prevent nonspecific binding of the CBM by incubation of section with a buffer such as PBS with 3% (w/v) milk

protein (e.g., Marvel Milk) (PBS/MP) or 3% (w/v) bovine serum albumin (Sigma-Aldrich) in PBS (PBS/BSA).
3. Wash the section in PBS for 5 min.
4. Incubate the section with CBM-fluorophore in PBS/MP for 2 h. A good CBM concentration to start with is 10 μg/ml—but it is important to try a range of concentrations (say from 1 to 100 μg/ml) to determine which is the most effective concentration that generates the clearest signal.
5. Wash with at least three changes of PBS.
6. Use post-labeling treatments as appropriate (see Section 3.4).
7. Mount slide for microscopy using a small drop of fluorescence antifade reagent such as Citifluor glycerol/PBS AF1 (Agar Scientific).
8. Examine in a fluorescence microscope.

3.3. Indirect fluorescence imaging methods for *in situ* cellulose detection with CBMs

This procedure is for the indirect immunofluorescence labeling of sections of plant material. Always ensure that there is a no-CBM-control to assess the extent of cell wall autofluorescence, while this will also indicate if there is any nonspecific binding of the fluorophore-containing secondary reagents. In our hands, the tertiary antibody system described below is most effective—but reagents are available (such as antipolyhistidine coupled to Alexa Fluor 488 (Serotec, Kidlington, UK or Invitrogen)) to use a two-step system rather than the three-step system described here.

1. Use a hydrophobic pen (Super PAP hydrophobic pen; Agar Scientific) for marking buffer incubation regions on a glass slide around the sections to be labeled (if not using eight-well slides). The regions will contain the incubation solutions.
2. Block the section to prevent nonspecific binding of the CBM by incubation of section with a buffer such as PBS with 3% (w/v) milk protein (e.g., Marvel Milk) or 3% (w/v) bovine serum albumin (Sigma-Aldrich) in PBS.
3. Wash the section in PBS for 5 min.
4. *First stage*. Incubate the section with the his-tagged CBM of interest in PBS/MP for 2 h. A good CBM concentration to start with is 10 μg/ml of—but it is important to try a range of concentrations (say from 1 to 100 μg/ml) to determine which is the most effective concentration that generates the clearest signal.
5. Wash with at least three changes of PBS.
6. *Second stage*. Incubate with an antipolyhistidine antibody (the secondary antibody) such as mouse antipolyhistidine (Sigma-Aldrich) diluted in the range of 1000-fold in PBS/MP for at least 1 h at RT.
7. Wash with three changes of PBS.

8. *Third stage.* Incubate with the secondary antibody (e.g., anti-mouse coupled to FITC (Sigma-Aldrich) at 100-fold dilution) in MP/PBS for at least 1 h.
9. Wash with three changes of PBS.
10. Consider post-CBM-labeling treatments with either Calcofluor White or Toluidine Blue O or equivalent as appropriate (see Section 3.4).
11. Mount slides using a small drop of antifade reagent such as Citifluor AF1 mountant solution (Agar Scientific) and examine.

3.4. Post-CBM-labeling treatments prior to fluorescence imaging

After the final wash of the CBM-labeling procedures, there are a couple of options that can be useful for the subsequent imaging of the FITC or equivalent fluorescence.

3.4.1. Counterstaining with Calcofluor White for UV-induced fluorescence of cell walls

The first of these is to counterstain with Calcofluor White which can be useful to see all cell walls in plant materials and, with UV excitation (excitation 365 nm and emission $\sim$440 nm, can use DAPI filter set), can provide images alongside FITC or Alexa Fluor visible light-excited fluorescence images.

1. After final washing step, add 10-fold dilution into PBS of the Calcofluor White stock solution in water. The stock solution (0.25% (w/v) Calcofluor White/Fluorescent Brightener 28, Sigma) can be stored at 4 °C for many months.
2. Incubate for 5 min at RT.
3. Wash slide thoroughly with PBS and then mount in preparation for fluorescence microscopy imaging.

3.4.2. Post-CBM labeling with Toluidine Blue O to block cell wall autofluorescence

If the plant material under examination displays a high level of autofluorescence (e.g., to cell wall phenolics) in the same emission regions as the fluorophore that is being used, this can be specifically quenched by a post-CBM labeling with Toluidine Blue O. In this treatment, the TBO will bind strongly to the cell wall glycans and it cannot be used in conjunction with the use of Calcofluor White.

1. After the final washing step of the CBM-labeling procedure, cover the plant material in an aliquot of Toluidine Blue O solution (Sigma, 0.1% (w/v) in sodium phosphate buffer, pH 5.5).

2. Incubate for 5 min at RT.
3. Wash slide thoroughly with PBS and then mount in preparation for fluorescence microscopy imaging.

4. Summary

A range of methods and protocols are now in place for the detailed examination of diverse cellulose structures within the context of plant materials. Such methods are capable of revealing precise details of the occurrence of specific cellulose structures in relation to both architectural contexts within cell walls and also in relation to cell development. The use of cellulose-directed CBMs in conjunction with molecular probes for other cell wall glycans will be useful in revealing the nature of biomaterials, their diversity, and their functional properties. There is also considerable potential for the use of CBM in *in situ* detection procedures for the analysis of cellulose substructures and supramolecular structures in the context of the processing and industrial use of cellulose-based materials such as plant fibers.

ACKNOWLEDGMENT

I thank Susan Marcus for advice on technical aspects and Thomas Benians for providing the micrographs of cotton fibers.

REFERENCES

Anderson, C. T., Carroll, A., Akhmetova, L., and Somerville, C. (2010). Real-time imaging of cellulose reorientation during cell wall expansion in Arabidopsis roots. *Plant Physiol.* **152,** 787–796.

Berg, R. H., Erdos, G. W., Gritzali, M., and Brown, R. D., Jr. (1988). Enzyme-gold affinity labelling of cellulose. *J. Electron Microsc. Tech.* **8,** 371–379.

Blake, A. W., McCartney, L., Flint, J. E., Bolam, D. N., Boraston, A. B., Gilbert, H. J., and Knox, J. P. (2006). Understanding the biological rationale for the diversity of cellulose-directed carbohydrate-binding modules in prokaryotic enzymes. *J. Biol. Chem.* **281,** 29321–29329.

Bonfante-Fasolo, P., Vian, B., Perotto, S., Faccio, A., and Knox, J. P. (1990). Cellulose and pectin localization in roots of mycorrhizal *Allium porrum*: Labelling continuity between host cell wall and interfacial material. *Planta* **180,** 537–547.

Boraston, A. B., Bolam, D. N., Gilbert, H. J., and Davies, G. J. (2004). Carbohydrate-binding modules: Fine-tuning polysaccharide recognition. *Biochem. J.* **382,** 769–781.

Hervé, C., Rogowski, A., Gilbert, H. J., and Knox, J. P. (2009). Enzymatic treatments reveal differential capacities for xylan recognition and degradation in primary and secondary plant cell walls. *Plant J.* **58,** 413–422.

Hervé, C., Rogowski, A., Blake, A. W., Marcus, S. E., Gilbert, H. J., and Knox, J. P. (2010). Carbohydrate-binding modules promote the enzymatic deconstruction of intact plant cell walls by targeting and proximity effects. *Proc. Natl. Acad. Sci. USA* **107**, 15293–15298.

Hervé, C., Marcus, S. E., and Knox, J. P. (2011). Monoclonal antibodies, carbohydrate-binding modules, and the detection of polysaccharides in plant cell walls. *In* "The Plant Cell Wall: Methods and Protocols," (Z. A. Popper, ed.), Vol. 715, pp. 103–113. Springer/Humana Press, New York, USA.

Hildén, L., Daniel, G., and Johansson, G. (2003). Use of a fluorescence labelled carbohydrate-binding module from *Phanerochaete chrysosporium* Cel7D for studying wood cell wall ultrastructure. *Biotechnol. Lett.* **25**, 553–558.

Kljun, A., Benians, T. A. S., Goubet, F., Meulewaeter, F., Knox, J. P., and Blackburn, R. S. (2011). Comparative analysis of crystallinity changes in cellulose I polymers using ATR-FTIR, X-ray diffraction, and carbohydrate-binding module (CBM) probes. *Biomacromolecules* **12**, 4121–4126.

Knox, J. P. (2008). Revealing the structural and functional diversity of plant cell walls. *Curr. Opin. Plant Biol.* **11**, 308–313.

Marcus, S. E., Verhertbruggen, Y., Hervé, C., Ordaz-Ortiz, J. J., Farkas, V., Pedersen, H. L., Willats, W. G. T., and Knox, J. P. (2008). Pectic homogalacturonan masks abundant sets of xyloglucan epitopes in plant cell walls. *BMC Plant Biol.* **8**, 60.

McCartney, L., Gilbert, H. J., Bolam, D. N., Boraston, A. B., and Knox, J. P. (2004). Glycoside hydrolase carbohydrate-binding modules as molecular probes for the analysis of plant cell wall polymers. *Anal. Biochem.* **326**, 49–54.

McCartney, L., Blake, A. W., Flint, J., Bolam, D. N., Boraston, A. B., Gilbert, H. J., and Knox, J. P. (2006). Differential recognition of plant cell walls by microbial xylan-specific carbohydrate-binding modules. *Proc. Natl. Acad. Sci. USA* **103**, 4765–4770.

Seibert, G. R., Benjaminson, M. A., and Hoffman, H. (1978). A conjugate of cellulase with fluorescein isothiocyanate: A specific stain for cellulose. *Stain Technol.* **53**, 103–106.

Steedman, H. F. (1957). A new ribboning embedding medium for histology. *Nature* **179**, 1345.

Stone, B. (2005). Cellulose: structure and distribution. eLS. John Wiley & Sons Ltd, Chichester. http://www.els.net [doi: 10.1038/npg.els.0003892].

CHAPTER THIRTEEN

Interactions Between Family 3 Carbohydrate Binding Modules (CBMs) and Cellulosomal Linker Peptides

Oren Yaniv, Felix Frolow, Maly Levy-Assraf, Raphael Lamed, *and* Edward A. Bayer

Contents

1. Introduction	248
2. Enzyme-Linked Immunosorbent Assay (ELISA) to Check Possible Interactions Between CBM3 and Cellulosomal Linkers	250
2.1. Materials	250
2.2. Cloning procedure	250
2.3. Protein expression	251
2.4. Protein purification	252
2.5. ELISA protocol	252
2.6. Results	253
3. Isothermal Titration Calorimetry for Analysis of Interactions Between CBM3 and Cellulosomal Linkers	253
3.1. Principle	253
3.2. Linker analysis	254
3.3. Materials	256
3.4. Cloning procedure	256
3.5. Expression and purification of CBM3	256
3.6. ITC protocol	257
3.7. Results	257
3.8. Notes	258
Acknowledgments	258
References	258

Department of Molecular Microbiology and Biotechnology, The Daniella Rich Institute for Structural Biology, Tel Aviv University, Ramat Aviv, Israel

Abstract

Family 3 carbohydrate-binding modules (CBM3s) are among the most distinctive, diverse, and robust. CBM3s, which are numerous components of both free cellulases and cellulosomes, bind tightly to crystalline cellulose, and thus play a key role in cellulose degradation through their substrate targeting capacity. In addition to the accepted cellulose binding surface of the CBM3 molecule, a second type of conserved face (the "shallow groove") is retained on the opposite side of the molecule in all CBM3 subfamilies, irrespective of the loss or modification of the cellulose-binding function. The exact function of this highly conserved shallow groove is currently unknown.

The cellulosomal system contains many linker segments that interconnect the various modules in long polypeptides chains. These linkers are varied in length (5–700 residues). The long linkers are commonly composed of repeated sequences that are often rich in Ser, Pro, and Thr residues. The exact function of the linker segments in the cellulosomal system is currently unknown, although they likely play several roles.

In this chapter, we document the binding interaction between the conserved shallow-groove region of the CBM3s with selected cellulosomal linker segments, which may thus induce conformational changes in the quaternary structure of the cellulosome. These conformational changes would presumably promote changes in the overall arrangement of the cellulosomal enzymes, which would in turn serve to enhance cellulosome efficiency and degradation of recalcitrant polysaccharide substrates.

Here, we describe two different methods for determining the interactions between a model CBM3 and cellulosomal linker peptides.

1. Introduction

Cellulosomes are multienzyme complexes designed for efficient degradation of the plant cell wall in general, and cellulose in particular (Bayer et al., 2004, 2008; Doi and Kosugi, 2004; Fontes and Gilbert, 2010). A central feature of cellulosomes is the noncatalytic "scaffoldin" subunit(s) that plays a multiplicity of important functions, including the incorporation of the various enzymes into the complex by specific cohesin–dockerin interactions, the anchoring of the; complex onto the cell surface of the bacterium, and the targeting of the complex and the bacterial cell to the insoluble cellulosic substrate (Gerngross et al., 1993). Substrate targeting is thus mediated by a family 3 cellulose-specific carbohydrate-binding module (CBM3), a component of the primary scaffoldin (Poole et al., 1992). The CBM3s include two defined surfaces located on opposite sides of the molecule. A planar linear strip of aromatic and polar residues, which are proposed to interact with crystalline cellulose, dominates one of these faces. The other conserved residues are contained in a shallow groove of yet unknown function (Tormo et al., 1996).

Other types of CBM3 (known as CBM3c) are fused to the catalytic module of some family 9 glycoside hydrolases (GH9s), and serve to alter the enzymatic characteristics of the parent protein from a standard endoglucanase to a processive enzyme (Bayer et al., 1998; Irwin et al., 1998; Sakon et al., 1997). Additional types of CBM3s have also been described, which may play alternative roles, although they have yet to be properly characterized (Jindou et al., 2006). In these latter types of CBM3, the cellulose-binding surface appears to be altered.

Linker segments are sequences in a given protein that interconnect the various modules within the same protein. In the cellulosomal system, the linkers are of various lengths and are characteristic of the multimodular cellulosomal subunits, both structural and enzymatic. Some linkers are only five residues long, while others can consist of up to 700 amino acids. Medium-length linkers commonly comprise repeated sequences that are frequently rich in Ser, Pro, and Thr residues. The very long linkers can contain abundant repeats of specific amino acid motifs; the structural and functional consequence of such an arrangement is still an enigma.

The exact function of the linker segments in the cellulosomal system is currently unknown. It has been proven that Ser and Thr residues can serve as oligosaccharide attachment sites (Xu et al., 2003). Glycosylation may modulate cellulosome action in various ways, including protection against protease attack and contribute to protein–protein or protein–carbohydrate recognition (Schwarz, 2001). Moreover, the spatial flexibility of the scaffoldin, imparted in part by the linker segments, may allow precise positioning of the complexed enzymes according to the topography of the cellulosic substrate (Hammel et al., 2005)

In case of the cellulosomal system of *Clostridium thermocellum* (discovered by Lamed et al., 1983), the high incidence of Pro residues might suggest that the linkers form extended structures in the polypeptide chains, which would separate the various scaffoldin subunits. Such spatial separation of the cellulolytic subunits may facilitate intersubunit protein–protein interactions and promote dynamic synergistic interactions among the catalytic domains.

The first hypothesis regarding the function of the shallow groove of the CBM3s was suggested by Tormo et al. (1996), who determined the first 3D crystal structure of a CBM (i.e., CBM3 from the *C. thermocellum* CipA scaffoldin; PDB entry 1NBC). The CBM3 was crystallized together with seven linker residues at its C terminus, and it was found that the linker segment interacts with the conserved shallow groove of a neighboring molecule in the crystal. Furthermore, it has been shown (Shimon et al., 2000) that the shallow-groove region of the scaffoldin-borne CBM3 from a related bacterium, *C. cellullolyticum*, is capable of binding Pro residues from the linker segment of an adjacent symmetry-related molecule (Petkun et al., 2010). We therefore examined whether the shallow-groove region of CBM3s would interact with the Pro/Thr-containing cellulosomal linker segments of the scaffoldin, and thereby induce conformational changes in the quaternary structure of the cellulosome. The consequences of such conformational changes could generate differences in the overall enzyme arrangement within the cellulosome and might assist in enhancing the carbohydrate degradation profile of the cellulosome.

In this chapter we describe two experimental methods that demonstrate interactions between the CBM3's shallow-groove region and cellulosomal linker segments.

2. Enzyme-Linked Immunosorbent Assay (ELISA) to Check Possible Interactions Between CBM3 and Cellulosomal Linkers

2.1. Materials

Prepare all solutions using analytical grade reagents and ultrapure water (prepared by purifying deionized water to attain a sensitivity of 18 MΩ cm at 25 °C). Follow all waste disposal regulations carefully when disposing waste materials.

Washing buffer: TBS (1.37 M NaCl, 20 mM KCl, 250 mM Tris–HCl, pH 7.4) containing 10 mM CaCl$_2$ and 0.05% Tween 20
Blocking buffer: TBS, 10 mM CaCl$_2$, 2% BSA, 0.05% Tween 20
Anti-MBP monoclonal antibody (HRP-conjugated) (New England Biolabs, Ipswich, MA)
TMB (3,3′,5,5′-tetramethylbenzidine) reagent (Sigma Chemical Co., St. Louis, MO)
1 M H$_2$SO$_4$

The procedure in this chapter describes a method using protein modules and a linker segment from *C. thermocellum* (strain ATCC27405) CipA (accession number Q06851.2; Fig. 13.1A). The linkers are fused to a carrier protein, that is, the *Escherichia coli* 42.5-kDa maltose-binding protein (MBP).

Constructs used for this assay are as follows (Fig. 13.1B):

(i) Linker-free CBM3
(ii) MBP-Linker-Coh3, which includes the third cohesin (Coh3) in CipA and the scaffoldin-derived linker segment between the CBM3 and Coh3
(iii) MBP-Coh3, which includes the Coh3 without the linker segment

2.2. Cloning procedure

A DNA fragment encoding CBM3a (CipA residues 368–522) was amplified by PCR from *C. thermocellum* ATCC27405 genomic DNA using the following primers:

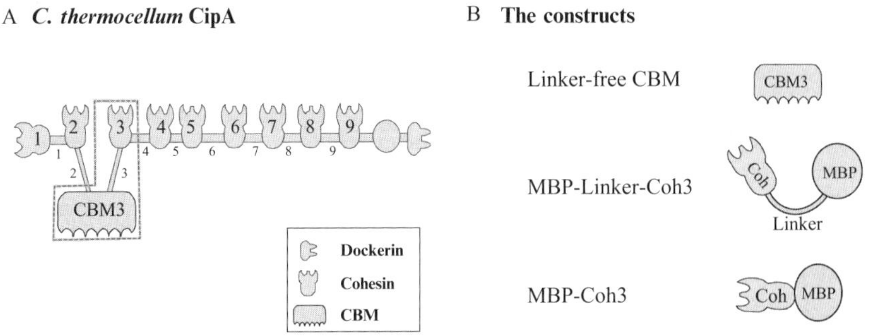

Figure 13.1 CBM3 and cellulosomal linkers of the scaffoldin subunit. (A) Schematic representation of *C. thermocellum* CipA scaffoldin subunit, showing the location of the CBM3, cohesin, and intervening linker (in dashed rectangle) used in this chapter. Cohesins and linkers are numbered according to their position on the scaffoldin. (B) Schematic representation of the protein constructs used for the assay described in this chapter (MBP, maltose-binding protein).

F-CATATGGCAAATACACCGGTATC
R-GGATCCTATTTACCCCATACAAGAACACC

PCR products were purified and cleaved with the restriction enzymes *Nde*I and *Bam*HI and inserted into the pET-28a(+) expression vector (Novagen, Madison, WI) together with an N-terminal hexahistidine tag.

DNA fragments encoding cohesin 3 with (Linker-Coh3) and without (Coh3) the adjacent linker (CipA residues 543–702 and residues 561–702, respectively) were amplified by PCR from *C. thermocellum* ATCC27405 genomic DNA, using the following primers:

F-GAATTCGAACCCGGTGGCAGTGTAG (Linker-Coh3)
F-GAATTCAATGCAATAAAGATTAAGGTGGACACAG (Coh3)
R-AAGCTTCTAATCTCCAACATTTACTCCACCGTC (for both clones)

PCR products were purified and cleaved with restriction enzymes *Eco*RI and *Hin*dIII and inserted into the pMAL-c2x expression vector (New England Biolabs), which contains the MBP gene immediately upstream of the *Eco*RI and *Hin*dIII restriction sites. The resultant plasmids encode, respectively, Linker-Coh3 and Coh3 fused to the C-terminal residue of MBP.

2.3. Protein expression

E. coli BL21 (DE3) harboring the recombinant expression vectors were aerated at 37 °C in 1 l Luria-Bertani (LB) growth medium, supplemented with 50 μg/ml kanamycin (for His-tagged protein) or 100 μg/ml ampicillin (for MBP-tagged protein). After 3–4 h, when the culture reached an A_{600} of 0.6, 0.5 mM isopropyl-β-D-1-thiogalactopyranoside (IPTG) was added to induce recombinant gene expression, and incubation at 37 °C was continued for an additional 12 h. Cells were then harvested by centrifugation (5000 × g for 10 min) at 4 °C and subsequently resuspended in 50 mM sodium phosphate buffer, pH 8.0, containing 300 mM NaCl at a ratio of 1 g wet pellet/4 ml buffer. A few micrograms of DNase powder were added before sonication. The suspension was kept in ice during sonication, after which it was centrifuged (20,000 × g at 4 °C for 30 min), and the supernatant fluid was collected.

2.4. Protein purification

The expressed His-tagged proteins were isolated by metal-chelate affinity chromatography using a Ni-IDA resin (Rimon Biotech, Israel). No attempts to remove the His tag were made. The expressed MBP-tagged proteins were isolated by amylose-resin affinity chromatography following the supplier's (New England Biolabs) specifications. Fast protein liquid chromatography (FPLC) was performed as a second purification step for all the expressed proteins, by using a Superdex 75pg column and an ÄKTA prime system (GE Healthcare).

2.5. ELISA protocol

1. Coat 96-well plates (Nunc-Immuno™ Plates MaxiSorp F96, NUNC A/S, Roskilde, Denmark) with 100 μl/well of 10 mg/ml Linker-free CBM3 in 100 mM Na_2CO_3, pH 9, overnight at 4 °C (Fig. 13.2A).
2. Wash three times using washing buffer to remove unabsorbed CBM (Fig. 13.2B).
3. Block wells using 100 μl/well blocking buffer for 1 h at 25 °C (Fig. 13.2C).
4. Add 100 μl/well of MBP-Linker-Coh3 (0.1 μg/ml to 0.1 mg/ml) in blocking buffer. Incubate for 1 h at 25 °C (Fig. 13.2D).
5. Wash three times using washing buffer to remove unbound protein (Fig. 13.2E).
6. Add 100 μl/well of monoclonal anti-MBP antibody conjugated to horseradish peroxidase (HRP), diluted 1:10,000 in blocking buffer. Incubate for 1 h at 25 °C (Fig. 13.2F).
7. Wash four times using wash buffer to remove unbound antibodies (Fig. 13.2G).
8. Detect using 100 μl/well TMB color reagent (add H_2O_2 according to manufacturer's instructions; Fig. 13.2H).
9. After 1–2 min (color should become blue), add 50 μl/well of 1 M H_2SO_4 to stop the reaction (color should become yellow).
10. Measure signal at $A_{450\ nm}$.

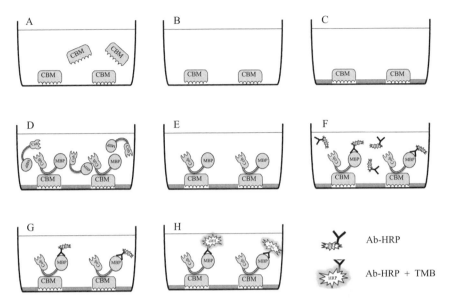

Figure 13.2 Schematic illustration of the ELISA assay. The schemes correspond to the description in Section 2.2—ELISA protocol. Schematic representations of proteins are displayed in Figure 13.1. Symbols for HRP-conjugated antibody (Ab-HRP), with and without the TMB reagent, are shown at the bottom right-hand side of the figure.

2.6. Results

Figure 13.3 represents the results of the described ELISA assay. The data showed that the CBM3 interacted with the linker-containing protein MBP-Linker-Coh3, while very little binding to the control protein (MBP-Coh3) which lacks the linker, was observed. These data indicate that the CBM3 recognizes the CipA-derived linker sequence.

3. Isothermal Titration Calorimetry for Analysis of Interactions Between CBM3 and Cellulosomal Linkers

3.1. Principle

Isothermal titration calorimetry (ITC) is a technique widely used for measuring the binding energetics of biological macromolecular interactions, including protein–ligand binding, protein–protein binding, etc. (Velazquez-Campoy et al., 2004).

The ITC instrument contains two identical coin-shaped cells, that is, the sample and reference cells, which are enclosed in an adiabatic shield (jacket;

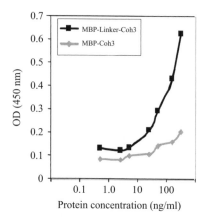

Figure 13.3 Results of ELISA assay, demonstrating a defined interaction between the CBM3 and the linker segment.

Fig. 13.4). It also contains an injection syringe that sits inside the sample cell and also acts as a stirring device. Each injection of the syringe solution triggers the binding reaction, and if such an interaction takes place, then a certain amount of complex is formed. Complex formation is accompanied by the release (exothermic reaction) or the absorption (endothermic reaction) of heat that causes a difference in temperature between the two cells. Then, a feedback system either lowers or raises the thermal power applied to compensate such temperature imbalance. After each injection, the system reaches equilibrium, and the temperature balance is restored. Therefore, the recorded signal shows a typical deflection pattern in the form of a peak (Velazquez-Campoy et al., 2004).

The procedure here describes a method to measure interactions between the CipA CBM3 and protein modules that contain a linker segment from the *C. thermocellum* (strain ATCC27405) cellulosomal scaffoldin (accession number Q06851.2; Fig. 13.1A, enumerated). The peaks that are formed represent the release of heat in the formation of a CBM3-linker complex.

3.2. Linker analysis

C. thermocellum CipA scaffoldin contains nine linker segments (between adjacent cohesins and CBM3) that are highly homologous and contain multiple Pro and Thr residues (Fig. 13.5A and B). Multiple sequence alignment was conducted using the ClustalW (Larkin et al., 2007) website (http://www.ebi.ac.uk/Tools/msa/clustalw2/), in order to find a consensus linker sequence that could be used as a probe in the ITC assays. A nine-residue sequence was chosen as it appears in five of the nine linkers (and partially in three additional linkers; Fig. 13.5A, highlighted).

Interactions Between CBM3s and Cellulosomal Linker Peptides   255

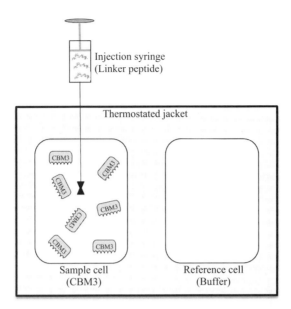

Figure 13.4 Schematic illustration of the ITC instrument. The two cells (sample and reference) are enclosed by a thermostated jacket, and an injection syringe also works as a stirring device. The CBM3 is placed into the sample cell and the linker peptide is loaded into the injection syringe.

A *C. thermocellum* CipA linker alignment

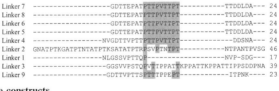

B **The constructs**

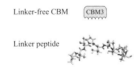

Figure 13.5 Linker analysis of *C. thermocellum* CipA scaffoldin. (A) Multiple sequence alignment of CipA linker segments (created using ClustalW). Linker numbering, corresponding to Fig. 13.1A, is indicated at the beginning of each line. Linker length is indicated at the end of each line. The consensus sequence that was chosen for calorimetric assays is highlighted. (B) Schematic representation of the constructs was used in the ITC assays.

3.3. Materials

ITC buffer: 250 mM Tris–HCl, pH 7.5, 0.25% triethylamine (TEA)
Microcrystalline cellulose, type 50 (Sigma)
1 M sodium bicarbonate
200 mM sodium bicarbonate
1 M Tris–HCl, pH 7.0

Constructs used for this assay (Fig. 13.4C):

(i) Linker-free CBM3 (CipA residues 361–520)
(ii) Consensus *C. thermocellum* CipA Linker peptide. The linker peptide does not contain any additional tag. It was synthesized chemically (Genemed Synthesis, Inc., TX, USA) and purified by HPLC (purity of more than 95%).

3.4. Cloning procedure

A DNA fragment encoding CBM3 (CipA residues 361–520) was amplified by PCR from *C. thermocellum* ATCC27405 genomic DNA using the following primers:

F-CCATGGGCAATTTGAAGGTTG
R-CTCGAGTCAACCCCATACAAGAACACCG

PCR products were purified and cleaved with restriction enzymes *Nco*I and *Xho*I and inserted into the pET-28a(+) expression vector (Novagen).

3.5. Expression and purification of CBM3

CBM3a was expressed according to the protocol described in Section 2.3 for His-tagged constructs, except that the growth medium was supplemented with 50 μg/ml kanamycin. All other steps were the same.

The expressed CBM3 was purified by cellulose affinity purification using microcrystalline cellulose. The recombinant cell-free extract was incubated with cellulose for 1 h with gentle stirring at 4 °C. The cellulose pellet was recovered by centrifugation and washed three times with 1 M sodium bicarbonate and three times with 200 mM sodium bicarbonate. The pH of the bicarbonate solution is about 8.5 and is a convenient high-ionic strength solution to minimize charge-mediated adsorption of extraneous proteins. The resulting cellulose pellet was resuspended in 1% (w/v) aqueous TEA solution to elute the CBM3 protein, and the cellulose powder was removed by centrifugation. The eluate was neutralized immediately to pH 7.5 using 1 M Tris–HCl buffer, pH 7.0. FPLC was performed as a second purification step, by using Superdex 75pg column and the ÄKTA prime system (GE Healthcare).

3.6. ITC protocol

1. Place 3 mg of purified CBM3 (total volume of 2 ml) in the sample cell and 0.27 mg linker peptide (total volume of 300 μl) in the injection syringe.
2. The assay parameters for the ITC are as follows:
   ITC instrument model: VP-ITC (MicroCal, LLC, Northhampton, MA, USA)
   Temperature: 25 °C
   Number of injections: 30(10 μl/injection)
   Time between injections: 240 s
   Stirring velocity: 372 rpm

3.7. Results

The results of the ITC analysis indicated a weak interaction between the *C. thermocellum* CipA CBM3 and the peptide corresponding to the consensus linker segment of the CipA scaffoldin (Fig. 13.6A). Clear peaks are evident, indicating heat release upon complex formation between the CBM and the linker. Note that the peaks are reduced upon successive injections. In this

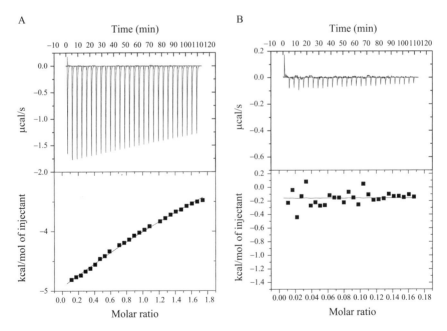

Figure 13.6 ITC results. (A) Interaction between the *C. thermocellum* CipA CBM3 and the consensus linker segment of the CipA scaffoldin. (B) Control experiment showing the CBM3 sample without the linker.

example, saturation was not achieved apparently due to the weak character of the interactions. In contrast, in the control experiments, in which buffer instead of linker is titrated into the cell containing the CBM3 sample the peaks are very small and uniform, indicating a lack of reaction. These results corroborate the ELISA data, shown in Fig. 13.3, thus supporting a defined interaction between the CBM3 and the scaffoldin-borne linker segments.

3.8. Notes

1. All protein samples must be dialyzed against the same buffer due to high sensitivity of measurements to variations in buffer conditions.
2. The lyophilized peptide can be resuspended in the dialysis buffer of the protein sample.
3. All solutions must be thoroughly degassed and devoid of any particulate matter.
4. Several control experiments should be conducted (O'Brien et al., 2001; Pierce et al., 1999). First, the buffer solution should be titrated into the cell containing the protein and the heat observed (Fig. 13.5B). Additionally, the reverse experiment where the peptide is titrated into the buffer solution containing no protein should also be performed if sufficient amounts of peptide are available. These control experiments determine the heats of dilution of the system and the results are deducted from the experimental data to obtain accurate values (Jacobs et al., 2004).

ACKNOWLEDGMENTS

This work was supported by the Israel Science Foundation (ISF; grant nos. 293/08 and 24/11). E.A.B. holds the Maynard I and Elaine Wishner Chair of Bio-Organic Chemistry at the Weizmann Institute of Science.
The authors are grateful to Mr. Oshik Segev (Tel Aviv University) for his assistance with the ITC data.

REFERENCES

Bayer, E. A., Morag, E., Lamed, R., Yaron, S., and Shoham, Y. (1998). Cellulosome structure: Four-pronged attack using biochemistry, molecular biology, crystallography and bioinformatics. In "Carbohydrases from Trichoderma reesei and Other Microorganisms," (M. Claeyssens, W. Nerinckx, and K. Piens, eds.), pp. 39–67. The Royal Society of Chemistry, London.

Bayer, E. A., Belaich, J. P., Shoham, Y., and Lamed, R. (2004). The cellulosomes: Multienzyme machines for degradation of plant cell wall polysaccharides. *Annu. Rev. Microbiol.* **58,** 521–554.

Bayer, E. A., Lamed, R., White, B. A., and Flint, H. J. (2008). From cellulosomes to cellulosomics. *Chem. Rec.* **8,** 364–377.

Doi, R. H., and Kosugi, A. (2004). Cellulosomes: Plant-cell-wall-degrading enzyme complexes. *Nat. Rev Microbiol.* **2,** 541–551.

Fontes, C. M., and Gilbert, H. J. (2010). Cellulosomes: Highly efficient nanomachines designed to deconstruct plant cell wall complex carbohydrates. *Annu. Rev. Microbiol.* **79,** 655–681.

Gerngross, U. T., Romaniec, M. P., Kobayashi, T., Huskisson, N. S., and Demain, A. L. (1993). Sequencing of a *Clostridium thermocellum* gene (*cipA*) encoding the cellulosomal S_L-protein reveals an unusual degree of internal homology. *Mol. Microbiol.* **8,** 325–334.

Hammel, M., Fierobe, H. P., Czjzek, M., Kurkal, V., Smith, J. C., Bayer, E. A., Finet, S., and Receveur-Brechot, V. (2005). Structural basis of cellulosome efficiency explored by small angle X-ray scattering. *J. Biol. Chem.* **280,** 38562–38568.

Irwin, D., Shin, D. H., Zhang, S., Barr, B. K., Sakon, J., Karplus, P. A., and Wilson, D. B. (1998). Roles of the catalytic domain and two cellulose binding domains of *Thermomonospora fusca* E4 in cellulose hydrolysis. *J. Bacteriol.* **180,** 1709–1714.

Jacobs, S. A., Fischle, W., and Khorasanizadeh, S. (2004). Assays for the determination of structure and dynamics of the interaction of the chromodomain with histone peptides. *Methods Enzymol.* **376,** 131–148.

Jindou, S., Xu, Q., Kenig, R., Shulman, M., Shoham, Y., Bayer, E. A., and Lamed, R. (2006). Novel architecture of family-9 glycoside hydrolases identified in cellulosomal enzymes of *Acetivibrio cellulolyticus* and *Clostridium thermocellum*. *FEMS Microbiol. Lett.* **254,** 308–316.

Lamed, R., Setter, E., and Bayer, E. A. (1983). Characterization of a cellulose-binding, cellulase-containing complex in *Clostridium thermocellum*. *J. Bacteriol.* **156,** 828–836.

Larkin, M. A., Blackshields, G., Brown, N. P., Chenna, R., McGettigan, P. A., McWilliam, H., Valentin, F., Wallace, I. M., Wilm, A., Lopez, R., Thompson, J. D., Gibson, T. J., *et al*. (2007). Clustal W and Clustal X version 2.0. *Bioinformatics* **23,** 2947–2948.

O'Brien, R., Chowdhry, B. Z., and Ladbury, J. (2001). Isothermal Titration Calorimetry of Biomolecules. Oxford University Press, London.

Petkun, S., Jindou, S., Shimon, L. J., Rosenheck, S., Bayer, E. A., Lamed, R., and Frolow, F. (2010). Structure of a family 3b' carbohydrate-binding module from the Cel9V glycoside hydrolase from *Clostridium thermocellum*: Structural diversity and implications for carbohydrate binding. *Acta Crystallogr. D* **66,** 33–43.

Pierce, M. M., Raman, C. S., and Nall, B. T. (1999). Isothermal titration calorimetry of protein-protein interactions. *Methods* **19,** 213–221.

Poole, D. M., Morag, E., Lamed, R., Bayer, E. A., Hazlewood, G. P., and Gilbert, H. J. (1992). Identification of the cellulose-binding domain of the cellulosome subunit S1 from *Clostridium thermocellum* YS. *FEMS Microbiol. Lett.* **78,** 181–186.

Sakon, J., Irwin, D., Wilson, D. B., and Karplus, P. A. (1997). Structure and mechanism of endo/exocellulase E4 from *Thermomonospora fusca*. *Nat. Struct. Biol.* **4,** 810–818.

Schwarz, W. H. (2001). The cellulosome and cellulose degradation by anaerobic bacteria. *Appl. Microbiol. Biotechnol.* **56,** 634–649.

Shimon, L. J., Pages, S., Belaich, A., Belaich, J. P., Bayer, E. A., Lamed, R., Shoham, Y., and Frolow, F. (2000). Structure of a family IIIa scaffoldin CBD from the cellulosome of *Clostridium cellulolyticum* at 2.2 A resolution. *Acta Crystallogr. D* **56,** 1560–1568.

Tormo, J., Lamed, R., Chirino, A. J., Morag, E., Bayer, E. A., Shoham, Y., and Steitz, T. A. (1996). Crystal structure of a bacterial family-III cellulose-binding domain: A general mechanism for attachment to cellulose. *EMBO J.* **15,** 5739–5751.

Velazquez-Campoy, A., Ohtaka, H., Nezami, A., Muzammil, S., and Freire, E. (2004). Isothermal titration calorimetry. *Curr. Protoc. Cell Biol.* **23,** 17.8.1–17.8.24. Chapter 17, Unit 17.8.

Xu, Q., Gao, W., Ding, S. Y., Kenig, R., Shoham, Y., Bayer, E. A., and Lamed, R. (2003). The cellulosome system of *Acetivibrio cellulolyticus* includes a novel type of adaptor protein and a cell surface anchoring protein. *J. Bacteriol.* **185,** 4548–4557.

CHAPTER FOURTEEN

Approaches for Improving Thermostability Characteristics in Cellulases

Michael Anbar *and* Edward A. Bayer

Contents

1. Construction of Cellulase Libraries	262
1.1. Random gene libraries by error-prone PCR	262
1.2. Alignment-guided consensus libraries	263
2. Screening Endoglucanases for Enhanced Thermostability	265
2.1. CMC-plate assay for enzyme activity	265
2.2. Thermostability screen of endoglucanase in 96-plate format	266
3. Screening β-Glucosidase for Enhanced Thermostability	267
3.1. Magenta-Glc-plate assay for activity combined with thermostability	267
3.2. Thermostability screen in 96-plate format	269
4. *In Vitro* Recombination Between Best Thermostable Mutants	269
References	271

Abstract

Many efforts have been invested to reduce the cost of biofuel production to substitute renewable sources of energy for fossil-based fuels. At the forefront of these efforts are the initiatives to convert plant-derived cellulosic material to biofuels. Although significant improvements have been achieved recently in cellulase engineering in both efficiency and cost reduction, complete degradation of lignocellulosic material still requires very long periods of time and high enzyme loads.

Thermostable cellulases offer many advantages in the bioconversion process, which include increase in specific activity, higher levels of stability, inhibition of microbial growth, increase in mass transfer rate due to lower fluid viscosity, and greater flexibility in the bioprocess. Besides rational design methods, which require deep understanding of protein structure–function relationship, two of the major methods for improvement in specific cellulase properties are directed evolution and knowledge-based library design based on multiple sequence

Department of Biological Chemistry, The Weizmann Institute of Science, Rehovot, Israel

alignments. In this chapter, we provide protocols for constructing and screening of improved thermostable cellulases. Modifications of these protocols may also be used for screening for other improved properties of cellulases such as pH tolerance, high salt, and more.

1. Construction of Cellulase Libraries

Due to the difficulties of and peculiarities inherent the various types of cellulosic substrates, the establishment of truly improved mutant forms of a given cellulase is a monumental task. In this context, one may improve the hydrolysis characteristics of a given cellulase (e.g., endoglucanase) on a given model substrate (Zhang *et al.*, 2006), but the true effect of its "improvement" is how the modified enzyme works in concert with other enzymes in a cellulase cocktail to efficiently degrade recalcitrant cellulosic substrates. There are no current high-throughput assays that would facilitate mutant selection on this basis. The story is different when one wants to improve the thermostability characteristics of a given enzyme, which is the topic of this chapter.

1.1. Random gene libraries by error-prone PCR

One of the most established methods for generating random amino acid substitutions for directed evolution of proteins is performing a PCR reaction under conditions that reduce the fidelity of the polymerase. Recently, error-prone PCR protocols use an enzyme blend, formulated to provide mutation with minimal mutational bias like Stratagen's GeneMorph II (Agilent Technologies Inc., Santa Clara, CA). The mutation load could be determined by altering either the number of PCR cycles or amount of DNA template. The number of mutations per kilobase should be decided experimentally for each enzyme by determining the activity profiles for several expressed libraries of varying error rate and the best library can be chosen for further screening. Usually, an error rate resulting in a library with about 30–50% of inactive enzyme mutants is desirable, but this value depends on the specific enzyme used. The endoglucanase Cel8A of *Clostridium thermocellum* was chosen as a model for thermostability enhancement. Interestingly, although this enzyme is already considered thermostable in its native state, it can be further improved without the expense of reduced cellulose hydrolysis.

Error-prone PCR reaction: The *cel8A* gene cloned into pET28a was prepared with concentrations varying from 1 to 50 ng of the template DNA. The primers used were T7 promoter primer and T7 terminator primer found upstream and downstream of the gene, respectively. Thermal cycling parameters were 95 °C for 3 min and 28 cycles of 95 °C for 40 s, 56 °C for 40 s, and 72 °C for 1.2 min. The resulting PCR product was treated with *Dpn*I to destroy the template plasmid, purified from agarose gel, and then used as a template for a nested PCR using high-fidelity PCR reaction and primers 5′-AAGAAGGAGATATA**CCATGG**-3′ and 5′-GTGGTGGTGGTG**CTCGAG**-3′ (boldface letters indicate *Nco*I and *Xho*I restriction sites, respectively). The amplified product was purified and ligated into the expression vector pET28a through the restriction sites and transformed into *Escherichia coli* DH5a cells, yielding about 10^5 transformants. Plasmid DNA was then extracted to obtain the library for subsequent transformations and screening. The diversity of the genes in the resulting library was examined by sequence analysis of 10 randomly selected colonies. Based on the subsequent screening results, mutation rates of 2–7 mutations per kilobase were chosen.

1.2. Alignment-guided consensus libraries

A complementary approach to random mutagenesis takes advantage of the large number of available protein sequences. This semi-rational "consensus approach" is a well-established strategy to improve thermostability and has been used successfully on both enzymatic and nonenzymatic proteins (Amin *et al.*, 2004; Lehmann *et al.*, 2002; Polizzi *et al.*, 2006). The approach is based on the substitution of specific amino acids in a particular protein with the most prevalent amino acid present at these positions among the homologous family members. Several studies have shown that applying this strategy frequently leads to the creation of thermostable protein variants. A possible explanation for the stabilizing effect of consensus mutations based on analogy with statistical thermodynamics has been proposed by Steipe *et al.* (1994). However, it was also shown that only some of the consensus mutations contribute to protein stability, while others destabilize the protein or are neutral (Lehmann *et al.*, 2002). It is therefore suggested that a selection should be made in order to include only the beneficial mutations.

Selecting consensus mutations: The amino acid sequence of Cel8A was used to identify 18 homologous sequences in GenBank. The sequences were selected based on amino acid identity values of 30–60%. The sequences were aligned using the ClustalW algorithm and either consensus positions or most abundant positions were determined. Overall the Cel8A gene differed in eight positions from the consensus sequence (Fig. 14.1).

Constructing the DNA library: Plasmid pET28aCel8A (Anbar *et al.*, 2010) was used to construct the library. The *cel8A* gene was amplified and digested with DNaseI. The resulting 50–200 bp fragments were assembled by PCR in the presence of an equimolar mixture of eight oligonucleotides encoding

```
 34                                         GVPFNTKYPYGPTSIAD    50
                                            ----------------

 51   NQSEVTAMLKAEWEDWKSKRITSNGAGGYKRVQRDASTNYDTVSEGMGYG        100
      --------------------------------------------------

101   LLLAVCFNEQALFDDLYRYVKSHFNGNGLMHWHIDANNNVTSHDGGDGAA        150
      M-----------G-------------------------------------

151   TDADEDIALALIFADKLWGSSGAINYGQEARTLINNLYNHCVEHGSYVLK        200
      ---------------------------------I----------------

201   PGDRWGGSSVTNPSYFAPAWYKVYAQYTGDTRWNQVADKCYQIVEEVKKY        250
      ----------------------F--F------------------------

251   NNGTGLVPDWCTASGTPASGQSYDYKYDATRYGWRTAVDYSWFGDQRAKA        300
      -------------------------------P---------Y--------

301   NCDMLTKFFARDGAKGIVDGYTIQGSKISNNHNASFIGPVAAASMTGYDL        350
      ---------------------L----------------------------

351   NFAKELYRETVAVKDSEYYGYYGNSLRLLTLLYITGNFPNPLSDLSGQPT        400
      --------------------------------------------------

401   PPSNPTPSLPPQVVYGDVNGDGNVNSTDLTMLKRYLLKSVTNINREAADV        450
      --------------------------------------------------

451   NRDGAINSSDMTILKRYLIKSIPHLPYLEHHHHHH 485
      ----------------------------------
```

Figure 14.1 Protein sequence of Cel8A from *C. thermocellum*. Consensus mutations are indicated.

the consensus mutations (total of 10 pmol) as previously described (Herman and Tawfik, 2007). The resulting library was cloned into the pET28 vector using the *Nco*I and *Xho*I restriction sites.

Primers used (the codon changes are underlined):

Primer name	Sequence 5′ → 3′
L101M	GGTATGGGATACGGAATGCTTTTGGCGGTTTGC
D115G	CAGGCTTTGTTTGACGGTTTATACCGTTACGTA
L187I	ACATTGATAAACAATATTTACAACCATTGTGTA
Y224F	GCATGGTACAAAGTGTTTGCTCAATATACAGG
Y227F (comp)	CTTGTGTCTCCTGTAAATTGAGCATACACTTTG
G283P	GATGCTACACGTTACCCGTGGAGAACTGCCGTG
F293Y	GTGGACTATTCATGGTATGGTGACCAGAGAGC
I323L	GTTGACGGATACACACTGCAAGGTTCAAAAATTAG

2. SCREENING ENDOGLUCANASES FOR ENHANCED THERMOSTABILITY

Reagents

E. coli strain appropriate for protein expression, for example, BL21 (DE3)

DNA library containing the endoglucanase (in this case *C. thermocellum* Cel8A) subcloned into an expression plasmid, for example, pET28a (Novagen, Madison, WI) and the same expression plasmid containing the wild-type Cel8A as control

LB media and LB agar plates with appropriate antibiotics, for example, kanamycin (50 μg/ml)

Isopropyl-β-D-1-thiogalactopyranoside (IPTG)

Replica plating apparatus (base and collar) and velveteen pads

CMC (carboxymethyl cellulose)-soft agar, containing 0.3% (w/v) CMC (medium viscosity; Sigma, St. Louis, MO), 0.7% agar, and 0.2 mM IPTG as an inducer, in 25 mM sodium acetate, pH 6.0

0.25% (w/v) Congo red (Acros Organics Geel, Belgium)

1 M NaCl

PopCulture reagent (Novagen) or other lysis buffer

50 mM sodium acetate, pH 6.0

High-throughput screening equipment: 96-deep-well plates (Nunc, Roskilde, Denmark), 96-well PCR plates (Axygen, Union City, CA, USA), 96-well microplates (Nunc) for screening, plate reader, etc.

CMC solution: 1% (w/v) CMC, 10 mM $CaCl_2$, and 50 mM sodium acetate, pH 6.0

3,5-Dinitrosalicylic acid (DNS) reagent: The procedure of Wood and Bhat was followed (Wood and Bhat, 1988):
DNS: 40 g (Sigma)
Phenol: 8 g
Sodium sulfite (Na_2SO_3): 2 g
Na-K tartarate (Rochelle salts): 800 g

All components are dissolved in 2 l of 2% (w/v) NaOH solution and then diluted to 4 l with distilled water. The container is covered with aluminum foil.

2.1. CMC-plate assay for enzyme activity

In order to screen for thermostable mutant Cel8A enzymes, a two-step screening strategy was employed. For the initial screening step, identification of active endoglucanase enzymes was performed using a medium-throughput screening procedure employing double layered CMC-containing plates. This assay takes advantage of small leakage of enzyme from the individual colonies, which express Cel8A to facilitate the detection of active enzymes.

Procedures

1. Transform *E. coli* cells with the plasmids containing the library of Cel8A variants and wild-type Cel8A as control.
2. Spread onto large LB plates (140 mm) containing kanamycin and incubate overnight at 37 °C (to facilitate the screening process, aiming for about 400 CFU per plate).
3. Replicate the plates using velveteen pads.
4. Melt CMC-agar by boiling. Then, overlay about 20 ml of the CMC-agar, cooled to about 45 °C, onto the LB plates.
5. Leave the plates at room temperature until the CMC-agar solidifies.
6. Incubate for 30 min at 37 °C to induce enzyme expression and 2 h at 60 °C to facilitate enzyme activity. It should be noted that for some enzymes IPTG at lower concentration could be added to the transformation plates and incubated overnight instead of adding it to the soft-agar overlay.
7. Stain the plates for 10 min with fresh 0.25% Congo red solution and destain with 1 M NaCl.
8. Positive clones form large halos around the colonies, which is indicative of their endoglucanase activity. Select the positive clones from replica plates for further study (Fig. 14.2).

2.2. Thermostability screen of endoglucanase in 96-plate format

Approximately 9000 colonies were screened for enzyme activity using the double layered CMC-containing plates of which approximately 30% showed clearing zones similar in size to the parental clones. Out of these, 2880 clones were selected for subsequent screening. In the second screen, the retention of endoglucanase activity was measured after heating the samples for 15 min at 82 °C. After incubation at this elevated temperature, the wild-type Cel8A enzyme retained only about 41% of its activity. Thus, any clone that showed a significant level of improvement after the heat treatment was considered a candidate for a thermostable enzyme.

Procedures

1. Culture the colonies that expressed active enzyme overnight at 37 °C in 96-deep-well plates containing 0.4 ml LB medium, 50 µg/ml kanamycin, and 0.1 mM IPTG.

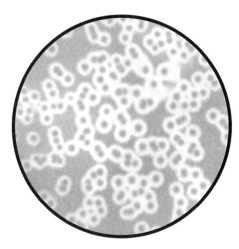

Figure 14.2 Halos formed around colonies of *E. coli* cells expressing an active Cel8A enzyme.

2. Lyse 0.2 ml of the culture using PopCulture reagent and dilute 10 μl of the lysate into 390 μl of 50 mM sodium acetate, pH 6.0.
3. Incubate 100 μl of diluted lysate at 82 °C for 15 min. Incubate 20 μl of the diluted lysate before and after the heat treatment and determine the residual activity in 60 μl of CMC solution for 2 h at 65 °C in 96-well PCR plates, and cool on ice for 1 min.
4. Determine the amount of reducing sugars released by the enzyme colorimetrically using DNS reagent (Miller, 1959) by adding 120 μl of DNS to the reaction mixture and boiling for 5–10 min. The absorbance is measured at 540 nm.

The clones that demonstrate higher than wild-type activity are picked from the 96-well stock and streaked again on an agar plate.

To avoid false positive clones, a few colonies are picked and reassayed in a 96-well format on CMC as a substrate.

3. Screening β-Glucosidase for Enhanced Thermostability

3.1. Magenta-Glc-plate assay for activity combined with thermostability

Screening of β-glucosidase enzymes for increased thermostability can be performed in a manner very similar to that described in the above sections for the endoglucanase. The thermostability of the enzyme β-glucosidase A

(BglA) from *C. thermocellum* was thus increased using both random mutagenesis and alignment-guided mutagenesis. The following protocol was used to screen and select the thermostable variants.

Reagents

E. coli strain appropriate for protein expression, for example, BL21 (DE3)
DNA library containing the *bglA* subcloned into an expression plasmid, for example, pET28a (Novagen) and a plasmid containing the wild-type *bglA* as control
LB media and LB agar plates with 50 μg/ml kanamycin and 4 μM IPTG, if required
IPTG
Replica plating apparatus (base and collar) and velveteen pads
Soft Magenta-Glc agar containing: 0.7% agar, 0.02% bromo-6-chloro-3-indolyl-β-D-glucopyranoside (Magenta-Glc) (Sigma) in 25 mM Na-citrate, pH 6.1
PopCulture reagent (Novagen) or other lysis buffer
50 mM Na-citrate, pH 6.1
High-throughput screening equipment: 96-deep-well plates (Nunc), 96-well PCR plates (Axygen), 96-well microplates (Nunc) for screening, plate reader, etc.
p-Nitrophenyl glucopyranoside (PNPG) (Sigma), 1 mM in 50 mM Na-citrate, pH 6.1
1 M Na-carbonate, pH 9.5

Procedures

1. Transform *E. coli* cells derived from the library of *bglA* variants and wild-type *bglA* as control.
2. Spread onto large LB plates (140 mm) containing kanamycin and incubate overnight at 37 °C (to facilitate the screening process, aiming for about 400 CFU per plate).
3. Replicate the plates using velveteen pads.
4. Perform a heat challenge of the variant plates in a preheated oven for 50 min at 70 °C. This temperature was selected, so that the wild-type *bglA* is inactivated. Different temperatures and incubation times should be selected for each enzyme.
5. Cool plates for 40 min at 4 °C and then 20 min at room temperature.
6. Prepare soft Magenta-Glc agar by boiling.
7. Overlay about 20 ml of the soft-agar, cooled to about 45 °C, onto the LB plates. Leave the plates at room temperature until the soft-agar solidifies.
8. Incubate 1–2 h at 60 °C to facilitate enzyme activity. Colonies that remain red after the heat treatment potentially contain thermostable

enzymes. Two control plates containing wild-type *bglA* should be prepared each time and used with and without the heat challenge.

3.2. Thermostability screen in 96-plate format

1. Culture the colonies that expressed active enzyme overnight at 37 °C in 96-deep-well plates containing 0.4 ml LB medium, containing kanamycin and IPTG.
2. Lyse 0.2 ml of the culture using the PopCulture reagent.
3. Dilute 10 μl of the lysate into 290 μl of sodium citrate buffer.
4. Incubate 100 μl of diluted lysate at 66 °C for 90 min and cool on ice for 1 min.
5. Incubate 15 μl of the diluted lysate before and after the heat treatment and determine the residual activity in 85 μl of PNPG solution at 60 °C for 45 min in 96-well PCR plates.
6. Stop the reaction with an equal volume of 1 M Na-carbonate solution.
7. Read OD at 415 nm in spectrophotometer. The clones that demonstrate higher than wild-type activity are selected from the 96-well stock and streaked again onto an agar plate.

To avoid false positive clones, a few colonies are selected and reassayed on the 96-well format plates using PNPG as a substrate.

4. *IN VITRO* RECOMBINATION BETWEEN BEST THERMOSTABLE MUTANTS

When screening for thermostable variants, it is common to identify several independent thermostable mutants that contain different mutations. It is possible to produce a better thermostable mutant by randomly combining mutations from the different variants. This type of DNA shuffling is usually done by partial digestion with DNaseI and reassembling the fragments using cycles of denaturation, annealing, and extension by a high-fidelity polymerase. Following the reassembly reaction, PCR amplification is performed with nested primers to generate full-length chimeras that contain different combinations of mutations. These PCR fragments are then cloned into expression vectors and the thermostability of the resulted enzyme is determined.

Reagents

Two sets of primers in a nested configuration
Pfu polymerase and 10 × buffer (Stratagene, La Jolla, CA)
dNTP (2.5 mM each)

Template DNA (gene of interest cloned in plasmid)
Gel and PCR purification kits (RBC Bioscience, Xindian City, Taiwan) or equivalent
1 M Tris–HCl, pH 7.5
100 mM MnCl$_2$
Thermal cycler
Agarose gel containing ethidium bromide
0.25 M EDTA
DNaseI (Sigma), 0.05 U/ml

Procedure

Obtaining the DNA for shuffling:
1. Amplify the genes for shuffling; use the outer primers (e.g., T7 promoter primer and T7 terminator primer found upstream and downstream of the gene, respectively, in pET28a).
2. Purify the genes using a PCR purification kit—usually 20–25 μg of DNA is required.
3. Prepare the solution for DNA fragmentation by mixing 120 μl of DNA solution containing at least 20 μg DNA, 15 μl of 1 M Tris–HCl, and 15 μl of 100 mM MnCl$_2$. Bring these solutions to 20 °C in a thermocycler.
4. At the same time, prepare 2% agarose gel containing ethidium bromide, heat the heating block to 90 °C, and prepare PCR tubes with 5 μl of 0.25 M EDTA solution. Prepare DNaseI solution 0.05 U/ml in DNase buffer. Prepare 1, 2, and 3 μl of DNaseI in Eppendorf tubes. Add 50 μl of the reaction mix to the DNaseI mix and incubate at 20 °C for 0.5–3 min. To stop the reaction, transfer the solution to the tube containing EDTA in the heating block, mix thoroughly, and incubate for 10 min.
5. Spin down the tubes after the incubation, run the sample on 2% agarose gel and excise the desired size range under UV light (usually an average size of 70–150 bp is desirable) (Fig 14.3), and purify.

Reassembly of digested fragments:
1. Prepare 42 μl of purified fragment DNA, 5 μl of 10× *Pfu* buffer, 2 μl of dNTP solution, and 1 μl of *Pfu*. Assembly PCR program: 94 °C 2 min, 35 cycles of 94 °C 30 s, 65 °C 1:30 min, 62 °C 1:30 min, 59 °C 1:30 min, 56 °C 1:30 min, 53 °C 1:30 min, 50 °C 1:30 min, 47 °C 1:30 min, 45 °C 1:30 min, 41 °C 1:30 min, 72 °C 45 s, end with 72 °C 7 min.
2. Perform PCR using nested primers with 1 μl of unpurified assembly mixture at three concentrations (1/1, 1/10, and 1/100).
3. Run the reactions on an agarose gel and purify the band at the molecular weight of the parent gene and clone into an expression vector.

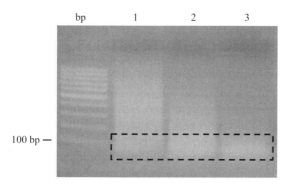

Figure 14.3 DNase-treated DNA library at different incubation times: (1) 0.5 min, (2) 1 min, (3) 2 min. The DNA fragments between 70 and 150 bp were excised.

REFERENCES

Amin, N., Roberge, M., Estabrook, M., Basler, J., Chin, R., Gualfetti, P., Liu, A., Wong, S. B., Rashid, M. H., Graycar, T., Babé, L., and Schellenberger, V. (2004). Construction of stabilized proteins by combinatorial consensus mutagenesis. *Protein Eng. Des. Sel.* **17,** 787–793.

Anbar, M., Anbar, M., Lamed, R., and Bayer, E. A. (2010). Thermostability enhancement of *Clostridium thermocellum* cellulosomal endoglucanase Cel8A by a single glycine substitution. *ChemCatChem.* **2,** 997–1003.

Herman, A., and Tawfik, D. S. (2007). Incorporating synthetic oligonucleotides via gene reassembly (ISOR): A versatile tool for generating targeted libraries. *Protein Eng. Des. Sel.* **20,** 219–226.

Lehmann, M., Loch, C., Middendorf, A., Studer, D., Lassen, S. F., Pasamontes, L., van Loon, A. P., and Wyss, M. (2002). The consensus concept for thermostability engineering of proteins: Further proof of concept. *Protein Eng.* **15,** 403–411.

Miller, G. L. (1959). Use of dinitrosalicylic acid reagent for determination of reducing sugar. *Anal. Biochem.* **31,** 426–428.

Polizzi, K. M., Chaparro-Riggers, J. F., Vazquez-Figueroa, E., and Bommarius, A. S. (2006). Structure-guided consensus approach to create a more thermostable penicillin G acylase. *Biotechnol. J.* **1,** 531–536.

Steipe, B., Schiller, B., Plückthun, A., and Steinbacher, S. (1994). Sequence statistics reliably predict stabilizing mutations in a protein domain. *J. Mol. Biol.* **240,** 188–192.

Wood, T., and Bhat, K. (1988). Methods for measuring cellulases activities. *Methods Enzymol.* **160,** 87–112.

Zhang, Y. H., Himmel, M. E., and Mielenz, J. R. (2006). Outlook for cellulase improvement: Screening and selection strategies. *Biotechnol. Adv.* **24,** 452–481.

CHAPTER FIFTEEN

Thermophilic Glycosynthases for Oligosaccharides Synthesis

Beatrice Cobucci-Ponzano, Giuseppe Perugino, Andrea Strazzulli, Mosè Rossi, *and* Marco Moracci

Contents

1. Introduction	274
2. Thermophilic β-Glycosynthases from GH1	276
2.1. Cloning, mutagenesis, and expression and purification of β-glycosynthases from GH1	276
2.2. Enzymatic assays and chemical rescue of β-glycosynthases from GH1	278
2.3. Transglycosylation reactions catalyzed by β-glycosynthases from GH1	279
2.4. Characterization of the β-glycosynthases from GH1	280
3. Thermophilic α-L-Fucosyntases from GH29	285
3.1. Cloning, mutagenesis, expression and purification	285
3.2. Enzymatic assays and chemical rescue of thermophilic α-fucosynthases	287
3.3. Synthesis and analysis of fucosylated oligosaccharides	287
3.4. Characterization of the thermophilic α-fucosynthases	288
4. Thermophilic α-D-Galactosynthase from GH36	292
4.1. Mutagenesis, expression and purification	292
4.2. Assays and chemical rescue of TmGalA wild type and D327G mutant	293
4.3. Synthesis and analysis of galactosylated oligosaccharides	294
4.4. Characterization of the thermophilic α-galactosynthase	295
5. Summary	295
Acknowledgments	296
References	296

Abstract

Glycosynthases are engineered glycoside hydrolases that in suitable reaction conditions promote the synthesis of oligosaccharides with exquisite stereoselectivity and enhanced regioselectivity, if compared to traditional chemical

Institute of Protein Biochemistry, National Research Council, Naples, Italy methods. This approach was demonstrated to be successful in a number of cases including β-glycosynthases acting at the termini or within an oligosaccharide chain (*exo-* and *endo-*glycosynthases, respectively) and, more recently, α-glycosynthases. This led to the production of a vast repertoire of products that include poly- and oligosaccharides, glycoconjugates, and glycopeptides. These molecules can be used as ligands of glycoside hydrolases, for the characterization of therapeutic enzymes, and as leads of drugs for the pharmaceutical industry. In this panorama, hyperthermophilic organisms, which thrive at temperatures as high as 80 °C, which usually impede the growth of other living forms, have been used in the development of interesting novel glycosynthases. In fact, the extreme stability of these catalysts to extremes of pH and high concentrations of organics has allowed the exploration of novel reaction conditions, revealing new avenues for enzyme-catalyzed oligosaccharide synthesis.

1. INTRODUCTION

In 1998, Withers and colleagues showed, for the first time, that it was possible to mutate the nucleophile of an *exo*-acting β-glucosidase, which followed a *retaining* reaction mechanism (Fig. 15.1A), and to restore the activity (transglycosylation activity) of the enzyme in the presence of a donor glycoside with a good leaving group and an anomeric configuration opposite to that of the original substrate (α-glycosyl-fluoride) (Fig. 15.1B) (Mackenzie *et al.*, 1998). In these conditions, the mutant could not hydrolyze the synthesized β-oligosaccharides, which accumulated in quantitative yields: these findings led to the invention of the *glycosynthases*.

In the same year, the group of Planas described for the first time an *endo*-acting *glycosynthase* (Malet and Planas, 1998), while Moracci and coworkers applied this methodology to a hyperthermophilic β-glycosidase from the archaeon *Sulfolobus solfataricus* (Moracci *et al.*, 1998). In this latter case, however, the approach involved, for the first time, β-glycoside donors with good leaving groups and external nucleophiles, such as sodium formate, which, by mimicking the functional role of the original nucleophile, formed meta-stable intermediates in the active site allowing the transfer of the glycoside moiety of the donor into the acceptor (Fig. 15.1C). Again, oligosaccharides were produced in high yields because the poor leaving ability of the sugar products prevented further hydrolysis.

Since then, the glycosynthase approach has been extended to several β-glycosidases from a variety of glycoside hydrolase (GH) families (for GH family classification, see Cantarel *et al.*, 2009), confirming the validity of the approach (Bojarová and Kren, 2009; Cobucci-Ponzano *et al.*, 2011a; Hancock *et al.*, 2006; Hrmova *et al.*, 2002; Jakeman and Withers, 2002; Mayer *et al.*, 2001; Perugino *et al.*, 2004). On the other hand, not many α-glycosidases have been engineered into α-glycosynthases, thus, an alternative approach,

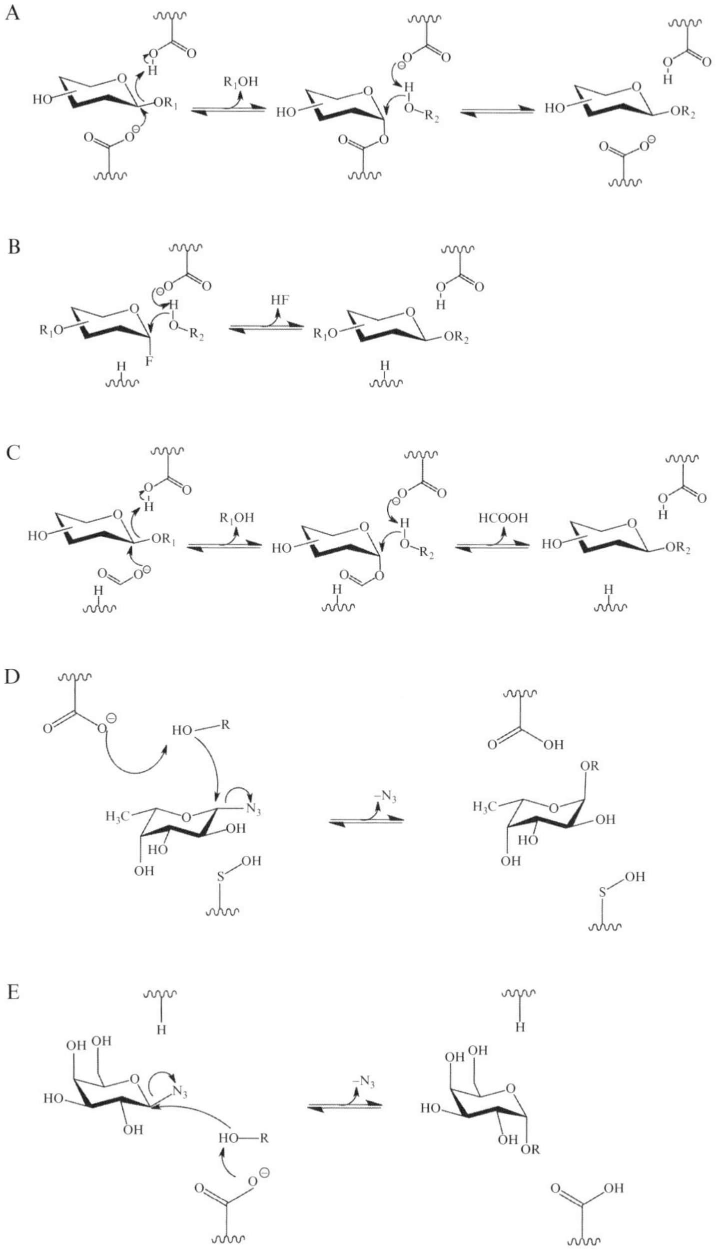

Thermophilic Glycosynthases

Figure 15.1 Reaction mechanisms of glycoside hydrolases and glycosynthases. Reaction mechanism of *retaining exo*-β-glycoside hydrolases (A); mesophilic β-glycosynthases (B); thermophilic β-glycosynthases; here, in the first step, the activated substrate shows a group with good leaving ability (R_1) and the intact carboxylic acid allows the formation of a

again using a hyperthermophilic GH, but exploiting β-glycosyl-azide donors, allowed the production in high yields of α-oligosaccharides (Fig. 15.1D and E) (Cobucci-Ponzano et al., 2009, 2011b).

The unusual stability to protein denaturants (extremes of pH, ionic strength, detergents, temperature, etc.) of thermophilic GHs, working at the interface between chemistry and biology, makes these catalysts excellent tools to explore novel reaction conditions for oligosaccharide synthesis (Zhu and Schmidt, 2009).

2. Thermophilic β-Glycosynthases from GH1

The β-glycosidase from the archaeon *S. solfataricus* (*Ss*βgly) (Aguilar et al., 1997; Ausili et al., 2004; Moracci et al., 1995; Moracci et al., 2001; Nucci et al., 1993; Pouwels et al., 2000; Trincone et al., 1994) was the first thermophilic glycosidase converted into a glycosynthase (Moracci et al., 1998). Later, the approach was extended to two other β-glycosynthases from the hyperthermophilic archaea, *Thermosphaera aggregans* and *Pyrococcus furiosus* (Taβ-gly and CelB, respectively) (Perugino et al., 2003).

2.1. Cloning, mutagenesis, and expression and purification of β-glycosynthases from GH1

2.1.1. Cloning and mutagenesis of *Ss*βgly

The *lacS* gene encoding for the *retaining S. solfataricus* β-glycosidase (E.C. 3.2.1.x), showing wide substrate specificity on β-gluco, -galacto, and -fucosides, was cloned in the pGEX-2TK (GE Healthcare) plasmid, in frame and downstream to the gene encoding the glutathione *S*-transferase (GST) tag, as reported previously (Moracci et al., 1996). Glu387, identified as the catalytic nucleophile of the enzyme (Moracci et al., 1996), was removed by site-directed mutagenesis, following the method of Mikaelian and Sergeant (1992), and replaced with a glycine (*Ss*E387G mutant) (Moracci et al., 1998).

2.1.2. Expression and purification of *Ss*βgly wild type and mutants

1. To express the wild type and *Ss*E387G mutant of *Ss*βgly as GST-fusion proteins, transform *Escherichia coli* RB791 with the appropriate plasmid.

---

glycoside-formate intermediate. In the second step, the glycoside is transferred to an acceptor (R_2) and the resulting transglycosylation product is not cleaved, even in the presence of the functional acid/base and external nucleophile, because the disaccharide obtained has a poor leaving group for the *exo*-acting enzyme (C); thermophilic α-fucosynthases (D); thermophilic α-galactosynthases (E). R = sugar; R_1 = leaving group; R_2 = H: hydrolysis; R_2 = sugar or alcohol: transglycosylation.

2. Grow 2 l of E. coli RB791 until 0.6 OD$_{600}$ and induce the culture with 1 mM isopropyl-β-D-1-thiogalactopyranoside (IPTG) overnight.
3. Resuspend the cells 1:3 g/ml in PBS buffer (20 mM sodium phosphate, pH 7.3; 150 mM NaCl) supplemented with 1% Triton X-100.
4. Lyse the cells using a French pressure cell and centrifuge for 30 min at 17,000 × g.
5. Load the cell-free crude extracts on a 2.5 ml of Glutathione-Sepharose 4B matrix (GE Healthcare) equilibrated with PBS buffer supplemented with 1% Triton X-100.
6. Wash the column with 10 column volumes of PBS buffer.
7. Elute the fusion protein with 0.5 M Tris–HCl, pH 8.0; 15 mM glutathione.
8. Perform a GST assay according to the manufacturer's instruction (GE Healthcare) to identify the active fractions.
9. To separate Ssβgly from the GST-tag, pool the active fractions and incubate overnight at 4 °C with thrombin protease (10 units/mg of protein).
10. Test the thrombin cleavage by SDS-PAGE inspection.
11. After thrombin cleavage, load proteins in a FPLC Superdex 200 size-exclusion column (GE Healthcare), equilibrated in PBS buffer supplemented with 1 mM DTT.
12. Pool and concentrate the fractions. SDS-PAGE showed that the purified Ssβgly was >95% pure.
13. Dialyze against 50 mM sodium phosphate buffer, pH 6.5, supplemented with 50% glycerol, and store at −20 °C. Alternatively, store the protein sample at 4 °C in the same buffer without glycerol.

The purification procedures were performed with affinity matrices dedicated only to the purification of the specific enzyme to exclude cross-contamination (Moracci et al., 1998).

2.1.3. Cloning and mutagenesis of Taβgly and CelB

The gene encoding for Taβgly was PCR amplified from genomic DNA by using specific oligonucleotides, and cloned in the pET9d vector (Novagen), in which Taβgly is under the control of the IPTG-inducible T7 RNA polymerase promoter as reported (Perugino et al., 2003). The celB gene was previously cloned into the same plasmid (Kaper et al., 2000). Primary structure comparison with other members of GH1 family, and 3D inspection of Taβgly (PDB ID 1QVB; Chi et al., 1999) and CelB (preliminary 3D data from 3.3 Å resolution crystal and a model obtained by using 3D structure of Ssβgly as template) led to the identification of their catalytic nucleophiles, namely glutamic acid residues in position 386 and 372, respectively (Chi et al., 1999; Kaper et al., 2002). PfE372A mutant was prepared as already described (Kaper et al., 2002), while the TaE386G mutant was prepared by site-directed mutagenesis following the method of Higuchi et al. (1988).

2.1.4. Expression and purification of Taβgly and CelB wild type and mutants

1. To express and purify wild-type and mutant enzymes, transform E. coli BL21(DE3)RIL cells and follow the steps 2–4 described in Section 2.1.2 above.
2. Incubate the free cell crude extracts for 30 min at 75 °C, for Taβgly and TaE386G, and 80 °C for CelB and PfE372A, respectively, to eliminate the heat-labile proteins of the bacterial host.
3. Centrifuge for 30 min at 30,000 × g.
4. Add ammonium sulfate to the supernatant to a final concentration of 1 M, and load the sample onto a Phenyl-Sepharose 26/10 FPLC column (GE Healthcare) equilibrated with 50 mM sodium phosphate buffer, pH 6.8, and 1 M ammonium sulfate.
5. Apply appropriate column volumes (usually three) of a linear gradient of this buffer against water for protein elution.
6. Wild-type and mutant enzymes elute in a single fraction in water (Perugino et al., 2003).
7. Add phosphate buffer pH 6.5 to obtain a final concentration of 50 mM and store as previously described (step 13, Section 2.1.2).

All the enzymes were >95% pure by SDS-PAGE analysis (Moracci et al., 1998; Perugino et al., 2003). Before the enzyme characterization of each mutant, perform the following procedure to eliminate any trace of wild-type contaminating activity in the preparations:

1. Incubate the enzymes overnight at 50 °C in phosphate buffer 50 mM (pH 6.5) with a 100-fold molar excess of 2,4-dinitrophenyl-2-deoxy-2-fluoro-β-D-glucoside (2,4DNP-2F-β-D-Glc) (Moracci et al., 1998; Perugino et al., 2003). This mechanism-based inhibitor forms a stable covalent bond with the glutamic acid that acts as the catalytic nucleophile.

2.2. Enzymatic assays and chemical rescue of β-glycosynthases from GH1

A common procedure to test if a carboxylic acid in a GH acts as the nucleophile of the reaction is to delete this amino acid by mutagenesis and assess whether there is a substantial reduction in activity (Shallom et al., 2002). Then, the function of the catalytic nucleophile can be unequivocally assigned by the so-called *chemical rescue* of the enzymatic activity of the mutant followed by the chemical analysis of the products. Small ions, such as sodium azide or sodium formate, can chemically restore the activity working as external nucleophiles and lead to the production of glycosides with anomeric configuration inverted compared to the substrate (Ly and Withers, 1999; McIntosh et al., 1996; Viladot et al., 1998, 2001). Chemical rescue is also a useful tool to measure the activity of nucleophile mutants of GHs.

1. To test the activity of the wild-type enzymes, prepare in duplicate the mixtures containing Buffer A (50 mM sodium phosphate buffer, pH 6.5) and 5 mM of 2-nitrophenyl-β-D-glucoside (2NP-β-D-Glc) or 4-nitrophenyl-β-D-glucoside (4NP-β-D-Glc) (commercially available substrates were purchased from Sigma-Aldrich).
2. Preheat the mixtures in quartz cuvettes by incubating for 2 min at 65 °C. One is used as the blank mixture to correct for spontaneous hydrolysis of the substrates.
3. Start the reaction by adding enzyme (ranging from 0.5 to 20.0 μg) to the assay mixture and keep the temperature constant during all activity measurements (Perugino et al., 2003).
4. Follow the change in absorbance at 405 nm with a Peltier thermally controlled spectrophotometer (Varian Cary 100 Scan).
5. To rescue the activity of SsE387G, TaE386G, and PfE372A mutants, add to Buffer A an external nucleophile (sodium azide or sodium formate) in the range of 0–4 M. Alternatively, perform the assay in Buffer B (50 mM sodium formate, pH 3.0 or 4.0). The values of the molar extinction coefficient at 405 nm (ε_M) for all the conditions used are listed in Table 15.1 and used for the calculation of the enzymatic activity (Moracci et al., 1998).
6. One unit of enzyme activity is the amount of enzyme catalyzing the hydrolysis of 1 μmol of substrate in 1 min at the conditions described.

2.3. Transglycosylation reactions catalyzed by β-glycosynthases from GH1

To determine the transglycosylation ability of the mutants prepare and analyze the reactions as follows:

1. In 1 ml final volume, add Buffer A (with sodium formate) or B, add a fixed amount of enzyme and 2NP-β-D-Glc as donor/acceptor substrate (1.29 μg of protein/μmol of substrate), and incubate at 65 °C for 3 h. Prepare appropriate blanks to correct the values for spontaneous hydrolysis of 2NP-β-D-Glc.
2. To test the transglycosylation activity of acceptor substrates that are different from the donor, prepare the reactions as described above and add the acceptors at appropriate molar ratios with the donor (usually 1:2 and 1:3 donor:acceptor).

Table 15.1 Molar extinction coefficients measured at 405 nm

Leaving group	Buffer[a]	pH	T (°C)	Nucleophile (M)	ε_M (M^{-1} cm^{-1})
2-Nitrophenol	A	6.5	65	–	1711
	B	3.0	65	Formate (0.05)	403
4-Nitrophenol	A	6.5	65	–	9340
	A	6.5	65	Azide (1–2)	8000
	C	6.0	60	–	5300
	C	6.0	60	Azide (1)	4600[b]
	B	3.0	65	Formate (0.05)	74
2-Chloro-4-nitrophenol	A	6.5	65	–	17,000
	A	6.5	65	Azide (2)	17,000
	C	6.0	60	–	11,200
	C	6.0	60	Azide (1)	15,600

[a] For the composition of buffer, see the text for details (Cobucci-Ponzano et al., 2009, 2011b; Moracci et al., 1998; Perugino et al., 2003).
[b] Measured at 400 nm.

3. Determine the complete consumption of the substrate analyzing aliquots of the reactions by the glucose oxidase-peroxidase enzymatic assay (GOD/POD), according to the manufacturer's instruction (Roche). This method measures the amount of free glucose in the reaction as a direct indication of the efficiency of the transfer reaction to an acceptor (Perugino et al., 2003).
4. To monitor the formation of oligosaccharides products promoted by SsE387G, TaE386G, and PfE372A mutants, spot aliquots of each reaction on analytical thin layer chromatography (TLC) silica gel plate (Merck). Load appropriate carbohydrate standards as reference.
5. Develop the TLC in EtOAc:MeOH:H$_2$O (70:20:10 by volume) and visualize the products under UV light (in case of glycosides containing nitrophenol groups) and by charring with α-naphthol reagent.
6. Identify the products of transglycosylation by NMR analysis as reported (Moracci et al., 1998; Perugino et al., 2003; Trincone et al., 2000).

2.4. Characterization of the β-glycosynthases from GH1

2.4.1. Kinetic characterization of the β-glycosynthases from GH1

The substitution of the glutamic acid residue acting as nucleophile with non-nucleophile small residues, such as glycine and alanine, led to the complete inactivation of the β-glycosidase mutants. Their reactivation (described in Fig. 15.1C), in terms of hydrolytic activity, depended on the concentrations of sodium formate in the standard reaction, as showed in the steady-state kinetic parameters listed in Table 15.2 (Moracci et al., 1998; Perugino et al., 2003). Interestingly, all three hyperthermophilic enzyme mutants showed a remarkable reactivation, using a commercially available substrate containing a 2-nitrophenol a group, which displays limited leaving ability. The reactivation rates of SsE387G and TaE387G mutants were comparable, whereas PfE372A mutant showed the lowest catalytic efficiency: this could be explained by the presence, in this mutant, of an alanine, which, in comparison to the glycine present in the other two mutants, offered limited access to the formate ion (Perugino et al., 2003).

2.4.2. Optimization of the hydrolytic activity of mutants at low pH

Kinetic parameters of mutants were measured by using 2NP-β-D-Glc substrate in diluted acid buffered sodium formate (Buffer B). As reported in Table 15.2, the chemical rescue by acidic sodium formate buffer greatly enhanced the catalytic activity of all the mutants; in particular, SsE387G and TaE387G reached levels comparable to the respective wild-type enzymes. Remarkably, these results were obtained by using concentrations of sodium formate as low as 50 mM, whereas, at neutral pH, the chemical rescue of the activity was lower even at molar concentration of external nucleophiles (Table 15.2) (Moracci et al., 1998; Perugino et al., 2003). No activity could be obtained for all the mutants in the presence of sodium citrate and sodium acetate buffers 50 mM, pH 3.0, indicating that the nature of the external nucleophile plays a critical role in the reactivation of these enzymes (Perugino et al., 2003).

2.4.3. Transglycosylation efficiency

To test if the hyperthermophilic glycosynthases could be exploited for the production of oligosaccharides, 50 mM sodium formate buffer (pH 4.0) was used in all synthetic reactions, as long incubations at pH 3.0 inactivated the mutants. SsE387G was the most efficient catalyst converting, after 0.5 h, almost all (91%) the 2NP-β-D-Glc substrate into products (Perugino et al., 2003). In addition, only ca. 8% of the free monosaccharide was present in the reaction mixture (Table 15.3). It is worth noting that this glycosynthetic approach at acidic conditions enabled a 50% reduction in enzyme used and a 50% decrease in reaction time (2.64 μg/μmol of substrate and 1 h incubation time), compared to the conditions reported previously (4 M sodium formate in Buffer A) (Moracci et al., 1998). The TaE386G mutant converted all the substrate into products after 1 h but a larger amount of free glucose was found in the reaction mixture (30% of the products). In the case of PfE372A, the efficiency was much lower; the substrate was still present after 7 h (Perugino et al., 2003).

The TLC analysis of aliquots of the reaction mixtures at various time intervals revealed that SsE387G produced a mixture of oligosaccharides,

Table 15.2 Kinetic constants of the β-glycosynthases from GH1

Enzyme	Buffer[a]	pH	Formate (M)	k_{cat} (s^{-1})	K_M (mM)	k_{cat}/K_M (s^{-1} mM^{-1})	% reactivation[b]	$^{mut}k_{cat}/^{wt}k_{cat}$
SsβgIy	A	6.5	—	538.0 ± 11.0	1.01 ± 0.24	533.0	—	—
SsE387G	A	6.5	—	ND[c]	ND	—	—	—
	A	6.5	2.00	53.0 ± 1.2	1.17 ± 0.12	45.0	8.44	0.10
	B	3.0	0.05	901.4 ± 32.9	16.4 ± 1.6	55.0	10.32	1.67
TaβgIy	A	6.5	—	309.0 ± 18.5	0.20 ± 0.06	1575.7	—	—
TaE386G	A	6.5	—	ND	ND	—	—	—
	A	6.5	2.00	72.9 ± 2.4	1.01 ± 0.10	72.9	4.60	0.23
	B	3.0	0.05	970.0 ± 39.9	6.6 ± 0.6	147.4	9.35	3.14
CelB	A	6.5	—	1796.9 ± 90.2	0.28 ± 0.06	6480.0	—	—
PfE372A	A	6.5	—	ND	ND	—	—	—
	A	6.5	2.00	3.9 ± 0.1	1.49 ± 0.12	2.6	0.04	0.002
	B	3.0	0.05	47.6 ± 0.9	4.3 ± 0.2	10.9	0.17	0.026

[a] For the composition of buffer, see the text for the details.
[b] % reactivation is calculated by taking as 100% the k_{cat}/K_M of the wild-type β-glycosidases assayed at standard conditions.
[c] ND: not detectable (Perugino et al., 2003).

Table 15.3 Comparison of the transglycosylation products of two hyperthermophilic glycosynthases

Enzymes	Substrate (donor and acceptor)	Products	Relative ratios (%)
SsE387G/TaE386G	2NP-β-D-Glc	β-Glc-(1–3)-β-Glc-2NP	40/50
		β-Glc-(1–6)-β-Glc-2NP	9/10
		β-Glc-(1–4)-β-Glc-2NP	1/24
		β-Glc-(1–6) | β-Glc-(1–3)-β-Glc-2NP	28/ND[a]
		β-Glc-(1–6) | β-Glc-(1–3)-β-Glc-2NP	6/ND
		β-Glc-(1–3)-β-Glc-2NP β-Glc-(1–3)-β-Glc-(1–6) | β-Glc-(1–3)-β-Glc-2NP	5/ND

[a] Not detected. Data from Moracci et al. (1998), Perugino et al. (2003), and Trincone et al. (2000).

with a high degree of polymerization (tri- and tetrasaccharides), similar to those obtained at neutral pH in 4 M sodium formate, while TaE386G did not generate products larger than the trisaccharide. The products were similar for the two glycosynthases, although the composition of regioisomers was different (Table 15.3) (Perugino et al., 2003). NMR analysis revealed the nature of disaccharide derivatives as glucose dimers β-O-linked to 2-nitrophenol with β-(1–3), β-(1–4), and β-(1–6) bonds. These products of autocondensation of 2NP-β-D-Glc accumulated during the reaction, becoming new acceptors for further cycles of transglycosylation reactions leading to linear and branched trisaccharides, which, in turn, became acceptors for the synthesis of tetrasaccharides (Moracci et al., 1998). As expected, the low transglycosylation efficiency of PfE372A resulted in the production of mainly β-(1–3) disaccharides at all pH values tested (Perugino et al., 2003).

SsE387G and TaE386G were also tested using 2NP-β-D-Glc donor and different acceptors, as reported in Table 15.4. Remarkably, both were able to recognize different glycosides in the acceptor binding site. In particular, SsE387G was also able to recognize α-glycosides acceptors, a result in some ways unexpected, considering the specificity of the parental wild-type enzyme for β-glycosides. The high yields observed with the α-anomers might result from their inability to compete with the donor in the active site (Trincone et al., 2005). Finally, the nature of the aglycon, other than its lipophilic character, is decisive: SsE387G produced β-(1–3)-disaccharides using acceptors containing the 2-nitrophenyl group, while β-(1–6)-disaccharides are predominant with 4-nitrophenyl-β-D-glucoside acceptors (Trincone et al., 2005).

Table 15.4 Synthetic products of β-glycosynthases

Reaction	Enzyme	Donor	Acceptor (m.e.)[a]	Products	Relative ratios of transfucosylation products (%)
I	SsE387G	2NP-β-D-Glc	4MUGlc (2)	β-Glc-(1–3)-β-Glc-4MU	60
			4MUαGlc (3)	β-Glc-(1–6)-α-Glc-4MU	33
			4NPαGlc (3)	β-Glc-(1–6)-α- Glc-4NP	31
			4NPαGal (3)	β-Glc-(1–6)-α-Gal-4NP	Quantitative yields
			4NPαMan (3)	β-Glc-(1–6)-α-Man-4NP	72
			2NPαGlcNAc (3)	β-Glc-(1–6)-α-GlcNAc-2NP	75
			2NPLam (3)	Branched trisaccharides	53.3
			4NPCell (2)	Branched trisaccharides	9
			4MUGlc (2)	β-Glc-(1–3)-β-Glc-4MU	24
				β-Glc-(1–6)-β-Glc-4MU	36
		2NP-β-D-Glc	Glc-(1–3)-β-Glc-4MU (2)	β-Glc-(1–4)	16
				|	
				Glc-(1–3)-β-Glc-β-4MU	
				β-Glc-(1–3)	
				|	
				Glc-(1–3)-β-Glc-β-4MU	
II	TaE386G	2NP-β-D-Gal	β-Xyl-4P	β-Gal-(1–3)-β-Xyl-4P	34

[a] Molar excess of acceptor with respect to the donor.

Abbreviations: 4MUGlc, 4-methyl umbelliferyl-β-glucopyranoside; 4MUαGlc, 4-methyl umbelliferyl-α-glucopyranoside; 2NPLam, 2-nitrophenyl-β-D-laminaribioside; 4MUCell, 4-methyl umbelliferyl-β-cellobioside; β-Xyl-4P, β-(pent-4-en-1-yl)-D-xylopyranoside (Moracci *et al.*, 1998; Perugino *et al.*, 2003; Trincone *et al.*, 2000, 2003, 2005).

3. THERMOPHILIC α-L-FUCOSYNTASES FROM GH29

The glycosynthetic methodology was applied on two hyperthermophilic α-L-fucosidases from the archaeon *S. solfataricus* (Ssαfuc) and the bacterium *Thermotoga maritima* (Tmαfuc), both belonging to GH29 (Cobucci-Ponzano et al., 2009). The recombinant Ssαfuc and Tmαfuc are well-characterized enzymes, showing high thermophilicity and thermostability as other enzymes from this source (Cobucci-Ponzano et al., 2003a; Rosano et al., 2004; Tarling et al., 2003). The catalytic residues were identified by site-directed mutagenesis and enzymological studies (Cobucci-Ponzano et al., 2003b, 2005a, 2008; Tarling et al., 2003), and the 3D structure of the enzyme from *Thermotoga* is deposited (PDB 1HL8) (Sulzenbacher et al., 2004).

3.1. Cloning, mutagenesis, expression and purification

The two enzymes were cloned and expressed as fusion proteins with tags for affinity chromatography purification.

3.1.1. Cloning and mutagenesis of Ssαfuc

Ssαfuc is encoded by two open reading frames (ORF) separated by a -1 frameshift and is expressed *in vivo* by a mechanism of translational regulation of gene expression called *programmed -1 frameshifting* (Cobucci-Ponzano et al., 2005b, 2006). A full-length enzyme was produced by merging the two ORFs with a mutation producing a single frame between them and then cloned in the pGEX-2TK vector (GE Healthcare), which introduces a GST-tag at the N-terminal of the protein (pGEX-frameFuc) (Cobucci-Ponzano et al., 2003a). This plasmid was used as template for site-directed mutagenesis (Gene-Tailor Site-directed Mutagenesis System, Invitrogen) to modify the nucleophile residue Asp242 into serine (D242S) and the gene containing the desired mutation was identified by complete sequencing of the gene (Cobucci-Ponzano et al., 2003b, 2009).

3.1.2. Cloning and mutagenesis of Tmα-fuc

The plasmid pDEST17/Tmα-fuc expressing Tmα-fuc fused to a histidine tag (Sulzenbacher et al., 2004) was used as template for the preparation, by site-directed mutagenesis (Gene-Tailor Site-directed Mutagenesis System, Invitrogen), of the nucleophile mutant TmD224G. The gene containing the desired mutation was identified by direct sequencing (Cobucci-Ponzano et al., 2009).

3.1.3. Expression and purification of Ssαfuc wild type and mutants

1. To express the Ssαfuc wild type and D242S mutant as GST-fusion proteins, transform *E. coli* RB791 with the appropriate plasmid.
2. Grow 2 l of *E. coli* RB791 until 0.6 OD_{600} and induce the culture with 0.5 mM IPTG overnight.
3. Resuspend the cells 1:3 g/ml in PBS buffer supplemented with 1% Triton X-100.
4. Lyse the cells using a French pressure cell and, after centrifugation for 30 min at 17,000 × g, load the cell-free crude extracts on 2.5 ml of Glutathione-Sepharose 4B matrix (GE Healthcare) equilibrated with PBS buffer supplemented with 1% Triton X-100.
5. Wash the column with 10 column volumes of PBS.
6. Load 60 units of thrombin protease onto the column in one column volume of PBS buffer and incubate overnight at 4 °C.
7. Elute the protein by five washings with one column volume of PBS each.
8. Pool the elutions and perform two heating steps for 30 min at 65 and 70 °C.
9. Centrifuge the sample for 30 min at 17,000 × g. Store the supernatant at 4 °C.

The purification procedures were performed with affinity matrices dedicated only to the purification of the specific enzyme to exclude cross-contamination. The enzymes, >95% pure by SDS-PAGE, are stored in PBS buffer and are stable for several months at 4 °C.

3.1.4. Expression and purification of Tmα-fuc wild type and mutants

1. To express Tmα-fuc wild type and TmD224G mutant as His-tag fusion proteins, transform *E. coli* Rosetta (DE3) with the appropriate plasmid.
2. Grow 2 l of *E. coli* Rosetta (DE3) until 1 OD_{600} and induce the culture with 1 mM IPTG for 4 h.
3. Resuspend the cells 1:3 g/ml in 50 mM NaH_2PO_4, 300 mM NaCl, pH 8.0 buffer (lysis buffer).
4. Lyse the cells using a French pressure cell and, after centrifugation, load the cell-free crude extracts on Protino Ni-TED 1000 packed columns (Macherey-Nagel).
5. After washing with lysis buffer, elute the proteins with lysis buffer supplemented with 250 mM imidazole.
6. Collect the fractions and dialyze against PBS buffer.

The enzymes, >95% pure by SDS-PAGE, are stored in PBS buffer and were stable for several months at 4 °C.

3.2. Enzymatic assays and chemical rescue of thermophilic α-fucosynthases

1. Prepare the assay mixtures in Buffer A (for wild-type and mutant enzymes from *S. solfataricus*) and in Buffer C (50 mM sodium citrate/phosphate buffer, pH 6.0) for wild-type and mutant enzymes from *T. maritima*. Add 4-nitrophenyl- (1 mM) or 2-chloro-4-nitrophenyl-α-L-fucopyranosides (3 mM) (4NP-α-L-Fuc and 2C4NP-α-L-Fuc, respectively) (Sigma-Aldrich and Carbosynth).
2. Preheat the mixtures by incubating for 2 min at 60 or 65 °C in quartz cuvettes for *T. maritima* and *S. solfataricus* enzymes, respectively.
3. Prepare a blank mixture for each reaction in order to correct for the spontaneous hydrolysis of the substrates.
4. Start the reaction by adding up to 20 μg of enzyme to the assay mixture and keep the temperature constant during all activity measurements.
5. Follow changes in absorbance at 405 nm, as described in Section 2.2. The molar extinction coefficients used are reported in Table 15.1.
6. To chemically rescue the activity of the nucleophile mutants, proceed as in Section 2.2, in Buffer A in the presence of sodium formate and sodium azide in the range of 0–2 M.
7. For all the enzymes, 1 unit of enzyme activity is defined as the amount of enzyme catalyzing the hydrolysis of 1 μmol of substrate in 1 min at the conditions described.

3.3. Synthesis and analysis of fucosylated oligosaccharides

3.3.1. Transfucosylation reactions

1. For standard transglycosylation reactions, incubate in 0.2 ml final volume D242S (50 μg) or TmD224G (38 μg) in the presence of 20 mM β-L-fucopyranosyl azide (β-L-Fuc-N$_3$), or in the presence of 0.1 M sodium azide and 20 mM 2C4NP-α-L-Fuc substrate, for 16 h at 65 and 60 °C, respectively.
2. For preparative analyses, incubate 94 μg of SsD242S for 16 h at 65 °C in 0.8 ml of Buffer A in the presence of 20 mM β-L-Fuc-N$_3$ as donor and different acceptors at molar ratios between 1:2 and 1:3.4 donor:acceptor. For TmD224G, incubate 38 μg of enzyme in the presence of 10 mM β-L-Fuc-N$_3$ and 100 mM 4NP-β-D-Xyl (1:10 donor:acceptor molar ratios) for 16 h at 70 °C.
3. To monitor the formation of oligosaccharides in the reactions, spot aliquots of the reactions and of the appropriate standards on analytical TLC and proceed as described in Section 2.3.

3.3.2. Analysis of the reaction products and determination of the transfucosylation efficiency

1. The isolation and the structural characterization of the transfucosylation products are described by Cobucci-Ponzano et al. (2009).
2. Dilute (10- to 100-fold) the preparative reaction mixtures with H_2O and 0.5 nmol of arabinose as internal standard.
3. Run a high-performance anion-exchange chromatography with pulsed amperometric detection (HPAEC-PAD) (Dionex) by loading 25 μl of the diluted mixture onto the CarboPac PA1 column (Dionex) and elute with 16 mM NaOH.
4. Determine the moles of fucose by integrating the peaks in the chromatogram, based on fucose and arabinose standard curves.
5. The transfucosylation efficiency is defined as the amount of fucose transferred to an acceptor that is different from water. Calculate the amount of fucose transferred by the enzyme to water by subtracting the amount of free fucose measured in the blank mixtures from that monosaccharide identified in the reaction mixtures.
6. To measure the total amount of fucose enzymatically transferred, incubate 1/10 of the reaction mixtures for 90 min at 65 °C in the presence of 1.2 μg of Ssαfuc wild type.
7. Dilute (10- to 100-fold) the preparative reaction mixtures with H_2O and add 0.5 nmol of arabinose as the internal standard.
8. Run the HPAEC-PAD as described above to measure the total amount of fucose.

The efficiency of the transfucosylation reaction was calculated as: [(total amount of fucose transferred − moles of fucose transferred to water)/total amount of fucose transferred] × 100 (Cobucci-Ponzano et al., 2009).

3.4. Characterization of the thermophilic α-fucosynthases

3.4.1. Kinetic characterization of the α-fucosynthases

The activity of D242S and TmD224G on 4NP-α-L-Fuc and 2C4NP-α-L-Fuc was chemically rescued in the presence of sodium azide. The steady-state kinetic parameters of the mutants on these substrates were measured in several conditions and reported in Table 15.5. Remarkably, the turnover number of SsαD242S on 2C4NP-α-L-Fuc in saturating sodium azide (2 M) was 1.8-fold higher than that of the wild type. However, the steady-state kinetic constants of TmD224G in 1 M sodium azide revealed specificity constants of the mutant that were 4- and 18-fold lower than those of the wild-type enzyme on 4NP-α-L-Fuc and 2C4NP-α-L-Fuc, respectively. The affinity for both substrates was not significantly impaired by the mutation (Table 15.5).

Table 15.5 Kinetic constants of α-glycosynthases

	k_{cat} (s^{-1})	K_M (mM)	k_{cat}/K_M (s^{-1} mM^{-1})
Wild type Ssαfuc			
4NP-α-L-Fuc	287 ± 11	0.028 ± 0.004	10,250
2C4NP-α-L-Fuc	157 ± 9	0.013 ± 0.004	11,602
D242S			
4NP-α-L-Fuc	0.08^a	–	–
2C4NP-α-L-Fuc	0.25^a	–	–
4NP-α-L-Fuc + NaN$_3$ (2 M)	47 ± 1	0.19 ± 0.02	247
2C4NP-α-L-Fuc + NaN$_3$ (2 M)	286 ± 35	0.3 ± 0.1	987
Wild type Tmα-fuc			
4NP-α-L-Fuc	80 ± 3	0.033 ± 0.005	2412
2C4NP-α-L-Fuc	88 ± 5	0.007 ± 0.003	12,287
D242G			
4NP-α-L-Fuc	0.05^a	–	–
2C4NP-α-L-Fuc	0.46^a	–	–
4NP-α-L-Fuc + NaN$_3$ (1 M)	9.2 ± 0.3	0.015 ± 0.002	613
2C4NP-α-L-Fuc + NaN$_3$ (1 M)	27 ± 1	0.040 ± 0.005	681
Wild type TmGalA			
4NP-α-D-Gal	65 ± 2	0.04 ± 0.02	1585
D327G			
4NP-α-D-Gal	0.060 ± 0.003	0.10 ± 0.05	0.6
4NP-α-D-Gal + NaN$_3$ (0.5 M)	2.07 ± 0.04	0.12 ± 0.01	17

^{a} k_{cat} was determined from the initial velocity at saturating concentration of substrate.

3.4.2. Transglycosylation efficiency of the α-fucosynthases

The characterization of the transfucosylation products of D242S, obtained by incubating the enzyme in the presence of 0.1 M sodium azide and 20 mM 2C4NP-α-L-Fuc, revealed the formation of β-L-Fuc-N$_3$ and the disaccharide α-L-fucopyranosyl-(1–3)-β-L-fucopyranosyl azide (α-L-Fuc-(1–3)-β-L-Fuc-N$_3$), suggesting that the newly formed β-L-Fuc-N$_3$ might become a novel donor and, as a result of its accumulation, also an acceptor. This was confirmed by the observation that D242S, incubated for 16 h at 65 °C with 20 mM β-L-Fuc-N$_3$, catalyzed the formation of the disaccharide with a transfucosylation efficiency of 40% and exclusive formation of the α-L-(1–3)-bond (see Fig. 15.1D for the reaction mechanism proposed). By contrast, TmD224G incubated with 20 mM 2C4NP-α-L-Fuc in sodium azide 0.05–1.0 M for 16 h at 60 °C synthesized only β-L-Fuc-N$_3$, and no disaccharide products were formed suggesting that no homo-condensation occurred.

The substrate specificities of the α-fucosynthases, using β-L-Fuc-N$_3$ as donor, in the presence of different glycosides as acceptors are summarized in Table 15.6. Reactions I and II show the products of D242S in 10 mM of

Table 15.6 Synthetic products of α-glycosynthases

Reaction	Enzyme	Donor	Acceptor	Products	Relative ratios of transfucosylation products (%)
I	D242S	β-L-Fuc-N$_3$ (10 mM)	4NP-β-D-Xyl (34 mM)	α-L-Fuc-(1–3)-β-L-Fuc-N$_3$ α-L-Fuc-(1–3)-β-D-Xyl-4NP α-L-Fuc-(1–4)-β-D-Xyl-4NP α-L-Fuc-(1–3)-β-D-Xyl-4NP — α-L-Fuc-(1–4)	ND 53 31 16
Total transfucosylation efficiency 50%					
II	D242S	β-L-Fuc-N$_3$ (10 mM)	4NP-β-D-Gal (20 mM)	α-L-Fuc-(1–3)-β-L-Fuc-N$_3$ α-L-Fuc-(1–6)-β-D-Gal-4NP α-L-Fuc-(1–4)-β-D-Gal-4NP α-L-Fuc-(1–3)-β-D-Gal-4NP α-L-Fuc-(1–2)-β-D-Gal-4NP — α-L-Fuc-(1–3)	ND 78 10 8 4
Total transfucosylation efficiency 48%					
III	D242S	β-L-Fuc-N$_3$ (5 mM)	4NP-β-D-GlcNAc (15 mM)	α-L-Fuc-(1–3)-β-L-Fuc-N$_3$ α-L-Fuc-(1–3)-β-D-GlcNAc–4NP	ND 100

	Enzyme	Donor	Acceptor	Product	%
	Total transfucosylation efficiency 86%	β-L-Fuc-N₃ (10 mM)	4NP-β-D-Xyl (100 mM)		
IV	TmD224G			α-L-Fuc-(1–4)-β-D-Xyl-4NP	45
				α-L-Fuc-(1–4)-β-D-Xyl-4NP	55
	Total transfucosylation efficiency 91%	β-D-Gal-N₃ (14 mM)	4NP-α-D-Glc (14 mM)		
V	D327G			α-D-Gal-(1–6)-α-D-Glc-4NP	100
	Total transgalactosylation efficiency 33%	β-D-Gal-N₃ (14 mM)	4NP-α-D-Xyl (14 mM)		
VI	D327G			α-D-Gal-(1–2)-α-D-Xyl-4NP	73
				α-D-Gal-(1–4)-α-D-Xyl-4NP	27
	Total transgalactosylation efficiency 40%	β-D-Gal-N₃ (14 mM)	4NP-β-D-Xyl (14 mM)		
VII	D327G			α-D-Gal-(1–4)-β-D-Xyl-4NP	100
	Total transgalactosylation efficiency 38%	β-D-Gal-N₃ (14 mM)	4NP-α-D-Man (14 mM)		
VIII	D327G			α-D-Gal-(1–6)-α-D-Man-4NP	92
				α-D-Gal-(1–3)-α-D-Man-4NP	8
	Total transgalactosylation efficiency 51%				

β-L-Fuc-N₃ donor with 4NP-β-D-Xyl and 4NP-β-D-Gal acceptors (34 and 20 mM, respectively), while in reaction III the mutant was incubated with 5 mM donor and 15 mM 4NP-β-D-GlcNAc acceptor. The regioselectivity of D242S depends on the acceptor used producing α-(1–3) and α-(1–4) linkages with 4NP-β-D-Xyl and mainly an α-(1–6) bond with 4NP-β-D-Gal. Instead, with 4NP-β-D-GlcNAc, the α-(1–3) regioselectivity prevails with the formation of a single product. Interestingly, with 4NP-β-D-Xyl and 4NP-β-D-Gal acceptors, the mutant also catalyzed the formation of branched trisaccharides. In all the reactions, the formation of the homo-condensation product also occurred. Reaction IV shows the results of incubation of TmD224G in the presence of 10 mM β-L-Fuc-N₃ and 100 mM 4NP-β-D-Xyl (1:10 donor:acceptor molar ratios) after 16 h at 70 °C; no homo-condensation occurred and the mutant synthesized disaccharides with α-(1–4) and α-(1–3) linkages, but at very high efficiency (91%) (Cobucci-Ponzano et al., 2009).

Interestingly, crystals of TmD224G in complex with β-L-Fuc-N₃ at 2.65 Å, obtained as reported for the native enzyme (Sulzenbacher et al., 2004) by adding 75 mM β-L-Fuc-N₃ to the crystallization buffer, revealed the presence of the homo-condensation product α-L-Fuc-(1–2)-β-L-Fuc-N₃ (PDB ID 2WSP) in contrast to what had been observed in solution. The 3D-structure of the mutant complexed with the transglycosylation product did not reveal major changes compared to the wild type, but it is worth mentioning that the complex with β-L-Fuc-N₃ substrate was never observed. Presumably, the mutation allowed the attack to the anomeric center of the donor by an acceptor leading to the transfucosylation product.

4. THERMOPHILIC α-D-GALACTOSYNTHASE FROM GH36

The use of β-glycosyl azide derivatives as possible donors for α-glycosynthases was further demonstrated by using β-galactosyl-azide (β-Gal-N₃) to convert the α-galactosidase from *T. maritima* (TmGalA), belonging to family GH36, into an α-galactosynthase (see the reaction mechanism proposed in Fig. 15.1E) (Cobucci-Ponzano et al., 2011b). The recombinant enzyme is a homodimeric protein, optimally active at pH 5.0 and 90 °C, and extremely stable (50% residual activity after 7 days at 70 °C). TmGalA is active on 4NP-α-D-Gal and on the di- and trisaccharides melibiose and raffinose (Miller et al., 2001). The catalytic residues were identified and the 3D structure at 2.3 Å resolution has also been deposited (PDB ID 1ZY9) (Comfort et al., 2007).

4.1. Mutagenesis, expression and purification

The *galA* gene from *T. maritima* MSB8 (ORF TM1192) cloned into pET24d (Novagen) was used as template to prepare, by site-directed mutagenesis, the nucleophile mutant D327G (Comfort et al., 2007).

1. To express wild type and the D327G mutant of TmGalA, transform *E. coli* BL21 (DE3) with the appropriate plasmid.
2. Grow *E. coli* Rosetta (DE3) until 0.8 OD_{600} and induce the culture with 0.5 mM IPTG overnight.
3. Resuspend the cells 1:3 g/ml in 20 mM Tris–HCl, pH 8.0.
4. Lyse the cells using a French pressure cell and centrifuge for 30 min at $17,000 \times g$.
5. Incubate the clarified supernatant for 20 min at 80 °C in order to eliminate the heat-labile proteins of the bacterial host and centrifuge for 30 min at $17,000 \times g$.
6. Apply the sample on a Q-Sepharose anion-exchange column (GE Healthcare), equilibrated with 20 mM Tris–HCl, pH 8.0.
7. Elute the proteins with a linear gradient of NaCl (0–1.0 M) in 20 mM Tris–HCl, pH 8.0. TmGalA activity was found in the fractions eluting at about 0.5 M NaCl.
8. Pool the fractions containing TmGalA activity and add 5.0 M NaCl.
9. Load the sample onto a Phenyl-Sepharose 26/10 (GE Healthcare) and elute with a linear gradient of NaCl (5.0–0 M) in 20 mM Tris–HCl, pH 8.0. TmGalA activity is found in the fractions eluting at about 1.5 M NaCl.
10. Desalt active fractions by dialysis against 25 mM phosphate buffer, pH 7.5 (Cobucci-Ponzano *et al.*, 2011b; Comfort *et al.*, 2007).

The enzyme, >95% pure by SDS-PAGE, stored at 4 °C is stable for several months.

4.2. Assays and chemical rescue of TmGalA wild type and D327G mutant

1. Prepare the assay mixtures in 50 mM sodium acetate buffer, pH 5.0 (Buffer D), by using concentrations of 4NP-α-D-Gal (Sigma-Aldrich or Carbosynth) ranging from 0.01 to 2.5 mM in a final volume of 0.2 ml, preheat them for 2 min at 65 °C.
2. Prepare a blank mixture for each reaction in order to correct the spontaneous hydrolysis of the substrates.
3. Start the reaction by adding 1–10 μg of enzyme to the assay mixture and keep the temperature constant during all activity measurements.
4. After 1 min, stop the reactions by adding 0.8 ml of ice-cold 1 M Na_2CO_3.
5. Measure the changes in absorbance at 420 nm. The molar extinction coefficient of 4NP at 25 °C at 420 nm is 17,200. One enzymatic unit is defined as the amount of enzyme catalyzing the conversion of 1 μmol of substrate into product in 1 min, under the indicated conditions.

6. To rescue the activity of the D327G mutant, proceed as above, but prepare a reaction mixture containing Buffer D, 2.5 mM 4NP-α-D-Gal, and 0.5 M sodium azide (Cobucci-Ponzano et al., 2011b).

4.3. Synthesis and analysis of galactosylated oligosaccharides

4.3.1. Transgalactosylation reactions

1. In 0.1 ml final volume in Buffer D, incubate D327G (10 μg) for 16 h at 65 °C in the presence of 1–14 mM of β-Gal-N$_3$ (donor) and using 4NP-α-D-Glc (glucopyranoside), -Man (mannopyranoside), -Xyl (xylopyranoside), and 4NP-β-D-Xyl as acceptors at 1:1 molar ratio with the donor (Cobucci-Ponzano et al., 2011b).
2. To monitor the formation of oligosaccharides and the degree of polymerization in the reaction, spot aliquots of the reactions, and of the appropriate standards, on analytical TLC and proceed as described in Section 2.3.

4.3.2. Analysis of the reaction products and determination of the transgalactosylation efficiency

1. Scale up the reactions described in Section 4.3.1 to 2 ml.
2. The isolation and the structural characterization of the transgalactosylation products are described in Cobucci-Ponzano et al. (2011b).
3. Dilute (10- to 100-fold) the reaction mixtures described above with H$_2$O and 0.5 nmol of arabinose as internal standard.
4. Run an HPAEC-PAD (Dionex) by loading 25 μl of the diluted reaction mixture onto the CarboPac PA200 column (Dionex) and elute with 20 mM NaOH at a flow rate of 0.5 ml/min.
5. Determine the moles of galactose by integrating the peaks from the chromatogram, based on galactose and arabinose standard curves.
6. The transgalactosylation efficiency is defined as the amount of galactose transferred to an acceptor different from water. Calculate the amount of galactose transferred by subtracting the amount of free galactose measured in the blank mixtures from that identified in the reaction mixtures.
7. To measure the total amount of galactose enzymatically transferred, incubate 1/10 of the reaction mixtures for 90 min at 65 °C in the presence of 5.2 μg of TmGalA wild type.
8. Dilute (10- to 100-fold) the reactions with H$_2$O and 0.5 nmol of arabinose as internal standard and analyze by HPAEC-PAD as described above to measure the total amount of galactose.

The efficiency of the transgalactosylation reaction was calculated as: [(total amount of galactose transferred − moles of galactose transferred to water)/total amount of galactose transferred] × 100 (Cobucci-Ponzano et al., 2011b).

4.4. Characterization of the thermophilic α-galactosynthase

The removal by site-directed mutagenesis of the catalytic nucleophile (D327G) resulted in a decrease of the k_{cat} and of the k_{cat}/K_M of 10^3- and 2.6×10^3-fold, respectively. The chemical rescue of D327G in the presence of 0.5 M sodium azide was 30-fold (Table 15.5).

The inspection of the transgalactosylation reactions of D327G was obtained by incubating the enzyme in the presence of β-Gal-N_3 as donor revealed that no homo-condensation reaction occurred. Transgalactosylation products were obtained by including different glycosides in the reaction as possible acceptors (Table 15.6). The mutant is rather selective in the −1 subsite of the catalytic center and products were observed only by using 4NP-α-D-Glc, -Man, -Xyl, and 4NP-β-D-Xyl as acceptors at 1:1 molar ratio with the donor (Cobucci-Ponzano et al., 2011b).

The enzyme showed an exquisite regioselectivity, and the transgalactosylation products identified are summarized in Table 15.6. In the presence of 4NP-α-Glc, only the compound α-Gal-(1–6)-α-Glc-4NP (Hada et al., 2006) was synthesized in 33% yield (reaction V), demonstrating that the enzyme transgalactosylated exclusively the primary alcohol at the C6. Higher yields were obtained when 4NP-α-Xyl and 4NP-β-Xyl were used as acceptors (40% and 38% for reactions VI and VII, respectively), and the mutant showed also good regioselectivity with transfer of the galactose moiety exclusively onto a single hydroxyl group. The regioselectivity on xyloside acceptors differs based on their anomeric configuration: transgalactosylation occurred on the OH at the C2 and C4 groups using 4NP-α-Xyl as acceptor, while only the C4 was recognized with 4NP-β-Xyl (Table 15.6). In reaction VIII, with 4NP-α-Man as acceptor, two products were synthesized, α-Gal-(1–6)-α-Man-4NP and, interestingly, α-Gal-(1–3)-α-Man-4NP (Galili, 2001; Macher and Galili, 2008), respectively, with total yields of 51% (Cobucci-Ponzano et al., 2011b).

5. Summary

Substantial evidence exists that (hyper) thermophilic glycosynthases are interesting and powerful alternatives in the rich repertoire of glycosynthases. The extreme stability to heat and to other common denaturants (i.e., high concentrations of organics, high ionic strength, and extremes of pH) makes these biocatalysts simple to manipulate and handle. In fact, the intrinsic stability of the (hyper) thermophilic glycosynthases can afford mutations in the nucleophile, such as Asp/Glu → Gly, that can weaken the protein backbone of the more fragile enzymes from mesophiles. In addition, the incubation at high temperature of the *E. coli* extracts facilitates simple purification of the recombinant glycosynthase from the proteins of the host, while inactivating other *E. coli* GHs that, even in tiny amounts, can prevent glycosynthetic reactions.

Due to their characteristics, (hyper) thermophilic glycosynthases have been exploited in the development of new reaction conditions that significantly improved the glycosynthesis of oligosaccharides. In fact, β-glycosynthases from GH1 could use high concentrations (up to 4 M) of external nucleophiles and buffers at a pH as low as 3.0 to synthesize oligosaccharides. Such reactions would not be practical for conventional glycosynthases. Similarly, α-glycosynthases from families GH29 and GH36 were able to use β-glycosyl azides as donors, compounds considered very stable in mild conditions. Presumably, the high temperatures of the reactions weakened these substrates allowing their use in α-glycosynthetic reactions that, previously, were known only in one case using conventional α-glycosyl fluorides (Okuyama *et al.*, 2002; Wada *et al.*, 2008).

In conclusion, thermophilic glycosynthases are interesting catalysts that have significant biotechnological applications in oligosaccharide synthesis (Galili, 2001; Macher and Galili, 2008; Ma *et al.*, 2006; Vanhooren and Vandamme, 1999).

ACKNOWLEDGMENTS

This work was supported by the Agenzia Spaziale Italiana (project MoMa n. 1/014/06/0) and by the project "Nuove glicosidasi d'interesse terapeutico nella salute umana" within the exchange program Galileo 2009–10 of the Università Italo-Francese.

REFERENCES

Aguilar, C., Sanderson, I., Moracci, M., Ciaramella, M., Nucci, R., Rossi, M., and Pearl, L. H. (1997). Crystal structure of the beta-glycosidase from the hyperthermophilic archeon *Sulfolobus solfataricus*: Resilience as a key factor in thermostability. *J. Mol. Biol.* **271,** 789–802.

Ausili, A., Di Lauro, B., Cobucci-Ponzano, B., Bertoli, E., Scirè, A., Rossi, M., Tanfani, F., and Moracci, M. (2004). Two-dimensional IR correlation spectroscopy of mutants of the beta-glycosidase from the hyperthermophilic archaeon *Sulfolobus solfataricus* identifies the mechanism of quaternary structure stabilization and unravels the sequence of thermal unfolding events. *Biochem. J.* **384,** 69–78.

Bojarová, P., and Kren, V. (2009). Glycosidases: A key to tailored carbohydrates. *Trends Biotechnol.* **27**, 199–209.
Cantarel, B. L., Coutinho, P. M., Rancurel, C., Bernard, T., Lombard, V., and Henrissat, B. (2009). The Carbohydrate-Active EnZymes database (CAZy): An expert resource for Glycogenomics. *Nucleic Acids Res.* **37**, D233–D238.
Chi, Y. I., Martinez-Cruz, L. A., Jancarik, J., Swanson, R. V., Robertson, D. E., and Kim, S. H. (1999). Crystal structure of the β-glycosidase from the hyperthermophile *Thermosphaera aggregans*: Insights into its activity and thermostability. *FEBS Lett.* **445**, 375–383.
Cobucci-Ponzano, B., Trincone, A., Giordano, A., Rossi, M., and Moracci, M. (2003a). Identification of an archaeal alpha-L-fucosidase encoded by an interrupted gene. Production of a functional enzyme by mutations mimicking programmed − 1 frameshifting. *J. Biol. Chem.* **278**, 14622–14631.
Cobucci-Ponzano, B., Trincone, A., Giordano, A., Rossi, M., and Moracci, M. (2003b). Identification of the catalytic nucleophile of the family 29 alpha-L-fucosidase from *Sulfolobus solfataricus* via chemical rescue of an inactive mutant. *Biochemistry* **42**, 9525–9531.
Cobucci-Ponzano, B., Mazzone, M., Rossi, M., and Moracci, M. (2005a). Probing the catalytically essential residues of the alpha-L-fucosidase from the hyperthermophilic archaeon *Sulfolobus solfataricus*. *Biochemistry* **44**, 6331–6342.
Cobucci-Ponzano, B., Rossi, M., and Moraccim, M. (2005b). Recoding in archaea. *Mol. Microbiol.* **55**, 339–348.
Cobucci-Ponzano, B., Conte, F., Benelli, D., Londei, P., Flagiello, A., Monti, M., Pucci, P., Rossi, M., and Moracci, M. (2006). The gene of an archaeal alpha-L-fucosidase is expressed by translational frameshifting. *Nucleic Acids Res.* **34**, 4258–4268.
Cobucci-Ponzano, B., Conte, F., Mazzone, M., Bedini, E., Corsaro, M. M., Rossi, M., and Moracci, M. (2008). Design of new reaction conditions for characterization of a mutant thermophilic a-L-fucosidase. *Biocat. Biotrans.* **26**, 18–24.
Cobucci-Ponzano, B., Conte, F., Bedini, E., Corsaro, M. M., Parrilli, M., Sulzenbacher, G., Lipski, A., Dal Piaz, F., Lepore, L., Rossi, M., and Moracci, M. (2009). β-Glycosyl azides as substrates for α-glycosynthases: Preparation of efficient α-L-fucosynthases. *Chem. Biol.* **16**, 1097–1108.
Cobucci-Ponzano, B., Strazzulli, A., Rossi, M., and Moracci, M. (2011a). Glycosynthases in biocatalysis. *Adv. Synth. Catal.* **353**, 2177–2548.
Cobucci-Ponzano, B., Zorzetti, C., Strazzulli, A., Carillo, S., Bedini, E., Corsaro, M. M., Comfort, D. A., Kelly, R. M., Rossi, M., and Moracci, M. (2011b). A novel alpha-D-galactosynthase from *Thermotoga maritima* converts beta-D-galactopyranosyl azide to alpha-galacto-oligosaccharides. *Glycobiology* **21**, 448–456.
Comfort, D. A., Bobrov, K. S., Ivanen, D. R., Shabalin, K. A., Harris, J. M., Kulminskaya, A. A., Brumer, H., and Kelly, R. M. (2007). Biochemical analysis of *Thermotoga maritima* GH36 α-galactosidase TmGalA confirms the mechanistic commonality of clan GH-D glycoside hydrolases. *Biochemistry* **46**, 3319–3330.
Galili, U. (2001). The α-gal epitope Galα1-3Galβ1-4GlcNAc-R in xenotransplantation. *Biochimie* **83**, 557–563.
Hada, N., Oka, J., Nishiyama, A., and Takeda, T. (2006). Stereoselective synthesis of 1,2-cis galactosides: Synthesis of a glycolipid containing Galα1-6Gal component from *Zygomycetes* species. *Tetrahedron Lett.* **47**, 6647–6650.
Hancock, S. M., Vaughan, M. D., and Withers, S. G. (2006). Engineering of glycosidases and glycosyltransferases. *Curr. Opin. Chem. Biol.* **10**, 509–519.
Higuchi, R., Krummel, B., and Saiki, R. K. (1988). A general method of *in vitro* preparation and specific mutagenesis of DNA fragments: Study of protein and DNA interactions. *Nucleic Acids Res.* **16**, 7351–7367.

Hrmova, M., Imai, T., Rutten, S. J., Fairweather, J. K., Pelosi, L., Bulone, V., Driguez, H., and Fincher, G. B. (2002). Mutated varley (1,3)-β-D-glucan endohydrolases synthesize crystalline (1,3)-β-D-glucans. *J. Biol. Chem.* **277**, 30102–30111.

Jakeman, D. L., and Withers, S. G. (2002). Glycosynthases: New tools for oligosaccharide synthesis. *Trends Glycosci. Glycotechnol.* **14**, 13–25.

Kaper, T., Lebbink, J. H., Pouwels, J., Kopp, J., Schulz, G. E., van der Oost, J., and de Vos, W. M. (2000). Comparative structural analysis and substrate specificity engineering of the hyperthermostable beta-glucosidase CelB from *Pyrococcus furiosus*. *Biochemistry* **39**, 4963–4970.

Kaper, T., van Heusden, H. H., van Loo, B., Vasella, A., van der Oost, J., and de Vos, W. M. (2002). Substrate specificity engineering of beta-mannosidase and beta-glucosidase from Pyrococcus by exchange of unique active site residues. *Biochemistry* **41**, 4147–4155.

Ly, H. D., and Withers, S. G. (1999). Mutagenesis of glycosidases. *Annu. Rev. Biochem.* **68**, 487–522.

Ma, B., Simala-Grant, J. L., and Taylor, D. E. (2006). Fucosylation in prokaryotes and eukaryotes. *Glycobiology* **12**, 158R–184R.

Macher, B. A., and Galili, U. (2008). The Galα1,3Galβ1,4GlcNAc-R (α-Gal) epitope: A carbohydrate of unique evolution and clinical relevance. *Biochim. Biophys. ACTA Gen. Subjects* **1780**, 75–88.

Mackenzie, L. F., Wang, Q., Warren, R. A. J., and Withers, S. G. (1998). Glycosynthases: Mutant glycosidases for oligosaccharide synthesis. *J. Am. Chem. Soc.* **120**, 5583–5584.

Malet, C., and Planas, A. (1998). From beta-glucanase to beta-glucansynthase: Glycosyl transfer to alpha-glycosyl fluorides catalyzed by a mutant endoglucanase lacking its catalytic nucleophile. *FEBS Lett.* **440**, 208–212.

Mayer, C., Jakeman, D. L., Mah, M., Karjala, G., Gal, L., Warren, R. A., and Withers, S. G. (2001). Directed evolution of new *glycosynthases* from *Agrobacterium* β-glucosidase: A general screen to detect enzymes for oligosaccharide synthesis. *Chem. Biol.* **5**, 437–443.

McIntosh, L. P., Hand, G., Johnson, P. E., Joshi, M. D., Korner, M., Plesniak, L. A., Ziser, L., Wakarchuk, W. W., and Withers, S. G. (1996). The pKa of the general acid/base carboxyl group of a glycosidase cycles during catalysis: A 13^C NMR study of *Bacillus circulans* xylanase. *Biochemistry* **35**, 9958–9966.

Mikaelian, I., and Sergeant, A. (1992). A general and fast method to generate multiple site directed mutations. *Nucleic Acids Res.* **20**, 376.

Miller, E. S., Jr., Kimberley, E. S., Parker, N., Liebl, W., Lam, D., Callen, W., Snead, M. A., Mathur, E. J., Short, J. M., and Kelly, R. M. (2001). Alpha-D-galactosidases from Thermotoga species. *Methods Enzymol.* **330**, 246–260.

Moracci, M., Nucci, R., Febbraio, F., Vaccaro, C., Vespa, N., La Cara, F., and Rossi, M. (1995). Expression and extensive characterization of a beta-glycosidase from the extreme thermoacidophilic archaeon *Sulfolobus solfataricus* in *Escherichia coli*: Authenticity of the recombinant enzyme. *Enzyme Microb. Technol.* **17**, 992–997.

Moracci, M., Capalbo, L., Ciaramella, M., and Rossi, M. (1996). Identification of two glutamic acid residues essential for catalysis in the β-glycosidase from the thermoacidophilic archaeon *Sulfolobus solfataricus*. *Protein Eng.* **9**, 1191–1195.

Moracci, M., Trincone, A., Perugino, G., Ciaramella, M., and Rossi, M. (1998). Restoration of the activity of active-site mutants of the hyperthermophilic β-glycosidase from *Sulfolobus solfataricus*: Dependence of the mechanism on the action of external nucleophiles. *Biochemistry* **37**, 17262–17270.

Moracci, M., Trincone, A., Cobucci-Ponzano, B., Perugino, G., Ciaramella, M., and Rossi, M. (2001). Enzymatic synthesis of oligosaccharides by two glycosyl hydrolases of *Sulfolobus solfataricus*. *Extremophiles* **3**, 145–152.

Nucci, R., Moracci, M., Vaccaro, C., Vespa, N., and Rossi, M. (1993). Exo-glucosidase activity and substrate specificity of the beta-glycosidase isolated from the extreme thermophile *Sulfolobus solfataricus*. *Biotechnol. Appl. Biochem.* **17**, 239–250.

Okuyama, M., Mori, H., Watanabe, K., Kimura, A., and Chiba, S. (2002). α-Glucosidase mutant catalyzes "α-glycosynthase"-type reaction. *Biosci. Biotechnol. Biochem.* **66**, 928–933.

Perugino, G., Trincone, A., Giordano, A., van der Oost, J., Kaper, T., Rossi, M., and Moracci, M. (2003). Activity of hyperthermophilic glycosynthases is significantly enhanced at acidic pH. *Biochemistry* **42**, 8484–8493.

Perugino, G., Trincone, A., Rossi, M., and Moracci, M. (2004). Oligosaccharide synthesis by glycosynthases. *Trends Biotechnol.* **22**(1), 31–37.

Pouwels, J., Moracci, M., Cobucci-Ponzano, B., Perugino, G., van der Oost, J., Kaper, T., Lebbink, J. H., de Vos, W. M., Ciaramella, M., and Rossi, M. (2000). Activity and stability of hyperthermophilic enzymes: A comparative study on two archaeal β-glycosidases. *Extremophiles* **3**, 157–164.

Rosano, C., Zuccotti, S., Cobucci-Ponzano, B., Mazzone, M., Rossi, M., Moracci, M., Petoukhov, M. V., Svergun, D. I., and Bolognesi, M. (2004). Structural characterization of the nonameric assembly of an archaeal alpha-L-fucosidase by synchrotron small angle X-ray scattering. *Biochem. Biophys. Res. Commun.* **320**, 176–182.

Shallom, D., Belakhov, V., Solomon, D., Shoham, G., Baasov, T., and Shoham, Y. (2002). Detailed kinetic analysis and identification of the nucleophile in α-L-arabinofuranosidase from *Geobacillus stearothermophilus* T-6, a family 51 glycoside hydrolase. *J. Biol. Chem.* **277**, 43667–43673.

Sulzenbacher, G., Bignon, C., Nishimura, T., Tarling, C. A., Withers, S. G., Henrissat, B., and Bourne, Y. (2004). Crystal structure of *Thermotoga maritima* alpha-L-fucosidase. Insights into the catalytic mechanism and the molecular basis for fucosidosis. *J. Biol. Chem.* **279**, 13119–13128.

Tarling, C. A., He, S., Sulzenbacher, G., Bignon, C., Bourne, Y., Henrissat, B., and Withers, S. G. (2003). Identification of the catalytic nucleophile of the family 29 alpha-L-fucosidase from *Thermotoga maritima* through trapping of a covalent glycosyl-enzyme intermediate and mutagenesis. *J. Biol. Chem.* **278**, 47394–47399.

Trincone, A., Improta, R., Nucci, R., Rossi, M., and Gambacorta, A. (1994). Enzymatic synthesis of carbohydrate derivatives using β-glycosidase of *Sulfolobus solfataricus*. *Biocatalysis* **10**, 195–210.

Trincone, A., Perugino, G., Rossi, M., and Moracci, M. (2000). A novel thermophilic glycosynthase that effects branching glycosylation. *Bioorg. Med. Chem. Lett.* **4**, 365–368.

Trincone, A., Giordano, A., Perugino, G., Rossi, M., and Moracci, M. (2003). Glycosynthase-catalysed syntheses at pH below neutrality. *Bioorg. Med. Chem. Lett.* **13**, 4039–4042.

Trincone, A., Giordano, A., Perugino, G., Rossi, M., and Moracci, M. (2005). Highly productive autocondensation and transglycosylation reactions with *Sulfolobus solfataricus* glycosynthase. *Chembiochem* **6**, 1431–1437.

Vanhooren, P., and Vandamme, E. J. (1999). L-Fucose: Occurrence, physiological role, chemical, enzymatic and microbial synthesis. *J. Chem. Technol. Biotechnol.* **74**, 479–497.

Viladot, J. L., de Ramon, E., Durany, O., and Planas, A. (1998). Probing the mechanism of *Bacillus* 1,3-1,4-beta-D-glucan 4-glucanohydrolases by chemical rescue of inactive mutants at catalytically essential residues. *Biochemistry* **37**, 11332–11342.

Viladot, J. L., Canals, F., Barllori, X., and Planas, A. (2001). Long-lived glycosyl-enzyme intermediate mimic produced by formate re-activation of a mutant endoglucanase lacking its catalytic nucleophile. *Biochem. J.* **355**, 79–86.

Wada, J., Honda, Y., Nagae, M., Kato, R., Wakatsuki, S., Katayama, T., Taniguchi, H., Kumagai, H., Kitaoka, M., and Yamamoto, K. (2008). 1,2-alpha-l-Fucosynthase: A glycosynthase derived from an inverting alpha-glycosidase with an unusual reaction mechanism. *FEBS Lett.* **582,** 3739–3743.

Zhu, X., and Schmidt, R. R. (2009). New principles for glycoside-bond formation. *Angew. Chem. Int. Ed.* **48,** 1900–1934.

CHAPTER SIXTEEN

Engineering Cellulase Activity into *Clostridium acetobutylicum*

Henri-Pierre Fierobe,* Florence Mingardon,* *and* Angélique Chanal[†]

Contents

1. Introduction	302
2. Electrotransformation of *C. acetobutylicum* ATCC824 and Storage of Recombinant Strains	303
2.1. Preparation of the plasmid	303
2.2. Electrotransformation of *C. acetobutylicum*	305
2.3. Storage of recombinant strains	305
3. Detection/Quantification of the Secreted Heterologous Cellulosomal Protein	306
3.1. Heterologous scaffoldins	306
3.2. Heterologous cellulosomal enzymes	307
3.3. Integrity of heterologous proteins secreted by *C. acetobutylicum*	308
4. Recombinant Strain of *C. acetobutylicum* Secreting a Heterologous Minicellulosome	308
5. Secretion Issue and the Use of "Carrier" Modules	311
5.1. Introduction of modules from the *C. cellulolyticum* scaffoldin CipC	312
5.2. Introduction of modules from the *C. acetobutylicum* scaffoldin CipA	313
5.3. Impact of carrier modules on cellulase activity	314
6. Outlook	314
Acknowledgments	315
References	315

Abstract

Clostridium acetobutylicum produces substantial amounts of butanol, and an engineered cellulolytic strain of the bacterium would be an attractive candidate for biofuel production using consolidated bioprocessing. Recent studies have

* Laboratoire de Chimie Bactérienne, UPR 9043, CNRS IMM, Marseille, France
[†] Total Gas and Power R & D Biotechnology Team, Paris, France

shown that this solventogenic bacterium can be used as a host for heterologous production and secretion of individual cellulosomal components, termed the minicellulosome. Their secretion yields range from 0.3 to 15 mg/L. Nevertheless, it appeared that key cellulosomal enzymes such as family GH48 processive enzymes and members of the large family of GH9 cellulases probably necessitate specific chaperone(s) for translocation and secretion, that is/are absent in the solventogenic bacterium. Heterologous secretion of the latter enzymes, however, can be obtained by grafting specific combinations of scaffoldin modules at the N-terminus of these cellulases, which are then used as cargo domains.

1. INTRODUCTION

Solvent-producing *Clostridia*, especially the best known, *Clostridium acetobutylicum*, efficiently converts various mono and disaccharides as well as some polysaccharides like starch into acids and solvents (acetone–butanol–ethanol fermentation). For instance, during solventogenesis, *C. acetobutylicum* can produce up to 13 g/L of butanol (Lutke-Eversloh and Bahl, 2011). This property makes this anaerobic bacterium and closely related microorganisms like *Clostridium beijerinckii*, attractive candidates for biofuel production using separate hydrolysis and fermentation (SHF) process (Qureshi et al., 2007). In the longer term, *C. acetobutylicum* can also be envisioned as a putative bacterium for consolidated bioprocessing (CBP) in which a single microorganism carries out all major steps (enzyme production, saccharification of plant biomass, and fermentation into products of industrial interest). The conversion of *C. acetobutylicum* into a suitable microorganism for CBP requires major modifications of the wild-type strain, above all the introduction of an efficient extracellular cellulolytic system, since the bacterium is unable to utilize native (crystalline) or chemically treated (amorphous) cellulose as a source of carbon. However, *C. acetobutylicum* displays several advantages. First of all, various genetic tools including replicative vectors, constitutive or inducible promoters (Girbal et al., 2003) and gene "knock out" systems are available (Heap et al., 2007, 2010). The bacterium can ferment all the major monosaccharides (glucose, xylose, arabinose, galactose, and mannose) generated from plant cell wall depolymerization (Servinsky et al., 2010) and can secrete xylan-degrading enzymes (Lee et al., 1987), although *C. acetobutylicum* does not utilize this polysaccharide efficiently (Lopez-Contreras et al., 2001). Finally, the solvent-producer is not phylogenetically distant from cellulolytic anaerobic bacteria like *Clostridium cellulolyticum*, *Clostridium josui*, or *Clostridium cellulovorans* and exhibits a similar codon bias. The cellulolytic system of these clostridia consists of numerous plant cell wall degrading enzymes assembled into high molecular mass complexes called cellulosomes. These complexes are composed of catalytic subunits exhibiting a "dockerin" module that strongly binds to complementary "cohesin" modules hosted by a backbone protein called scaffoldin. These scaffoldins produced by mesophilic *Clostridia* usually contain several copies of the cohesin module, a cellulose-specific carbohydrate binding module (CBM), and 100-residue modules of unknown function(s) called the X-modules. The *C. acetobutylicum* genome also contains a cluster of genes encoding a cellulosome, which, however, was shown to be inactive toward crystalline cellulose (Sabathe *et al.*, 2002). Taken together, these characteristics make *C. acetobutylicum* an appropriate host for the production and secretion of heterologous active cellulosomes or cellulosomal components.

2. Electrotransformation of *C. acetobutylicum* ATCC824 and Storage of Recombinant Strains

Since degradation of cellulosic insoluble substrates usually requires significant amounts of cellulases, heterologous overexpression of cellulosomal genes should preferably be performed by cloning the gene(s) of interest in a multicopy expression vector downstream of a strong promoter. Several expression vectors are available, and one of the most suitable for overexpression is the *Escherichia coli/Clostridium* shuttle vector pSOS95 (ColE1, pIM13 *repL*, *bla*, *ermB*) and its derivatives carrying the strong and constitutive clostridial promoter P_{thl} and the *adc* transcriptional terminator. This 7-kb vector (Girbal *et al.*, 2003) originally contained the acetone operon (*ctfA-ctfB-adc*, cloned at *Bam*HI/*Nar*I) and can carry up to 4–5 kb of foreign DNA downstream of P_{thl}. In the solventogenic host, pSOS95 is compatible with the shuttle vector *E. coli/C. acetobutylicum* pMTL500E vector (and derivatives) carrying the replication origin from pAMβ1.

2.1. Preparation of the plasmid

The successful cloning of a DNA encoding the N-terminal region of the *C. acetobutylicum* endogenous scaffoldin CipA (encompassing the leader peptide, the CBM3a, three X-modules, and two cohesins) in pSOS95 was earlier reported, and it led to the secretion of the truncated scaffoldin by *C. acetobutylicum* (Sabathe and Soucaille, 2003). It should, however, be noted that P_{thl} is functional in *E. coli* leading to heterologous production of protein precursors appended with a typical Gram + (SEC) leader peptide. The accumulation of these precursors may be harmful for *E. coli* and preventing the formation of colonies (Perret *et al.*, 2004b). The cause of this effect remains unknown, but a disruption or a jamming of the *E. coli* secretion machinery is hypothesized.

To avoid this technical issue, the vector pSOS95 can be modified by the introduction of two *lac* operators at the 5′ and 3′ extremities of P_{thl}, respectively, as shown in Fig. 16.1 (vector pSOS952). The presence of the operators does not notably diminish the strength of P_{thl} in *C. acetobutylicum*, whereas in *E. coli* strains, harboring the repressor gene *lacI* or *lacI^q*, the expression of the heterologous genes is low and is not detrimental to the enteric bacterium in the absence of lactose. The use of the *E. coli* strain SG-13009 containing the pREP4 repressor plasmid (Qiagen) is recommended as the recipient strain for this initial step (Perret *et al.*, 2004b).

C. acetobutylicum possesses a restriction endonuclease *Cac*824I, whose recognition sequence is 5′-GCNGC-3′, which constitutes a major barrier to electrotransformation of the solventogenic bacterium. No methyltransferase that could protect DNA from restriction by this endonuclease is commercially available. Nevertheless, in 1993, Mermelstein and Papoutsakis showed that methylation by ϕ3T 1 methyltransferase encoded by *Bacillus substilis* phage ϕ3T provides efficient protection (Mermelstein and Papoutsakis, 1993). Methylation of the vector should be performed *in vivo* by transforming the *E. coli* strain ER2275 containing the vector pAN1, which encodes the ϕ3T 1 methyltransferase (p15A origin; Cm^r, *ϕ3tI*) (Mermelstein and Papoutsakis, 1993). The methylation of the purified vector can be verified by incubation with the Cac824I isoschizomer Fnu4HI (Ozyme).

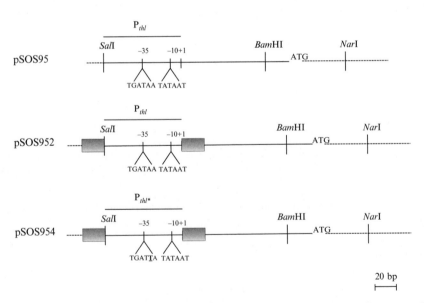

Figure 16.1 Schematic representation of the cloning region of pSOS95 and position of the two *lac* operators (gray boxes) inserted upstream and downstream of the promoter P_{thl} thereby generating the plasmid pSOS952. The mutation in the −35 box to attenuate the P_{thl} promoter (pSOS954) is underlined. Adapted from Perret *et al.* (2004b).

This last step is indeed superfluous if a modified strain of *C. acetobutylicum* exhibiting a disruption of the gene CA_C1502 encoding the restriction endonuclease Cac824I (Dong et al., 2010) is employed.

2.2. Electrotransformation of *C. acetobutylicum*

The following procedure was inspired by reports published by Mermelstein and Papoutsakis (1993) as well as Nakotte et al. (1998).

All manipulations are carried out in anaerobic media in an anaerobic chamber equipped with a centrifuge, electroporation device, and incubator. Resazurine (1 mg/L) can be added to media and buffers to detect any oxygen contamination. Plates can be poured aerobically but must be stored at least 24 h before use in the anaerobic chamber to dry and remove traces of oxygen.

1. Ten milliliters of spore germination medium CGM (0.625% yeast extract, 5.4 mM K$_2$HPO$_4$, 6.9 mM KH$_2$PO$_4$, 2 mM MgSO$_4$·7H$_2$O, 75 µM MnSO$_4$·H$_2$O, 45 µM FeSO$_4$·7H$_2$O, 21 mM NaCl, 19 mM (NH$_4$)$_2$SO$_4$, 16.5 mM asparagine, 4 mM cysteine, 0.15 M glucose, and pH adjusted to 6.5 with 10 M NaOH) is inoculated with 0.5–1 mL of *C. acetobutylicum* spore suspension. Germination is obtained by incubation at 80 °C for 15 min, followed by overnight growth at 37 °C.
2. The CGM culture is used to inoculate 50 mL of 2× YT glucose medium (16 g/L bactotryptone, 10 g/L yeast extract, 70 mM NaCl, 28 mM glucose, and pH adjusted to 5.2 with 1 M HCl). Cells are grown until mid-log phase (A_{620} of 0.6–0.8) and cooled down in ice.
3. The cells are harvested (3000 × g, 30 min, 4 °C) and washed twice with 10 and 1 mL of ice-cold electrotransformation buffer (270 mM saccharose, 3 mM NaH$_2$PO$_4$/Na$_2$HPO$_4$, pH 7.4), respectively.
4. Electrotransformation is performed by thoroughly mixing 500 µL of cells with 0.5–5 µg of methylated plasmid (maximum volume of vector solution of 10 µL) in an electroporation cuvette with 4 mm gap width (Eurogentec). Immediately after the pulse (2.5 kV, ∞Ω, 25 µF) (Bio-Rad), the cell suspension is diluted in 10 mL of ice cold 2× YT medium supplemented with 28 mM glucose and incubated for 4 h at 37 °C.
5. Cells are harvested (3000 × g, 30 min) and resuspended in 500 µL of 2× YT medium supplemented with 28 mM glucose. 250 µL is spread on each RCA plate (38 g/L Reinforced Clostridial Medium from Biokar and 15 g/L agar) supplemented with the appropriate antibiotic. Colonies appear after 48–96 h of incubation at 37 °C. Typically, 5×10^2 colonies are obtained per µg of plasmid.

2.3. Storage of recombinant strains

One individual colony is used to inoculate 10 mL of 2× YT medium supplemented with 28 mM glucose and the appropriate antibiotic. After overnight growth at 37 °C, 10 mL of MS medium (0.33 M glucose, 2.9 mM K$_2$HPO$_4$,

3.7 mM KH$_2$PO$_4$, 1 mM MgSO$_4$·7H$_2$O, 40 mM acetic acid, 45 μM FeSO$_4$·7H$_2$O, 60 μM of p-aminobenzoic acid, 80 μM biotin, and 4 mM cysteine, pH adjusted to 6.5 with NH$_3$) supplemented with the appropriate antibiotic is inoculated with 1 mL of 2 × YT glucose culture. After 1 week of incubation at 37 °C, spore formation is verified by microscopic observation, and the suspension can be stored at 4 or −20 °C.

3. Detection/Quantification of the Secreted Heterologous Cellulosomal Protein

The genes encoding cellulosome components were generally cloned in pSOS952. Heterologous cellulosomal components are usually secreted at yields ranging from 0.3 to 15 mg/L. The recombinant strain should be grown in 10–100 mL of 2 × YT medium supplemented with 5 g/L glucose or cellobiose and the appropriate antibiotic, up to late log phase (typically A_{620} 2.5–3.5) or early stationary phase. The culture is then centrifuged aerobically for 15 min at 8000 × g (4 °C), and the pH of the supernatant (usually around 4) may be increased to 6–7 by addition of 1 M phosphate pH 7.0, or 1 M Tris–HCl pH 8.0 buffers.

3.1. Heterologous scaffoldins

Scaffoldins appended with a cellulose-specific CBM, such as the CBM3a, can be easily detected and/or purified by loading the culture supernatant on a column containing 5–10 g of Avicel (Flucka PH 101) equilibrated in 50 mM phosphate buffer pH 7.0 (Perret et al., 2004b). The column is subsequently washed with 50 and 10 mM phosphate buffer pH 7.0. The scaffoldin is then eluted with distilled water or 1% (v/v) triethylamine. In the conditions described above, nearly no endogenous *C. acetobutylicum* protein should be retained on cellulose, and the heterologous scaffoldin eluted from Avicel should be at least 95% pure and easily quantified by absorbance at 280 nm.

Scaffoldins can also be detected using the far Western blot technique by exploiting the ability of their cohesin modules to specifically bind to the dockerin modules with high affinity. The supernatant of the recombinant strain should be concentrated approximately 100 times by ultrafiltration (using non cellulosic membrane, such as polyethersulfone membrane, Millipore) and blotted after denaturing polyacrylamide gel electrophoresis (SDS-PAGE), onto a nitro-cellulose membrane. After saturation with 2% (w/v) milk in 50 mM Tris–HCl and 150 mM NaCl pH 7.5, the membrane is overlaid with the same solution but containing 1–5 μg/mL of cellulosomal enzyme appended with the cognate dockerin. The cohesin/dockerin interaction is calcium dependent, but the calcium content of milk is sufficient to trigger the binding between the complementary modules. The scaffoldin-bound enzyme is subsequently detected using a specific antiserum raised against the enzyme. This experiment allows controlling the functionality of the cohesin(s) in the scaffoldin secreted by *C. acetobutylicum*.

Heterologous truncated scaffoldins called miniscaffoldins (shortened version of a native scaffoldin), composed of *C. cellulolyticum* and/or *Clostridium thermocellum* modules, are usually secreted at 10–15 mg/L in the culture supernatant (Perret et al., 2004b).

3.2. Heterologous cellulosomal enzymes

Small and/or soluble cellulosomal cellulases (Cel5A, Cel8C, and Cel9M) or hemicellulase (Man5K) from *C. cellulolyticum* were previously shown to be secreted by *C. acetobutylicum* at yields ranging from 0.5 to 5 mg/L, using either their native (SEC) signal peptide or the leader peptide displayed by the scaffoldin CipC from *C. cellulolyticum* (Mingardon et al., 2005, 2011). The pSOS952 expression vector was used except for the mannanase encoding gene in which overexpression through wild-type P_{thl} prevented the formation of colonies on selective medium. To circumvent this issue, a modified promoter, mutated in the -35 box (Fig. 16.1) and presumably leading to lower expression levels, was selected (pSOS954) (Mingardon et al., 2005). As for scaffoldins, the presence of the heterologous enzymes in the supernatant can indeed be verified by Western blot analysis using a specific antiserum. Alternatively, dockerin-containing enzymes can be probed with a scaffoldin harboring a complementary cohesin. Biotinylated scaffoldin and subsequent incubation with streptavidin–peroxidase conjugate or the far Western blot technique with a scaffoldin-specific antiserum can be used to probe the heterologous cellulosomal enzymes.

Detection of the extracellular endoglucanase activity can be easily performed using carboxymethyl cellulose (CMC) plates (Mingardon et al., 2011). The medium is composed of 1.5% (w/v) agar, 0.5% (w/v) CMC in 0.1 M potassium phosphate, pH 7.0. Twenty to fifty microliters of culture supernatant (prior to concentration) are loaded in 5-mm-diameter wells and incubated overnight at 37 °C. The plates are stained afterward for 15 min with 1% (w/v) Congo red and washed several times with 1 M NaCl. A clear halo surrounding the wells that contain some endoglucanase activity is observed. The size is roughly proportional to the enzyme concentration and its specific activity on CMC. A halo is sometimes observed with the supernatant of the control strain (containing the pSOS952 without any cellulase-encoding gene), but its size should be much smaller than that generated by strains secreting even low amounts (< 1 mg/L) of heterologous endoglucanase.

Quantification of the secreted enzyme is usually performed by measuring the activity of the 100-fold concentrated supernatant of the recombinant strain on a specific substrate, assuming the specific activity of the pure enzyme is known. On cellulosic substrates such as phosphoric acid swollen cellulose or Avicel, the concentrated supernatant of the control strain does not display any detectable activity, in contrast to CMC as mentioned above. C. acetobutylicum does not produce extracellular activity on mannan (locust bean gum, LBG, Fluka), which can therefore be used as the substrate to detect and quantify a secreted heterologous mannanase (Mingardon et al., 2005).

3.3. Integrity of heterologous proteins secreted by C. acetobutylicum

Miniscaffoldins secreted by *C. acetobutylicum*, containing up to four different modules (one CBM, one X-module, and two cohesins) connected by linkers, were found to be mainly secreted as full length forms (Perret et al., 2004b). In contrast, a large fraction of the secreted cellulosomal enzymes were truncated, displaying a molecular mass reduced by 8–10 kDa, compared to the expected mass of the mature protein (Fig. 16.2). In some cases, over 90% of the extracellular enzyme is in the truncated form. Western blot analyses performed simultaneously using either a specific antiserum or a biotinylated scaffoldin clearly indicated that the truncated forms lack the dockerin module, since they fail to interact with the scaffoldin probe. This phenomenon was observed for enzymes appended with either an N- or a C-terminus dockerin module (Mingardon et al., 2005, 2011). The proteolysis is likely to occur after the translocation step across the cytoplasmic membrane, since the same Western blot analyses performed on the cellular fractions indicate that the enzyme precursor still contains the dockerin module (Mingardon et al., 2011). In general, the loss of the dockerin module has no or moderate impact on the activity of the catalytic module. In one case, however, the mannanase Man5K from *C. cellulolyticum*, the dockerin-less form is 5.5 times more active on mannan (LBG) than the full length hemicellulase (Perret et al., 2004a).

The proteolysis can be almost completely avoided by coexpression of a scaffoldin-encoding gene as described in Section 4.

4. Recombinant Strain of *C. acetobutylicum* Secreting a Heterologous Minicellulosome

To date, only one case of successful production and secretion of a heterologous two-component minicellulosome by *C. acetobutylicum* has been reported (Mingardon et al., 2005). The genes encoding a truncated scaffoldin (miniCipC1) containing the three N-terminal modules of CipC and the gene *man5K* encoding a mannanase from *C. cellulolyticum* were cloned as an operon in the expression vector. The expression of the heterologous genes was under the control of a weakened form of

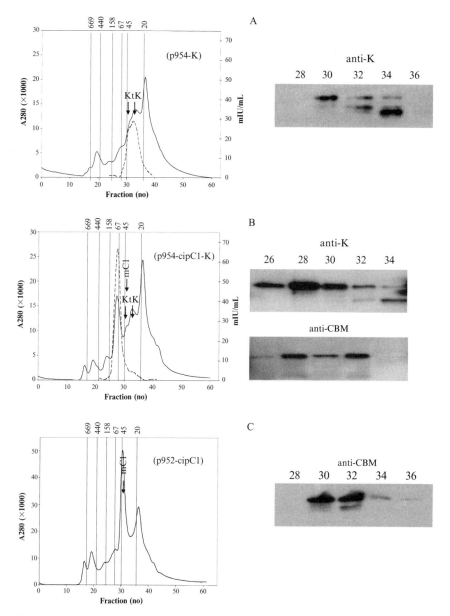

Figure 16.2 Examination of the supernatants of *C. acetobutylicum* recombinant strains producing Man5K (A), miniCipC1-Man5K (B) and miniCipC1 (C). The concentrated supernatants were analyzed by gel filtration (left panel) and subsequent Western blot analyses with polyclonal antibodies raised against Man5K (anti-K) or against CBM3a of miniCipC1 (anti-CBM) (right panel). Protein elution was followed absorbance at 280 nm (solid lines) and activity (dashed line) on 0.8% locust bean gum. Vertical lines (left panel) indicate the elution volume of molecular mass markers: tyroglobulin, 669 kDa; ferritin, 440 kDa; aldolase, 158 kDa; bovine serum albumin, 67 kDa; ovalbumin, 45 kDa; and trypsin inhibitor, 20 kDa. Arrows indicate the elution volume of purified miniCipC1 (mC1), entire (K) and truncated (tK) Man5K. The numbers on top of the Western blots (right panel) indicate fraction numbers. Adapted from Mingardon *et al.* (2005).

the promoter P_{thl} (pSOS954-cipC1-K). The demonstration that both components form the complex required: (i) large culture volumes of the recombinant strain and (ii) parallel similar experiments using two control strains carrying the expression vectors encoding either gene alone (pSOS952-cipC1; encodes miniCipC and pSOS954-K; encodes the mannanases ManK).

Batch fermentations of the various strains were performed in a 2-L fermentor at 37 °C in 2 × YT medium supplemented with 5 g/L cellobiose and 40 μg of erythromycin/mL. The culture was maintained anaerobic by a continuous flow of sterile N_2 gas and agitated at an impeller speed of 50 rpm. The pH was maintained at 5.5 with controlled additions of 2 M NaOH. The cultures were stopped at the late exponential phase and centrifuged at 10,000 × g for 30 min. Analyses were performed on 90 mL of supernatants by combining gel filtration separation, mannanase activity assay, and Western blot analyses.

1. Tris–HCl (pH 8.0) and potassium acetate are added to 90 mL of culture supernatants at final concentrations of 25 mM and 1 M, respectively.
2. The sample is loaded on 4 mL of phenyl-sepharose resin (GE Healthcare) equilibrated with the same buffer. The bound proteins are eluted with 50 mL of distilled water.
3. The sample is concentrated by ultrafiltration to a final volume of 0.5 mL.
4. One hundred microliters of concentrated sample is subjected to gel filtration chromatography by injection onto a 30-mL Superdex 200 GL column (GE Healthcare) equilibrated in 30 mM potassium phosphate (pH 7.0), 150 mM NaCl with an Akta FPLC system (GE Healthcare). The flow rate is set at 0.5 mL/min and 0.5-mL fractions are collected.
5. The fractions are subjected to Western blot analysis using polyclonal antibodies raised against Man5K or miniCipC1. Mannanase activity assay is performed using 0.8% LBG as the substrate.

As shown in Fig. 16.2, the complex formation is revealed by a shift of the mannanase activity toward a smaller elution volume compared to the control strain, secreting only Man5K. Western blot analyses also confirmed that when both genes are coexpressed, miniCipC1 as well as entire Man5K are detected in the most active fraction and shifted toward smaller elution volume compared to the supernatants of the control strains secreting either protein alone. Some free miniCipC1 is detected, indicating that the miniscaffoldin is produced and/or secreted in excess compared to Man5K. This experience also reveals that complexation of the mannanase dockerin module to miniCipC1 protected the enzyme from proteolysis, since the unmodified form is the dominant species in the supernatant, whereas in the case of the strain carrying the pSOS954-K, the truncated form was the most abundant.

5. Secretion Issue and the Use of "Carrier" Modules

The secretion by *C. acetobutylicum* of some *C. cellulolyticum* cellulosomal enzymes was found to be more difficult. The cloning in the solventogenic strain of the *C. cellulolyticum* genes encoding the large family 9 cellulases such as Cel9G or Cel9E, or the pivotal processive enzyme Cel48F in the expression vector pSOS952, failed to generate recombinant colonies on selective medium, suggesting a deleterious effect. This phenomenon was also observed when the gene encoding the endogenous Cel48A was cloned in the expression vector (Mingardon *et al.*, 2011). The toxicity was found to be related to the secretion step, since the cloning of a gene coding for the mature form of Cel48F generated multiple *C. acetobutylicum* clones that accumulated the heterologous cellulase in the cytoplasm, indicating that the enzyme is not intrinsically toxic for the solventogenic bacterium.

In contrast to *man5K* (see Section 4), the use of a weakened P_{thl} promoter did not alleviate the toxicity. The replacement of the leader peptide of the deleterious Cel48F by secreted heterologous (miniCipC1) or endogenous (AmyP) proteins, or the coexpression of a scaffoldin-encoding gene, also failed to circumvent this issue. The replacement of the constitutive promoter P_{thl} by the inducible promoter P_{xyl} from *Staphylococcus aureus*, which was previously shown to be functional in *C. acetobutylicum* (Girbal *et al.*, 2003) also failed to generate colonies. Most probably, this promoter generates deleterious levels of expression of the *cel48F* gene even in the absence of the inducer (xylose).

One possible explanation is that a specific, and as yet undiscovered, chaperone(s) involved in the secretion of the key family-48 cellulase (and possibly Cel9G and Cel9E), is missing or insufficiently produced in *C. acetobutylicum*. In the absence of the chaperone, at 37 °C, the Cel48F precursor (and possibly unprocessed Cel9G and Cel9E) would rapidly adopt a conformation incompatible with secretion, inducing obstruction of the SEC complex, which in turn inhibits the growth of the solventogenic bacterium. This hypothesis is consistent with the fact that if electrotransformation of *C. acetobutylicum* with pSOS952-cel48F is performed using the general procedure described in Section 2.2, except that all steps are carried out at 30 °C instead of 37 °C, some colonies that secrete very low amounts of Cel48F are obtained (Mingardon *et al.*, 2011).

It is also worth noting that the mature forms of the scaffoldins and enzymes described in Section 3 are produced in a soluble form in the *E. coli* cytoplasm at 37 °C, whereas the overproduction of Cel48F (Reverbel-Leroy *et al.*, 1996, 1997), Cel9E (Gaudin *et al.*, 2000), and Cel9G (Gal *et al.*, 1997) generates inclusion body unless the expression of their genes is induced at a lower temperature (15–18 °C). Thus, the translation and folding of these large enzymes need to be slowed down to prevent aggregation and obtain soluble forms of these cellulases in *E. coli*.

5.1. Introduction of modules from the *C. cellulolyticum* scaffoldin CipC

The putative chaperone(s) specifically involved in the secretion process of the large cellulases in *C. cellulolyticum* has not been identified yet. Another approach has therefore been selected to slow down the folding of the enzymes and facilitate their secretion by *C. acetobutylicum*: the introduction of carrier (or cargo) modules at the N-terminus of the mature forms of the cellulases. As the scaffoldin miniCipC1 (which contains the family-3a CBM, one X2 module, and the first cohesin of CipC from *C. cellulolyticum*) was efficiently secreted by *C. acetobutylicum* (Perret et al., 2004b), the gene encoding the miniCipC1 (including its leader peptide) was fused to DNA encoding the mature form of Cel48F and cloned in pSOS952 (see Fig. 16.3) (Chanal et al., 2011). To avoid any self-complexation (due to simultaneous presence of complementary cohesin and dockerin modules in the same protein), the DNA encoding the C-terminus dockerin of the cellulase was deleted. Recombinant clones were obtained and Western blot analyses performed on the supernatant and the cell fraction of one clone indicated that the protein is secreted. Activity assays on amorphous cellulose provided a value of 0.3 mg/L for the secretion yield, based on the specific activity of the same fusion enzyme purified from an *E. coli* overproducing strain. It also appeared that the fusion enzyme was mainly produced in a soluble form by *E. coli* at 37 °C, in contrast to wild-type mature Cel48F.

Removal of the cohesin module in the fusion enzyme (3a-Xc-48F and 3a-Xc-48F-do, Fig. 16.3) had no impact on the secretion yield of the modified cellulase by *C. acetobutylicum*, which remained in the 0.3–0.5 mg/L range. Nevertheless, subsequent deletion of the DNA encoding the N-terminal Family 3a CBM (protein Xc-48F-do) notably diminished the secretion yield, since the resulting chimeric protein is barely detectable in the supernatant of the recombinant strain. Nevertheless, electrotransformation of *C. acetobutylicum* with the expression vector encoding this protein generated multiple colonies on selective medium, indicating the protein precursor is nonlethal to the bacterium (Chanal et al., 2011).

The combination of carrier modules (3a-Xc) that gave the best secretion yield of Cel48F was assayed with another deleterious cellulase: Cel9G. The chimeric cellulase was found to be secreted at similar yield (0.4 mg/L) to the engineered Cel48F appended with the same carrier modules. In the case of Cel9G, the deletion of the DNA coding for the X2 module was performed, but no colonies were obtained after electrotransformation with the expression vector encoding 3a-9G. This observation indicates that removal of the X2

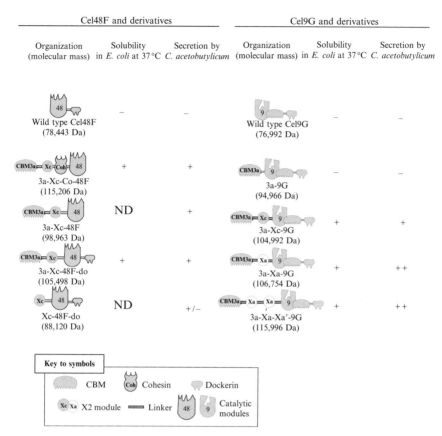

Figure 16.3 Schematic representation of the various cellulases engineered to display scaffoldin modules at the N-terminus. Gray (*C. cellulolyticum*) and white (*C. acetobutylicum*) denote the source of the respective modules or domains (see key to symbols). Solubility in *E. coli* was examined by overproduction of the various enzymes in the cytoplasm at 37 °C. The CipC leader peptide was omitted for production/secretion in *C. acetobutylicum* (pSOS952 expression vector). ND, not determined. Adapted from Chanal *et al.* (2011).

module restored the toxicity of Cel9G despite the presence of the CBM3a module at the N-terminus. When overproduced at 37 °C in the cytoplasm of *E. coli*, the chimeric 3a-9G also generated inclusion bodies at 37 °C.

5.2. Introduction of modules from the *C. acetobutylicum* scaffoldin CipA

As mentioned above, the X2 module is an efficient cargo module (in combination with the scaffoldin Family-3a CBM) facilitating the heterologous secretion of deleterious cellulases. In this respect, it is worth noting that the scaffoldin produced by *C. acetobutylicum* is exceptionally rich in X2 modules since it contains six modules (for only five cohesin modules). In contrast, the scaffoldins of *C. cellulolyticum* or *C. josui* only harbor two and one X2 modules, respectively. The X2 module from *C. cellulolyticum* (Xc) in the combination of carrier domains grafted to Cel9G was therefore replaced by one (3a-Xa-9G) or two (3a-Xa-Xa′-9G) homologous modules (Xa) found in the *C. acetobutylicum* scaffoldin CipA (Fig. 16.3). After electrotransformation of *C. acetobutylicum* with the corresponding expression vectors, batch fermentations at pH 5.5 of the recombinant strains producing 3a-Xc-9G, 3a-Xa-9G, and 3a-Xa-Xa′-9G were performed as described in Section 4. The secretion yield was estimated by incubating 10-mL aliquots of supernatants with 40 mg of Avicel for 10 min at room temperature. The suspensions were centrifuged at $10,000 \times g$ for 10 min (at 4 °C) and the cellulose-bound proteins were eluted by incubating the pellet with 50–100 μL of SDS-PAGE loading buffer during 10 min at 100 °C. The samples were centrifuged for 10 min at $10,000 \times g$, and the supernatants were analyzed by SDS-PAGE and also quantified by microfluidic automated electrophoresis performed on an Experion device (Bio-Rad) (Chanal *et al.*, 2011), and using known concentrations of the corresponding fusion enzymes (purified from overproducing strains of *E. coli*) as the standards. Secretion yields of 0.4, 2.6, and 3.9 mg/L were obtained for 3a-Xc-9G, 3a-Xa-9G, and 3a-Xa-Xa′-9G, respectively. These data indicate that the use of several endogenous X2 modules improves, nearly 10-fold, the secretion of the fusion enzyme.

5.3. Impact of carrier modules on cellulase activity

The presence of a family-3a CBM (and X2 module(s)) at the N-terminus of the GH48 or GH9 fusion cellulases modifies their activity on cellulosic substrates. In general, it improves the specific activity on microcrystalline cellulose such as Avicel, whereas the degradation of acid swollen cellulose (amorphous cellulose) is diminished, especially in the case of Cel9G (Mingardon *et al.*, 2007).

6. OUTLOOK

To our knowledge, the various recombinant strains of *C. acetobutylicum* secreting cellulase(s) (or minicellulosomes) are unable to grow on a medium containing cellulose as the sole source of carbon. The achievement of a cellulolytic *C. acetobutylicum* still needs further investigation and will require the expression of several genes encoding complementary and synergistic cellulases, optionally bound onto a scaffoldin to improve their activity. The genetic tools recently developed by the Minton laboratory

(ClosTron, Heap et al., 2010) will indeed be helpful to clone and express a number of heterologous DNAs. Another challenge concerns the secretion yield of the heterologous cellulosomal components which also need to be significantly improved for achieving a truly cellulolytic strain of *C. acetobutylicum*. Some chaperones specifically required for the secretion of key cellulosomal enzymes in the natural host are probably missing in *C. acetobutylicum* and have to be discovered. The coexpression of the corresponding genes with that encoding the critical cellulases in the solventogenic bacterium would probably improve their secretion yield beyond that attained by grafting "cargo" modules.

ACKNOWLEDGMENTS

This work was supported by Total SA and by grant ANR-05-BLAN-0259-01 from the Agence Nationale de la Recherche.

REFERENCES

Chanal, A., Mingardon, F., Bauzan, M., Tardif, C., and Fierobe, H. P. (2011). Scaffoldin modules serving as "cargo" domains to promote the secretion of heterologous cellulosomal cellulases by *Clostridium acetobutylicum*. *Appl. Environ. Microbiol.* **77,** 6277–6280.

Dong, H., Zhang, Y., Dai, Z., and Li, Y. (2010). Engineering *Clostridium* strain to accept unmethylated DNA. *PLoS One* **5,** e9038.

Gal, L., Gaudin, C., Belaich, A., Pages, S., Tardif, C., and Belaich, J. P. (1997). CelG from *Clostridium cellulolyticum*: A multidomain endoglucanase acting efficiently on crystalline cellulose. *J. Bacteriol.* **179,** 6595–6601.

Gaudin, C., Belaich, A., Champ, S., and Belaich, J. P. (2000). CelE, a multidomain cellulase from *Clostridium cellulolyticum*: A key enzyme in the cellulosome? *J. Bacteriol.* **182,** 1910–1915.

Girbal, L., Mortier-Barriere, I., Raynaud, F., Rouanet, C., Croux, C., and Soucaille, P. (2003). Development of a sensitive gene expression reporter system and an inducible promoter-repressor system for *Clostridium acetobutylicum*. *Appl. Environ. Microbiol.* **69,** 4985–4988.

Heap, J. T., Pennington, O. J., Cartman, S. T., Carter, G. P., and Minton, N. P. (2007). The ClosTron: A universal gene knock-out system for the genus *Clostridium*. *J. Microbiol. Methods* **70,** 452–464.

Heap, J. T., Kuehne, S. A., Ehsaan, M., Cartman, S. T., Cooksley, C. M., Scott, J. C., and Minton, N. P. (2010). The ClosTron: Mutagenesis in *Clostridium* refined and streamlined. *J. Microbiol. Methods* **80,** 49–55.

Lee, S. F., Forsberg, C. W., and Rattray, J. B. (1987). Purification and characterization of two endoxylanases from *Clostridium acetobutylicum* ATCC 824. *Appl. Environ. Microbiol.* **53,** 644–650.

Lopez-Contreras, A. M., Smidt, H., van der Oost, J., Claassen, P. A., Mooibroek, H., and de Vos, W. M. (2001). *Clostridium beijerinckii* cells expressing *Neocallimastix patriciarum* glycoside hydrolases show enhanced lichenan utilization and solvent production. *Appl. Environ. Microbiol.* **67,** 5127–5133.

Lutke-Eversloh, T., and Bahl, H. (2011). Metabolic engineering of *Clostridium acetobutylicum*: Recent advances to improve butanol production. *Curr. Opin. Biotechnol.* **22**, 634–647.

Mermelstein, L. D., and Papoutsakis, E. T. (1993). In vivo methylation in *Escherichia coli* by the *Bacillus subtilis* phage phi 3T I methyltransferase to protect plasmids from restriction upon transformation of *Clostridium acetobutylicum* ATCC 824. *Appl. Environ. Microbiol.* **59**, 1077–1081.

Mingardon, F., Perret, S., Belaich, A., Tardif, C., Belaich, J. P., and Fierobe, H. P. (2005). Heterologous production, assembly, and secretion of a minicellulosome by *Clostridium acetobutylicum* ATCC 824. *Appl. Environ. Microbiol.* **71**, 1215–1222.

Mingardon, F., Chanal, A., Tardif, C., Bayer, E. A., and Fierobe, H. P. (2007). Exploration of new geometries in cellulosome-like chimeras. *Appl. Environ. Microbiol.* **73**, 7138–7149.

Mingardon, F., Chanal, A., Tardif, C., and Fierobe, H. P. (2011). The issue of secretion in heterologous expression of *Clostridium cellulolyticum* cellulase-encoding genes in *Clostridium acetobutylicum* ATCC 824. *Appl. Environ. Microbiol.* **77**, 2831–2838.

Nakotte, S., Schaffer, S., Bohringer, M., and Dürre, P. (1998). Electroporation of, plasmid isolation from and plasmid conservation in *Clostridium acetobutylicum* DSM 792. *Appl. Microbiol. Biotechnol.* **50**, 564–567.

Perret, S., Belaich, A., Fierobe, H. P., Belaich, J. P., and Tardif, C. (2004a). Towards designer cellulosomes in *Clostridia*: Mannanase enrichment of the cellulosomes produced by *Clostridium cellulolyticum*. *J. Bacteriol.* **186**, 6544–6552.

Perret, S., Casalot, L., Fierobe, H. P., Tardif, C., Sabathe, F., Belaich, J. P., and Belaich, A. (2004b). Production of heterologous and chimeric scaffoldins by *Clostridium acetobutylicum* ATCC 824. *J. Bacteriol.* **186**, 253–257.

Qureshi, N., Saha, B. C., and Cotta, M. A. (2007). Butanol production from wheat straw hydrolysate using *Clostridium beijerinckii*. *Bioprocess Biosyst. Eng.* **30**, 419–427.

Reverbel-Leroy, C., Belaich, A., Bernadac, A., Gaudin, C., Belaich, J. P., and Tardif, C. (1996). Molecular study and overexpression of the *Clostridium cellulolyticum celF* cellulase gene in *Escherichia coli*. *Microbiology* **142**, 1013–1023.

Reverbel-Leroy, C., Pages, S., Belaich, A., Belaich, J. P., and Tardif, C. (1997). The processive endocellulase CelF, a major component of the *Clostridium cellulolyticum* cellulosome: Purification and characterization of the recombinant form. *J. Bacteriol.* **179**, 46–52.

Sabathe, F., and Soucaille, P. (2003). Characterization of the CipA scaffolding protein and in vivo production of a minicellulosome in *Clostridium acetobutylicum*. *J. Bacteriol.* **185**, 1092–1096.

Sabathe, F., Belaich, A., and Soucaille, P. (2002). Characterization of the cellulolytic complex (cellulosome) of *Clostridium acetobutylicum*. *FEMS Microbiol. Lett.* **217**, 15–22.

Servinsky, M. D., Kiel, J. T., Dupuy, N. F., and Sund, C. J. (2010). Transcriptional analysis of differential carbohydrate utilization by *Clostridium acetobutylicum*. *Microbiology* **156**, 3478–3491.

CHAPTER SEVENTEEN

Transformation of *Clostridium Thermocellum* by Electroporation

Daniel G. Olson[*,§] and Lee R. Lynd[*,†,‡,§]

Contents

1. Introduction	318
2. Materials	318
2.1. Media	318
2.2. Stock solutions of selective agents	320
2.3. Strains	320
2.4. Plasmids	320
3. Methods	320
3.1. Transformation protocol	320
3.2. Plasmid construction	322
3.3. Gene disruption by allelic replacement with a selective marker	322
3.4. Markerless deletion using removable marker system	325
3.5. Use of selective markers	327
4. Troubleshooting	328
4.1. Arcing during electric pulse application	328
4.2. No colonies after transformation	328
4.3. No colonies after selection with Tm and negative selection (5FOA, FUDR, or 8AZH)	328
4.4. Colony PCR fails	329
4.5. Lawn of colonies after selection	329
References	329

Abstract

In this work, we provide detailed instructions for transformation of *Clostridium thermocellum* by electroporation. In addition, we describe two schemes for genetic modification: allelic replacement—where the gene of interest is replaced by an antibiotic marker and markerless gene deletion—where the gene of interest is removed and the selective markers are recycled. The

---

[*] Thayer School of Engineering at Dartmouth College, Hanover, New Hampshire, USA
[†] Mascoma Corporation, Lebanon, New Hampshire, USA
[‡] Department of Biology, Dartmouth College, Hanover, New Hampshire, USA
[§] BioEnergy Science Center, Oak Ridge, Tennessee, USA

Methods in Enzymology, Volume 510
ISSN 0076-6879, DOI: 10.1016/B978-0-12-415931-0.00017-3

© 2012 Elsevier Inc.
All rights reserved.

markerless gene deletion technique can also be used for insertion of genes onto the *C. thermocellum* chromosome.

1. INTRODUCTION

Clostridium thermocellum has attracted attention due to its ability to rapidly solubilize cellulose. This ability appears to be due to a specialized multienzyme cellulase complex known as the cellulosome (Demain *et al.*, 2005). Since its first discovery (Lamed *et al.*, 1983), it has served as the model for complexed cellulase systems. Targeted gene deletion is a useful tool for understanding complex biological systems. Several improvements have been made to the *C. thermocellum* transformation protocol since it was first reported in our lab in 2004 (Olson, 2009; Tripathi *et al.*, 2010; Tyurin *et al.*, 2004). In addition, a variety of different schemes for gene disruption have been recently described (Argyros *et al.*, 2011; Olson *et al.*, 2010; Tripathi *et al.*, 2010). Here, we describe the transformation protocol currently in use in our lab and review the various engineering schemes that have been used for gene disruption.

Electrotransformation, the technique of introducing DNA into a cell by application of an electric field, is a technique that has existed for several decades. Most electrotransformation protocols require five steps: cell growth, harvest and washing, pulsing, recovery, and selection. Cells are typically harvested at some point in the growth curve between mid-log and early exponential phase. Then, cells are washed to remove the media, which typically contains charged ions that can interfere with electrical pulse application. Next, cells and DNA are mixed, and an electric field is applied. The electric field interacts with the cell membrane causing the temporary formation of holes (or pores) that allow the DNA to enter the cells. Pore size is affected by the amplitude and duration of the applied electrical field. Stronger field strength and longer duration leads to the formation of larger pores, which may allow DNA to enter the cell more readily; however, pores that are too large and may lead to cell death. There are several good references that describe the development and theory of the electrotransformation technique (Chang *et al.*, 1992; Dower et al., 1988).

2. MATERIALS

2.1. Media

A variant of DSM122 media referred to as CTFUD is used. The composition of this medium is described in Table 17.1. To prepare chemically defined medium (CTFUD-NY), yeast extract was replaced by a vitamin

Table 17.1 Recipe for CTFUD media

Reagent	FW (g/mol)	Amount (g/l)
Water (to 1L final volume)		
Sodium citrate tribasic dihydrate, $Na_3C_6H_5O_7 \cdot 2H_2O$	294.10	3.000
Ammonium sulfate, $(NH_4)_2SO_4$	132.14	1.300
Potassium phosphate monobasic, KH_2PO_4	136.09	1.500
Calcium chloride dihydrate, $CaCl_2 \cdot 2H_2O$	147.01	0.130
L-Cysteine–HCl	175.60	0.500
MOPS sodium salt[a]	231.20	11.560
Magnesium chloride hexahydrate, $MgCl_2 \cdot 6H_2O$	203.30	2.600
Ferrous sulfate heptahydrate, $FeSO_4 \cdot 7H_2O$	278.01	0.001
Cellobiose	342.30	5.000
Yeast extract		4.500
Resazurin 0.2% (w/v)		0.5 ml/l

[a] Adjust pH to 7.0 after addition of MOPS.

Table 17.2 Vitamin solution

25 ml 1000 × vitamin solution recipe	
Component	Amount
Water (final volume) (ml)	25
Pyridoxamine HCl (mg)	50
Biotin (mg)	5
p-Aminobenzoic acid (PABA) (mg)	10
Vitamin B_{12} (mg)	5

solution (Table 17.2). The vitamin solution was originally developed by Johnson et al. (1981). For solid media, 0.8% (w/v) agar was added to the medium. Solid media can be stored at 55 °C after autoclaving to avoid solidification. Alternatively, it can be allowed to solidify at room temperature and remelted by microwaving.

Washing of cells during preparation for electrotransformation is performed with a wash buffer consisting of autoclaved reverse-osmosis-purified water with a resistivity of 18 MΩ cm. The purpose of autoclaving the wash buffer is to remove oxygen in addition to sterilization. Alternatively, an aqueous wash buffer containing 10% glycerol and 250 mM sucrose can be used.

2.2. Stock solutions of selective agents

1. *Thiamphenicol (Tm)*: 10 mg/ml stock solution in either dimethyl sulfoxide (DMSO) or a 50% ethanol:50% water mixture, used at a range of final concentrations from 5 to 48 μg/ml. *Note*: DMSO is used for experiments where added ethanol could interfere with results.
2. *Neomycin (Neo)*: 50 mg/ml stock solution in water, filter sterilized and used at 250 μg/ml final concentration.
3. *5-Fluoroorotic acid (5FOA)*: 100 mg/ml stock solution in DMSO (Zymo Research Inc., part number F9003), used at 500 μg/ml final concentration.
4. *5-Fluoro-2'-deoxyuradine (FUDR)* (Sigma, F0503): 10 mg/ml stock solution in water, filter sterilized and used at a final concentration of 10 μg/ml.
5. *8-Azahypoxanthine (8AZH)* (Acros Organics, 202590010): 50 mg/ml stock solution in 1 M NaOH or KOH, filter sterilized and used at a final concentration of 500 μg/ml.

2.3. Strains

C. thermocellum strain DSM1313 (Deutsche Sammlung von Mikroorganismen und Zellkulturen GmbH, Germany) is used for routine transformations. Although we have previously reported transformation of *C. thermocellum* strain ATCC27405, we do not use it for routine transformation due to low and highly variable transformation efficiency.

2.4. Plasmids

Plasmid pMU102 (Fig. 17.1) (GenBank accession number JF423903) is derived from plasmid pNW33N (Olson, 2009), which is available from the Bacillus Genetic Stock Center (Cleveland, OH). It contains an origin of replication that functions in *C. thermocellum* and is the basis for all other replicating plasmids described here.

3. METHODS

3.1. Transformation protocol

1. Cell growth (anaerobic)
   a. Grow 50–500 ml of culture to an $OD_{600} = 0.6$–1.0. Stirring is optional, although not necessary.
   b. Place on ice for 20 min (optional).
   c. (Optional pause point)—at this point, cells can be divided into 50 ml plastic conical tubes and stored at $-80\ °C$. To resume the protocol, thaw tubes at room temperature.

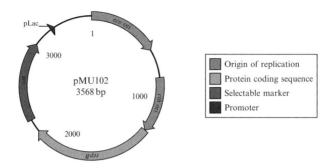

Name	Type	Region	Orientation	Description
eco ori	Replication origin	0001..0589	fwd	pUC19 origin of replication for plasmid propagation in *E. coli*
cth ori	Replication origin	0826..1280	fwd	pNW33N origin of replication for plasmid propagation in *C. thermocellum*
repB	CDS	1281..2285	fwd	Replication protein for pNW33N origin of replication
cat	CDS	2392..3042	fwd	Chloramphenicol acetyl-transferase, provides resistance to chloramphenicol and thiamphenicol
pLac	Promoter	3208..3244	fwd	Lactose inducible promoter from pUC19

Figure 17.1 Diagram of plasmid pMU102. (See Color Insert.)

2. Cell harvesting and washing (aerobic). The goal of washing is to remove as much of the media as possible to ensure the resulting cell suspension has the required electrical properties (i.e., low conductivity). There is no need to resuspend the pellet during washes.
    a. Harvest cells by centrifugation (aerobic or anaerobic) at $6500 \times g$ for 12 min at 4 °C. *Note*: cell harvest and washing can also be performed at room temperature.
    b. Decant supernatant being careful not to disturb pellet. *Note*: we typically observe 5–15% of cells are lost during each decanting step.
    c. Refill centrifuge bottle with electroporation wash buffer without disturbing the pellet.
    d. Centrifuge at $6500 \times g$ for 12 min at 4 °C.
    e. Decant supernatant.
    f. Repeat wash (steps 2c–e) a second time.
3. Pulse application (anaerobic)
    a. Bring cells into an anaerobic chamber (we use Coy Laboratory Products Inc. vinyl anaerobic chambers).
    b. Resuspend cells with 100 µl anaerobic wash buffer by gentle pipetting.
    c. In a standard 1 mm electroporation cuvette, add 20 µl cell suspension and 1–4 µl DNA (100–1000 ng resuspended in water), mix by tapping the cuvette.

d. Apply a square electrical pulse with an amplitude of 1500 V and duration of 1.5 ms (Bio-Rad Gene Pulser Xcell Microbial System part number 165-2662). *Note*: if arcing occurs, prepare a new cuvette and apply a pulse with amplitude decreased by 100 V. If arcing still occurs, repeat with yet another lower-amplitude pulse. Continue incrementally reducing amplitude until no arcing is detected.
4. Recovery (anaerobic)
   a. Resuspend pulsed cells in 3–5 ml media.
   b. Incubate at 51 °C for 6–18 h. Typically, we perform recovery in a stationary dry bath incubator. *Note*: an alternative recovery scheme is to plate cells directly onto selective media with no recovery period and incubate at 51 °C.
5. Select for transformants (anaerobic)
   a. Melt CTFUD agar (Table 17.1) and cool to ~55 °C before addition of 6 µg/ml thiamphenicol. Mix 1 ml and 50 µl aliquots of recovered cells into 20 ml medium.
   b. Pour into Petri dish and allow to solidify at room temperature for 25–30 min.
   c. Incubate at 55 °C. *Note*: secondary containment is necessary to prevent plates from drying out (Remel/Mitsubishi 2.5L AnaeroPak Jar).
   d. Colonies should appear in 3–5 days.

3.2. Plasmid construction

Construction of plasmids is performed using standard cloning methods (Maniatis *et al.*, 1982) and yeast-mediated recombination (Shanks *et al.*, 2006, 2009). Selection in *Escherichia coli* is performed with 5 µg/ml chloramphenicol or 100 µg/ml carbenicillin, selection in yeast is performed with SD ura$^-$ medium according to the protocol described in Shanks *et al.* (2006).

3.3. Gene disruption by allelic replacement with a selective marker (Olson et al., 2010; Tripathi et al., 2010)

3.3.1. Plasmid design

Plasmids designed for gene disruption by allelic replacement have two regions that are homologous to regions flanking the gene target on the *C. thermocellum* chromosome. These homologous flanks should be 500–1000 bp in length. A positive-selectable marker is placed between the homologous flanks, and a negative-selectable marker is placed outside of the homologous flanks. In practice, we have used the *cat* marker driven by the *gapDH* promoter as the positive-selectable marker and *pyrF* driven by the *cbp* promoter as the negative-selectable marker. Use of this selection

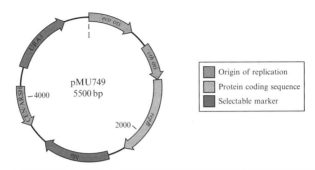

Name	Type	Region	Orientation	Description
eco ori	Replication origin	0001..0589	fwd	pUC19 origin of replication for plasmid propagation in *E. coli*
cth ori	Replication origin	0826..1280	fwd	pNW33N origin of replication for plasmid propagation in *C. thermocellum*
repB	CDS	1281..2285	fwd	Replication protein for pNW33N origin of replication
bla	CDS	2505..3362	fwd	Beta lactamase, provides resistance to ampicillin
CEN6/ARSH	Replication origin	3631..4144	rev	Yeast origin of replication, used for yeast mediated ligation cloning
URA3	CDS	4383..5186	fwd	Provides uracil prototrophy for selection of plasmids during yeast mediated ligation cloning

Figure 17.2 Diagram of plasmid pMU749. (See Color Insert.)

requires a strain of *C. thermocellum* with an inactive *pyrF* gene. Plasmid pMU749 (GenBank accession number JN880474) is a backbone that, with the addition of an upstream (5′), a downstream (3′) flanking region, and the appropriate selectable markers, can be used for this gene disruption scheme (Fig. 17.2) (Olson *et al.*, 2010).

Cumulative marked genetic modifications via allelic replacement are limited by the number of positive-selectable markers available (currently three: *cat*, *neo*, and *pyrF*). Another limitation is that the insertion of a genetic marker onto the chromosome may affect expression of neighboring regions. For example, this scheme is unsuitable for deleting a single domain of a multidomain protein. The benefit of this scheme is that it requires fewer selection steps than markerless gene deletion.

3.3.2. Protocol for allelic replacement by Tm and 5FOA selection (Fig.17.3, allelic replacement, panels A–E)

1. Follow transformation protocol as described above.
2. Screen colonies by PCR to confirm presence of plasmid DNA.

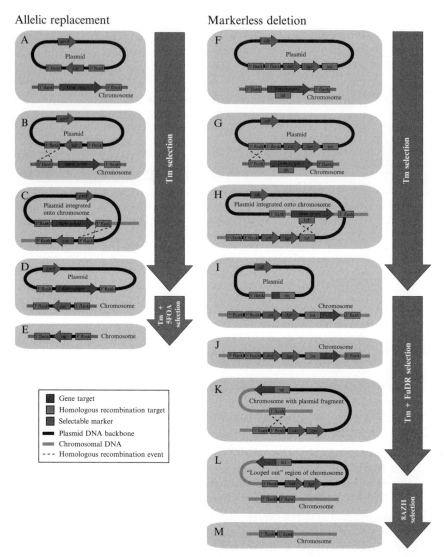

Figure 17.3 Selection scheme diagram. The steps involved in gene disruption by allelic replacement or markerless gene deletion are described. (See Color Insert.)

3. Inoculate confirmed plasmid-containing colony into liquid medium containing positive selection (i.e., Tm) (Fig. 17.3, panels B and C).
4. Plate four 100-fold serial dilutions of the culture on solid medium containing positive and negative selection such as Tm and 5FOA, respectively (Fig. 17.3, panels D and E).
5. Colonies should appear in 2–4 days.

6. Screen colonies by PCR to confirm the allelic replacement has been successful. *Note*: colony PCR is performed by resuspending 5 µl of a fresh colony in 100 µl water and using 0.02–2 µl of the resulting suspension as template for a PCR reaction. The PCR cycling protocol is modified to include a-10 min incubation at either 95 or 98 °C to lyse the cells.
7. Dilution plate to isolate individual colonies on media without selection. This final round of colony purification is recommended to ensure purity of colonies picked from within solid plates.

3.4. Markerless deletion using removable marker system (Argyros *et al.*, 2011)

3.4.1. Plasmid design

Plasmids designed for gene disruption by markerless deletion have two regions that are homologous to regions flanking the gene target on the *C. thermocellum* chromosome. These homologous flanks should be 500–1000 bp in length and are referred to as "5′ flank" and "3′ flank." Additionally, there is a third region with homology to the gene target. Typically, this region referred to as the "int region" is internal to the gene of interest and is also 500–1000 bp in length. *Note*: if the gene target is <500 bp in length, the int region can overlap with the 5′ or 3′ flank. The 3′ flank is chosen from a region downstream of the gene target.

The 5′ and 3′ flanks are inserted on one side of the P*gapDH-cat-hpt* cassette, and the int region is inserted on the other side. The relative orientation of the 5′–3′ flanks and the int region affects the resulting intermediate chromosomal configuration. If the orientation of the flanks is the same as that shown in Fig. 17.3 (i.e., the order is 5′ flank, 3′ flank, cat, hpt, and int region), the intermediate chromosomal configuration will contain a version of the target gene missing the sequence *upstream* of the int region. The flanking regions can be placed in an alternate order (i.e., int region, cat, hpt, 5′ flank, and 3′ flank), and the intermediate chromosomal configuration will contain a version of the target gene missing the sequence *downstream* of the int region. In most cases, the difference between these two possibilities will be unimportant and in that case, we recommend placing the flanks in the order shown in Fig. 17.3. Note that orientations of homologous flanks must be in the same direction or the desired recombination events will not occur.

For this type of gene deletion strategy, we typically use a plasmid with a configuration similar to pAMG258 (Fig. 17.4) (GenBank accession number JN880475). Typically, the 5′ and 3′ flanking regions are inserted near the *Bam*HI site, and the int region is inserted near the *Asc*I site.

Although markerless gene deletion requires more rounds of selection than allelic replacement, it is currently considered the preferred method of chromosomal modification in *C. thermocellum*.

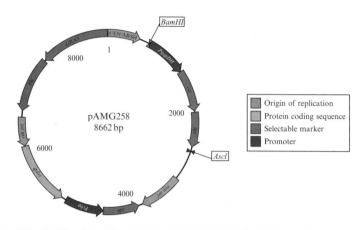

Name	Type	Region	Orientation	Description
CEN6/ARSH4	Replication origin	0001..0519	fwd	Yeast origin of replication, used for yeast mediated ligation cloning
PgapD	Promoter	0692..1339	fwd	Promoter from glyceraldehyde-3-phosphate dehydrogenase gene
cat	CDS	1340..1990	fwd	Chloramphenicol acetyl-transferase, provides resistance to chloramphenicol and thiamphenicol
hpt	CDS	2008..2553	fwd	Marker used for counterselection in *C. thermocellum* with 8AZH
eco ori	Replication origin	3099..3772	fwd	pUC19 origin of replication for plasmid propagation in *E. coli*
tdk	CDS	3833..4411	rev	Marker used for counterselection in *C. thermocellum* with FuDR
Pcbp	Promoter	4412..5032	rev	Promoter from cellobiose phosphorylase (cbp) gene of *C. thermocellum*
repB	CDS	5108..6112	rev	Replication protein for pNW33N origin of replication
cth ori	Replication origin	6113..6567	rev	pNW33N origin of replication for plasmid propagation in *C. thermocellum*
bla	CDS	6670..7530	rev	Beta lactamase, provides resistance to ampicillin
URA3	CDS	7548..8650	rev	Provides uracil prototrophy for selection of plasmids during yeast mediated ligation cloning

Figure 17.4 Diagram of plasmid pAMG258. (See Color Insert.)

3.4.2. Protocol for markerless gene deletion by TM/FUDR/8AZH selection (Fig. 17.3, panels F–M)

1. Follow transformation protocol as described above (Fig. 17.3, panel F).
2. Screen colonies by PCR to confirm presence of plasmid DNA.
3. Inoculate confirmed plasmid-containing colony into liquid medium containing positive selection (i.e., Tm) (Fig. 17.3, panels G–I).
4. When the culture has grown up, plate four 100-fold serial dilutions on solid medium containing positive and first negative selection (Tm and FUDR, respectively) (Fig. 17.3, panel J).

5. Colonies should appear in 2–4 days.
6. Screen colonies by PCR to confirm first and second recombination events (i.e., Fig. 17.3, panels G and H).
7. Inoculate confirmed colony into liquid medium without selection.
8. When the culture has grown up, plate four 100-fold serial dilutions on solid medium containing the second negative selection (i.e., 8AZH) (Fig. 17.3, panels L and M).
9. Screen colonies by PCR to confirm the marker removal has been successful.
10. Dilution plate to isolate individual colonies on media without selection. This final round of colony purification is recommended to ensure purity of colonies picked from within solid plates.

3.5. Use of selective markers

1. *cat*—Preferred positive selection marker. Can be used with either CTFUD or CTFUD-NY media. Addition of Tm to the medium selects for the presence of a functional *cat* gene. Tm concentrations from 5 to 48 μg/ml are routinely used for selection, and increased concentrations have only a marginal effect on growth rate. See Olson thesis Figure 7 for details (Olson, 2011).
2. *neo*—Less-preferred compared with *cat*, since it inhibits growth at the concentrations required for selection. See Olson thesis Figure 14 for details (Olson, 2011). Can be used with either CTFUD or CTFUD-NY media. Addition of *neo* to the medium selects for presence of a functional *neo* gene.
3. *pyrF*—Can be used for both positive and negative selection in *ΔpyrF* backgrounds. Strains with nonfunctional *pyrF* gene need to be grown in media supplemented with at least 40 μg/ml uracil; however, excess uracil can inhibit growth (Tripathi *et al.*, 2010). This selection works best with actively growing cells in mid-log phase and with a final concentration of $< 10^6$ cells/ml. *Positive selection*: Growth in CTFUD-NY media selects for presence of functional *pyrF* gene. *Negative selection*: Addition of 5FOA to CTFUD medium selects for the absence of a functional *pyrF* gene.
4. *tdk*—Preferred negative-selectable marker due to the ability to use it in the wild-type strain background. Either CTFUD or CTFUD-NY media can be used. Addition of FUDR to the medium selects for absence of a functional *tdk* marker. This selection works best with actively growing cells in mid-log phase and with a final concentration of $< 10^6$ cells/ml.
5. *hpt*—Negative-selectable marker which requires a *Δhpt* background. Addition of 8AZH to CTFUD-NY medium selects for the absence of

the *hpt* marker. Since *C. thermocellum* contains a functional *hpt* gene (locus tag Cthe_2254 in the *C. thermocellum* strain ATCC 27405 genome), selection must be performed in a strain where this gene has been inactivated. This selection works best with actively growing cells in mid-log phase and with a final concentration of $< 10^6$ cells/ml.

4. Troubleshooting

4.1. Arcing during electric pulse application

1. Make sure DNA was eluted in water. Common DNA elution buffers such as EB and TE contain salts that can interfere with electroporation. We use the clean and concentrate kit for this purpose (Zymo Research Inc., part number D4003).
2. Reduce the quantity of DNA added.
3. Reduce pulse amplitude by 100 V and try pulsing a freshly prepared cuvette. If arcing is still observed, lower the amplitude and try again with a new cuvette. Repeat as necessary. Although transformation efficiency declines with decreasing pulse amplitude, transformation has been observed at amplitudes as low as 6 kV/cm (600 V pulse amplitude with a 1 mm cuvette).
4. Check resistivity of wash buffer, it should be 18 MΩ cm.
5. Make sure to remove as much media as possible during wash steps. Residual media may contain salts that can interfere with electroporation.

4.2. No colonies after transformation

1. Check to make sure recovery temperature is $< 51\ °C$, for details see Olson thesis Figure 3 (Olson, 2011).
2. Concentrate DNA. DNA should be at a concentration of > 200 ng/μl for best results.
3. If possible, use plasmid pMU102 as a positive control. We have occasionally observed sequence-dependent reductions in transformation efficiency. Plasmid pMU102 is known to transform *C. thermocellum* at high efficiency.

4.3. No colonies after selection with Tm and negative selection (5FOA, FUDR, or 8AZH)

1. Function of *cat* in single copy on the chromosome requires that it be driven by a strong promoter such as *gapDH* (Tripathi et al., 2010). Check plasmid design and construction.

4.4. Colony PCR fails

1. If no PCR product was detected, try repeating the PCR with three 10-fold serial dilutions of the resuspended colony. Frequently PCR reactions fail as a result of *too much* cell material.
2. If the PCR generates a smear instead of a distinct band, try raising the annealing temperature. If you still get a smear, this may be the result of using an old culture. Try subculturing your cells and repeat the PCR with fresh culture.

4.5. Lawn of colonies after selection

1. All selections (except Tm) can be "overwhelmed" by plating concentrations of cells that are too high. Plate serial dilution to ensure a cell concentration to $<10^6$ cells/ml is achieved.
2. CTFUD media occasionally develops a precipitate that can be mistaken for colony growth. If this is the case, an uninoculated plate can be prepared for comparison. To eliminate media precipitation, make sure components are added in the order listed. Also, when media is reheated, occasionally a precipitate forms that disappears upon cooling.
3. Confirm that genetic markers are still present in plasmid.

REFERENCES

Argyros, D. A., Tripathi, S. A., Barrett, T. F., Rogers, S. R., Feinberg, L. F., Olson, D. G., Foden, J. M., Miller, B. B., Lynd, L. R., Hogsett, D. A., and Caiazza, N. C. (2011). High ethanol titers from cellulose using metabolically engineered thermophilic, anaerobic microbes. *Appl. Environ. Microbiol.* **77,** 8288–8294.

Chang, D. C., Chassy, B. M., and Saunders, S. (1992). Guide to Electroporation and Electrofusion. Academic Press, San Diego, CA.

Demain, A. L., Newcomb, M., and Wu, J. H. D. (2005). Cellulase, clostridia, and ethanol. *Microbiol. Mol. Biol. Rev.* **69,** 124–154.

Dower, W. J., Miller, J. F., and Ragsdale, C. W. (1988). High efficiency transformation of *E. coli* by high voltage electroporation. *Nucleic Acids Res.* **16,** 6127–6145.

Johnson, E. A., Madia, A., and Demain, A. L. (1981). Chemically defined minimal medium for growth of the anaerobic cellulolytic thermophile *Clostridium thermocellum*. *Appl. Environ. Microbiol.* **41,** 1060–1062.

Lamed, R., Setter, E., and Bayer, E. A. (1983). Characterization of a cellulose-binding, cellulase-containing complex in *Clostridium thermocellum*. *J. Bacteriol.* **156,** 828–836.

Maniatis, T., Fritsch, E. F., and Sambrook, J. (1982). Molecular Cloning: A Laboratory Manual. Cold Spring Harbor Laboratory Press, Cold Spring Harbor, NY.

Olson, D. G. (2009). Electrotransformation of Gram-positive, Anaerobic, Thermophilic Bacteria. (W. I. P. Orginazation, ed.). Mascoma Corporation, Geneva, Switzerland.

Olson, D. G. (2011). Genetic investigations of the *Clostridium thermocellum* cellulosome. Thayer School of Engineering, Ph.D., Dartmouth College, Hanover, NH, 2011, p. 118.

Olson, D. G., Tripathi, S. A., Giannone, R. J., Lo, J., Caiazza, N. C., Hogsett, D. A., Hettich, R. L., Guss, A. M., Dubrovsky, G., and Lynd, L. R. (2010). Deletion of the Cel48S cellulase from *Clostridium thermocellum*. *Proc. Natl. Acad. Sci. USA* **107,** 17727–17732.

Shanks, R. M. Q., Caiazza, N. C., Hinsa, S. M., Toutain, C. M., and O'Toole, G. A. (2006). *Saccharomyces cerevisiae*-based molecular tool kit for manipulation of genes from gram-negative bacteria. *Appl. Environ. Microbiol.* **72,** 5027–5036.

Shanks, R. M. Q., Kadouri, D. E., MacEachran, D. P., and O'Toole, G. A. (2009). New yeast recombineering tools for bacteria. *Plasmid* **62,** 88–97.

Tripathi, S. A., Olson, D. G., Argyros, D. A., Miller, B. B., Barrett, T. F., Murphy, D. M., McCool, J. D., Warner, A. K., Rajgarhia, V. B., Lynd, L. R., Hogsett, D. A., and Caiazza, N. C. (2010). Development of pyrF-based genetic system for targeted gene deletion in *Clostridium thermocellum* and creation of a pta mutant. *Appl. Environ. Microbiol.* **76,** 6591–6599.

Tyurin, M. V., Desai, S. G., and Lynd, L. R. (2004). Electrotransformation of *Clostridium thermocellum*. *Appl. Environ. Microbiol.* **65,** 883–890.

CHAPTER EIGHTEEN

Genetic and Functional Genomic Approaches for the Study of Plant Cell Wall Degradation in *Cellvibrio japonicus*

Jeffrey G. Gardner *and* David H. Keating

Contents

1. Introduction	332
2. Growth and Maintenance of *C. japonicus*	333
2.1. Acquisition of strain	333
2.2. Culture and storage conditions for *C. japonicus*	333
2.3. Medium for routine growth of *C. japonicus*	334
2.4. Plant cell wall-containing media for screening and propagation	335
3. Genetic Manipulation of *C. japonicus*	336
3.1. Plasmid constructs used in *C. japonicus*	336
3.2. Introduction of DNA into *C. japonicus*	337
4. Construction of Mutations	338
4.1. Random mutagenesis by the use of transposons	338
4.2. Gene inactivation via vector integration	339
5. Transcriptomic Analysis of *C. japonicus*	341
5.1. Transcription analysis via Northern blots	341
5.2. Real-time PCR	341
5.3. β-Glucuronidase (GUS) transcriptional fusions	342
6. Global Expression Profiling	342
6.1. Cell capture during growth in the presence of soluble substrates	344
6.2. Cell capture during growth in insoluble carbon sources	344
6.3. Extraction of RNA	344
6.4. cDNA synthesis, labeling, and microarray analysis	344
7. Summary and Future Directions	345
Acknowledgment	345
References	345

DOE Great Lakes Bioenergy Research Center, University of Wisconsin-Madison, Madison, Wisconsin, USA

Methods in Enzymology, Volume 510   © 2012 Elsevier Inc.
ISSN 0076-6879, DOI: 10.1016/B978-0-12-415931-0.00018-5   All rights reserved.

Abstract

Microbial degradation of plant cell walls is a critical contributor to the global carbon cycle, and enzymes derived from microbes play a key role in the sustainable biofuels industry. Despite its biological and biotechnological importance, relatively little is known about how microbes degrade plant cell walls. Much of this gap in knowledge has resulted from difficulties in extending modern molecular tools to the study of plant cell wall-degrading microbes. The bacterium *Cellvibrio japonicus* has recently emerged as a powerful model system for the study of microbial plant cell wall degradation. *C. japonicus* is unique among microbial model systems in that it possesses the ability to carry out the complete degradation of plant cell wall polysaccharides. Furthermore, an extensive array of genetic and molecular tools exists for functional genomic analysis. In this review, we describe progress in the development of methodology for the functional genomic study of plant cell wall degradation by this microbe, and discuss future directions for research.

1. INTRODUCTION

The plant cell wall is a chemically and physically complex structure composed of an elaborate array of interconnected structural polysaccharides. The major component of plant cell walls is cellulose, a β-1,4-linked crystalline polymer of glucose residues that comprises 35–50% of plant cell walls by weight (Gilbert, 2010). The cellulose is present in a complex matrix with hemicellulose, which makes up 20–35% of plant cell walls, and consists of three components: xylan, mannan, and xyloglucan. In addition, the molecule pectin is present and is found in three distinct varieties: homogalacturonan, rhamnogalacturonan I, and rhamnogalacturonan II (Mohnen, 2008). Finally, secondary cell walls are modified by linkage to lignin, a highly hydrophobic molecule composed of hydroxycinnamyl, coniferyl, sinapyl, and *p*-coumaryl alcohols (Vanholme et al., 2010).

The cell walls of plants represent a major reservoir of global carbon (Leschine, 1995), and microbial turnover of plant cell walls contributes significantly to the carbon balance (Heimann and Reichstein, 2008; Wilson, 2011). In addition to their role in the carbon cycle, microbial enzymes involved in the degradation of plant cell walls have recently emerged as critical components of the sustainable biofuels industry. Despite the abundance of carbon available in plant cell walls, the combination of cellulose crystallinity, hemicellulose complexity, and lignin-mediated enzyme inhibition makes these structures highly recalcitrant to microbial degradation (Lynd et al., 2002). This recalcitrance reduces turnover of plant cell walls in natural environments and is a major impediment to efforts to develop cost-effective bioenergy. Currently, overcoming this recalcitrance requires the use of large quantities of enzymes and chemical pretreatments, which significantly increase the cost of lignocellulosic biofuels (Mosier et al., 2005). Capturing the energy in plant cell walls for fuel production requires an improved understanding of the mechanism of microbial degradation. Furthermore, because plant cell wall degradation occurs within the context of a microbial cell, a systems-level understanding of the cellular state during plant cell wall breakdown, including the genes and proteins involved, and the regulatory networks that control their expression, will be required to fully understand this process and ultimately exploit it for the improvement of bioenergy.

In this review, we will describe progress in the development of *Cellvibrio japonicus* as a model system for the study of plant cell wall degradation. This Gram-negative soil dwelling bacterium possesses several traits critical for systems-biological analysis. First, it contains one of the most extensive plant cell wall-degrading systems known, with the capacity to completely depolymerize polysaccharides from diverse plant sources, including bioenergy-relevant switchgrass and corn stover. Second, a significant number of plant cell wall hydrolases from this bacterium have been subjected to detailed biochemical characterization, and the crystal structure of 13 of these enzymes have been determined. Third, the microbe possesses an impressive array of genetic and molecular tools. We will describe the current biochemical, genetic, and genomic tools available in *C. japonicus*, and how they are being used to understand the mechanism by which this organism carries out degradation of plant cell walls.

2. GROWTH AND MAINTENANCE OF *C. JAPONICUS*

2.1. Acquisition of strain

C. japonicus sp. *nov.* strain Ueda107 (Ueda et al., 1952), formerly known as *Pseudomonas fluorescens* subsp. *cellulosa*, can be obtained from the National Collections of Industrial, Marine and Food Bacteria (http://www.ncimb.com).

2.2. Culture and storage conditions for *C. japonicus*

C. japonicus is not known to be resistant to any antibiotics, although a rifampicin resistant mutant has been reported (Beylot et al., 2001). The bacterium has been reported to grow at temperatures between 20 and 37 °C (Humphry et al., 2003). However, in our experience, growth of *C. japonicus* is most rapid at 30 °C and in the presence of high aeration (225 rpm). Although the strain has been reported to grow under anaerobic conditions (Humphry et al., 2003), we have not observed growth in the absence of oxygen. For short-term storage, plates containing *C. japonicus* can be stored on the bench (22 °C) for up to 2 weeks but storage at 4 °C results in loss of viability. For long-term storage, freezing of cultures at −80 °C in the presence of 25% glycerol (v/v, final) led to the greatest retention of culture viability.

2.3. Medium for routine growth of *C. japonicus*

C. japonicus strains have been reported to grow in the defined minimal medium M9 (Beylot *et al.*, 2001; Miller, 1975) supplemented with $MgSO_4$ (1 m*M*) and $CaCl_2$ (0.1 m*M*). However, in our experience, *C. japonicus* strains display a more rapid growth rate in MOPS (morpholinopropane sulfonic acid) minimal medium (Fig. 18.1; Neidhardt *et al.*, 1974). In addition, growth in MOPS leads to improved viability during long-term storage at −80 °C. The reason for the improved growth and viability in MOPS medium has not been investigated, but may result from the presence of iron and other trace elements that are added to MOPS media, but lacking in conventional M9 recipes. Despite the difference in growth rate, the cells reach a similar final cell density in both media (Fig. 18.1). There are multiple reports involving the use of lysogenic broth (LB; Bertani, 1951, 2004) as a medium for propagation of *C. japonicus* (Beylot *et al.*, 2001). In our experience, however, the bacterium grows relatively poorly in LB, with the cells only reaching a final cell density of only 0.3–0.5 (as measured by OD_{600}), whereas cells grown in glucose MOPS medium reach a final cell density of

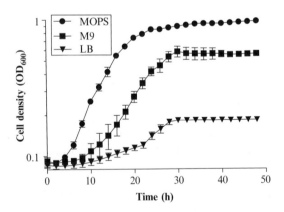

Figure 18.1 Comparison of growth of *C. japonicus* in M9 and MOPS minimal media, as well as LB medium. *C. japonicus* was cultured in a baffled shake flask at 30 °C, and cell density measured by OD_{600} during growth in M9 (squares), MOPS (circles) minimal media, and LB medium (triangles). MOPS and M9 medium contained glucose (10 g/l) as a carbon source.

1.3 (Fig. 18.1). In addition, culturing of cells in LB led to a reduction in viability upon freezing.

2.4. Plant cell wall-containing media for screening and propagation

Genetic and systems-biological approaches for the study of *C. japonicus* require culturing cells in the presence of plant cell walls, as well as their monosaccharide, oligosaccharide, and polysaccharide components. Plant cell walls and purified cellulose present a particular challenge due to their insolubility and physical/chemical heterogeneity. Despite these challenges, both solid and liquid media containing these substrates have been developed for culture of *C. japonicus*. In the case of solid medium, two distinct types have been developed: a traditional agar-based solid medium and an alternative solid medium that uses acrylamide as a solidifying agent (Gardner *et al.*, 2011).

2.4.1. Plant cell wall liquid medium

Derivatives of M9 and MOPS media were developed with corn stover or switchgrass as a sole carbon source (Gardner and Keating, 2010). The plant material (corn stover or switchgrass) was ground to a small particle size (~0.5 mm) by the use of a porcelain mortar and pestle and then was subjected to a 15-min autoclave cycle with 30 min of drying. Finally, the sterile plant material was washed five times with sterile water and added at a concentration of 10 g/l to M9 or MOPS media (Neidhardt *et al.*, 1974; prepared without the addition of glucose or any other carbon source).

2.4.2. Plant cell wall agar-solidified medium

To facilitate the high-throughput screening and propagation of *C. japonicus*, a solid phase derivative of MOPS medium was developed with plant cell walls as the sole carbon source (Gardner *et al.*, 2011). The plant material was prepared in the same manner as described for the preparation of liquid medium and transferred aseptically to 5 ml of molten MOPS agar (Neidhardt *et al.*, 1974) at a concentration of 10 g/l in a 50-ml conical tube. The tube was mixed briefly via the use of a vortex mixer and the suspension overlaid onto a previously prepared M9 agar plate that lacked a carbon source.

2.4.3. Plant cell wall acrylamide-solidified medium

The use of acrylamide as a solidifying agent required special polymerization conditions, which increased the efficiency of polyacrylamide polymerization, and reduced the concentration of unpolymerized acrylamide in the medium. To a 125 ml flask was added 25 ml of 40% acrylamide:bis-acrylamide solution (29:1 ratio with 3.3% cross-linking) in combination with 1 g of mixed-bed ion-exchange resin and stirred for 30 min at room temperature. The acrylamide:bis-acrylamide solution was then passed through a 0.45-μm syringe filter to remove the mixed-bed ion-exchange resin and sterilize the solution. The acrylamide:bis-acrylamide solution was added to 20 ml of sterile 5 × M9 (or MOPS), 100 μl of sterile 1 M MgSO$_4$, 10 μl of sterile 1 M CaCl$_2$, and 10 μl of thiamine (10 μg/ml) in a sterile glass bottle, and then added to the autoclaved and washed plant material in a sterile conical tube. The mixture was equilibrated overnight in an anaerobic chamber (H$_2$ 10%/CO$_2$ 10%/N$_2$ 80%, mol/mol), along with the plastic materials used for preparation of the plates (Petri dishes, 50 ml conical tubes, and pipetman tips). Then 200 μl of freshly prepared sterile 10% ammonium persulfate and 100 μl of TEMED were added to the acrylamide:bis-acrylamide/plant material mixture and rapidly poured into Petri dishes.

Once polymerized (approximately 5 min), the plates were removed from the anaerobic chamber and subjected to washing to remove residual unpolymerized acrylamide and other growth inhibitory compounds. The lids were removed from the Petri dishes, the plates sterilized via immersion in 95% ethanol for 10 min, transferred via the use of sterile forceps to a beaker containing 1 l of sterile M9 or MOPS medium, and incubated at room temperature for 12 h in a covered container. The plates were then removed and placed in a sterile hood with a pair of sterile forceps and dried for 15 min.

3. Genetic Manipulation of *C. japonicus*

The development of *C. japonicus* as a system for systems-biological studies requires the ability to manipulate its genotype. *C. japonicus* possesses all of the attributes necessary for efficient genetic manipulation, namely the ability to introduce foreign DNA, and the presence of systems for gene overexpression and gene disruption via random and targeted methods. Below, we will describe each of these methods. First, we will describe the plasmids known to replicate in *C. japonicus* and protocols for introduction of DNA into this organism. Methods for random gene disruption via the use of transposon mutagenesis, targeted gene disruption, and gene overexpression will then be described.

3.1. Plasmid constructs used in *C. japonicus*

Although an extensive characterization of plasmid stability has not been carried out in *C. japonicus*, stable replication has been reported for a derivative of plasmid pVS1, an IncP plasmid originally isolated from *Pseudomonas aeruginosa* (Itoh and Haas, 1985). In addition, our laboratory has shown that the gentamycin resistant broad host range plasmid pBBR1-MCS5 (Kovach et al., 1994) replicates in *C. japonicus* (Gardner and Keating, 2010). The ability to introduce plasmids into *C. japonicus* allows for the engineering of bacterial metabolism for the production of molecules of biotechnological interest. For example, the *pdc* and *adhB* genes from *Zymomonas mobilis*, encoding pyruvate decarboxylase and alcohol dehydrogenase, respectively, were cloned behind the *lac* promoter in the plasmid pBBR1-MCS5 (Kovach et al., 1994). Introduction of the plasmid into *C. japonicus* resulted in production of ethanol, indicating that the *pdc* and *adhB* genes were expressed and functional (Gardner and Keating, 2010).

3.2. Introduction of DNA into *C. japonicus*

Introduction of DNA is a critical component for overexpression of genes, construction of transcriptional fusions, and manipulation of the *C. japonicus* genome. Plasmid DNA can be introduced via two approaches, electroporation and conjugation. Both will be discussed in the following sections.

3.2.1. Electroporation

Plasmids can be introduced into *C. japonicus* by electroporation (Emami et al., 2002). However, in our experience, the efficiency of the method is relatively low, likely due to the presence of a restriction system that leads to the degradation of plasmid DNA (analysis of the *C. japonicus* genome suggested the presence of multiple Type I restriction systems).

Protocol

C. japonicus was cultured in LB (400 ml) to mid-exponential phase, and the bacterial cells were recovered via centrifugation at $6000 \times g$ for 10 min at 4 °C. The cell pellet was then washed three times with a 400 ml solution of cold 5% (v/v) glycerol, and the cells resuspended in 1 ml of the 5% glycerol solution. The washed cells were frozen as 50 μl aliquots and stored at -80 °C. To introduce DNA into cells, 50 μl of a thawed aliquot of electrocompetent cells and 200–500 ng of plasmid DNA were placed in Gene Pulser cuvettes (0.1-cm gap), and an electrical current applied with the following parameters (1.6 kV cm, 100 Ω, pulse time of 4 ms). The bacterial cells were then diluted with 250 μl of LB and incubated 1 h at 37 °C with aeration and plated on media containing appropriate antibiotics.

3.2.2. Conjugation

Multiple reports have indicated that plasmids can be moved into *C. japonicus* via the use of binary conjugation systems (Beylot et al., 2001; Emami et al., 2002; Gardner and Keating, 2010), in which a "helper" plasmid provides the conjugation functions necessary for plasmid mobilization into the bacterium. The helper functions can be provided by the plasmid pRK2013 (Figurski and Helinski, 1979), which has been widely used for conjugation with diverse hosts. To select against the *Escherichia coli* strains (a step referred to as counterselection), previous studies have employed a spontaneous rifampin resistant mutant of *C. japonicus* (Emami et al., 2002). An alternative approach takes advantage of the valine sensitivity of *E. coli* K12 strains (Gardner and Keating, 2010), which result from mutations in the *ilvG* gene, and engender an inability to produce sufficient isoleucine for bacterial growth when valine is present in the medium (Leavitt and Umbarger, 1961, 1962). This method can, in principle, be used with any *C. japonicus* strain, as well as other types of cell wall-degrading bacteria that are valine resistant.

Protocol

An *E. coli* S17 λ_{PIR} strain harboring the plasmid to be mobilized, an *E. coli* strain harboring plasmid pRK2013, and the *C. japonicus* strain were streaked into a common region of an LB plate (or a MOPS plate in the presence of supplements necessary for growth of all three strains), followed by incubation for 48 h at 30 °C. The cells from the mating region were then streaked on M9 or MOPS minimal glucose medium containing antibiotic and 50 μg/ml valine. Finally, the *C. japonicus* colonies were purified by streaking twice on the same medium.

4. Construction of Mutations

Systems biology requires the capacity to construct mutations in candidate genes identified via transcriptomic, metabolomic, or proteomic approaches, or from bioinformatic analysis of genome sequence. Methodology now exists for the construction of targeted and random mutations in *C. japonicus*, which will be described in the following sections.

4.1. Random mutagenesis by the use of transposons

The first reported genetic tool in *C. japonicus* was the development of a system for random mutagenesis via a Tn*10*-derived mini transposon (a Tn*5*-derived system was also developed but proved ineffective due to nonrandom transposition; Emami et al., 2002). The Tn*10* mini transposon was present on a plasmid pLOFKm that can be introduced into *C. japonicus* via conjugation (Herrero et al., 1990). Because this plasmid cannot replicate within *C. japonicus*, the transposon-encoded kanamycin resistance can only be retained in the organism by transposition of the mini-Tn*10* into the chromosome. In addition, because the gene encoding the transposase is expressed from the plasmid, the transposon is stable upon integration into the host chromosome. Using this approach, the authors of the study were able to identify over 30,000 transposition events, including mutations in the xylanases *xyn10A* and *xyn10B*, as well as the regulator *abfS* (Emami et al., 2002)

Protocol

The transposon Tn*10* was introduced into the *C. japonicus* genome by culturing an *E. coli* S17-1λpir strain and *C. japonicus* overnight in LB supplemented with antibiotics. The cells were then recovered by centrifugation, washed, and resuspended in phosphate-buffered saline. The cell suspension was applied in a ratio of 5:1 (*C. japonicus*:*E. coli*), and the 6 ml was allowed to undergo filter mating (Beylot et al., 2001) overnight on LB plates, to which 1 mM isopropyl-β-D-1-thiogalactopyranoside (IPTG) was added (which serves to upregulate expression of the transposase gene). Transposition events were selected on medium with 50 μg/ml rifampicin and 50 μg/ml kanamycin.

4.2. Gene inactivation via vector integration

Although the ability to identify random mutations was a significant advance in understanding the genetics of plant cell wall breakdown, efficient systems biology requires the construction of targeted mutations in candidate genes. Targeted gene disruption in *C. japonicus* can be carried out using the plasmid pK18*mobsacB* (Schafer et al., 1994), which can be mobilized into diverse Gram-negative strains. The plasmid cannot replicate within *C. japonicus* (Gardner and Keating, 2010). However, if the plasmid contains an internal region of the target gene, then this region can undergo recombination with its cognate region of the *C. japonicus* chromosome, leading to plasmid integration, and gene disruption (Fig. 18.2A). The first test of this system involved the inactivation of the gene *gspD* of *C. japonicus*, which encoded the critical secretin component of the Type II secretion system, and is responsible for secretion of plant cell wall-degrading enzymes from *C. japonicus* (Gardner and Keating, 2010). The *gspD* plasmid integration mutant grew in a manner similar to wild type when glucose was used as a carbon source, but displayed a decrease in endoglucanase secretion, and a greatly reduced ability to utilize plant cell wall containing substrates such as corn stover (Fig. 18.2B).

Protocol

A derivative of plasmid pK18*mobsacB* was introduced into S17 λ$_{PIR}$ by electroporation, followed by selection for kanamycin resistance, and then introduced into *C. japonicus* by conjugation. After incubation for 48 h at

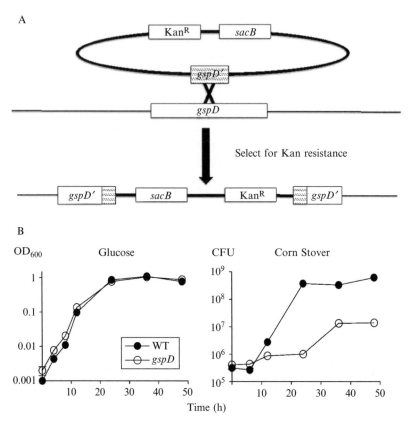

Figure 18.2 Construction of targeted mutations in *C. japonicus*. An insertion mutation was constructed in *gspD*, which was then subjected to phenotypic analysis. (A) Strategy for insertional inactivation of *gspD* gene. Upon introduction into *C. japonicus* by conjugation, a region of the plasmid containing DNA identical to an internal 500 bp portion of the *gspD* gene recombines with the cognate full-length *gspD* gene on the chromosome. The result is integration of the plasmid into the genome, and disruption of *gspD*. (B) Phenotypic characterization of the *gspD* mutant. Wild type *C. japonicus* and a strain containing the insertionally inactivated *gspD* gene were subjected to growth in M9 medium containing either 10 g/l glucose or 10 g/l corn stover. Growth was measured by optical density (glucose) or by viable cell counting (corn stover). Data is reprinted from Gardner and Keating (2010).

30 °C, the mating mixture was streaked onto MOPS medium containing 25 μg/ml kanamycin and 50 μg/ml valine. The colonies that arose after 4–5 days of incubation at 30 °C were purified by streaking two times on MOPS medium containing 25 μg/ml kanamycin and 50 μg/ml valine, and the integration event verified by PCR analysis.

5. Transcriptomic Analysis of *C. japonicus*

The genomes of plant cell wall-degrading microbes often encode large numbers of glycoside hydrolases, polysaccharide lyases, and carbohydrate esterases, often with unassigned functions. Measurement of transcription of these genes can identify conditions under which a given candidate gene is expressed, which can provide insight into its function. Multiple methods have been developed for the measurement of transcription in *C. japonicus*. Methods such as Northern blots, real time-PCR (RT-PCR), and transcriptional fusions are most useful for the transcriptional characterization of individual genes or small groups of genes, whereas global expression profiling methods such as microarrays are used to measure expression of larger groups of genes.

5.1. Transcription analysis via Northern blots

The first study of transcription of plant cell wall-degrading genes in *C. japonicus* (Emami *et al.*, 2002) made use of the well-established Northern blot approach. The studies demonstrated that the *xyn10* and *xyn11* genes can be monitored during growth of *C. japonicus* in glucose and soluble xylan. Critically, temporal analysis of expression using this approach identified differences in transcription of *xyn11A* and *xyn11B* genes, which provided insight into their role during xylan breakdown.

Protocol

RNA was isolated from cultures of *C. japonicus* via the use of RNAZOL as previously described (Beylot *et al.*, 2001). The extracted RNA (5 µg) was then fractionated by electrophoresis on 1.2% (w/v) agarose-formaldehyde denaturing gels (Sambrook *et al.*, 1989) and transferred to nylon membranes (Hybond N; Amersham Pharmacia) with 300 mM sodium citrate (pH 7.2), 3 M NaCl as a buffer, followed by immobilization of the RNA by UV cross-linking. The immobilized RNA was then probed with radiolabeled DNA corresponding to the genes of interest, and the hybridized DNA detected by the use of a phosphorimager.

5.2. Real-time PCR

A rapid method for quantification of transcription in *C. japonicus* involves the use of RT-PCR (Emami *et al.*, 2002). In this approach, RNA is extracted and converted to cDNA via the use of reverse transcriptase. The cDNA is then amplified by PCR, and the amount of amplified DNA quantified by fluorescence of the reagent SYBR green.

Protocol

C. japonicus total RNA (1 µg), prepared as described above, was used as a template for synthesis of cDNA via the use of random primers and a Superscript preamplification cDNA synthesis kit (Gibco BRL). The cDNA was then used as a template for amplification via PCR with transcript-specific primers (200 µM), 3 mM MgCl$_2$, and amplification mix (Light Cycler-DNA SYBR Green 1; Roche Diagnostics). Amplifications were carried out in a Roche Light Cycler instrument according to the following temperature protocol: 95 °C for 20 min, 50 cycles of 95 °C for 15 s, 60 °C for 5 s, and 72 °C for 10 s. The intensity of SYBR Green 1 fluorescence, which measures dsDNA, was measured at the end of each elongation reaction. Fluorescent signals from primer dimers and other nonspecific background were controlled for as described by the manufacturer's instructions.

5.3. β-Glucuronidase (GUS) transcriptional fusions

A rapid method for evaluating expression involves the transcriptional fusion of a promoter of interest to a gene encoding an easily assayed protein product. A previous study (Emami *et al.*, 2002) employed the use of transcriptional fusions to *gusA*, which encodes β-glucuronidase (Roberts *et al.*, 1989; Wilson *et al.*, 1991), using the broad host range promoter fusion vector pRG960SD (Van den Eede *et al.*, 1992). The fusion vector can be introduced by electroporation or conjugation, and the copy number of this plasmid has been reported to be approximately 1 per cell, which reduces the titration effects of regulatory proteins that often accompany the use of higher copy plasmids.

Protocol

A reaction mixture (500 µl) composed of 50 mM sodium phosphate buffer (pH 7.0), 1 mM 4-methylumbelliferyl-β-D-glucuronide, 1 mg of bovine serum albumin (BSA)/ml, and dilutions of cell extracts were incubated at 37 °C for 10 min and was terminated by the addition of a stop solution (2.5 ml of 50 mM glycine–NaOH buffer, pH 10.4). Activity was then quantified by measurement of fluorescence associated with release of the 4-methylumbelliferone via the use of an RF-1501 spectrofluorophotometer (Shimadzu) using excitation and emission wavelengths of 365 and 460 nm, respectively.

6. Global Expression Profiling

Generation of a systems-level understanding of plant cell wall degradation requires characterization of the complete cellular transcriptome during growth in the presence of artificial growth substrates, as well as plant cell walls. Plant cell walls and their constituents lead to significant challenges for global expression analysis, due their insolubility and polysaccharide/lignin composition, which inhibits downstream applications with the RNA. To circumvent these challenges, our laboratory developed a two step protocol for extraction of RNA from cells grown in the presence of plant cell wall containing substrates. In the first step, a solution of phenol and ethanol is added to the bacterial culture, which serves to quench RNA turnover (Rhodius and Gross, 2011). The second step involves the physical separation of the cells and the carbon source. In the case of soluble substrates such as monosaccharides, disaccharides, and soluble polysaccharides such as xylan and carboxymethylcellulose, the cells undergo a simple washing step. In the case of insoluble substrates, such as plant cell walls and purified cellulose, the material is removed by fractionation through a course filter or low-speed centrifugation. The cells are then subjected to RNA extraction via a hot phenol separation method (Rhodius and Gross, 2011), the RNA converted to cDNA, labeled, and subjected to microarray analysis on a custom NimbleGen microarray. Using this approach, expression of the xylanases *xyn11A* and *xyn11B* during growth in xylan during exponential and stationary phase growth (Fig. 18.3) were observed to be similar to previously reported expression results (Emami *et al.*, 2002). We expect that variations of this method should be broadly applicable to a number of cellulose-degrading microorganisms. The methods for sample collection, removal of insoluble material, and RNA extraction are listed in the following sections.

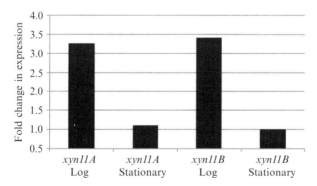

Figure 18.3 Expression of *xyn11A* and *xyn11B* during growth in the presence of glucose and xylan. *C. japonicus* was cultured in the presence of glucose and xylan (10 g/l). The cells were then subjected to phenol–ethanol treatment, RNA extracted from cell pellets, converted to cDNA, labeled, hybridized to NimbleGen microarrays, and the label quantified. Values represent fold changes in expression of *xyn11A* and *xyn11B* in log phase and stationary phase cells cultured in the presence of xylan, with respect to growth in the presence of glucose.

6.1. Cell capture during growth in the presence of soluble substrates

RNA quenching was carried out by the rapid collection of 35 ml of cell culture into a prechilled 50 ml conical tube that already contained 5 ml of ice cold stop solution (95:5, v/v, absolute ethanol:saturated unbuffered phenol; Rhodius and Gross, 2011). In the case of soluble substrates such as monosaccharides, disaccharides, and soluble forms of cellulose such as CMC, the cells were then collected by centrifugation at $10,000 \times g$ for 5 min at 4 °C to pellet the cells. The cell pellet then underwent an additional wash with a 50 ml solution of MOPS medium and phenol/ethanol (mixed in the same ratios as above). The supernatant was removed and the pellets flash frozen in a dry ice/ethanol bath for 5 min before being stored at -80 °C.

6.2. Cell capture during growth in insoluble carbon sources

6.2.1. Removal of cells by filtration

Glass wool was placed at the bottom of a 50-ml syringe. The suspension of cells/phenol:ethanol stop solution was then forced through the syringe, trapping the insoluble carbon source in the glass wool, and the cells collected in a tube. The cells were then recovered by centrifugation as described above. A limitation of this method was the low flow rate resulting from the presence of the insoluble material.

6.2.2. Centrifugation

An alternative method for separation of cells from insoluble material involves the use of a very low speed spin at $500 \times g$ for 5 min at 4 °C to pellet the insoluble carbon sources, but not the cells. The cell suspension was then transferred to a fresh 50 ml conical tube and subjected to a subsequent high-speed spin at $10,000 \times g$ for 5 min at 4 °C to recover the cells. The cells were then flash frozen and subjected to RNA extraction as discussed in the following sections.

6.3. Extraction of RNA

RNA was extracted by hot phenol extraction as described previously (Rhodius and Gross, 2011).

6.4. cDNA synthesis, labeling, and microarray analysis

Purified RNA was converted to cDNA, labeled, and subjected to microarray analysis as described previously (Nimblegen, 2010).

7. Summary and Future Directions

The efforts of multiple laboratories have produced the tools necessary for sophisticated functional genomic analysis in *C. japonicus*. However, additional technical hurdles remain. A key challenge in the study of plant cell wall-degrading organisms such as *C. japonicus* is the large number of predicted genes that appear to encode redundant activities. Bioinformatic analysis of these genes often provides little insight into their function, in that they appear to encode highly similar proteins. Understanding the function of these genes in plant cell wall degradation will require a combination of existing tools, such as global transcriptional analysis, as well as the development of new tools. In particular, the study of these genes will require the development of new strategies for mutant construction. The plasmid pK18*mobsacB* used for the targeted gene disruption contains the *sacB* gene, which encodes sensitivity to sucrose. The *sacB* gene allows for selection against strains in which the plasmid has integrated into the genome (Gay *et al.*, 1985), allowing for identification of double recombinants that have undergone allelic replacement of the gene of interest. Thus, this construct facilitates the construction of marked and unmarked deletions, as well as "in-frame" deletions, which disrupt the gene without affecting the transcription of downstream genes. We expect that this system will also allow the construction of strains that contain multiple deletions, which will be critical for understanding the function of genes that encode redundant functions in plant cell wall degradation. Finally, this system can be used to introduce point mutations into candidate genes in *C. japonicus*, allowing for testing of mutations affecting key catalytic or regulatory residues in their native chromosomal context. We expect that the development of these tools will lead to a significant acceleration in the pace of research into plant cell wall breakdown, opening up new avenues for the study of global carbon recycling, and facilitating the development of cost-effective biofuels.

ACKNOWLEDGMENT

This work was funded by the DOE Great Lakes Bioenergy Research Center (DOE BER Office of Science DE-FC02-07ER64494).

REFERENCES

Bertani, G. (1951). Studies on lysogenesis. I. The mode of phage liberation by lysogenic *Escherichia coli*. *J. Bacteriol.* **62,** 293–300.

Bertani, G. (2004). Lysogeny at mid-twentieth century: P1, P2, and other experimental systems. *J. Bacteriol.* **186,** 595–600.

Beylot, M. H., Emami, K., McKie, V. A., Gilbert, H. J., and Pell, G. (2001). *Pseudomonas cellulosa* expresses a single membrane-bound glycoside hydrolase family 51 arabinofuranosidase. *Biochem. J.* **358,** 599–605.

Emami, K., Nagy, T., Fontes, C. M., Ferreira, L. M., and Gilbert, H. J. (2002). Evidence for temporal regulation of the two *Pseudomonas cellulosa* xylanases belonging to glycoside hydrolase family 11. *J. Bacteriol.* **184,** 4124–4133.

Figurski, D. H., and Helinski, D. R. (1979). Replication of an origin-containing derivative of plasmid RK2 dependent on a plasmid function provided in trans. *Proc. Natl. Acad. Sci. USA* **76,** 1648–1652.

Gardner, J. G., and Keating, D. H. (2010). Requirement of the type II secretion system for utilization of cellulosic substrates by *Cellvibrio japonicus*. *Appl. Environ. Microbiol.* **76,** 5079–5087.

Gardner, J. G., Zeitler, L. A., Wigstrom, W. J., Engel, K. C., and Keating, D. H. (2011). A high-throughput solid phase screening method for identification of lignocellulose-degrading bacteria from environmental isolates. *Biotechnol. Lett.* **3.4,** 81–89.

Gay, P., Le Coq, D., Steinmetz, M., Berkelman, T., and Kado, C. I. (1985). Positive selection procedure for entrapment of insertion sequence elements in gram-negative bacteria. *J. Bacteriol.* **164,** 918–921.

Gilbert, H. J. (2010). The biochemistry and structural biology of plant cell wall deconstruction. *Plant Physiol.* **153,** 444–455.

Heimann, M., and Reichstein, M. (2008). Terrestrial ecosystem carbon dynamics and climate feedbacks. *Nature* **451,** 289–292.

Herrero, M., de Lorenzo, V., and Timmis, K. N. (1990). Transposon vectors containing non-antibiotic resistance selection markers for cloning and stable chromosomal insertion of foreign genes in gram-negative bacteria. *J. Bacteriol.* **172,** 6557–6567.

Humphry, D. R., Black, G. W., and Cummings, S. P. (2003). Reclassification of '*Pseudomonas fluorescens* subsp. *cellulosa*' NCIMB 10462 (Ueda *et al.* 1952) as *Cellvibrio japonicus* sp. nov. and revival of *Cellvibrio vulgaris* sp. nov., nom. rev. and *Cellvibrio fulvus* sp. nov., nom. rev. *Int. J. Syst. Evol. Microbiol.* **53,** 393–400.

Itoh, Y., and Haas, D. (1985). Cloning vectors derived from the *Pseudomonas* plasmid pVS1. *Gene* **36,** 27–36.

Kovach, M. E., Phillips, R. W., Elzer, P. H., Roop, R. M., 2nd, and Peterson, K. M. (1994). pBBR1MCS: A broad-host-range cloning vector. *Biotechniques* **16,** 800–802.

Leavitt, R. I., and Umbarger, H. E. (1961). Isoleucine and valine metabolism in *Escherichia coli*. X. The enzymatic formation of acetohydroxybutyrate. *J. Biol. Chem.* **236,** 2486–2491.

Leavitt, R. I., and Umbarger, H. E. (1962). Isoleucine and valine metabolism in *Escherichia coli*. XI. Valine inhibition of the growth of *Escherichia coli* strain K-12. *J. Bacteriol.* **83,** 624–630.

Leschine, S. B. (1995). Cellulose degradation in anaerobic environments. *Annu. Rev. Microbiol.* **49,** 399–426.

Lynd, L. R., Weimer, P. J., van Zyl, W. H., and Pretorius, I. S. (2002). Microbial cellulose utilization: Fundamentals and biotechnology. *Microbiol. Mol. Biol. Rev.* **66,** 506–577.

Miller, J. H. (1975). *Experiments in Molecular Genetics*. Cold Spring Harbor Laboratory Press, Cold Spring Harbor, NY.

Mohnen, D. (2008). Pectin structure and biosynthesis. *Curr. Opin. Plant Biol.* **11,** 266–277.

Mosier, N., Wyman, C., Dale, B., Elander, R., Lee, Y. Y., Holtzapple, M., and Ladisch, M. (2005). Features of promising technologies for pretreatment of lignocellulosic biomass. *Bioresour. Technol.* **96,** 673–686.

Neidhardt, F. C., Bloch, P. L., and Smith, D. F. (1974). Culture medium for enterobacteria. *J. Bacteriol.* **119,** 736–747.

Nimblegen (2010). Gene expression arrays. NimbleGen Arrays User's Guide Version 5.1.

Rhodius, V. A., and Gross, C. A. (2011). Using DNA microarrays to assay part function. *Methods Enzymol.* **497,** 75–113.

Roberts, I. N., Oliver, R. P., Punt, P. J., and van den Hondel, C. A. (1989). Expression of the *Escherichia coli* beta-glucuronidase gene in industrial and phytopathogenic filamentous fungi. *Curr. Genet.* **15,** 177–180.

Sambrook, J., Fritsch, E. F., and Maniatis, T. (1989). *Molecular Cloning: A Laboratory Manual.* 2nd edn Cold Spring Harbor Laboratory Press, Cold Spring Harbor, NY.

Schafer, A., Tauch, A., Jager, W., Kalinowski, J., Thierbach, G., and Puhler, A. (1994). Small mobilizable multi-purpose cloning vectors derived from the *Escherichia coli* plasmids pK18 and pK19: Selection of defined deletions in the chromosome of *Corynebacterium glutamicum. Gene* **145,** 69–73.

Ueda, K., Ishikawa, S., Itami, T., and Asai, T. (1952). Studies on the aerobic mesophilic cellulose-decomposing bactera. Taxonomical study of genus *Pseudomonas. J. Agric. Chem. Soc. Jpn.* **26,** 35–41.

Van den Eede, G., Deblaere, R., Goethals, K., Van Montagu, M., and Holsters, M. (1992). Broad host range and promoter selection vectors for bacteria that interact with plants. *Mol. Plant Microbe Interact.* **5,** 228–234.

Vanholme, R., Demedts, B., Morreel, K., Ralph, J., and Boerjan, W. (2010). Lignin biosynthesis and structure. *Plant Physiol.* **153,** 895–905.

Wilson, D. B. (2011). Microbial diversity of cellulose hydrolysis. *Curr. Opin. Microbiol.* **14,** 259–263.

Wilson, K. J., Giller, K. E., and Jefferson, R. A. (1991). P-Glucuronidase (GUS) operon fusions as a tool for studying plant-microbe interactions. *In* "Advances in Molecular Genetics of Plant-Microbe Interactions," (H. Hennecke and D. P. S. Verma, eds.), pp. 226–229. Kluwer Academic Publishers, Dordrecht, The Netherlands, Vol. 1.

CHAPTER NINETEEN

METHODS FOR THE ISOLATION OF CELLULOSE-DEGRADING MICROORGANISMS

James E. McDonald,* David J. Rooks,[†] and Alan J. McCarthy[†]

Contents

1. Introduction	350
2. Cellulosic Substrates	351
2.1. Filter paper and ball-milled cellulose	351
2.2. Acid-treated cellulose	351
2.3. Dewaxed cotton string	352
2.4. Bacterial cellulose	352
2.5. Avicel	353
2.6. Carboxymethyl cellulose (CMC)	353
2.7. Cellobiose	353
2.8. Xylan, wood fractions, wood pulp, and plant material	353
3. Isolation and Cultivation of Cellulolytic Microorganisms	354
3.1. The rumen	354
3.2. Soils, sediments, and aquatic habitats	361
3.3. Landfill sites, waste digesters, fermentors, and activated sludge	363
3.4. Termite guts	365
3.5. Human feces	365
4. Methods for the Detection of Cellulose Degradation in Cultures	366
4.1. Clear and transparent zones around growing colonies	366
4.2. Cellulose azure dye-release assay	366
4.3. Congo red	367
4.4. Stable isotope probing	367
4.5. Filter paper strips	367
4.6. Cellulase activity assays	367
5. Future Prospects	368
Acknowledgments	369
References	369

* School of Biological Sciences, Bangor University, Bangor, Gwynedd, Wales, United Kingdom
[†] Microbiology Research Group, Institute of Integrative Biology, Biosciences Building, University of Liverpool, Liverpool, United Kingdom

Abstract

The biodegradation of lignocellulose, the most abundant organic material in the biosphere, is a feature of many aerobic, facultatively anaerobic and obligately anaerobic bacteria and fungi. Despite widely recognized difficulties in the isolation and cultivation of individual microbial species from complex microbial populations and environments, significant progress has been made in recovering cellulolytic taxa from a range of ecological niches including the human, herbivore, and termite gut, and terrestrial, aquatic, and managed environments. Knowledge of cellulose-degrading microbial taxa is of significant importance with respect to nutrition, biodegradation, biotechnology, and the carbon-cycle, providing insights into the metabolism, physiology, and functional enzyme systems of the cellulolytic bacteria and fungi that are responsible for the largest flow of carbon in the biosphere. In this chapter, several strategies employed for the isolation and cultivation of cellulolytic microorganisms from oxic and anoxic environments are described.

1. INTRODUCTION

Cellulose is the most abundant organic polysaccharide in the biosphere, and the microorganisms responsible for the colonization and degradation of this recalcitrant substrate drive the largest carbon flow in the terrestrial biosphere. To date, the majority of isolated cellulolytic bacteria belong to the largely aerobic order *Actinomycetales* (phylum *Actinobacteria*) and the anaerobic order *Clostridiales* (phylum *Firmicutes*) (Lynd *et al.*, 2002). However, within the domain *Eukarya*, cellulose hydrolysis is distributed across the entire kingdom of filamentous fungi. Anaerobic Chytridiomycetes are the most primitive fungal group and potent degraders of cellulosic biomass in the anaerobic herbivore gut (Orpin, 1975; Trinci *et al.*, 1994). Aerobic fungi of the Ascomycetes, Basidiomycetes, and Deuteromycetes contain large numbers of cellulolytic species, some of which have received significant attention with respect to their cellulose-degrading capabilities. In the third domain of life, the *Archaea*, cellulases have recently been identified (Ando *et al.*, 2002; Birbir *et al.*, 2007; Sunna *et al.*, 1997). Cellulolytic microorganisms display substantial differences in their cellulose hydrolysis strategies, but there are two primary approaches: the extracellular synergistic model characterized in *Trichoderma* spp. (Saloheimo and Pakula, 2012) and suited to the colonization of solid substrates by mycelial microorganisms in a terrestrial setting; and the cell associated "cellulosome" extensively studied in clostridia and regarded as typical of unicellular anaerobic bacteria or Chytridiomycetes (Gilbert, 2007). There are modifications to typical cellulosomal structures, exemplified by the cell surface cellulolytic structures identified in the marine bacterium *Saccharophagus degradans* (Weiner *et al.*, 2008), and the processive cellulose degradation model recently proposed for fibrobacters (Suen et al., 2011; Wilson, 2009).

2. Cellulosic Substrates

Essentially, the choice is between insoluble forms of cellulose, the degradation of which is a true indication of cellulolytic activity, or soluble derivatives such as carboxymethyl cellulose (CMC), which is convenient to incorporate in isolation media, but can be degraded by many microorganisms that produce endoglucanases in the absence of significant activity against native cellulose. A number of factors influence the choice of cellulose substrate: (1) the nature of the medium, (2) if the enrichment of certain organisms is desired, (3) the target activity or cellulosic substrate, (4) if downstream screening/activity assays are required, and (5) when cultures are to be serially diluted for the isolation of individual colonies. Specific examples and issues are discussed throughout the chapter.

2.1. Filter paper and ball-milled cellulose

The detection of cellulolytic activity in microbial cultures has been achieved by the addition of filter paper strips to culture medium. Mann (1968) incorporated strips of filter paper into broth in anaerobic culture tubes and recorded pitting and disintegration. This method enabled estimates of cellulolytic bacterial counts, confirmed the cellulolytic activities of pure cultures and allowed the morphological characteristics and other properties of cellulolytic bacteria to be determined. Recently, filter paper strips have been included in culture medium to visually confirm the cellulolytic activity of freshwater *Micromonospora* isolates (de Menezes et al., 2008). Filter paper is often added to culture medium as a powder, and Whatman no. 1 filter paper or indeed cotton can be ground for 72 h in a pebble mill to produce a powder suspension of cellulose (Hungate, 1950). Acid-treatment of Whatman no. 1 filter paper prior to ball-milling generates a finer suspension; however, cellulose powders are now directly available from commercial suppliers.

2.2. Acid-treated cellulose

Hungate (1950) produced finely dispersed cellulose from absorbent cotton:

1. Absorbent cotton is packed into a 1-l conical flask containing 270 ml concentrated HCl diluted to 300 ml.
2. Stand at room temperature for 24 h until the fibers break apart and a fine cellulose suspension is obtained.

3. Collect cellulose by filtration at low pressure, wash with distilled water, and air dry.
4. The cellulose is combined with 600 ml water and ground in a pebble mill for 72 h, producing a finely divided suspension of cellulose at ca. 4% (w/v).

Bose (1963) described a simple protocol for the rapid preparation of modified cellulose from Whatman no. 1 filter paper:

1. Immerse Whatman no. 1 filter paper (3%, w/v) in concentrated HCl (specific gravity, 1.18).
2. Incubate with occasional shaking at 25–27 °C for 3 h.
3. Stir the mixture in excess distilled water and stand for 2–3 h.
4. Pour off the supernatant and wash cellulose on a Buchner funnel until acid-free.
5. Wash with 95% ethanol and air dry.

The treated cellulose can be combined with a small volume of medium or mineral/salts solution in a pestle and mortar to produce a paste, which can be added to liquid or solid medium as a uniform suspension. The formation of transparent zones and partial clearing around colonies on agar plates unequivocally indicates cellulose decomposition.

2.3. Dewaxed cotton string

Natural waxes associated with cotton can be removed by repeated extraction with chloroform and ethanol to produce crystalline cellulose in the form of dewaxed cotton string (Wood, 1988). Dewaxed cotton string has been used as a "bait" for the *in situ* enrichment of cellulolytic actinomycetes and their subsequent isolation from aquatic environments (de Menezes *et al.*, 2008) and the generation of cellulolytic biofilms for metagenomic analysis (Edwards *et al.*, 2010), and can also be added to liquid culture medium for visual confirmation of cellulose hydrolysis by isolates in axenic culture.

2.4. Bacterial cellulose

Cellulose mats generated by the cellulose-producing bacterium *Gluconacetobacter xylinus* and its relatives may also be added to isolation medium; *G. xylinus* is grown in a sugar-rich medium to produce mats of bacterial cellulose (Mikkelsen *et al.*, 2009). This cellulose can be purified for addition to microbial isolation media (Dickerman and Starr, 1951) as follows:

1. Wash harvested cellulose in running water for 8 h.
2. Place cellulose in 5% NaOH for two 24-h periods (for removal of bacterial protein from cellulose mats).
3. Remove NaOH by washing in running water for 8 h.
4. Immerse in 3 M HCl for 24 h.

5. Rinse cellulose with running water for 8 h, followed by washing in distilled water until neutral with phenol red.
6. Drain the cellulose on cheesecloth for 24 h.
7. Cellulose suspensions are produced by processing the purified bacterial cellulose in a blender.

2.5. Avicel

Avicel powder is an insoluble form of microcrystalline cellulose that has been partially hydrolyzed by acid-treatment and is available commercially.

2.6. Carboxymethyl cellulose (CMC)

CMC is a derivative of cellulose, containing carboxymethyl groups that are generated via the reaction of cellulose with chloroacetate in alkali to produce substitutions in the C2, C3, or C6 positions of glucose units (Gelman, 1982). As a result, CMC is water soluble and more amenable to the hydrolytic activity of cellulases. CMC is therefore a useful additive to both liquid and solid medium for the detection of cellulase activity, and its hydrolysis can be subsequently determined by the use of the dye Congo red, which binds to intact β-D-glucans. Zones of clearing around colonies growing on solid medium containing CMC, subsequently stained with Congo red, provides a useful assay for detecting hydrolysis of CMC and therefore, β-D-glucanase activity (Teather and Wood, 1982). The inoculation of isolates onto membrane filters placed on the surface of CMC agar plates is a useful modification of this technique, as the filter may subsequently be removed allowing visualization of clear zones in the agar underneath cellulolytic colonies.

2.7. Cellobiose

The repeating unit of cellulose is cellobiose, a disaccharide of glucose in which there is a 180° rotation of each monomer in relation to its neighbor. The hydrolysis of cellobiose is an indicator of β-glucosidase activity, which, although universally produced by cellulolytic microorganisms intra- or extracellularly, is also produced by a number of organisms that do not attack cellulose *per se*. However, the addition of cellobiose as the sole carbon source has proven useful for the isolation and purification of anaerobic rumen bacteria involved in cellulose degradation, as described in Section 3.1.2.7.

2.8. Xylan, wood fractions, wood pulp, and plant material

Cellulose rarely occurs in pure form (cotton) in nature and is almost always associated with hemicellulose and lignin. Hemicelluloses are linear decorated heteropolymers of D-xylose, L-arabinose, D-mannose, D-glucose, D-galactose, and D-glucuronic acid (Leschine, 1995). The backbone of the most abundant hemicellulose, xylan, contains acetyl, methylglucuronyl, and arabinofuranosyl side chains, the latter of which can be further esterified by aromatic acids to generate hemicellulose–lignin cross-linkages. The degree of association of cellulose with lignin and hemicellulose therefore complicates the microbial degradation of cellulose, as hemicellulose occupying the spaces between fibrils must be at least partly degraded before cellulose hydrolysis can begin.

Consequently, cellulolytic microorganisms invariably produce hemicellulases, but the reverse is certainly not true. Any cellulose-containing organic matter, primarily plant material, can be incorporated into enrichment or isolation medium, but dried substrates in powdered form are the most useful because they test the organism's ability to degrade the cellulosic component. Essentially, any visual evidence of clearing around a colony on an agar plate containing insoluble lignocellulose is evidence that the isolate can degrade cellulose. Wood can be chipped and ball-milled to produce a fine powder for incorporation into medium, and this has been used to directly isolate cellulolytic fungi, in particular (Rodriguez et al., 1996). Access to a vibratory ball mill, as opposed to a rolling ball mill, is advantageous because it reduces processing time dramatically and provides more efficient disruption of lignified substrates to give a product that can be attacked by cellulolytic microorganisms to produce visible "clearing." Cereal straw can be used in the same way, and zones of clearing require the degradation of cellulose rather than simply attack of the hemicellulose components or indeed lignin (McCarthy and Broda, 1984). These powdered lignocellulosic substrates should be added to isolation of agar medium at 0.2% (w/v), and in order to produce agar plates with finely dispersed lignocellulose, the powders should be autoclaved separately as suspensions in a small volume of distilled water, added to the sterile molten agar, and gently agitated to produce an evenly distributed suspension just prior to pouring the agar plates (Ball and McCarthy, 1988; McCarthy and Broda, 1984).

The use of lignocellulose in culture medium is therefore an attractive option to researchers who prefer to utilize a substrate that is more similar to native forms of cellulose. Hungate (1950) utilized dried grass and malt sprouts in anaerobic culture medium for the isolation of saccharolytic rumen bacteria. Delignified wood pulp derived predominantly from softwood was also used to enrich and isolate cellulolytic aerobic bacteria from soils and aquatic habitats (Mullings and Parish, 1984).

3. Isolation and Cultivation of Cellulolytic Microorganisms

3.1. The rumen

The rumen is the largest of the pre-gastric chambers of ruminant herbivores and forms a large fermentation vessel with the reticulum (Trinci et al., 1994). Within the rumen, a dense population of bacteria, fungi, and protozoa are maintained at ~39 °C in a continuous culture system, relentlessly supplied with plant polysaccharides that are fermented by microorganisms to yield predominantly volatile fatty acids that are the primary energy source for the ruminant animal. Cellulolytic microorganisms are therefore key to the rumen ecosystem. Microbial concentrations in the rumen have been estimated at 10^9–10^{10} ml^{-1} for bacteria, 10^5–10^6 ml^{-1} for protozoa (Hungate, 1966), and anaerobic fungi can represent ca. 8% of the rumen microbial biomass (Kemp et al., 1984). Over 200 species of rumen bacteria have been described, highlighting the astonishing diversity of this ecosystem. Any oxygen is rapidly consumed by facultative anaerobic bacteria, maintaining the redox potential of rumen digesta between -250 and -400 mV (Hobson and Wallace, 1982). Consequently, the majority of cellulolytic microbes are obligate anaerobes, and the preparation of growth medium and manipulation of strains under strictly anoxic conditions is critical to successful isolation and cultivation.

3.1.1. Methods for the isolation of cellulolytic rumen microorganisms

Two key methodological approaches are used for the cultivation of cellulolytic rumen microorganisms. In the "direct" method, colony counts of cellulolytic bacteria are estimated from the number of visible clearings, produced in thin films of a selective agar medium containing cellulose and inoculated with high dilutions of rumen fluid (e.g., Hungate, 1969). The "indirect" approach utilizes a nonselective agar medium to isolate the widest possible range of rumen bacteria, and each isolate is subsequently screened for cellulolytic activity (e.g., Bryant and Burkey, 1953). Both approaches have advantages and disadvantages, and both demonstrated similar results when tested on the same samples; the reader is directed to the study of Vangylsw (1970) for a wider discussion of these methods.

The "direct" isolation technique of Hungate is the most widely applied method for recovering anaerobic rumen cellulolytic microorganisms, including the three major cellulolytic rumen bacterial species: *Fibrobacter succinogenes*, *Ruminococcus albus*, and *Ruminococcus flavefaciens*. The technique is described in detail in the following section.

3.1.2. The Hungate technique for cultivating cellulolytic obligate anaerobes

This method was first described by Hungate in 1947 and is still extensively used today, and widely regarded as the best method for the isolation of obligate anaerobic cellulose-degrading bacteria and fungi that cannot be cultivated on solid medium in Petri dishes. Briefly, for inoculation and subculture, anoxic conditions are maintained in Hungate tubes by degassing with oxygen-free CO_2 via gas hooks which hang over the rim of the tubes to displace oxygen. Butyl rubber bungs or septa are used to seal the tubes and maintain anaerobiosis during incubation. Both liquid and solid media can be used, allowing serial dilutions of cellulolytic cultures for colony counting and the isolation of individual colonies from Hungate agar roll tubes. For full descriptions and modifications of the technique, the reader is directed to the following articles (Bryant, 1972; Hungate, 1947, 1950, 1966, 1969).

3.1.2.1. The culture medium Several culture media have been reported for the isolation of cellulolytic rumen microbes. Typically, they comprise an inorganic salts solution, rumen fluid, resazurin (a colorimetric indicator of reducing conditions), a cellulosic substrate, and a reducing agent such as cysteine hydrochloride. Oxygen-free CO_2 gas is bubbled through the medium during preparation as described below.

3.1.2.2. Rumen fluid The addition of rumen fluid to isolation medium has proven critical (Hobson and Mann, 1961; Hungate, 1966) and is obtained from either a recently killed or live fistulated animal. The rumen fluid is clarified by sterilization, followed by centrifugation and straining through cheesecloth or gauze to remove coarse particles (Stewart and Bryant, 1988).

3.1.2.3. Preparation of culture medium Preparation of the anaerobic cultivation medium of Hungate (1950) is described here; however, numerous minor modifications in medium composition have subsequently been reported. Concentrations of medium constituents listed below represent the final concentration when all medium components are combined.

Inorganic mineral medium: 0.05% (w/v) $(NH_4)_2SO_4$, 0.05% (w/v) K_2HPO_4, 0.02% (w/v) KH_2PO_4, 0.005% (w/v) $CaCl_2$, 0.005% (w/v) $MgSO_4$, and 0.1% (w/v) NaCl.

Buffer: Carbon dioxide (oxygen-free) is the most widely utilized gas for the maintenance of anoxic conditions and is used in combination with sodium bicarbonate (0.5%, w/v). Where nitrogen or hydrogen gas is used to maintain anaerobiosis, phosphate is the principal buffer and the final concentration of K_2HPO_4 and KH_2PO_4 is increased to 0.5–1% (w/v). A pH of approximately 7.0 has been satisfactory for the growth of pure cultures.

Reducing agent: 0.02% (w/v) sodium thioglycolate or 0.01% (w/v) cysteine hydrochloride. Sodium sulfide may also be used but must be added following sterilization.

Redox indicator: Resazurin (0.0001%, w/v) is used as an indicator and is a dark blue dye that becomes reduced to the pink-colored resorufin when boiled or sterilized in an oxygen-free environment. A second, but reversible, reduction step converts resorufin to colorless hydroresorufin at a redox potential of ca. −110 mV. Pink coloration will be regained if the redox potential increases above -51 mV, indicating an oxidized medium that would inhibit the growth of many rumen anaerobes.

Cellulose substrate: Hungate (1950) used 1.2% (w/v) acid-treated cellulose as prepared in Section 2.2. However, any cellulosic substrate can be incorporated into the isolation medium.

Agar: It is used at 2% (w/v) for agar roll tubes and 0.75% (w/v) for sloppy agar tubes.

For preparation of 50 ml of the culture medium, 20 ml of inorganic mineral medium (at 2.5 times the concentration described above) is added to a glass bottle or flask. If solid medium is required, the agar is added and dissolved by boiling. The resazurin and 15 ml of a 4% (w/v) acid-treated and milled cellulose suspension prepared as in Section 2.2 are added (final concentration 1.2%, w/v). The medium is again boiled to remove oxygen. CO_2 gas is also bubbled through the medium during this process via a Pasteur pipette on rubber tubing connected to the gas source. Finally, 5 mg of cysteine hydrochloride (reducing agent) and 15 ml clarified rumen fluid are added. At this stage, a butyl rubber stopper may be used to seal the medium bottle, as the gas source is withdrawn for sterilization. Once the medium has been autoclaved and cooled to 50 °C, a sterile Pasteur pipette is used to bubble CO_2 through the medium whilst 2.5 ml of sterile 10% (w/v) bicarbonate solution is added with shaking, followed by 2 min without shaking to equilibrate the medium to ca. pH 7.0. Other heat labile medium constituents such as sugar solutions and beef blood serum may also be added at this point. The medium is transferred to sterile culture tubes (usually 18 mm × 150 mm glass tubes and screw caps with a 9-mm opening to house a butyl rubber stopper) aseptically and under anaerobiosis. Alternatively, the medium may be aliquoted into individual Hungate tubes under anaerobic conditions (as described below) prior to sterilization.

3.1.2.4. Procedure for the transfer of medium, inoculation, and dilution of strains under anaerobic conditions

Metal gas hooks or Pasteur pipettes with bent tips connected to the gas source (typically CO_2) via rubber tubing are inserted over the neck of the culture tube; CO_2 is more dense than air, thus displacing air from culture tubes, preventing its access, and reducing contamination. Y-shaped connectors are typically used to allow two or more gas lines to be used, allowing the simultaneous maintenance of anaerobiosis in both the culture tube/flask from which medium or culture is removed and the fresh tube to be inoculated. Medium and liquid or solid microbial cultures may be added or removed via rapid transfer from one tube to the next via the use of a sterile glass Pasteur pipette. Addition of the last drop of liquid from the pipette is avoided to limit the introduction of air. Upon withdrawal of the gas hook, a butyl rubber stopper is flamed and rapidly inserted to prevent the access of air and maintain anoxic conditions within the culture tube. All of the above is carried out aseptically, in close proximity to a Bunsen flame (Fig. 19.1).

Dilution of cultures: A Pasteur pipette or inoculation wire is used to pick a single colony from an agar tube or transfer liquid culture. Dilutions may be prepared directly into agar tubes or diluted before inoculation into an agar

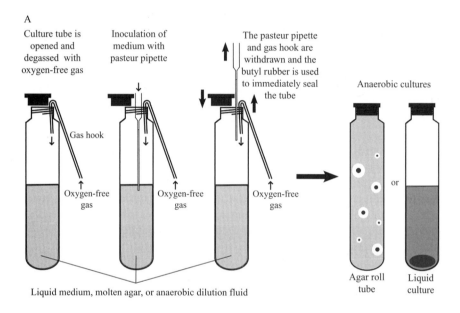

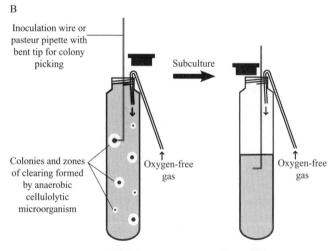

Figure 19.1 Schematic diagram of Hungate's anaerobic technique. Method for the inoculation and manipulation of culture tubes under anoxic conditions (A), and picking of individual colonies from agar roll tubes (B).

tube. The gas supply is withdrawn from the inoculation culture, whilst a butyl rubber stopper is used to seal the tube. The gas hook is transferred to the second dilution tube, and the first dilution is sealed with a stopper and the contents mixed. The first dilution tube is reopened with the addition of a gas hook/pipette, and a sterile 1 ml pipette filled with oxygen-free gas is used to transfer the required volume (0.1–1 ml). Dilution tube one is aseptically sealed and dilution three is inoculated from dilution two in the same manner, and so on (Fig. 19.1). Agar tubes are kept molten in a water bath at 42 °C during preparation of a dilution series.

3.1.2.5. Agar roll tubes Originally, agar roll tubes were cooled by placing under running cold water, with repeated inversion until the agar is almost solid. Modifications to this method include spinning the tubes horizontally, or using a mechanical roller, on a bed of ice for rapid and uniform deposition of the agar on the wall of the tube. Following incubation (ca. 39 °C for rumen strains), colonies are visible throughout the agar films on the walls of the tubes, making smaller colonies easier to detect and enabling the visualization of zones of clearing where cellulose has been degraded. Some rumen bacteria, for example, *Ruminococcus* spp. produce distinct colonies in agar roll tubes, with visible zones of cellulose hydrolysis around the colony; the cellulolytic rumen bacterium *F. succinogenes* is able to migrate through the agar, producing clearings in the agar without the observation of visible colonies.

3.1.2.6. Picking and purification of individual colonies Colonies in thinner areas of the agar are most suitable for picking and subculture. A sterile Pasteur pipette or inoculating wire, with 5 mm of the tip bent at a right angle, is used to pick the colony from the agar tube, whilst anoxia is maintained via the use of the gas hooks, and the colony is inoculated into a fresh culture tube (Fig. 19.1B). Researchers have noted significant difficulties in the removal of contaminating bacteria associated with cellulose-digesting colonies, and it is critical that care is taken during colony picking; in this context, a microscope can be used to ensure that there are no adjacent microcolonies. Minimizing the transfer of agar surrounding a cellulolytic colony and performing several serial transfers of the culture with high dilution factors will improve the chances of successful purification.

A serum bottle modification of the Hungate technique has also been described (Miller and Wolin, 1974); butyl rubber stoppers and crimped metal seals are used to seal the bottles, which are subsequently inoculated with a needle and syringe. This simple approach allows the preparation of larger culture volumes in addition to greater flexibility and separation of colonies when attempting to isolate pure cultures (Miller and Wolin, 1974).

3.1.2.7. Determining the purity of cultures The presence of insoluble cellulose substrates in culture medium often makes it difficult to identify the presence of contaminants that do not digest cellulose or grow slowly. Consequently, the isolate should be inoculated from the solid cellulose medium into a dilution series of a clear solid sugar medium, allowing all colony types to be visualized. Cellobiose is the sugar substrate of choice (McBee, 1950). A single colony from a high dilution should be inoculated into a second dilution series of solid sugar-containing medium until two successive dilution series produce colony types of the same form, confirmed microscopically. A colony from the second sugar dilution series should give rise to cellulose-digesting colonies on reintroduction to a dilution series of cellulose medium, somewhat analogous to Koch's postulates. If all of the above criteria are met, the culture can be considered pure, but the application of molecular biological techniques (i.e., PCR, cloning, and sequencing of taxonomic marker genes) now provide very effective means to track the purity and identity of colonies.

3.1.2.8. Isolation of cellulolytic anaerobic fungi The Hungate roll tube method is also the method of choice for the cultivation of anaerobic cellulolytic fungi (Joblin, 1981), which were first identified in the rumen in 1975 (Orpin, 1975; Orpin and Joblin, 1988). The addition of combinations of benzyl penicillin, ampicillin, and streptomycin sulfate suppresses the growth of rumen bacteria (Bauchop, 1979; Joblin, 1981), and fungal growth was not observed in roll tubes where antibiotics were not included (Joblin, 1981). This method allowed the direct isolation of fungi from rumen fluid, and enumeration of fungal zoospores, without the requirement for enrichment. In addition, strains could be maintained in roll tubes for up to 7 months at 39 °C, with the addition of glucose to reactivate cultures within 2–3 days (Joblin, 1981). Theodorou *et al.* (1990) described a similar anaerobic culture technique based on the most probable number method for the enumeration of anaerobic fungi based on thallus forming units (Davies *et al.*, 1993; Theodorou *et al.*, 1990), and this approach has successfully been used in the isolation of new anaerobic fungal genera (Ozkose *et al.*, 2001). To date, all described anaerobic fungal species are cellulolytic, obviating the necessity to observe clear zones in cellulose agar upon isolation. Anaerobic fungal cellulase systems are highly active and an obvious target for biotechnological application (Wood *et al.*, 1986); however, successful cultivation of these strictly anaerobic fungi requires close adherence to anaerobic procedure and significant expertise in the technique. Consequently, the cultivation of anaerobic fungi is not widely practiced, but despite this, novel species of anaerobic fungi are still being described (Chen *et al.*, 2007). Recent sequence-based analysis of phylogenetic marker genes, including the internal transcribed spacer region 1 (ITS1) and ribosomal RNA (rRNA) gene, have proven an effective method for the analysis of anaerobic fungal diversity and could be applied to the identification of isolates. A recent pyrosequencing study with more than a quarter of a million anaerobic fungal ITS1/rRNA gene sequence data points determined that 38% of the sequence reads could not be ascribed to a known genus, representing eight novel lines of decent within the phylum *Neocallimastigomycota* (Liggenstoffer *et al.*, 2010), and this provides an impetus to persevere with research focused on their isolation and cultivation.

3.2. Soils, sediments, and aquatic habitats

Soils and sediments represent a major reservoir of global carbon. The microbial-mediated degradation of organic matter in soil is a key feature of the global carbon cycle, and cellulose represents up to 14% of soil organic material (Chidthaisong and Conrad, 2000). Consequently, a taxonomically diverse range of aerobic and anaerobic bacteria, and aerobic fungi, have been isolated from soils and sediments.

3.2.1. Isolation of anaerobic bacteria

Several species of anaerobic cellulolytic bacteria, and particularly cellulolytic *Clostridium* spp., have been isolated from soils (He *et al.*, 1991; Monserrate *et al.*, 2001; Skinner, 1960; Warnick *et al.*, 2002), the sediments of freshwater (Leschine and Canaleparola, 1983), and estuarine systems (Madden *et al.*, 1982; Murray *et al.*, 1986). Typically, the Hungate method (Hungate, 1950) is again used in combination with an appropriate medium for the enrichment of cellulolytic soil bacteria (Skinner, 1960). In contrast to isolations from rumen fluid, where carbon dioxide gas is used, 100% nitrogen gas (oxygen-free), or a gas mix comprised of ca. 80% N_2 with CO_2 and H_2 is used in isolations from soil. In some studies, the growth of isolated strains could be maintained on solid agar in Petri dishes in an anaerobic cabinet (Ozkan *et al.*, 2001; Warnick *et al.*, 2002) or anaerobic jars (Skinner, 1960).

3.2.2. Isolation of aerobic bacteria

In many cases, cellulolytic bacteria have been isolated from soils and sediments by simply incorporating a cellulosic substrate into the required growth medium. Psychrophilic cellulolytic bacteria were isolated and enumerated from Antarctic benthic sediment by the addition of Avicel as the sole carbon source in the isolation medium (Reichardt, 1988), and a similar approach was used to isolate several cellulolytic *Cellulomonas* spp. from soils (Elberson *et al.*, 2000; Malekzadeh *et al.*, 1993). A solid medium containing acid-washed cellulose and Congo red, which indicates cellulolytic bacteria via zones of clearing, has also been described (Hendricks *et al.*, 1993). Mullings and Parish (1984) used a combination of methods to isolate predominantly Gram-negative bacteria identified as belonging to the genera *Cellvibrio*, *Pseudomonas*, and *Cytophaga* from soils and aquatic sediments, in addition to a smaller number of Gram-positive actinomycetes, including *Streptomyces* spp. In this study, wood pulp was added to a basal salt medium to enrich cellulolytic bacteria; the enrichment was subsequently plated onto agar-containing filter paper and bacterial cellulose mats for the screening of cellulolytic activity. Cycloheximide (50 µg/ml) can be included in the isolation medium to prevent the growth of cellulolytic soil fungi.

Lignocellulose-degrading actinomycetes are well established as degraders of plant biomass in soils, and although their contribution to polysaccharide hydrolysis in aquatic environments is yet to be fully established, their distribution in these environments suggests a potential role in cellulose hydrolysis. Again, the incorporation of a cellulosic substrate into enrichment and isolation medium is adequate for the isolation of actinomycetes from soils (Ball and Crawford, 2006; Li and Gao, 1996; Phelan *et al.*, 1979), compost (Ball and McCarthy, 1988; McCarthy and Broda, 1984), freshwater (de Menezes *et al.*, 2008), and marine sediments (Veiga *et al.*, 1983). The isolation media are typically nutrient limited to suppress the growth of unicellular bacteria, for example, basal salts media with the addition of low concentrations of yeast extract or casamino acids and vitamins (McCarthy and Williams, 1990). However, the inclusion of antifungal agents such as cycloheximide, the heat treatment of samples at temperatures where only actinomycete spores can survive, and the use of an Andersen sampler to detect and isolate actinomycete spores from the air are also useful methods for the selective isolation of actinomycetes (Ball and McCarthy, 1988; McCarthy, 1987; McCarthy and Broda, 1984). The latter approach utilizes a sedimentation chamber to create an aerosol of actinomycete spores from dried samples such as compost, fodders, and soil, and the resulting spore cloud is sampled with an Andersen sampler (McCarthy and Broda, 1984).

There are few reports on the recovery of cellulolytic microorganisms from the water column of aquatic environments, possibly due to the fact that the majority of cellulose hydrolysis is expected to take place at, or close to, the sediment. However, several cellulolytic *Micromonospora* strains were recently isolated from the water column and sediment of freshwater lakes, and these actinomycetes were obtained via the incubation of dewaxed cotton string *in situ* for the enrichment and isolation of cellulolytic strains (de Menezes *et al.*, 2008). Samples were heat-treated, or treated with tunicamycin to inactive non-*Micromonospora* Gram-positive bacteria, to promote the survival and enrichment of actinomycete spores prior to plating on solid medium. The ability of these isolates to degrade cellulose in pure culture was determined by incubation in minimal salts medium or sterile lake water supplemented with dewaxed cotton string and filter paper strips, followed by scanning electron microscopy and visual observations of degradation, respectively. Endoglucanase activity of the isolates was also determined by measuring reducing sugars generated from culture supernatants derived from cultures supplemented with fibrous filter paper powder (de Menezes et al., 2008).

3.2.3. Isolation of aerobic fungi

Warcup (1950) described a soil-plate method for the isolation of soil fungi, in which a thin layer of nutrient agar medium is poured over soil crumbs dispersed in a Petri dish. However, the isolation of cellulolytic fungi was hampered by the rapid overgrowth of the "sugar fungi," which rapidly utilized the simple sugar constituents and outcompeted the slower growing degraders of recalcitrant plant biomass. Eggins and Pugh (1962) subsequently modified Warcup's technique for the specific isolation of cellulolytic soil fungi, using cellulose as the sole carbon source to keep the sugar fungi in check:

1. Soil crumbs added to Petri dishes are overlaid with a shallow layer of solid medium containing powdered cellulose, 0.5 g ammonium sulfate, 0.5 g L-asparagine, 1 g potassium dihydrogen phosphate, 0.5 g potassium chloride, 0.2 g magnesium sulfate, 0.1 g calcium chloride, 0.5 g Difco yeast extract, 20 g agar, 10 g ball-milled cellulose, and water to 1 l. Adjust pH to 6.2.
2. When added to Petri dishes, the cellulose becomes evenly distributed, forming a white and opaque medium on setting.
3. Fungal growth is evident after 5–7 days incubation at 25 °C due to the clearing of cellulose in the medium, due to cellulase activity.

Further modifications of this technique include the addition of acid-treated cellulose (method described in Section 2.2), which can be prepared more rapidly and without the requirement for specialized equipment such as a ball mill (Bose, 1963). Recently, a method for the rapid isolation of cellulolytic fungi and bacteria from soils has been described. Soil suspensions were inoculated onto a mineral salts agar medium and overlaid with a 70-mm diameter Whatman no. 5 filter paper. Following incubation for 2 weeks at 30 °C, fungal and bacterial colonies were visible on the filter papers in a ratio of 4:1, respectively (Fujii et al., 2011). It is noteworthy that the original study of Warcup (1950) demonstrated the isolation of a greater number of fungal species when the actual soil residue was incorporated in the medium rather than a soil suspension.

3.3. Landfill sites, waste digesters, fermentors, and activated sludge

Cellulose and hemicellulose represent the most abundant biodegradable organic polymers in landfill, comprising ca. 40–50% and ca. 12% of the biodegradable waste content, respectively (Barlaz et al., 1989). Cellulose hydrolysis is therefore the rate-limiting step of waste degradation, yet surprisingly there is little information on the ecology and identity of the indigenous anaerobic cellulolytic community in landfill sites. Molecular biological techniques suggest that members of the genera *Clostridium* and *Eubacterium* are the predominant degraders of cellulose in landfill sites and municipal solid waste digesters (Burrell *et al.*, 2004; McDonald *et al.*, 2012; Van Dyke and McCarthy, 2002). This assertion has been supported by the isolation of several *Clostridium* and *Eubacterium* species from landfill sites (Westlake *et al.*, 1995), methanogenic bioreactors (Palop *et al.*, 1989; Shiratori *et al.*, 2006; Sleat *et al.*, 1984), and municipal solid waste digesters (Benoit *et al.*, 1992; Cailliez *et al.*, 1993). As with soils and sediments, the isolation of anaerobic bacteria from landfill sites and digesters typically relies on Hungate's anaerobic technique (Hungate, 1969) as described by Bryant (1972). A gas mix comprised either entirely or primarily of nitrogen gas (oxygen-free) is used, in combination with CO_2 and sometimes H_2.

Recently, molecular evidence has supported the presence of novel lineages of the *Fibrobacteres* phylum in landfill sites (McDonald *et al.*, 2008, 2012; Ransom-Jones *et al.*, 2012) and freshwater lakes (McDonald *et al.*, 2009), classified within the genus *Fibrobacter*, as currently defined. The *Fibrobacteres* phylum contains only two cultivated isolates, *F. succinogenes* and *F. intestinalis*, which are widely accepted as the major bacterial degraders of cellulose in the rumen. Due to recent advances in molecular biological techniques, the molecular phylogeny of the *Fibrobacteres* phylum has been extended to include lineages derived from termite gut contents (Hongoh *et al.*, 2005, 2006) where lignocellulose is the primary carbon source for growth (see Section 3.4) and the role of *Fibrobacteres* in lignocellulose hydrolysis in the termite gut has been confirmed using metagenomic and proteomic approaches (Warnecke *et al.*, 2007). In general, molecular studies have confirmed that the *Fibrobacteres* phylum has a broad ecological range and appears to be circumscribed by cellulolytic activity (Ransom-Jones *et al.*, 2012). Considering the functional and ecological importance of this group, it is lamentable that for an entire phylum there are only two cultivated isolates (*F. succinogenes* and *F. intestinalis*), derived from one ecological niche, the herbivore gut. Therefore, targeted cultivation-based studies are now required to isolate *Fibrobacteres* strains from other environments, to provide phenotypic profiles for the uncultivated taxa. *Fibrobacter* spp. are difficult to isolate into pure culture; problems associated with the isolation and cultivation of *F. succinogenes* from the rumen are mainly due to a lack of visible clearing of cellulose caused by gliding motility in agar and an apparent symbiosis with other rumen bacteria (Hungate, 1966). In one particular study, only one of several hundred isolated rumen strains was *F. succinogenes* (Latham *et al.*, 1971). It may be possible to isolate novel *Fibrobacter* spp. from environments beyond the herbivore gut using Hungate technique coupled with the application of *Fibrobacter*-specific molecular probes and PCR primers to confirm enrichment. Lack of progress may simply reflect research effort, but it is conceivable that an absolute dependence on syntrophy could render the successful isolation of these species improbable. Similarly, members of the cellulolytic anaerobic fungal order *Neocallimastigales* have never been isolated and cultivated from non-rumen environments, despite their molecular detection in landfill sites (Lockhart *et al.*, 2006; McDonald *et al.*, 2012).

3.4. Termite guts

The intriguing ability of saprotrophic termites to digest wood is important to its turnover in the environment (Varma *et al.*, 1994). The efficient degradation of lignocellulose is mediated by a nutritional symbiosis between termites and their hindgut microbiota, the hindgut representing a unique ecological niche where the depolymerization of complex biopolymers by specialized lignocellulose-degrading microorgansisms is a key feature (Slaytor, 1992; Varma *et al.*, 1994; Warnecke *et al.*, 2007), notwithstanding the relatively recent description of cellulose degradation by invertebrates (Watanabe and Tokuda, 2010). Numerous cellulolytic and xylanolytic microorganisms have been isolated from the termite intestine, with oxygen relationships that range from aerobic to microaerophilic, facultatively anaerobic, and anaerobic. A range of cultivation techniques are therefore available. Cellulolytic actinomycetes belonging to the genera *Micromonospora* and *Streptomyces* were isolated aerobically from termite gut enrichment cultures using solid medium in Petri dishes, supplemented with cellobiose, CMC, Avicel, and filter paper (Pasti and Belli, 1985). A similar approach was used to isolate 119 strains of aerobic and facultatively anaerobic bacteria belonging to 23 groups of bacteria, including members of the orders *Acinomycetales* and *Bacillales*, and relatives of several other genera including *Agrobacterium/Rhizobium*, *Pseudomonas*, *Sphingomonas/Zymomonas*, and *Brucella/Ochrobactrum* (Wenzel *et al.*, 2002). Thayer (1976) isolated facultative anaerobic wood-degrading bacterial strains belonging to the genera *Bacillus*, *Arthrobacter*, *Alcaligenes*, and *Serratia* under aerobic conditions on trypticase soy agar and demonstrated cellulolysis of these isolates on a range of media supplemented with CMC, cellulose, and mesquite wood as carbon source. Strictly anaerobic bacteria have also been isolated from the termite gut using the Hungate technique (Hethener *et al.*, 1992); for more detailed description of the diversity of lignocellulose-degrading bacteria in the termite gut, the reader is directed to Varma *et al.*, (1994).

3.5. Human feces

Cellulolytic and hemicellulolytic polysaccharides derived from plant cell walls are an important source of dietary fiber in humans. Consequently, several human colonic cellulolytic and xylanolytic bacteria have been isolated from human feces (Betian *et al.*, 1977; Chassard *et al.*, 2007, 2008;

Duncan *et al.*, 2007; Robert *et al.*, 2007; Wedekind *et al.*, 1988). As with the rumen bacteria, human colonic bacteria are strictly anaerobic, making the absence of oxygen a critical requirement for successful isolation (Duncan *et al.*, 2007). The cultivation of human colonic isolates is achieved using the anaerobic cultivation methods of Hungate, often with the addition of clarified rumen fluid to culture medium, and dilutions of fecal samples as the inoculum; several xylanolytic *Butyrivibrio*, *Clostridium*, *Bacteroides*, and *Roseburia* species have been isolated in this way (Chassard *et al.*, 2007, 2008; Wedekind *et al.*, 1988). Cellulolytic isolates include *Bacteroides*, *Clostridium*, and *Ruminococcus* spp. (Betian *et al.*, 1977; Robert *et al.*, 2007; Wedekind *et al.*, 1988). Polysaccharide substrates used in the isolation media in these studies included cellobiose, starch, CMC, ball-milled filter paper, spinach, wheat straw, larchwood xylan, birchwood xylan, and oat spelts.

4. Methods for the Detection of Cellulose Degradation in Cultures

4.1. Clear and transparent zones around growing colonies

Cellulose degradation can often be directly observed via the formation of zones of clearing around colonies in solid medium containing an insoluble cellulose substrate. Typically, cellulose concentrations of 1–1.5% (w/v) are used, providing sufficient opacity for the detection of zones of clearing, without altering the buffering capacity of the medium (Hungate, 1950). Several alternatives for the detection of cellulolysis are available, and the most widely used methods are described in the following sections.

4.2. Cellulose azure dye-release assay

Several studies have utilized dyed celluloses as substrates for the detection of cellulase activity in enzyme preparations and growing cultures; in these experiments, the release of dye from the substrate is measured as a proxy for cellulose hydrolysis. Cellulose azure is one such example of a commercially available dyed cellulose substrate, and a rapid tube test for the detection of cellulase production by fungal cultures has been described (Smith, 1977). Briefly, an initial layer of solid medium (2 ml/tube) without a carbon source is added to a screw capped tube and allowed to solidify following sterilization. A second layer of the same medium supplemented with 2% (w/v) cellulose azure (0.5 ml/tube) is subsequently added. Tubes are inoculated with cultures and incubated with regular inspection. The release of the blue colored dye from the top cellulose azure layer into the basal agar layer confirms cellulolysis (Smith, 1977). This assay has also been successfully used to assess cellulose degradation by thermophilic actinomycete bacteria (Cresswell *et al.*, 1988).

4.3. Congo red

The interaction of Congo red dye with intact β-D-glucans is a useful tool for the detection of β-D-glucanase enzyme activity of cellulolytic isolates. The soluble derivative of cellulose, CMC (Section 2.6), is added to isolation medium at a concentration of 0.1% (w/v) and incubated until growth has occurred. The surface of the agar is subsequently flooded with a 1 mg/ml solution of Congo red for 15 min. The Congo red is poured off and unbound dye removed via two washes with 1 M NaCl for 15 min (Teather and Wood, 1982). Regions of the agar where the hydrolysis of CMC has occurred are clear, whereas intact CMC will retain the Congo red dye.

4.4. Stable isotope probing

Stable isotope probing (SIP) has proved an effective technique to track specific functional groups of microorganisms that utilize a substrate of interest labeled with a stable isotope, such as ^{13}C. Microorganisms that utilize the labeled substrate assimilate the isotope into their nucleic acids, which may be subsequently detected using molecular techniques (Chen and Murrell, 2011; Lu and Conrad, 2011; O'Donnell *et al.*, 2011). Although SIP has never been applied directly to cellulolytic cultures, it has successfully been used to identify cellulose-degrading communities in soils (Haichar *et al.*, 2007) and may provide a tangible approach for the confirmation of cellulolytic phenotypes in pure culture. ^{13}C cellulose may be produced via the incorporation of ^{13}C glucose into bacterial cellulose mats produced by *G. xylinus* (Section 2.4) (Haichar *et al.*, 2007) or by refining plant biomass produced under an atmosphere of ^{13}CO$_2$ (Lu and Conrad, 2011).

4.5. Filter paper strips

As discussed in Section 2.1, the addition of filter paper strips to culture medium is useful for the visible detection of the cellulolytic activity of isolates (de Menezes *et al.*, 2008; Mann, 1968).

4.6. Cellulase activity assays

Several enzyme assays are available for the detection and quantification of cellulolytic activity in isolated strains. Three main approaches are used:

(1) *Measurement of hydrolysis products such as reducing sugars and total sugars*: Reducing sugar assays include the dinitrosalicyclic acid (DNS) method

(Miller, 1959) and the Nelson–Somogyi method (Somogyi, 1952). The anthrone–H_2SO_4 (Viles and Silverman, 1949) and phenol–H_2SO_4 (Dubois *et al.*, 1956) methods can be used to measure total soluble sugars released by extracellular enzymes in culture supernatants.

(2) *Measurement of reductions in substrate quantity via total carbohydrate assays*: The anthrone–H_2SO_4 (Viles and Silverman, 1949) and phenol–H_2SO_4 (Dubois *et al.*, 1956) methods are the most widely used. However, these techniques are limited to the study of pure celluloses, as derivatives of other carbohydrates may interfere with the quantification of cellodextrins (glucose equivalents).

(3) *Measuring changes in the physical properties of substrates*: Historically, physical characteristics including turbidity, viscosity, swollen factor, disruption of cellulose structure, and strength of cellulose fibers have also been used to assess cellulase activity.

For a comprehensive description of these methods, the reader is referred to the review of Zhang *et al.* (2006).

5. Future Prospects

Despite significant recent advances in the application of molecular biological and "omics" techniques for characterizing the structure and function of entire microbial communities, the isolation of microbial strains into pure culture remains important. While molecular studies have provided an exponential increase in the number of known taxa at every taxonomic level, many of these newly detected microbial lineages are based entirely on rRNA gene phylogeny and do not contain cultivated representatives. Attempts to isolate and cultivate representatives of these new lineages must be made; the isolation of individual species is a valuable tool for detailed studies of the metabolism and physiology of isolates as well as enabling genome sequencing to reveal the portfolio of enzymatic activities and perhaps discover new mechanistic approaches to the degradation of cellulose.

Often, the isolation of individual species from complex microbial assemblages/environments is not straightforward, or almost impossible, due to difficulties in providing the exact nutrient requirements for growth, reproducing environmental conditions in the laboratory, or most important of all, the interdependency of microbial taxa within the mixed communities that are the natural state. Despite this, however, significant progress in the isolation of cellulose-degrading microorganisms has been made, and developments in isolation techniques and medium formulations are providing fresh alternatives to isolate previously uncultivated microorganisms. Often, it may be just a reflection of the lack of effort and activity that is being applied to conventional culture-based microbiology. A significant proportion of cellulose-degrading taxa in natural environments undoubtedly await isolation and characterization.

ACKNOWLEDGMENTS

Research on the isolation of cellulolytic microorganisms by the authors has been funded by the UK Natural Environment Research Council (NERC) and the Systematics Association's SynTax award scheme, supported by the Linnean Society of London, BBSRC and NERC, to J.E.M.

REFERENCES

Ando, S., Ishida, H., Kosugi, Y., and Ishikawa, K. (2002). Hyperthermostable endoglucanase from *Pyrococcus horikoshii*. *Appl. Environ. Microbiol.* **68,** 430–433.

Ball, C. L., and Crawford, R. L. (2006). Bacterial diversity within the planktonic community of an artesian water supply. *Can. J. Microbiol.* **52,** 246–259.

Ball, A. S., and McCarthy, A. J. (1988). Saccharification of straw by actinomycete enzymes. *J. Gen. Microbiol.* **134,** 2139–2147.

Barlaz, M. A., Schaefer, D. M., and Ham, R. K. (1989). Bacterial population development and chemical characteristics of refuse decomposition in a simulated sanitary landfill. *Appl. Environ. Microbiol.* **55,** 55–65.

Bauchop, T. (1979). Rumen anaerobic fungi of cattle and sheep. *Appl. Environ. Microbiol.* **38,** 148–158.

Benoit, L., Cailliez, C., Petitdemange, E., and Gitton, J. (1992). Isolation of cellulolytic mesophilic clostridia from a municipal solid-waste digester. *Microb. Ecol.* **23,** 117–125.

Betian, H. G., Linehan, B. A., Bryant, M. P., and Holdeman, L. V. (1977). Isolation of a cellulolytic *Bacteroides* sp from human feces. *Appl. Environ. Microbiol.* **33,** 1009–1010.

Birbir, M., Calli, B., Mertoglu, B., Bardavid, R. E., Oren, A., Ogmen, M. N., and Ogan, A. (2007). Extremely halophilic Archaea from Tuz Lake, Turkey, and the adjacent Kaldirim and Kayacik salterns. *World J. Microbiol. Biotechnol.* **23,** 309–316.

Bose, R. G. (1963). A modified cellulosic medium for isolation of cellulolytic fungi from infected materials and soils. *Nature* **198,** 505–506.

Bryant, M. P. (1972). Commentary on Hungate technique for culture of anaerobic bacteria. *Am. J. Clin. Nutr.* **25,** 1324–1328.

Bryant, M. P., and Burkey, L. A. (1953). Cultural methods and some characteristics of some of the more numerous groups of bacteria in the bovine rumen. *J. Dairy Sci.* **36,** 205–217.

Burrell, P. C., O'Sullivan, C., Song, H., Clarke, W. P., and Blackall, L. L. (2004). Identification, detection, and spatial resolution of *Clostridium* populations responsible for cellulose degradation in a methanogenic landfill leachate bioreactor. *Appl. Environ. Microbiol.* **70,** 2414–2419.

Cailliez, C., Benoit, L., Gelhaye, E., Petitdemange, H., and Raval, G. (1993). Solubilization of cellulose by mesophilic cellulolytic clostridia isolated from a municipal solid-waste digester. *Bioresource Technol.* **43,** 77–83.

Chassard, C., Goumy, V., Leclerc, M., Del'homme, C., and Bernalier-Donadille, A. (2007). Characterization of the xylan-degrading microbial community from human faeces. *FEMS Microbiol. Ecol.* **61,** 121–131.

Chassard, C., Delmas, E., Lawson, P. A., and Bernalier-Donadille, A. (2008). *Bacteroides xylanisolvens* sp nov., a xylan-degrading bacterium isolated from human faeces. *Int. J. Syst. Evol. Microbiol.* **58**, 1008–1013.

Chen, Y., and Murrell, J. C. (2011). DNA stable isotope probing. *In* "Stable Isotope Probing and Related Technologies," (J. C. Murrell and A. S. Whiteley, eds.), pp. 3–24. ASM Press, Washington, DC.

Chen, Y.-C., Tsai, S.-D., Cheng, H.-L., Chien, C.-Y., Hu, C.-Y., and Cheng, T.-Y. (2007). *Caecomyces sympodialis* sp nov., a new rumen fungus isolated from *Bos indicus*. *Mycologia* **99**, 125–130.

Chidthaisong, A., and Conrad, R. (2000). Pattern of non-methanogenic and methanogenic degradation of cellulose in anoxic rice field soil. *FEMS Microbiol. Ecol.* **31**, 87–94.

Cresswell, M. A., Attwell, R. W., and Dempsey, M. J. (1988). Detection of cellulolytic actinomycetes using cellulose-azure. *J. Microbiol. Meth.* **8**, 299–302.

Davies, D. R., Theodorou, M. K., Lawrence, M. I. G., and Trinci, A. P. J. (1993). Distribution of anaerobic fungi in the digestive-tract of cattle and their survival in feces. *J. Gen. Microbiol.* **139**, 1395–1400.

de Menezes, A. B., Lockhart, R. J., Cox, M. J., Allison, H. E., and McCarthy, A. J. (2008). Cellulose degradation by *Micromonosporas* recovered from freshwater lakes and classification of these actinomycetes by DNA gyrase B gene sequencing. *Appl. Environ. Microbiol.* **74**, 7080–7084.

Dickerman, J. M., and Starr, T. J. (1951). A medium for the isolation of pure cultures of cellulolytic bacteria. *J. Bacteriol.* **62**, 133–134.

Dubois, M., Gilles, K. A., Hamilton, J. K., Rebers, P. A., and Smith, F. (1956). Colorimetric method for determination of sugars and related substances. *Anal. Chem.* **28**, 350–356.

Duncan, S. H., Louis, P., and Flint, H. J. (2007). Cultivable bacterial diversity from the human colon. *Lett. Appl. Microbiol.* **44**, 343–350.

Edwards, J. L., Smith, D. L., Connolly, J., McDonald, J. E., Cox, M. J., Joint, I., Edwards, C., and McCarthy, A. J. (2010). Identification of carbohydrate metabolism genes in the metagenome of a marine biofilm community shown to be dominated by *Gammaproteobacteria* and *Bacteroidetes*. *Genes* **1**, 371–384.

Eggins, H. O. W., and Pugh, G. J. F. (1962). Isolation of cellulose-decomposing fungi from soil. *Nature* **193**, 94–95.

Elberson, M. A., Malekzadeh, F., Yazdi, M. T., Kameranpour, N., Noori-Daloii, M. R., Matte, M. H., Shahamat, M., Colwell, R. R., and Sowers, K. R. (2000). *Cellulomonas persica* sp nov and *Cellulomonas iranensis* sp nov., mesophilic cellulose-degrading bacteria isolated from forest soils. *Int. J. Syst. Evol. Microbiol.* **50**, 993–996.

Fujii, K., Kuwahara, A., Nakamura, K., and Yamashita, Y. (2011). Development of a simple cultivation method for isolating hitherto-uncultured cellulase-producing microbes. *Appl. Microbiol. Biotechnol.* **91**, 1183–1192.

Gelman, R. A. (1982). Characterization of carboxymethylcellulose—Distribution of substituent groups along the chain. *J. Appl. Polym. Sci.* **27**, 2957–2964.

Gilbert, H. J. (2007). Cellulosomes: Microbial nanomachines that display plasticity in quaternary structure. *Mol. Microbiol.* **63**, 1568–1576.

Haichar, F. E. Z., Achouak, W., Christen, R., Heulin, T., Marol, C., Marais, M.-F., Mougel, C., Ranjard, L., Balesdent, J., and Berge, O. (2007). Identification of cellulolytic bacteria in soil by stable isotope probing. *Environ. Microbiol.* **9**, 625–634.

He, Y. L., Ding, Y. F., and Long, Y. Q. (1991). Two cellulolytic clostridium species: *Clostridium cellulosi* sp nov and *Clostridium cellulofermentans* sp nov. *Int. J. Syst. Evol. Microbiol.* **41**, 306–309.

Hendricks, C. W., Doyle, J. D., and Hugley, B. M. (1993). A new medium for enumerating cellulose-utilizing bacteria in soil. *Abstr. Gen. Meet. Am. Soc. Microbiol.* **93**, 307.

Hethener, P., Brauman, A., and Garcia, J. L. (1992). *Clostridium termitidis* sp nov, a cellulolytic bacterium from the gut of the wood-feeding termite, *Nasutitermes lujae*. *Syst. Appl. Microbiol.* **15,** 52–58.

Hobson, P. N., and Mann, S. O. (1961). Isolation of glycerol-fermenting and lipolytic bacteria from rumen of sheep. *J. Gen. Microbiol.* **25,** 227–240.

Hobson, P. N., and Wallace, R. J. (1982). Microbial ecology and activities in the rumen: Part 1. *CRC Crit. Rev. Microbiol.* **9,** 165–225.

Hongoh, Y., Deevong, P., Inoue, T., Moriya, S., Trakulnaleamsai, S., Ohkuma, M., Vongkaluang, C., Noparatnaraporn, N., and Kudol, T. (2005). Intra- and interspecific comparisons of bacterial diversity and community structure support coevolution of gut microbiota and termite host. *Appl. Environ. Microbiol.* **71,** 6590–6599.

Hongoh, Y., Deevong, P., Hattori, S., Inoue, T., Noda, S., Noparatnaraporn, N., Kudo, T., and Ohkuma, M. (2006). Phylogenetic diversity, localization, and cell morphologies of members of the candidate phylum TG3 and a subphylum in the phylum *Fibrobacteres*, recently discovered bacterial groups dominant in termite guts. *Appl. Environ. Microbiol.* **72,** 6780–6788.

Hungate, R. E. (1947). Studies on cellulose fermentation: 3. The culture and isolation of cellulose-decomposing bacteria from the rumen of cattle. *J. Bacteriol.* **53,** 631–645.

Hungate, R. E. (1950). The anaerobic mesophilic cellulolytic bacteria. *Bacteriol. Rev.* **14,** 1–49.

Hungate, R. E. (1966). The Rumen and Its Microbes. Academic Press, New York, USA.

Hungate, R. E. (1969). A roll tube method for cultivation of strict anaerobes. In "Methods in Microbiology," (J. R. Norris and D. W. Ribbons, eds.), pp. 117–132. Academic Press, London, England and New York, Vol. 1.

Joblin, K. N. (1981). Isolation, enumeration, and maintenance of rumen anaerobic fungi in roll tubes. *Appl. Environ. Microbiol.* **42,** 1119–1122.

Kemp, P., Lander, D. J., and Orpin, C. G. (1984). The lipids of the rumen fungus *Piromonas communis*. *J. Gen. Microbiol.* **130,** 27–37.

Latham, M. J., Sharpe, M. E., and Sutton, J. D. (1971). Microbial flora of rumen of cows fed hay and high cereal rations and its relationship to rumen fermentation. *J. Appl. Bacteriol.* **34,** 425–434.

Leschine, S. B. (1995). Cellulose degradation in anaerobic environments. *Annu. Rev. Microbiol.* **49,** 399–426.

Leschine, S. B., and Canaleparola, E. (1983). Mesophilic cellulolytic clostridia from freshwater environments. *Appl. Environ. Microbiol.* **46,** 728–737.

Li, X., and Gao, P. (1996). Isolation and partial characterization of cellulose-degrading strain of *Streptomyces* sp LX from soil. *Lett. Appl. Microbiol.* **22,** 209–213.

Liggenstoffer, A. S., Youssef, N. H., Couger, M. B., and Elshahed, M. S. (2010). Phylogenetic diversity and community structure of anaerobic gut fungi (phylum *Neocallimastigomycota*) in ruminant and non-ruminant herbivores. *ISME J.* **4,** 1225–1235.

Lockhart, R. J., Van Dyke, M. I., Beadle, I. R., Humphreys, P., and McCarthy, A. J. (2006). Molecular biological detection of anaerobic gut fungi (*Neocallimastigales*) from landfill sites. *Appl. Environ. Microbiol.* **72,** 5659–5661.

Lu, Y., and Conrad, R. (2011). Stable isotope probing and plants. In "Stable Isotope Probing and Related Technologies," (J. C. Murrell and A. S. Whiteley, eds.), pp. 151–163. ASM Press, Washington, DC.

Lynd, L. R., Weimer, P. J., van Zyl, W. H., and Pretorius, I. S. (2002). Microbial cellulose utilization: Fundamentals and biotechnology. *Microbiol. Mol. Biol. Rev.* **66,** 506–577.

Madden, R. H., Bryder, M. J., and Poole, N. J. (1982). Isolation and characterization of an anaerobic, cellulolytic bacterium, *Clostridium papyrosolvens* sp nov. *Int. J. Syst. Evol. Microbiol.* **32,** 87–91.

Malekzadeh, F., Azin, M., Shahamat, M., and Colwell, R. R. (1993). Isolation and identification of three *Cellulomonas* spp from forest soils. *World J. Microbiol. Biotechnol.* **9,** 53–55.

Mann, S. O. (1968). An improved method for determining cellulolytic activity in anaerobic bacteria. *J. Appl. Bacteriol.* **31,** 241–244.
McBee, R. H. (1950). The anaerobic thermophilic cellulolytic bacteria. *Bacteriol. Rev.* **14,** 51–63.
McCarthy, A. J. (1987). Lignocellulose-degrading actinomycetes. *FEMS Microbiol. Rev.* **46,** 145–163.
McCarthy, A. J., and Broda, P. (1984). Screening for lignin-degrading actinomycetes and characterization of their activity against C^{14} lignin-labeled wheat lignocellulose. *J. Gen. Microbiol.* **130,** 2905–2913.
McCarthy, A. J., and Williams, S. T. (1990). Methods for studying the ecology of actinomycetes. *Method Microbiol.* **22,** 533–563.
McDonald, J. E., Lockhart, R. J., Cox, M. J., Allison, H. E., and McCarthy, A. J. (2008). Detection of novel *Fibrobacter* populations in landfill sites and determination of their relative abundance via quantitative PCR. *Environ. Microbiol.* **10,** 1310–1319.
McDonald, J. E., de Menezes, A. B., Allison, H. E., and McCarthy, A. J. (2009). Molecular biological detection and quantification of novel *Fibrobacter* populations in freshwater lakes. *Appl. Environ. Microbiol.* **75,** 5148–5152.
McDonald, J. E., Houghton, J. N. I., Rooks, D. J., Allison, H. E., and McCarthy, A. J. (2012). The microbial ecology of anaerobic cellulose degradation in municipal waste landfill sites: Evidence of a role for fibrobacters. *Environ. Microbiol.* doi: 10.1111/j.1462-2920.2011.02688.x.
Mikkelsen, D., Flanagan, B. M., Dykes, G. A., and Gidley, M. J. (2009). Influence of different carbon sources on bacterial cellulose production by *Gluconacetobacter xylinus* strain ATCC 53524. *J. Appl. Microbiol.* **107,** 576–583.
Miller, G. L. (1959). Use of dinitrosalicylic acid reagent for determination of reducing sugar. *Anal. Chem.* **31,** 426–428.
Miller, T. L., and Wolin, M. J. (1974). Serum bottle modification of hungate technique for cultivating obligate anaerobes. *Appl. Microbiol.* **27,** 985–987.
Monserrate, E., Leschine, S. B., and Canale-Parola, E. (2001). *Clostridium hungatei* sp nov., a mesophilic, N-2-fixing cellulolytic bacterium isolated from soil. *Int. J. Syst. Evol. Microbiol.* **51,** 123–132.
Mullings, R., and Parish, J. H. (1984). Mesophilic aerobic Gram-negative cellulose degrading bacteria from aquatic habitats and soils. *J. Appl. Bacteriol.* **57,** 455–468.
Murray, W. D., Hofmann, L., Campbell, N. L., and Madden, R. H. (1986). *Clostridium lentocellum* sp nov, a cellulolytic species from river sediment containing paper-mill waste. *Syst. Appl. Microbiol.* **8,** 181–184.
O'Donnell, A. G., Jenkins, S. N., and Whiteley, A. S. (2011). RNA-radioisotope probing for studying carbon metabolism in soils. *In* "Stable Isotope Probing and Related Technologies," (J. C. Murrell and A. S. Whiteley, eds.), pp. 317–332. ASM Press, Washington, DC.
Orpin, C. G. (1975). Studies on rumen flagellate *Neocallimastix frontalis*. *J. Gen. Microbiol.* **91,** 249–262.
Orpin, C. G., and Joblin, K. N. (1988). The rumen anaerobic fungi. *In* "The Rumen Microbial Ecosystem," (P. N. Hobson, ed.), pp. 129–150. Elsevier Applied Science, London, UK.
Ozkan, M., Desai, S. G., Zhang, Y., Stevenson, D. M., Beane, J., White, E. A., Guerinot, M. L., and Lynd, L. R. (2001). Characterization of 13 newly isolated strains of anaerobic, cellulolytic, thermophilic bacteria. *J. Ind. Microbiol. Biotechnol.* **27,** 275–280.
Ozkose, E., Thomas, B. J., Davies, D. R., Griffith, G. W., and Theodorou, M. K. (2001). *Cyllamyces aberensis* gen.nov sp. nov., a new anaerobic gut fungus with branched sporangiophores isolated from cattle. *Can. J. Bot.* **79,** 666–673.

Palop, M. L., Valles, S., Pinaga, F., and Flors, A. (1989). Isolation and characterization of an anaerobic, celluloytic bacterium, *Clostridium celerecrescens* sp nov. *Int. J. Syst. Evol. Microbiol.* **39**, 68–71.

Pasti, M. B., and Belli, M. L. (1985). Cellulolytic activity of actinomycetes isolated from termites (*termitidae*) gut. *FEMS Microbiol. Lett.* **26**, 107–112.

Phelan, M. B., Crawford, D. L., and Pometto, A. L. (1979). Isolation of lignocellulose-decomposing actinomycetes and degradation of specifically C-14-labeled lignocelluloses by six selected streptomyces strains. *Can. J. Microbiol.* **25**, 1270–1276.

Ransom-Jones, E., Jones, D. L., McCarthy, A. J., and McDonald, J. E. (2012). The *Fibrobacteres*: An important phylum of cellulose-degrading bacteria. *Microb. Ecol.* doi: 10.1007/s00248-011-9998-1.

Reichardt, W. (1988). Impact of the antarctic benthic fauna on the enrichment of biopolymer degrading psychrophilic bacteria. *Microb. Ecol.* **15**, 311–321.

Robert, C., Chassard, C., Lawson, P. A., and Bernalier-Donadille, A. (2007). *Bacteroides cellulosilyticus* sp nov., a cellulolytic bacterium from the human gut microbial community. *Int. J. Syst. Evol. Microbiol.* **57**, 1516–1520.

Rodriguez, A., Perestelo, F., Carnicero, A., Regalado, V., Perez, R., DelaFuente, G., and Falcon, M. A. (1996). Degradation of natural lignins and lignocellulosic substrates by soil-inhabiting fungi imperfecti. *FEMS Microbiol. Ecol.* **21**, 213–219.

Saloheimo, M., and Pakula, T. M. (2012). The cargo and the transport system: Secreted proteins and protein secretion in *Trichoderma reesei* (*Hypocrea jecorina*). *Microbiology* **158**, 46–57.

Shiratori, H., Reno, H., Ayame, S., Kataoka, N., Miya, A., Hosono, K., Beppu, T., and Ueda, K. (2006). Isolation and characterization of a new *Clostridium* sp that performs effective cellulosic waste digestion in a thermophilic methanogenic bioreactor. *Appl. Environ. Microbiol.* **72**, 3702–3709.

Skinner, F. A. (1960). The isolation of anaerobic cellulose-decomposing bacteria from soil. *J. Gen. Microbiol.* **22**, 539–554.

Slaytor, M. (1992). Cellulose digestion in termites and cockroaches—What role do symbionts play. *Comp. Biochem. Phys. B* **103**, 775–784.

Sleat, R., Mah, R. A., and Robinson, R. (1984). Isolation and characterization of an anaerobic, cellulolytic bacterium, *Clostridium cellulovorans* sp nov. *Appl. Environ. Microbiol.* **48**, 88–93.

Smith, R. E. (1977). Rapid tube test for detecting fungal cellulase production. *Appl. Environ. Microbiol.* **33**, 980–981.

Somogyi, M. (1952). Notes on sugar determination. *J. Biol. Chem.* **195**, 19–23.

Stewart, C. S., and Bryant, M. P. (1988). The rumen bacteria. *In* "The Rumen Microbial Ecosystem," (P. N. Hobson, ed.), pp. 21–75. Elsevier Applied Science, New York, USA.

Suen, G., Weimer, P. J., Stevenson, D. M., Aylward, F. O., Boyum, J., Deneke, J., Drinkwater, C., Ivanova, N. N., Mikhailova, N., Chertkov, O., Goodwin, L. A., Currie, C. R., *et al.* (2011). The complete genome sequence of *Fibrobacter succinogenes* S85 reveals a cellulolytic and metabolic specialist. *PLoS One* **6**, e18814.

Sunna, A., Moracci, M., Rossi, M., and Antranikian, G. (1997). Glycosyl hydrolases form hyperthermophiles. *Extremophiles* **1**, 2–13.

Teather, R. M., and Wood, P. J. (1982). Use of Congo red polysaccharide interactions in enumeration and characterization of cellulolytic bacteria from the bovine rumen. *Appl. Environ. Microbiol.* **43**, 777–780.

Thayer, D. W. (1976). Facultative wood-digesting bacteria from hind-gut of termite *Reticulitermes hesperus*. *J. Gen. Microbiol.* **95**, 287–296.

Theodorou, M. K., Gill, M., Kingspooner, C., and Beever, D. E. (1990). Enumeration of anaerobic chytridiomycetes as thallus-forming units: Novel method for quantification of fibrolytic fungal populations from the digestive-tract ecosystem. *Appl. Environ. Microbiol.* **56,** 1073–1078.

Trinci, A. P. J., Davies, D. R., Gull, K., Lawrence, M. I., Nielsen, B. B., Rickers, A., and Theodorou, M. K. (1994). Anaerobic fungi in herbivorous animals. *Mycol. Res.* **98,** 129–152.

Van Dyke, M. I., and McCarthy, A. J. (2002). Molecular biological detection and characterization of *Clostridium* populations in municipal landfill sites. *Appl. Environ. Microbiol.* **68,** 2049–2053.

Vangylsw, N. O. (1970). A comparison of two techniques for counting cellulolytic rumen bacteria. *J. Gen. Microbiol.* **60,** 191–197.

Varma, A., Kolli, B. K., Paul, J., Saxena, S., and Konig, H. (1994). Lignocellulose degradation by microorganisms from termite hills and termite guts: A survey on the present state-of-art. *FEMS Microbiol. Rev.* **15,** 9–28.

Veiga, M., Esparis, A., and Fabregas, J. (1983). Isolation of cellulolytic actinomycetes from marine-sediments. *Appl. Environ. Microbiol.* **46,** 286–287.

Viles, F. J., and Silverman, L. (1949). Determination of starch and cellulose with anthrone. *Anal. Chem.* **21,** 950–953.

Warcup, J. H. (1950). The soil-plate method for isolation of fungi from soil. *Nature* **166,** 117–118.

Warnecke, F., Luginbuhl, P., Ivanova, N., Ghassemian, M., Richardson, T. H., Stege, J. T., Cayouette, M., McHardy, A. C., Djordjevic, G., Aboushadi, N., Sorek, R., Tringe, S. G., *et al.* (2007). Metagenomic and functional analysis of hindgut microbiota of a wood-feeding higher termite. *Nature* **450,** 560–565.

Warnick, T. A., Methe, B. A., and Leschine, S. B. (2002). *Clostridium phytofermentans* sp nov., a cellulolytic mesophile from forest soil. *Int. J. Syst. Evol. Microbiol.* **52,** 1155–1160.

Watanabe, H., and Tokuda, G. (2010). Cellulolytic systems in insects. *Annu. Rev. Entomol.* **55,** 609–632.

Wedekind, K. J., Mansfield, H. R., and Montgomery, L. (1988). Enumeration and isolation of cellulolytic and hemicellulolytic bacteria from human feces. *Appl. Environ. Microbiol.* **54,** 1530–1535.

Weiner, R. M., Taylor, L. E., II, Henrissat, B., Hauser, L., Land, M., Coutinho, P. M., Rancurel, C., Saunders, E. H., Longmire, A. G., Zhang, H., Bayer, E. A., Gilbert, H. J., *et al.* (2008). Complete genome sequence of the complex carbohydrate-degrading marine bacterium, *Saccharophagus degradans* strain 2-40(T). *PLoS Genet.* **4**(5), e1000087.

Wenzel, M., Schonig, I., Berchtold, M., Kampfer, P., and Konig, H. (2002). Aerobic and facultatively anaerobic cellulolytic bacteria from the gut of the termite *Zootermopsis angusticollis*. *J. Appl. Microbiol.* **92,** 32–40.

Westlake, K., Archer, D. B., and Boone, D. R. (1995). Diversity of cellulolytic bacteria in landfill. *J. Appl. Bacteriol.* **79,** 73–78.

Wilson, D. B. (2009). Evidence for a novel mechanism of microbial cellulose degradation. *Cellulose* **16,** 723–727.

Wood, T. M. (1988). Preparation of crystalline, amorphous, and dyed cellulase substrates. *Methods Enzymol.* **160,** 19–25.

Wood, T. M., Wilson, C. A., McCrae, S. I., and Joblin, K. N. (1986). A highly-active extracellular cellulase from the anaerobic rumen fungus *Neocallimastix frontalis*. *FEMS Microbiol. Lett.* **34,** 37–40.

Zhang, Y. H. P., Himmel, M. E., and Mielenz, J. R. (2006). Outlook for cellulase improvement: Screening and selection strategies. *Biotechnol. Adv.* **24,** 452–481.

CHAPTER TWENTY

METAGENOMIC APPROACHES TO THE DISCOVERY OF CELLULASES

David J. Rooks,* James E. McDonald,[†] and Alan J. McCarthy*

Contents

1. Introduction	376
2. Nucleic Acid Extraction from Environmental Samples	377
2.1. CTAB protocol for the coextraction of DNA and RNA from environmental samples	377
2.2. DNA extraction for the creation of large-insert metagenomic libraries	378
3. Metagenomic Libraries	379
3.1. Generation of large-insert libraries	380
3.2. Screening metagenomes for cellulases	382
4. Metatranscriptomics	385
4.1. mRNA enrichment	385
4.2. cDNA synthesis	388
5. Outlook	389
Acknowledgment	390
References	390

Abstract

Most of the microorganisms responsible for nutrient cycling in the environment have yet to be cultivated, and this could include those species responsible for the degradation of cellulose. Known cellulases are well defined at the protein sequence level, but gene variants are difficult to amplify from environmental DNA. The identification of novel cellulase genes independent of DNA amplification is made possible by adopting a direct metagenome sequencing approach to provide genes that can be cloned, expressed, and characterized prior to potential exploitation, all in the absence of any information on the species from which they originated. In this chapter, emerging strategies and methods that will enable the identification of novel cellulase genes and provide an unbiased perspective on gene expression *in situ* are presented.

* Microbiology Research Group, Institute of Integrative Biology, University of Liverpool, Liverpool, United Kingdom
[†] School of Biological Sciences, Bangor University, Bangor, United Kingdom

1. INTRODUCTION

The vast majority of microorganisms in the biosphere have yet to be cultivated and remain an untapped source of enzymes for biotechnological applications. The current impetus to find novel sources of cellulases for application in biomass refining derives from the importance of utilizing cellulose, one of nature's most abundant forms of carbon, as a substrate for second-generation biofuel production. The enzymology of cellulose degradation has only been investigated in a few model systems, primarily *Clostridium thermocellum* (Garcia-Alvarez et al., 2011) and *Trichoderma reesei* (Martinez et al., 2008). Metagenomics, essentially sequencing environmental DNA coupled with the interrogation of clone libraries for the presence of target genes, is still in its infancy, but there are already a number of examples of its efficacy, including the discovery of novel polysaccharide hydrolases in the microbial communities that inhabit termite and bovine intestinal tracts (Langer et al., 2006; Warnecke and Hess, 2009). While particularly suited to habitats that are intractable to conventional isolation and screening of cultivable microbial species, metagenomics for cellulase discovery can be applied to any environment in which cellulose-degrading microorganisms reside.

The landscape of environmental microbiology changed significantly with the advent of molecular biological tools. Most of the effort has been directed toward describing the true diversity in natural microbial communities via PCR amplification of taxonomic and, where applicable, functional marker genes. Cellulase genes, although well defined at the protein sequence level, can rarely be simply amplified from environmental DNA because nucleotide sequence variation does not lend itself well to the design of appropriate PCR primers. There are, however, some examples where this has been achieved for subsets of cellulases (Kataeva et al., 1999; Mayer et al., 2011). An alternative strategy has been to apply small subunit ribosomal RNA gene PCR primers directed at taxa that are known to be cellulolytic, for example, anaerobic fungi of the *Neocallimastigales* (Lockhart et al., 2006), subgroups of clostridia (Van Dyke and McCarthy, 2002), and fibrobacters (McDonald et al., 2008, 2009), but having demonstrated their presence and even importance in cellulose degradation in an environment (McDonald et al., 2012), the problem of isolating and cultivating these potentially novel sources of cellulases remains.

Identifying cellulase genes directly in metagenomic libraries circumvents this issue and provides genes that can then be cloned, expressed, and characterized prior to potential exploitation, in the absence of any information on the species from which they originated. Again, there are some examples (Duan et al., 2009; Ferrer et al., 2005; Liu et al., 2011), but this approach is set to gain considerable momentum fuelled by the arrival of pyrosequencing technology, which has just begun to be applied to the identification of cellulases in metagenomes (Edwards *et al.*, 2010; Graham *et al.*, 2011; Tasse *et al.*, 2010). It is the aim of this chapter to provide descriptions of the methods that are currently available in this rapidly developing technological field.

2. Nucleic Acid Extraction from Environmental Samples

A number of protocols are available for the coextraction of DNA and RNA from environmental materials. One of the most common is the CTAB (cetyltrimethylammonium bromide) extraction method described by Griffiths *et al.* (2000), which is based on the use of physical disruption (bead beating) to maximise the release of cellular DNA and RNA. The method incorporates a quaternary ammonium salt, CTAB, to remove contaminating organic matter, particularly humic material that interferes with DNA/RNA purification. Although originally designed for use with soil, it can be applied to virtually any material. One limitation of this method is shearing of the DNA, a consequence of bead beating, and although this step can be omitted for the preparation of large-insert libraries, an alternative approach may be required (see Section 2.2), especially if a limited amount of cellular material is present, for example biofilms.

2.1. CTAB protocol for the coextraction of DNA and RNA from environmental samples (Griffiths *et al.*, 2000)

1. Add the sample (0.5 g) to a lysing matrix tube (type E; MP Biochemicals). Add 500 µl of 5% CTAB in 120 mM sodium phosphate buffer, pH 8.0, and 500 µl phenol/chloroform/isoamyl alcohol (25:24:1) (Sigma).
2. Process each tube in a RiboLyser (Hybaid) and bead beat for 30 s at a speed setting of 5.5.
3. Centrifuge at 4 °C for 5 min at 14,000 rpm.
4. Extract the top aqueous layer into a 1.5-ml Eppendorf tube, and add an equal volume of chloroform/isoamyl alcohol (24:1) (Sigma). Vortex mix to form an emulsion.
5. Centrifuge at 4 °C for 5 min at 14,000 rpm.
6. Transfer the aqueous phase to a new Eppendorf tube and precipitate the nucleic acids in 2 volumes of 30% polyethylene glycol (PEG) 6000/1.6 M NaCl for 1–2 h at room temperature. This can be left overnight at 4 °C to increase nucleic acid yield. Pellet the nucleic acids by centrifugation at 10,000 rpm for 10 min.

7. Remove the supernatant and wash the pellet with 200 μl of 70% (v/v) ice-cold ethanol.
8. Air-dry the pellet and resuspend in 50 μl RNase-free water.

Each extract can be split into two aliquots, and DNase or RNase treated to provide purified RNA and DNA from the same sample. The RNA can be reverse transcribed to produce cDNA, or enriched for mRNA and used for the production of a metatranscriptome. The DNA can be used for the preparation of clone libraries, or for PCR amplification of target genes. It should be noted that if the DNA/RNA is not going to be used immediately, the extraction protocol should be stopped at step 6, and the sample stored at $-80\,^{\circ}\text{C}$ in the PEG solution until required. *Note:* All solutions and glassware should be rendered RNase-free by diethylpyrocarbonate (DEPC) and RNase Zap treatments (see Purdy, 2005). For Tris buffers, DEPC must be used prior to the addition of the Tris salt.

2.2. DNA extraction for the creation of large-insert metagenomic libraries (modified from Neufeld *et al.*, 2007)

The construction of metagenomic libraries derived from environmental samples for subsequent identification and cloning of functional genes requires a DNA extract of high quality, as the enzymatic modifications required during the construction of the libraries are sensitive to contamination by various biotic and abiotic components. High molecular weight environmental DNA (20–40 kb) is required for large-insert libraries (Simon and Daniel, 2010). The method below, modified from Neufeld *et al.* (2007), can be used to produce high molecular weight DNA of suitable purity for library production:

1. Add 1.6 ml SET buffer (0.75 M sucrose, 40 mM EDTA, 50 mM Tris–HCl, pH 8.0) and 180 μl freshly prepared lysozyme solution (990 μl sterile water, 9 mg lysozyme, and 9 μl 1 M Tris–HCl pH 8) to the sample and incubate with gentle rotation in a hybridization oven at 37 °C for 30 min.
2. Add 200 μl 10% (w/v) SDS and 55 μl proteinase K solution (95 μl sterile water, 50 μl of 1 M Tris–HCl pH 8, and 20 mg proteinase K) and incubate at 55 °C for 2 h with gentle rotation.
3. Withdraw the lysate using a 5-ml syringe.
4. Centrifuge a pellet phase lock gel (5 Prime) at $1500 \times g$ for 2 min and add the previously harvested lysate, 2 ml phenol:chloroform:isoamyl alcohol (25:24:1), and 2.0 ml CTAB/NaCl solution (mix equal volumes of 10% CTAB in 0.7 M NaCl with 240 mM potassium phosphate buffer pH 8.0). Mix thoroughly and incubate for 10 min at 65 °C.
5. Centrifuge for 5 min at $1500 \times g$ and transfer the 2–3 ml aqueous phase into a 25-ml centrifuge tube. Add 20 μg ml^{-1} glycogen, 0.5 volumes of

7.5 M ammonium acetate, and 2 volumes of 95% (v/v) ethanol and incubate overnight at $-20\,^\circ\mathrm{C}$.
6. Centrifuge the sample at $48{,}000 \times g$ to pellet the nucleic acid, remove the supernatant, and air-dry at room temperature for 30 min.
7. Resuspend DNA in 200 µl sterile distilled H_2O and quantify the DNA spectrophotometrically by measuring $A_{260/280}$ using a NanoDrop (Thermo Scientific) or a QuBit fluorometer (Invitrogen). DNA quality and molecular weight should be checked at this stage using pulsed-field gel electrophoresis.

3. Metagenomic Libraries

The construction and screening of metagenomic libraries of environmental DNA has proven to be a powerful tool for the recovery of novel biomolecules for biotechnological applications (Steele et al., 2009). Two types of libraries with respect to average insert size can be generated: small-insert libraries in plasmid vectors (< 10 kb) and large-insert libraries in fosmid vectors (up to 40 kb) or bacterial artificial chromosome (BAC) vectors (> 40 kb). Small-insert metagenomic libraries are useful for the isolation of single genes or small operons encoding novel biomolecules. Numerous vectors are suitable for the construction of small-insert metagenomic libraries, and small pieces of DNA can be efficiently transformed into a host cell using a variety of techniques, including heat shock, electroporation, conjugation, and phage transfection. Successful combinations of vector transformation protocols include p2ErO-2 with electroporation (Gabor et al., 2004) and Lambda Zap with phage transfection (Robertson et al., 2004). However, to identify complex pathways encoded by large gene clusters or large DNA fragments for the partial genomic characterization of uncultured microorganisms, the generation of large-insert libraries is the appropriate method. In recent years, fosmid cloning systems have been commonly used for the identification of novel genes, and this has aided the discovery of a wide range, including those encoding chitinase, dehydrogenase, oxidoreductase, amylase, esterase, endoglucanase, glycoside hydrolases (GHs), and cyclodextrinase (Ammiraju et al., 2005; Beloqui et al., 2010; Feng et al., 2007; Hu et al., 2008; Kim et al., 2008; Pang et al., 2009; Suenaga et al., 2007; Voget et al., 2006). Fosmids and BAC vectors both replicate using the F-factor replicon, which conveys high stability to the plasmid when carrying large inserts, enabling successful cloning of unstable DNA sequences and genes for expressed toxic proteins. Although construction of metagenomic libraries may be performed with BAC or fosmid vectors, the latter are invariably used. Genomic library construction kits are commercially available, and the "CopyControl™" Fosmid Library

Production Kit" (Epicentre) is a good example; the following protocol is adapted from the manufacturer's instructions.

3.1. Generation of large-insert libraries

3.1.1. Size selection

1. Separate the blunt-ended DNA via electrophoresis on a 1.5% low melting point agarose gel prepared with $1 \times$ TAE buffer and a DNA size marker in each of the outside lanes of the gel. *Note:* Do not include ethidium bromide in the gel.
2. Following electrophoresis, excise the outer lanes of the gel containing the DNA ladder and stain with ethidium bromide. Visualize the DNA ladder with UV light and mark the position of the desired fragment sizes on both DNA ladders. After removing the gel slices from the transilluminator, reassemble the gel and cut out a gel slice containing DNA of the desired fragment size range (30–40 kb).
3. Place the gel slice into a sterile Eppendorf tube and weigh.
4. Add GELase buffer (3 μl of buffer per mg of gel) and incubate at room temperature for 1 h.

3.1.2. End repair

1. The following reagents are included in the CopyControl™ Fosmid Library Production Kit and should be added to 50 μl of the size-fractionated GELase-treated DNA solution from Section 3.1.1: 8 μl end-repair buffer ($10\times$), 8 μl of 2.5 mM dNTP mix, 8 μl of 10 mM ATP (*Note:* ATP does not survive freeze thawing), and 4 μl end-repair enzyme mix. Make the final volume up to 80 μl with dH$_2$O.
2. Incubate at room temperature for 2 h, followed by enzyme inactivation at 70 °C for 10 min. The blunt-ended DNA should then be purified with SureClean (provided with kit) and used according to the manufacturer's instructions.
3. Resuspend the DNA pellet in 20–30 μl dH$_2$O. The DNA can be stored at -20 °C at this stage.

3.1.3. Ligation

1. The following reagents are included in the CopyControl™ Fosmid Library Production Kit and should be added to the end-repaired insert DNA (~ 600 ng): 1 μl Fast-Link ligation buffer ($10\times$), 1 μl 10 mM ATP, 1 μl CopyControl™ pCC1FOS Vector (0.5 μg μl^{-1}), and 1 μl Fast-Link DNA ligase (2 U μl^{-1}). Add sterile distilled H$_2$O to a final volume of 10 μl.
2. Incubate overnight at 16 °C.

3. Add 0.5 µl Fast-Link DNA ligase to the reaction mixture and incubate for an additional 1.5 h at room temperature.
4. Stop the reaction by incubating at 70 °C for 10 min. The ligation reaction can be stored at -20 °C until required.

3.1.4. Packaging of fosmids

1. The day before performing the lambda packaging reaction, inoculate 5 ml of Luria-Bertani (LB) broth containing 10 mM MgSO$_4$ and 0.2% (w/v) maltose with 0.5 ml of an overnight culture of EPI300-T1^R host cells. Incubate at 37 °C with shaking at 200 rpm to an A_{600} of 0.8–1.0.
2. Thaw one vial of the MaxPlax Lambda packaging extract on ice for each ligation you wish to perform.
3. Add 10 µl of the ligation reaction to 25 µl of the MaxPlax packaging extract and mix the solution without producing any air bubbles.
4. Incubate the reaction mixture for 90 min at 30 °C.
5. Add phage dilution buffer to a final volume of 1 ml and mix gently. Add 25 µl chloroform and mix gently. At this stage, the preparation can be stored at 4 °C for up to 2 days, as this does not have any effect on subsequent steps.

3.1.5. Host cell transduction

1. Add 10, 20, 30, 40, and 50 µl of the packaged phage particles individually to Eppendorf tubes, each containing 100 µl of the prepared EPI300-T1^R cells.
2. Incubate for 45 min at 37 °C.
3. Spread the infected EPI100-T1^R cells on an LB plate containing 12.5 µg ml^{-1} chloramphenicol and incubate overnight at 37 °C. Colonies are counted and the phage particle titer is calculated.
4. The fosmid clones obtained are then grown in 96-well culture plates (or 384-well format plates for robotic processing), containing LB broth and subsequently stored at -80 °C in the presence of 8% (v/v) DMSO.

It is recommended that the fosmid clones are subsequently induced to produce a higher copy number in order to achieve high fosmid DNA yields for sequencing, fingerprinting, or other downstream applications. Fosmid clones are cultured overnight at 37 °C in 96-well plates containing LB and 12.5 µg ml^{-1} chloramphenicol. Add 5 µl of the CopyControl^{TM} Induction Solution (1000 ×) (Epicentre) to each well and shake each plate at 37 °C for 5 h at 200 rpm, as aeration is critical for high copy number induction. The fosmid DNA is extracted, digested, and analyzed using standard techniques to ensure that the fosmid library contains metagenomic DNA. Clones with confirmed inserts are again stored at -80 °C in the presence of 8% DMSO.

3.2. Screening metagenomes for cellulases

The creation of clone libraries is no longer essential for identifying the *presence* of putative cellulases in metagenomes, due to the arrival of DNA pyrosequencing technology. The environmental DNA (or cDNA) preparations can be directly sheared and pyrosequenced using a number of sequencing platforms to generate large datasets of relatively short sequences, for assembly into contigs that can be identified by interrogation of appropriate databases. In this way, protein-encoding genes can be identified, although, of course, in environmental metagenomes, most will be hypothetical or of unknown function. The primary consideration in selection of the sequencing platform is read length and for this reason, 454 pyrosequencing (Rodrigue *et al.*, 2010) is the most popular. Datasets of predominantly short sequences, while not a significant problem in genome and transcriptome sequencing of pure cultures, where a genome sequence can be used as a scaffold, are of limited use in metagenomics. That said, pyrosequencing alternatives to 454 are continuing to develop increased read lengths and these include Illumina/Solexa Genome Analyzer, ABI SOLiD platform and Ion Torrent (Niedringhaus *et al.*, 2011), which can offer advantages in terms of convenience, volume of data generated, and cost. The second issue is depth of sequencing and the decision is largely dictated by cost. In metagenomes enriched in cellulose-degrading organisms, cellulases may be identified in relatively small pyrosequence datasets (Edwards *et al.*, 2010). It is possible to produce a small sequence dataset and apply rarefaction analysis (Foster and Dunstan, 2010) to estimate the extent of metagenome coverage that has been obtained and then decide whether resequencing to greater depth and to what extent is required to represent a credible cellulase gene discovery project. Primers for qPCR can be designed from these small datasets, or for genes known to be present, and applied to the environmental DNA preparation to verify and/or provide estimations of sequence coverage (Rooks *et al.*, 2010).

3.2.1. *In silico* screens

In environmental metagenomics, determining the true microbial community structure and discovering new taxa have been an important driver, and there are bioinformatics tools and approaches available in this context. MEGAN (Huson *et al.*, 2007) is a data management program used in the taxonomic analysis of large sequencing datasets, processing results of sequence comparisons between a known database and metagenome-derived sequences. Famously, the program has been applied to existing metagenome data by reanalyzing the original dataset from the Sargasso Sea (Venter *et al.*, 2004). The small subunit 16S or 18S ribosomal RNA genes are the taxonomic marker targets of choice, and in combination with the Ribosomal Database Project (Cole *et al.*, 2007) and Greengenes (DeSantis *et al.*, 2006), can be used to provide parallel information on taxonomic composition. Furthermore, the sequence dataset can be uploaded to the SEED (Overbeek *et al.*, 2005) and MG-RAST (Meyer *et al.*, 2008) to provide taxonomic affiliations for all of the known functional and hypothetical protein-encoding genes present. There are two important caveats here: first, identifications based on homology are only meaningful if the e-level cutoff is <0.001 (Meyer *et al.*, 2008); second, it is likely that the majority of the protein-encoding genes in the metagenome will be unidentifiable. While this taxonomic information is not the primary purpose here, it is a useful indicator of the presence of known groups of cellulose degraders in the sample used to generate the metagenomic DNA sample.

For identification of cellulase genes *per se*, this can be achieved by interrogating sequence databases to identify homologies or, more ambitiously, to look for new types or classes of cellulases among the genes of unknown function that invariably predominate in metagenome sequence datasets. The former is facilitated by the CAZy database (Cantarel *et al.*, 2009) which contains known polysaccharide-degrading enzymes, including both catalytic and carbohydrate-binding modules, and can be customized to search for members of established enzyme families (Edwards *et al.*, 2010), with follow-up using BlastX or BlastN (Altschul *et al.*, 1997) to verify homologies. However, although the collection of carbohydratase data in CAZy provides a very useful resource for enzymologists, including those engaged in interrogating genomic datasets, it is important to be able to distinguish annotations based on biochemical verification from the much less reliable computational predictions of function (see Sukharnikov *et al.*, 2011). In pyrosequenced metagenome datasets, these identifications can be used to design PCR primers or oligonucleotide probes to obtain intact genes for cloning and further characterization from clone libraries prepared from the same environmental DNA (or cDNA) sample. To find completely novel cellulases, more sophisticated bioinformatics approaches are required to search for domains and motifs indicative of enzymes with truly novel cellulose binding and/or catalytic functions. Sequence comparisons among proteins with suggestive domain architectures or genomic contexts in metagenomic DNA have the potential to identify novel cellulases; the discovery of a new carbohydrate-binding module in metagenomic DNA by Mello *et al.* (2010) is a particularly good example of what can be achieved by the continuing development of new bioinformatics tools.

3.2.2. Sequence-based screens
Traditionally, identification of genes from clone libraries is achieved through PCR screening. Here, PCR primers for cellulase family identification could be designed against known cellulase sequences in the CAZy database and then used to screen fosmid libraries. However, cellulases are GH enzymes, and to date, a total of 130 different GH families have been classified on CAZy (Sukharnikov *et al.*, 2011). The degree of nucleotide

sequence variation is such that it is not possible, or at best challenging, to construct alignments that enable the design of universal primers for specific families containing genes annotated as containing cellulases. A preferred method would be to use pyrosequenced datasets from parallel samples to identify the presence of potentially novel cellulases (modular enzymes that contains cellulose binding CBMs or in a locus that contains known cellulase genes) and then design PCR primers and, more importantly, oligonucleotide probes that can be used to recover and clone intact genes from libraries produced from the same sample.

3.2.3. Expression-based screens

The probability (hit rate) of identifying a certain gene depends on multiple factors that are inextricably linked: the host–vector system, size of the target gene, its abundance in the source metagenome, the assay method, and the efficiency of heterologous gene expression in a surrogate host (Uchiyama and Miyazaki, 2009; van Sint Fiet *et al.*, 2006). Functional screening has successfully recovered cellulases from metagenomes prepared from environments as varied as soil (Voget *et al.*, 2006), buffalo rumen (Duan *et al.*, 2009), and the termite hindgut (Zhang *et al.*, 2006); all using carboxymethyl cellulose (CMC) incorporation in agar plates and staining with Congo red to identify clones expressing endoglucanase (Teather and Wood, 1982). This approach has been used in our laboratory to recover and subsequently clone an endoglucanase from a metagenomic fosmid library prepared from estuarine sediment DNA (J. L. Edwards, A. Houlden, A. J. McCarthy, unpublished data). CMC is a derivative of cellulose containing carboxymethyl groups that render the substrate water-soluble. Although this method only directly identifies endoglucanases with activity against amorphous cellulose, it can also act as a proxy to identify fosmids that can be sequenced to find accessory modules, such as carbohydrate-binding modules, and genes for cellulases in addition to the endoglucanase responsible for the Congo red positive reaction.

Congo red interacts strongly with polysaccharides containing contiguous β-1,4-linked D-glucopyranosyl units, and zones of CMC hydrolysis around a metagenomic clone on an agar plate can be readily visualized. Q-Trays (Genetix), containing LB supplemented with agar (1.8% w/v), CMC (0.1%), and chloramphenicol (12.5 µg ml^{-1}), can be used to screen the fosmid library for endoglucanase activity. Clones are replicated onto agar with a 96-pin replicator (6 × 96 on each Q-Tray). Each plate is incubated overnight at 37 °C. The resultant colonies are blotted off the agar using damp Whatman paper, and the agar is subsequently stained by flooding with Congo red (0.1%) for 30 min, with gentle shaking. The Congo red is removed and each Q-Tray is then washed with 1 M NaCl for 30 min (twice). Clones positive for endoglucanase activity are identified by the presence of a clear yellowish zone around the clone, against a red background. This CMC/Congo red assay has previously been used to identify a single cellulase-positive clone in a metagenomic fosmid library prepared with DNA isolated from abalone intestine (Kim et al., 2011). The method was also used to screen four different environmental DNA libraries from forest soil, elephant dung, cattle rumen, and rotted wood, enabling the identification of seven independent clones specifying endoglucanase activities; subsequent sequence analysis of the clones revealed that the encoded products shared less than 50% identities and 70% similarities to cellulases in the CAZy databases (Wang et al., 2009).

4. METATRANSCRIPTOMICS

Environmental transcriptomics (metatranscriptomics) retrieves and sequences mRNAs from a microbial community to provide an unbiased perspective on gene expression *in situ*. This is important in ecological studies where the question of relevance to actual cellulose degradation in the environment is addressed by targeting expressed cellulases. It has also been inherent in some molecular ecological studies in which RNA has been extracted and reverse transcribed to cDNA for PCR determination of the presence, and hence activity, of known cellulose-degrading taxa (McDonald et al., 2012, 2009) circumventing problems of poor PCR amplification of rRNA genes themselves in, for example, fibrobacters (McDonald et al., 2008). Although for cellulase gene discovery the genetic capacity of the metagenome for cellulose degradation is the more important resource, the metatranscriptomic route does identify cellulases that are being expressed and therefore *bona fide*. Technologically, substantial progress has been made in the efficient analysis of complex expression profiles with the development of next-generation sequencing technologies, for example, Roche's 454, Illuminas' SOLEXA, and ABI's SOLiD platforms, which have been critically reviewed by Niedringhaus et al. (2011); these new technologies allow not only the direct sequencing of cDNA without the requirement for a cloning step (Medini et al., 2008) but also increase the throughput and reduce the cost. In the third generation of these technologies, the read length is in excess of 400 bp (Schadt et al., 2010) and will undoubtedly increase further.

4.1. mRNA enrichment

A major technical challenge for *de novo* transcriptome sequencing is the low relative abundance of mRNAs in total cellular RNA (1–5%), the bulk of which is rRNAs and tRNAs (Karpinets et al., 2006). Unlike eukaryotic mRNAs, which can be selectively reverse transcribed to cDNA by virtue of their polyA tails (Zhao et al., 1999), bacterial and archaeal cDNA libraries comprise predominantly rRNA sequences; it is clearly preferable to remove or at least reduce cDNA derived from prokaryotic rRNA prior to sequencing in order to facilitate mRNA detection at an acceptable level. For example, in the first study of a marine microbial metatranscriptome using 454-pyrosequencing technology, rRNA reads represented 53% of total sequences (Frias-Lopez et al., 2008). Although pyrosequencing reads of samples not enriched for mRNA are used for community structure analysis, it can be regarded as a waste of sequencing capacity if the goal is to study the functional expression profile of the community. Different methods have been used to eliminate prokaryotic rRNA, including subtractive hybridization (Stewart et al., 2010), digestion with exonuclease that preferentially acts on rRNA, poly(A) tail addition to discriminate against rRNA (Shi et al., 2009), and reverse transcription with rRNA-specific primers followed by RNase H digestion to degrade rRNA:DNA (Dunman et al., 2001).

4.1.1. Subtractive hybridization using sample-specific biotinylated rRNA probes (Stewart et al., 2010)

This is a sample-specific method for the subtraction of rRNA from total RNA. The method employs antisense rRNA probes transcribed *in vitro* from PCR products amplified from coupled DNA samples, thereby ensuring the specificity of the probe mix (Stewart et al., 2010). The method is summarized in Fig. 20.1.

1. Templates for probe generation are prepared using PCR, with universal primers that flank the entire length of the bacterial 16S rRNA gene and ∼85% of the 23S rRNA gene. The reverse primers are modified to contain the T7 RNA polymerase promoter sequence (DeLong et al., 1999; Hunt et al., 2006).
2. Each PCR mix (50 µl) contains 100 ng template DNA, 1 µl Herculase fusion DNA polymerase (Stratagene), 1× Herculase reaction buffer, 10 mM dNTP, and 10 µM forward and reverse primers.
3. Reaction conditions: 2 min at 92 °C; 35 cycles of 20 s at 39 °C (23S reactions) or 55 °C (16S reactions), 75 s (16S) or 90 s (23S) at 72 °C; 3 min at 72 °C.
4. Purify the resultant PCR products using the QIAquick PCR Purification kit (Qiagen). These products are then used as template for the generation of biotinylated antisense 16S and 23S RNA probes.
5. Probes for 16S and 23S RNA are generated separately in 20-µl reactions, each containing: 1× buffer, T7 RNA polymerase, SUPERase In RNase inhibitor (10 U), rATP (7.5 mM), rGTP (7.5 mM), rCTP (5.625 mM), rUTP (6.625 mM), biotin-11-rCTP (1.875 mM), and 16S/23S rDNA template (250–500 ng).
6. Each reaction is incubated at 37 °C for 4–5 h, followed by Turbo DNase digestion (Ambion) 15 min at 37 °C. Products are then purified using the MEGAclear kit (Ambion).

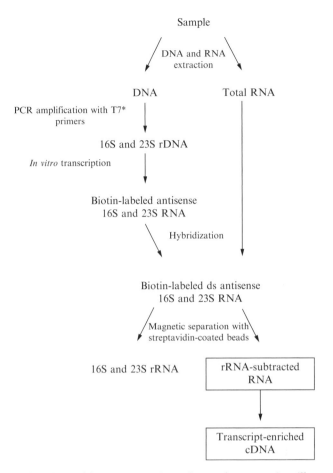

Figure 20.1 Flowchart of the metatranscriptomic sample preparation, illustrating steps for sample-specific subtraction of bacterial ribosomal RNA (16S and 23S subunits) from total RNA. *Primers contain the T7 RNA polymerase promoter sequence. Adapted from Stewart et al. (2010).

7. The biotinylated rRNA probes are hybridized to complimentary rRNA molecules in the total RNA sample. Each hybridization reaction (50 μl) contains formamide (20%), 1× SSC buffer (0.15 M sodium chloride, 0.015 M sodium citrate), SUPERase In RNase Inhibitor (20 U), template RNA (600 ng), and equal amounts of 16S and 23S rRNA probes, at a final template-to-probe ratio of 1:2.
8. The reaction is denatured at 70 °C for 5 min and then incubated at room temperature for 3 min. Biotinylated double-stranded rRNA can then be removed from the sample by hybridization to streptavidin-coated magnetic beads (New England BioLabs) according to the manufacturer's

instructions, and the resultant products are then purified using the RNeasy MinElute Cleanup kit (Qiagen).
9. The rRNA-subtracted RNA (~35–80 ng) can be amplified using the MessageAmp RNA amplification kit (Ambion) according to the manufacturer's instructions. The procedure consists of reverse transcription with an oligo(dt) primer bearing a T7 promoter, and *in vitro* transcription of the resulting DNA with T7 RNA polymerase to generate hundreds of thousands of antisense RNA copies of each mRNA in the sample. This mRNA is then converted to double-stranded cDNA (see Section 4.2).

4.1.2. Alternative mRNA enrichment methods

The MICROBExpress kit from Ambion mixes the total RNA with an optimized set of capture oligonucleotides that bind to bacterial 16S and 23S rRNA. The rRNA hybrids are then removed from the solution using derivatized magnetic microbeads. The mRNA remains in the supernatant and is recovered by ethanol precipitation. As this removal technique is sequence based, the efficiency of the removal varies between different communities. Gilbert *et al.* (2008) reported that using this technique to analyze a complete marine microbial metatranscriptome achieved enrichment to 99.92% mRNA.

Selective removal of rRNA utilizing *Escherichia coli* poly(A) polymerase, which preferentially polyadenylates mRNA, has also been used. The corresponding first-strand cDNA can then be synthesized using poly-dT primers included in the MessageAmp II-Bacteria Kit (Ambion). Frias-Lopez *et al.* (2008) used this technique to analyze the expression of genes across a microbial community. mRNA was enriched to 47% of total sequence output, and although many transcripts were similar to genes previously detected in ocean metagenomic surveys, a significant fraction (~50%) was unique.

4.2. cDNA synthesis

One of the most promising strategies for reverse transcription, without the introduction of an amplification bias, is the Just cDNA Double-Stranded cDNA Synthesis Kit from Agilent. This kit contains all the reagents necessary to make double-stranded cDNA from poly (A) + RNA, including random and oligo(dT) primers. AccuScript RT (in the Agilent kit) locates a site on the mRNA template where a specific primer has bound and begins to synthesize the first-strand cDNA. When using random primers, the primers anneal to various complimentary sites on the mRNA template. During second-strand synthesis, RNase H nicks the RNA bound to the first-strand cDNA to produce a multitude of fragments that serve as primers for DNA polymerase I. DNA polymerase I "nick translates" these RNA fragments into second-strand cDNA. The uneven termini of the double-stranded cDNA are then polished or filled in with cloned *pfu* DNA polymerase yielding blunt-ended double-stranded cDNA that can be used for downstream-sequencing processes.

Previous studies have demonstrated how high-throughput sequencing technologies can be applied to access information stored in transcripts isolated directly from complex environments, including marine and soil microbial communities. The first report of using 454 pyrosequencing for studying the metatranscriptome of a complex microbial community was published in 2006. Leininger and colleagues demonstrated that archaeal transcripts for the key enzyme (*amoA*) for ammonia oxidation were several orders of magnitude more abundant in soils than the bacterial equivalent of the gene, suggesting archaea as the numerically dominant ammonia oxidizers in soil (Leininger *et al.*, 2006). Gilbert *et al.* (2008) identified a large number of novel highly expressed sequence clusters of complex marine microbial communities, the majority of which were orphaned, demonstrating the utility of the approach for use in the discovery of novel genetic variants. At the time of writing, there are no reports of metatranscriptome preparation and screening aimed at the identification of expressed cellulases in RNA extracted from environmental samples.

5. Outlook

In the past 5 years, next-generation sequencing (NGS) has emerged as a revolutionary tool for the comprehensive characterization of DNA sequence variation, *de novo* sequencing of genomes, sequencing of microbiomes, quantitating transcript abundances, and identifying the degree to which mRNA transcripts are being actively transcribed. Currently, one of the major advantages of NGS in environmental metagenomic/metatranscriptomic studies is the small amounts of genetic starting material that are required. In addition to this recent and rapid advancement, a new generation of sequencing technologies is emerging (third-generation sequencing). This technology interrogates single molecules of DNA such that the need for amplification during the sequencing reaction is removed, with the promise of: higher throughput; faster turnaround time; longer read length; greater accuracy, enabling rare variant detection; even smaller amounts of starting material and reduced cost. These advances will allow the generation of much larger metagenome and metatranscriptome sequence datasets, which can then be fed into processing pipelines such as SEED and KEGG (Mitra *et al.*, 2011) to identify genes and proteins with known, putative, or predicted enzyme functions.

As the cost of sequencing continues to fall and the volume of data increases, there will be a continuing demand for novel analysis methods. The current "bioinformatics bottleneck" looks set to persist. Proper description of the protein universe requires that databases are functionally annotated, and ideally this should be done experimentally. Metaproteomics is technologically challenging and although progress is encouraging (Kong et al., 2011), its application in tandem with metagenomic and metatranscriptomic approaches to novel enzyme discovery is an ideal that is some way off. In the short term, however, it is undoubtedly the increased speed, capacity, and decreased cost of DNA sequencing, together with the ability to undertake this "in house" using benchtop platforms, such as Ion Torrent, that will promote dissemination of the metagenomic approach to novel cellulase discovery.

ACKNOWLEDGMENT

Research on the metagenomic analysis of cellulose-degrading communities by the authors has been funded by the UK Natural Environment Research Council (NERC).

REFERENCES

Altschul, S. F., Madden, T. L., Schaffer, A. A., Zhang, J., Zhang, Z., Miller, W., and Lipman, D. J. (1997). Gapped BLAST and PSI-BLAST: A new generation of protein database search programs. *Nucleic Acids Res.* **25**, 3389–3402.

Ammiraju, J. S., Yu, Y., Luo, M., Kudrna, D., Kim, H., Goicoechea, J. L., Katayose, Y., Matsumoto, T., Wu, J., Sasaki, T., and Wing, R. A. (2005). Random sheared fosmid library as a new genomic tool to accelerate complete finishing of rice (*Oryza sativa* spp. Nipponbare) genome sequence: Sequencing of gap-specific fosmid clones uncovers new euchromatic portions of the genome. *Theor. Appl. Genet.* **111**, 1596–1607.

Beloqui, A., Nechitaylo, T. Y., Lopez-Cortes, N., Ghazi, A., Guazzaroni, M. E., Polaina, J., Strittmatter, A. W., Reva, O., Waliczek, A., Yakimov, M. M., Golyshina, O. V., Ferrer, M., et al. (2010). Diversity of glycosyl hydrolases from cellulose-depleting communities enriched from casts of two earthworm species. *Appl. Environ. Microbiol.* **76**, 5934–5946.

Cantarel, B. L., Coutinho, P. M., Rancurel, C., Bernard, T., Lombard, V., and Henrissat, B. (2009). The Carbohydrate-Active EnZymes database (CAZy): An expert resource for glycogenomics. *Nucleic Acids Res.* **37**, D233–D238.

Cole, J. R., Chai, B., Farris, R. J., Wang, Q., Kulam-Syed-Mohideen, A. S., McGarrell, D. M., Bandela, A. M., Cardenas, E., Garrity, G. M., and Tiedje, J. M. (2007). The ribosomal database project (RDP-II): Introducing myRDP space and quality controlled public data. *Nucleic Acids Res.* **35**, D169–D172.

DeLong, E. F., Taylor, L. T., Marsh, T. L., and Preston, C. M. (1999). Visualisation and enumeration of marine planktonic archaea and bacteria by using polyribonucleotide probes and fluorescent in situ hybridization. *Appl. Environ. Microbiol.* **65**, 5554–5563.

DeSantis, T. Z., Hugenholtz, P., Larsen, N., Rojas, M., Brodie, E. L., Keller, K., Huber, T., Dalevi, D., Hu, P., and Andersen, G. L. (2006). Greengenes, a chimera-checked 16S rRNA gene database and workbench compatible with ARB. *Appl. Environ. Microbiol.* **72**, 5069–5072.

Duan, C. J., Xian, L., Zhao, G. C., Feng, Y., Pang, H., Bai, X. L., Tang, J. L., Ma, Q. S., and Feng, J. X. (2009). Isolation and partial characterization of novel genes encoding acidic cellulases from metagenomes of buffalo rumens. *J. Appl. Microbiol.* **107,** 245–256.

Dunman, P. M., Murphy, E., Haney, S., Palacios, D., Tucker-Kellogg, G., Wu, S., Brown, E. L., Zagursky, R. J., Shlaes, D., and Projan, S. J. (2001). Transcription profiling-based identification of *Staphylococcus aureus* genes regulated by the agr and/or sarA loci. *J. Bacteriol.* **183,** 7341–7353.

Edwards, J. L., Smith, D. L., Connolly, J., McDonald, J. E., Cox, M. J., Joint, I., Edwards, C., and McCarthy, A. J. (2010). Identification of carbohydrate metabolism genes in the metagenome of a marine biofilm community shown to be dominated by gammaproteobacteria and bacteroidetes. *Genes* **1,** 371–384.

Feng, Y., Duan, C. J., Pang, H., Mo, X. C., Wu, C. F., Yu, Y., Hu, Y. L., Wei, J., Tang, J. L., and Feng, J. X. (2007). Cloning and identification of novel cellulase genes from uncultured microorganisms in rabbit cecum and characterization of the expressed cellulases. *Appl. Microbiol. Biotechnol.* **75,** 319–328.

Ferrer, M., Golyshina, O. V., Chernikova, T. N., Khachane, A. N., Reyes-Duarte, D., Santos, V. A., Strompl, C., Elborough, K., Jarvis, G., Neef, A., Yakimov, M. M., Timmis, K. N., et al. (2005). Novel hydrolase diversity retrieved from a metagenome library of bovine rumen microflora. *Environ. Microbiol.* **7,** 1996–2010.

Foster, S. D., and Dunstan, P. K. (2010). The analysis of biodiversity using rank abundance distributions. *Biometrics* **66,** 186–195.

Frias-Lopez, J., Shi, Y., Tyson, G. W., Coleman, M. L., Schuster, S. C., Chisholm, S. W., and Delong, E. F. (2008). Microbial community gene expression in ocean surface waters. *Proc. Natl. Acad. Sci. USA* **105,** 3805–3810.

Gabor, E. M., de Vries, E. J., and Janssen, D. B. (2004). Construction, characterization, and use of small-insert gene banks of DNA isolated from soil and enrichment cultures for the recovery of novel amidases. *Environ. Microbiol.* **6,** 948–958.

Garcia-Alvarez, B., Melero, R., Dias, F. M. V., Prates, J. A. M., Fontes, C. M. G. A., Smith, S. P., Romao, M. J., Carvalho, A. L., and Llorca, O. (2011). Molecular architecture and structural transitions of a *Clostridium thermocellum* mini-cellulosome. *J. Mol. Biol.* **407,** 571–580.

Gilbert, J. A., Field, D., Huang, Y., Edwards, R., Li, W., Gilna, P., and Joint, I. (2008). Detection of large numbers of novel sequences in the metatranscriptomes of complex marine microbial communities. *PLoS One* **3,** e3042.

Graham, J. E., Clark, M. E., Nadler, D. C., Huffer, S., Chokhawala, H. A., Rowland, S. E., Blanch, H. W., Clark, D. S., and Robb, F. T. (2011). Identification and characterization of a multidomain hyperthermophilic cellulase from an archaeal enrichment. *Nat. Commun.* **2,** 375. Doi: 10.1038/ncomms1373.

Griffiths, R. I., Whiteley, A. S., O'Donnell, A. G., and Bailey, M. J. (2000). Rapid method for coextraction of DNA and RNA from natural environments for analysis of ribosomal DNA- and rRNA-based microbial community composition. *Appl. Environ. Microbiol.* **66,** 5488–5491.

Hu, Y., Zhang, G. M., Li, A. Y., Chen, J., and Ma, L. X. (2008). Cloning and enzymatic characterization of a xylanase gene from a soil-derived metagenomic library with an efficient approach. *Appl. Microbiol. Biotechnol.* **80,** 823–830.

Hunt, D. E., Klepac-Ceraj, V., Acinas, S. G., Gautier, C., Bertilsson, S., and Polz, M. F. (2006). Evaluation of 23S rRNA primers for use in phylogenetic studies of bacterial diversity. *Appl. Environ. Microbiol.* **72,** 2221–2225.

Huson, D. H., Auch, A. F., Qi, J., and Schuster, S. C. (2007). MEGAN analysis of metagenomic data. *Genome Res.* **17,** 377–386.

Karpinets, T. V., Greenwood, D. J., Sams, C. E., and Ammons, J. T. (2006). RNA: protein ratio of the unicellular organism as a characteristic of phosphorous and nitrogen stoichiometry and of the cellular requirement of ribosomes for protein synthesis. *BMC Biol.* **4**, 30.

Kataeva, I., Li, X. L., Chen, H., Choi, S. K., and Ljungdahl, L. G. (1999). Cloning and sequence analysis of a new cellulase gene encoding CelK, a major cellulosome component of *Clostridium thermocellum*: Evidence for gene duplication and recombination. *J. Bacteriol.* **181**, 5288–5295.

Kim, S. J., Lee, C. M., Han, B. R., Kim, M. Y., Yeo, Y. S., Yoon, S. H., Koo, B. S., and Jun, H. K. (2008). Characterization of a gene encoding cellulase from uncultured soil bacteria. *FEMS Microbiol. Lett.* **282**, 44–51.

Kim, D., Kim, S. N., Baik, K. S., Park, S. C., Lim, C. H., Kim, J. O., Shin, T. S., Oh, M. J., and Seong, C. N. (2011). Screening and characterization of a cellulase gene from the gut microflora of abalone using metagenomic library. *J. Microbiol.* **49**, 141–145.

Kong, T., Xia, Y., Seviour, R., He, M., McAllister, T., and Forster, R. (2011). In situ identification of carboxymethyl cellulose-digesting bacteria in the rumen of cattle fed alfalfa or triticale. *FEMS Microbiol. Ecol.* **80**, 159–167.

Langer, M., Gabor, E. M., Liebeton, K., Meurer, G., Niehaus, F., Schulze, R., Eck, J., and Lorenz, P. (2006). Metagenomics: An inexhaustible access to nature's diversity. *Biotechnol. J.* **1**, 815–821.

Leininger, S., Urich, T., Schloter, M., Schwark, L., Qi, J., Nicol, G. W., Prosser, J. I., Schuster, S. C., and Schleper, C. (2006). Archaea predominate among ammonia-oxidizing prokaryotes in soils. *Nature* **442**, 806–809.

Liu, J., Liu, W. D., Zhao, X. L., Shen, W. J., Cao, H., and Cui, Z. L. (2011). Cloning and functional characterization of a novel endo-beta-1,4-glucanase gene from a soil-derived metagenomic library. *Appl. Microbiol. Biotechnol.* **89**, 1083–1092.

Lockhart, R. J., Van Dyke, M. I., Beadle, I. R., Humphreys, P., and McCarthy, A. J. (2006). Molecular biological detection of anaerobic gut fungi (*Neocallimastigales*) from landfill sites. *Appl. Environ. Microbiol.* **72**, 5659–5661.

Martinez, D., Berka, R. M., Henrissat, B., Saloheimo, M., Arvas, M., Baker, S. E., Chapman, J., Chertkov, O., Coutinho, P. M., Cullen, D., Danchin, E. G., Grigoriev, I. V., et al. (2008). Genome sequencing and analysis of the biomass-degrading fungus *Trichoderma reesei* (syn. Hypocrea jecorina). *Nat. Biotechnol.* **26**, 553–560.

Mayer, W. E., Schuster, L. N., Bartelmes, G., Dieterich, C., and Sommer, R. J. (2011). Horizontal gene transfer of microbial cellulases into nematode genomes is associated with functional assimilation and gene turnover. *BMC Evol. Biol.* **11**, 13.

McDonald, J. E., Lockhart, R. J., Cox, M. J., Allison, H. E., and McCarthy, A. J. (2008). Detection of novel Fibrobacter populations in landfill sites and determination of their relative abundance via quantitative PCR. *Environ. Microbiol.* **10**, 1310–1319.

McDonald, J. E., de Menezes, A. B., Allison, H. E., and McCarthy, A. J. (2009). Molecular biological detection and quantification of novel Fibrobacter populations in freshwater lakes. *Appl. Environ. Microbiol.* **75**, 5148–5152.

McDonald, J. E., Houghton, J. N. I., Rooks, D. J., Allison, H. E., and McCarthy, A. J. (2012). The microbial ecology of anaerobic cellulose degradation in municipal waste landfill sites: Evidence of a role for fibrobacters. *Environ. Microbiol.* 10.1111/j.1462-2920.2011.02688.x.

Medini, D., Serruto, D., Parkhill, J., Relman, D. A., Donati, C., Moxon, R., Falkow, S., and Rappuoli, R. (2008). Microbiology in the post-genomic era. *Nat. Rev. Microbiol.* **6**, 419–430.

Mello, L. V., Chen, X., and Rigden, D. J. (2010). Mining metagenomic data for novel domains: BACON, a new carbohydrate-binding module. *FEBS Lett.* **584**, 2421–2426.

Meyer, F., Paarmann, D., D'Souza, M., Olson, R., Glass, E. M., Kubal, M., Paczian, T., Rodriguez, A., Stevens, R., Wilke, A., Wilkening, J., and Edwards, R. A. (2008). The metagenomics RAST server—A public resource for the automatic phylogenetic and functional analysis of metagenomes. *BMC Bioinformatics* **9**, 386.

Mitra, S., Rupek, P., Richter, D. C., Urich, T., Gilbert, J. A., Meyer, F., Wilke, A., and Huson, D. H. (2011). Functional analysis of metagenomes and metatranscriptomes using SEED and KEGG. *BMC Bioinformatics* **12**(Suppl 1), S21.

Neufeld, J. D., Schafer, H., Cox, M. J., Boden, R., McDonald, I. R., and Murrell, J. C. (2007). Stable-isotope probing implicates *Methylophaga* spp. and novel *Gammaproteobacteria* in marine methanol and methylamine metabolism. *ISME J.* **1**, 480–491.

Niedringhaus, T. P., Milanova, D., Kerby, M. B., Snyder, M. P., and Barron, A. E. (2011). Landscape of next-generation sequencing technologies. *Anal. Chem.* **83**, 4327–4341.

Overbeek, R., Begley, T., Butler, R. M., Choudhuri, J. V., Chuang, H. Y., Cohoon, M., de Crecy-Lagard, V., Diaz, N., Disz, T., Edwards, R., Fonstein, M., Frank, E. D., *et al.* (2005). The subsystems approach to genome annotation and its use in the project to annotate 1000 genomes. *Nucleic Acids Res.* **33**, 5691–5702.

Pang, H., Zhang, P., Duan, C. J., Mo, X. C., Tang, J. L., and Feng, J. X. (2009). Identification of cellulase genes from the metagenomes of compost soils and functional characterization of one novel endoglucanase. *Curr. Microbiol.* **58**, 404–408.

Purdy, K. J. (2005). Nucleic acid recovery from complex environmental samples. *Methods Enzymol.* **397**, 271–292.

Robertson, D. E., Chaplin, J. A., DeSantis, G., Podar, M., Madden, M., Chi, E., Richardson, T., Milan, A., Miller, M., Weiner, D. P., Wong, K., McQuaid, J., *et al.* (2004). Exploring nitrilase sequence space for enantioselective catalysis. *Appl. Environ. Microbiol.* **70**, 2429–2436.

Rodrigue, S., Materna, A. C., Timberlake, S. C., Blackburn, M. C., Malmstrom, R. R., Alm, E. J., and Chisholm, S. W. (2010). Unlocking short read sequencing for metagenomics. *PLoS One* **5**, e11840.

Rooks, D. J., Smith, D. L., McDonald, J. E., Woodward, M. J., McCarthy, A. J., and Allison, H. E. (2010). 454-Pyrosequencing: A molecular battiscope for freshwater viral ecology. *Genes* **1**, 210–226.

Schadt, E. E., Turner, S., and Kasarskis, A. (2010). A window into third-generation sequencing. *Hum. Mol. Genet.* **19**, R227–R240.

Shi, Y., Tyson, G. W., and DeLong, E. F. (2009). Metatranscriptomics reveals unique microbial small RNAs in the ocean's water column. *Nature* **459**, 266–269.

Simon, C., and Daniel, R. (2010). Construction of small-insert and large-insert metagenomic libraries. *Methods Mol. Biol.* **668**, 39–50.

Steele, H. L., Jaeger, K. E., Daniel, R., and Streit, W. R. (2009). Advances in recovery of novel biocatalysts from metagenomes. *J. Mol. Microbiol. Biotechnol.* **16**, 25–37.

Stewart, F. J., Ottesen, E. A., and DeLong, E. F. (2010). Development and quantitative analyses of a universal rRNA-subtraction protocol for microbial metatranscriptomics. *ISME J.* **4**, 896–907.

Suenaga, H., Ohnuki, T., and Miyazaki, K. (2007). Functional screening of a metagenomic library for genes involved in microbial degradation of aromatic compounds. *Environ. Microbiol.* **9**, 2289–2297.

Sukharnikov, L. O., Cantwell, B. J., Podar, M., and Zhulin, I. B. (2011). Cellulases: Ambiguous nonhomologous enzymes in a genomic perspective. *Trends Biotechnol.* **29**, 473–479.

Tasse, L., Bercovici, J., Pizzut-Serin, S., Robe, P., Tap, J., Klopp, C., Cantarel, B. L., Coutinho, P. M., Henrissat, B., Leclerc, M., Dore, J., Monsan, P., *et al.* (2010). Functional metagenomics to mine the human gut microbiome for dietary fiber catabolic enzymes. *Genome Res.* **20**, 1605–1612.

Teather, R. M., and Wood, P. J. (1982). Use of Congo red polysaccharide interactions in enumeration and characterization of cellulolytic bacteria from the bovine rumen. *Appl. Environ. Microbiol.* **43**, 777–780.

Uchiyama, T., and Miyazaki, K. (2009). Functional metagenomics for enzyme discovery: Challenges to efficient screening. *Curr. Opin. Biotechnol.* **20,** 616–622.

Van Dyke, M. I., and McCarthy, A. J. (2002). Molecular biological detection and characterization of *Clostridium* populations in municipal landfill sites. *Appl. Environ. Microbiol.* **68,** 2049–2053.

van Sint Fiet, S., van Beilen, J. B., and Witholt, B. (2006). Selection of biocatalysts for chemical synthesis. *Proc. Natl. Acad. Sci. USA* **103,** 1693–1698.

Venter, J. C., Remington, K., Heidelberg, J. F., Halpern, A. L., Rusch, D., Eisen, J. A., Wu, D., Paulsen, I., Nelson, K. E., Nelson, W., Fouts, D. E., Levy, S., *et al.* (2004). Environmental genome shotgun sequencing of the Sargasso Sea. *Science* **304,** 66–74.

Voget, S., Steele, H. L., and Streit, W. R. (2006). Characterization of a metagenome-derived halotolerant cellulase. *J. Biotechnol.* **126,** 26–36.

Wang, F., Li, F., Chen, G., and Liu, W. (2009). Isolation and characterization of novel cellulase genes from uncultured microorganisms in different environmental niches. *Microbiol. Res.* **164,** 650–657.

Warnecke, F., and Hess, M. (2009). A perspective: Metatranscriptomics as a tool for the discovery of novel biocatalysts. *J. Biotechnol.* **142,** 91–95.

Zhang, Y. H. P., Himmel, M. E., and Mielenz, J. R. (2006). Outlook for cellulase improvement: Screening and selection strategies. *Biotechnol. Adv.* **24,** 452–481.

Zhao, J., Hyman, L., and Moore, C. (1999). Formation of mRNA 3′ ends in eukaryotes: Mechanism, regulation, and interrelationships with other steps in mRNA synthesis. *Microbiol. Mol. Biol. Rev.* **63,** 405–445.

CHAPTER TWENTY-ONE

Escherichia coli Expression, Purification, Crystallization, and Structure Determination of Bacterial Cohesin–Dockerin Complexes

Joana L. A. Brás,* Ana Luisa Carvalho,[†] Aldino Viegas,[†]
Shabir Najmudin,* Victor D. Alves,* José A. M. Prates,*
Luís M. A. Ferreira,* Maria J. Romão,[†] Harry J. Gilbert,[‡]
and Carlos M. G. A. Fontes*

Contents

1. Introduction	396
2. Cloning of Cohesin and Dockerin Genes in Prokaryotic Expression Vectors	397
2.1. Cloning genes encoding dockerin and cohesin modules through PCR	397
2.2. Producing synthetic dockerin or cohesin genes	398
2.3. Producing gene constructs for protein coexpression	398
3. Expression and Purification of Cohesin–Dockerin Complexes in *E. coli*	400
4. The Dual Binding Mode and the Crystallization of Cohesin–Dockerin Complexes	402
5. X-ray Crystallography of Cohesin–Dockerin Complexes	402
5.1. Crystallization	404
5.2. X-ray data collection and reduction	407
5.3. Structure determination	409
5.4. Model building and structure refinement	410
6. Summary	413
Acknowledgments	414
References	414

* CIISA-Faculdade de Medicina Veterinária, Pólo Universitário do Alto da Ajuda, Avenida da Universidade Técnica, Lisboa, Portugal
[†] REQUIMTE/CQFB, Departamento de Química, Faculdade de Ciências e Tecnologia, Universidade Nova de Lisboa, Caparica, Portugal
[‡] Institute for Cell and Molecular Biosciences Medical School, Newcastle University, Framlington Place, United Kingdom

Abstract

Cellulosomes are highly efficient nanomachines that play a fundamental role during the anaerobic deconstruction of complex plant cell wall carbohydrates. The assembly of these complex nanomachines results from the very tight binding of repetitive cohesin modules, located in a noncatalytic molecular scaffold, and dockerin domains located at the C-terminus of the enzyme components of the cellulosome. The number of enzymes found in a cellulosome varies but may reach more than 100 catalytic subunits if cellulosomes are further organized in polycellulosomes, through a second type of cohesin–dockerin interaction. Structural studies have revealed how the cohesin–dockerin interaction mediates cellulosome assembly and cell-surface attachment, while retaining the flexibility required to potentiate catalytic synergy within the complex. Methods that might be applied for the production, purification, and structure determination of cohesin–dockerin complexes are described here.

1. INTRODUCTION

In anaerobic ecosystems, recycling of photosynthetically fixed plant cell wall carbon is mediated by an extensive repertoire of microbial modular enzymes that are organized in multienzyme complexes termed as cellulosomes (Fontes and Gilbert, 2010). Integration of cellulases and hemicellulases in these highly efficient nanomachines represents a powerful mechanism for targeting multienzymes to a localized region of the substrate, while also promoting enzyme synergy. The quaternary assembly of cellulosomes, exemplified by the *Clostridium thermocellum* complex, is dictated by highly ordered protein:protein interactions between cohesins and dockerins (Bayer *et al.*, 2004; Salamitou *et al.*, 1994). Cohesins are found as repetitive domains in a large noncatalytic molecular scaffold (defined as the scaffoldin), while dockerins are part of the cellulosomal catalytic subunits. The multifunctional scaffoldin contains, in addition to the various cohesin domains, a divergent dockerin that specifically interacts with cohesins found in polypeptides located at the bacterial surface (Leibovitz and Beguin, 1996). Thus, dockerins that are components of the enzymes only recognize scaffoldin cohesins. These interactions were termed as type I. In contrast, scaffoldin dockerins exclusively bind cell surface cohesins and these complexes were termed as type II. The structures of several cohesin–dockerin complexes have started to reveal the molecular determinants responsible for the high affinity and tight specificity displayed by these protein:protein interactions (Adams *et al.*, 2006; Carvalho *et al.*, 2003, 2007; Pinheiro *et al.*, 2008). The different methods that can be applied for this purpose are described below.

2. Cloning of Cohesin and Dockerin Genes in Prokaryotic Expression Vectors

Dockerins are usually present as a single copy at the C-terminus of cellulosomal cellulases and hemicellulases. These noncatalytic domains consist of ~70 amino acids that contain two duplicated segments, each of about 22 residues (Fontes and Gilbert, 2010). The first 12 residues of each duplicated segment resemble the calcium-binding loop of F-hand motifs in which the calcium-binding residues, asparagine or aspartate, are highly conserved (Pages et al., 1997). Calcium was shown to play a critical stabilizing role in dockerin structures (Choi and Ljungdahl, 1996). In addition, calcium is required for dockerin function (Fontes and Gilbert, 2010). Thus, in the presence of EDTA, dockerins are unable to interact with cohesins. Dockerins are highly unstable when produced as discrete entities in *Escherichia coli*, being very susceptible to proteolysis and degradation. However, high levels of dockerin expression in *E. coli* can be obtained when these unstable domains are coexpressed *in vivo* with their cognate cohesin partners (Carvalho et al., 2003, 2007). It is believed that the binding of dockerins to cohesins after the small recombinant domain (7–9 kDa) is properly folded fulfils a critical role in dockerin stabilization in *E. coli*. Coexpression of dockerins with cohesins might be obtained either through the cloning of both encoding genes in the same plasmid, or through the cloning of the two genes in different, but compatible, plasmids.

2.1. Cloning genes encoding dockerin and cohesin modules through PCR

The genes encoding dockerin and cohesin domains are amplified by PCR from bacterial genomic DNA using previously designed primers. Primers should contain engineered restriction sites for direct cloning into the appropriate vectors.

1. Set up a 50 µl PCR reaction using 50–200 ng of bacterial genomic DNA, 0.4 µM of each respective primer, 0.4 mM dNTPs, 2.5 U of a proofreading DNA polymerase in 1 × of the respective buffer as recommended by the thermostable polymerase manufacturer.
2. Run the PCR reaction in a conventional thermocycler using 30 amplification cycles with an annealing temperature that is 5 °C lower than the primers melting point.
3. When the amplification reaction is finished, subject 5–10 µl of each reaction to agarose electrophoresis, using a 1–1.5% (w/v) gel, to confirm that the amplification reaction was successful.

4. Purify the amplified genes from nucleotides and unincorporated primer dimers in silica columns following the manufacturer's instructions. Recover the purified PCR products in 50 μl of elution buffer (5–10 mM Tris–HCl, pH 8.5, or water).
5. Clone the genes into a blunt-ended prokaryotic vector following the manufacturer instructions. Sequence the cloned genes to verify that no mutations have accumulated during the amplification.

2.2. Producing synthetic dockerin or cohesin genes

Under some circumstances the lack of bacterial genomic DNA, or inappropriate codon usage for obtaining high levels of gene expression in *E. coli*, might entail the production of synthetic genes in addition to the strategy described in Section 2.1.

1. Select the primary sequences of the required cohesin and dockerin genes and design the genes encoding the respective proteins with a codon usage that is compatible with high level of expression in *E. coli* (gene design might be performed using Gene Designer by DNA2.0; https://www.dna20.com/genedesigner2; Villalobos *et al.*, 2006). This dedicated software excludes undesired internal restriction sites, repetitive regions, or putative regulatory sequences.
2. Divide the designed gene into overlapping oligonucleotides (20 bp overlap and 40 bp in length). One can use a dedicated software to design primers with overlapping regions with similar melting temperatures (e.g., Gene2Oligo http://berry.engin.umich.edu/gene2oligo; Rouillard *et al.*, 2004). Design upstream and downstream primers incorporating the engineered restriction sites that will be used for the subsequent cloning reactions.
3. Assemble a 50-μl PCR reaction using 25 pmol of the upstream and downstream primers and 0.25 pmol of the internal primers.
4. Perform the PCR reaction as described in Section 2.1 (point 2) using a proofreading thermostable DNA polymerase. Perform a standard PCR cycle using a 55 °C annealing temperature and an extension period of at least 1 min/kb.
5. Check the result of the PCR reaction through agarose gel electrophoresis. Clone the PCR product of the estimated size into a blunt-ended vector as described above.
6. Sequence the synthetic gene to confirm that no mutations have accumulated during the amplification.

2.3. Producing gene constructs for protein coexpression

Following our established strategy, cohesins and dockerins are cloned under the control of separated promoters in the same plasmid (Fig. 21.1), allowing the simultaneous expression of the two proteins in *E. coli*. Recombinant

Figure 21.1 DNA construct containing cohesin and dockerin genes cloned in tandem under the control of separate T7 promoter and terminator regions. (For the color version of this figure, the reader is referred to the Web version of this chapter.)

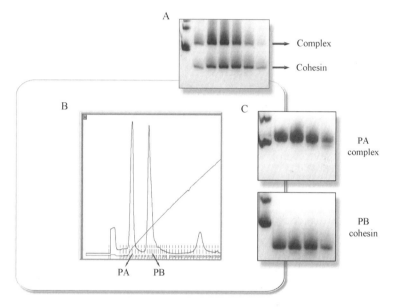

Figure 21.2 Purification of cohesin–dockerin complexes. In panel A, purified fractions collected after the first purification step through IMAC are analyzed through native acrylamide gel electrophoresis. Affinity chromatography generated both unbound cohesin and the cohesin–dockerin complex suggesting that the dockerin is expressed at a lower level. In panel B, ion exchange chromatography is used to separate the protein–protein complex from unbound cohesin. In panel C, purified fractions collected after ion exchange were analysed through native acrylamide gel electrophoresis. Fractions from Peak A (PA) contain the protein-protein complex. Fractions from Peak B (PB) contain unbound cohesin. (For the color version of this figure, the reader is referred to the Web version of this chapter.)

cohesins usually contain N- or C-terminal His_6-tags and coexpressed dockerins contain no additional vector-derived primary sequence. Following this approach, and assuming the *in vivo* binding of cohesins to dockerins, immobilized metal affinity chromatography (IMAC) can be used to purify the cohesin–dockerin complexes and unbound cohesins (Fig. 21.2). After the first chromatographic step, unbound dockerins are eliminated. A second purification step is, however, required to obtain purified cohesin–dockerin complexes (Fig. 21.2). Although we usually use this approach, we anticipate that the reverse strategy of including a His_6-tag in the dockerin might also be successful.

1. Subclone the cohesin gene into pET21a (Novagen Inc., Madison, WI, USA) or any other suitable vector so that the gene is under the control of a T7 promoter and a T7 terminator. The generated plasmid is termed as pET21a_coh. The cloned gene should have no 3′-end stop codon and should be in frame with the vector His_6-tag sequence. This strategy will ensure that the recombinant cohesin will contain a C-terminal His_6-tag.
2. Subclone the dockerin gene from the standard cloning vector into the prokaryotic expression vector pET3a (Novagen Inc.). Other vectors with similar properties can also be used. However, if using pET3a the gene can only be subcloned in *Nde*I and *Bam*HI restriction sites. The dockerin gene should contain a stop codon at the 5′-end, just before the engineered restriction site.
3. To allow excision of the dockerin gene with the T7 terminator sequences, use site-directed mutagenesis to engineer a *Bgl*II restriction sequence at the 5′-end of the T7 terminator sequence. Use a commercially available kit (e.g., NZYTech) and follow the manufacturers' instructions.
4. Subclone the dockerin gene under the control of T7 promoter and a T7 terminator by digesting the generated pET3a-dok-mut plasmid with *Bgl*II. Purify the digested gene following agarose gel electrophoresis. Ligate into pET21a_coh previously digested with *Bgl*II and dephosphorylated.
5. Perform a restriction analysis of the generated plasmids to evaluate the orientation of the Coh and Dok genes. However, gene orientation usually does not affect expression levels (Fig. 21.1).

3. Expression and Purification of Cohesin–Dockerin Complexes in *E. coli*

Expression of cohesin and dockerin genes under the control of T7 promoters requires the use of *E. coli* DE3 strains.

1. Plasmids containing the dockerin and cohesin genes organized in tandem and under the control of separate T7 promoter and terminator sequences are used to transform BL21 (DE3) cells.
2. Recombinant *E. coli* cells are grown in LB media supplemented with the appropriate antibiotic at 37 °C until OD_{550} 0.5. Gene expression is induced by the addition of 1 mM isopropyl-β-D-thiogalactopyranoside. Induced cells are further incubated for 16 h at 19 °C.
3. Centrifuge the cell suspension at $5000 \times g$ for 15 min at 4 °C.
4. Resuspend collected cells in 50 mM Na-Hepes buffer, pH 7.5, containing 1 M NaCl, 10 mM imidazole, and 5 mM $CaCl_2$. Disrupt bacterial membranes and cell wall through ultrasonication.
5. Collect cell-free extracts through centrifugation at $15,000 \times g$ for 30 min at 4 °C.

For crystallography, cohesin–dockerin complexes are usually purified through three purification steps using an FPLC chromatography system (Fig. 21.2). All procedures, unless otherwise indicated, are carried out at 4 °C.

1. Unbound cohesin and cohesin–dockerin complexes are initially purified through IMAC essentially as described by Pinheiro et al. (2008) in HisTrap^{TM} HP 5 ml columns (GE Healthcare, Uppsala, Sweden). The column is equilibrated with 50 mM Na-Hepes buffer, pH 7.5, containing 1 M NaCl, 10 mM imidazole, and 5 mM CaCl$_2$, and after loading the E. coli extracts the column is extensively washed with the same buffer. Proteins are eluted from the column in a gradient of the equilibration buffer and 50 mM Na-Hepes buffer, pH 7.5, containing 1 M NaCl, 300 mM imidazole, and 5 mM CaCl$_2$.
2. Fractions containing the protein–protein complexes are selected following 10% native gel electrophoresis and 15% SDS-PAGE (Fig. 21.2). The complex is usually copurified with unbound cohesin (Fig. 21.2). A control, consisting of purified cohesin, should be incorporated in the native gel to allow the identification of the cohesin–dockerin complex band.
3. IMAC purified proteins are buffer exchanged in PD-10 Sephadex G25M gel filtration columns (GE Healthcare) into 20 mM Tris–HCl buffer, pH 8.0, containing 2 mM CaCl$_2$.
4. Proteins are subjected to a further purification step by anion exchange chromatography using a column loaded with Source 30Q media (GE Healthcare).
5. Separation of the cohesin–dockerin complexes from the individual cohesin is achieved through the application of a 0–1 M NaCl elution gradient.
6. Protein fractions are analyzed through affinity gel electrophoresis. Fractions containing cohesin–dockerin complexes purified from isolated cohesin are selected and, if required, further purified by gel filtration chromatography (Fig. 21.2).
7. Before gel filtration chromatography, protein fractions are buffer exchanged into 20 mM Na-Hepes buffer, pH 7.5, containing 200 mM NaCl and 2 mM CaCl$_2$ as described above.
8. The protein is concentrated with an Amicon 10 kDa cut off molecular weight centrifugal membrane to approximately 25 mg/ml.
9. The protein–protein complex is loaded into an HiLoad 16/60 Superdex 75 column (GE Healthcare) previously equilibrated with 20 mM Na-Hepes buffer, pH 7.5, containing 200 mM NaCl and 2 mM CaCl$_2$.
10. Purity of eluted protein fractions is evaluated through SDS-PAGE and native gel electrophoresis, as mentioned before. Selected fractions are pooled and concentrated as described above. Pure complexes are buffer exchanged by washing with 2 mM CaCl$_2$ and concentrated to 6–12 mg/ml.

4. The Dual Binding Mode and the Crystallization of Cohesin–Dockerin Complexes

Dockerins are usually highly symmetrical molecules and generally contain two cohesin binding interfaces (Fontes and Gilbert, 2010). Dockerin primary sequence is a tandem duplication of a 22-residue segment that displays remarkable structural conservation. Thus, in C. thermocellum, the structure of the first duplicated segment, containing the N-terminal helix, can be precisely superimposed over the C-terminal helix (helix-3). Several mutagenesis studies informed by the structure of cohesin–dockerin complexes (Carvalho et al., 2003, 2007) revealed that the C. thermocellum type I dockerin contains two ligand binding sites that display similar affinity to its protein partner. Thus, in one complex, helix-3 dominates cohesin recognition with Ser-45 and Thr-46 playing a central role in the polar interactions between the two protein partners (Carvalho et al., 2003). In the second binding mode, the dockerin is rotated 180° relative to the cohesin and helix-1, rather than helix-3, plays a central role in complex formation (Carvalho et al., 2007). Thus, the equivalent residues to Ser-45 and Thr-46 in the N-terminal helix, Ser-11 and Thr-12, dominate the hydrogen-bonding interactions between the dockerin and its cohesin partner in this second binding mode. Similar observations were made in cohesin–dockerin interactions of *Clostridium cellulolyticum* (Pinheiro et al., 2008). Although dockerins present two cohesin-interacting surfaces, only one of these sites interacts with a cohesin at a defined moment.

The dual binding mode expressed by dockerins poses significant obstacles to cohesin–dockerin complex crystallization. Thus, initial attempts to crystallize the purified dockerin–cohesin complexes were unsuccessful. This observation likely reflects the dynamic binding of the two potential dockerin binding sites to the cohesin. To encourage a single binding mode between the protein partners, two variants of the dockerin should be constructed in which the function of site 1 or site 2 is disrupted by the introduction of mutations at the residue pairs that dominate cohesin recognition at each binding site. To achieve this, a commercial site-directed mutagenesis kit should be employed and the residues should be changed to alanine or larger amino acids such as glutamine.

5. X-ray Crystallography of Cohesin–Dockerin Complexes

Solving the three-dimensional structure of a macromolecule by X-ray crystallography involves several steps. Once a target protein is expressed and purified, it needs to be crystallized and single crystals must be obtained.

A single crystal is built-up by translationally repeating units, called unit cells. Each unit cell is characterized by three cell axes a, b, and c and by three angles α, β, and γ. Crystals are subjected to X-rays and diffraction data are measured, processed, and analyzed. Diffraction data are recorded as a range of diffraction patterns, containing spots (reflections), each attributed an hkl index. Each reflection hkl is produced by families of imaginary planes passing by all atoms in the crystal lattice. Thus, diffraction (the production of a spot in the diffraction pattern) only occurs when a constructive interference of the scattered radiation occurs, satisfying Bragg's law. The diffraction experiment produces a list of intensities (I_{hkl}) and associated errors for each reflection recorded. Each reflection can also be regarded as a scattered wave, with an amplitude, phase, and frequency (the same as the incident radiation). Mathematically, a wave can be described as a Fourier series, formulated as a function of the electron density $\rho(x, y, z)$ of all atoms in the unit cell, Eq. (21.1):

$$F_{hkl} = \int_V \rho(x, y, z) e^{[2\pi i(hx+ky+lz)]} dV \qquad (21.1)$$

where F_{hkl} (the structure factor) also possesses an associated amplitude, phase, and frequency. Applying the Fourier transform to Eq. (21.1), the electron density function is obtained:

$$\rho(x, y, z) = \frac{1}{V} \sum_h \sum_k \sum_l |F_{hkl}| e^{2\pi i \alpha_{hkl}} e^{[-2\pi i(hx+ky+lz)]} \qquad (21.2)$$

where α_{hkl} is the phase angle of reflection hkl, $|F_{hkl}|$ is the structure factor amplitude (proportional to $\sqrt{I}$), (x, y, z) are the fractional atomic coordinates in the unit cell, and V is the volume of the unit cell. The diffraction experiment provides the values of $|F_{hkl}|$; however, the phase information is lost, and this is known as the phase problem in crystallography. The Fourier transform marks the frontier between what is known as the reciprocal space (the space of the diffraction pattern, hkl reflections and structure factors) and the real space (the space of the electron density and atomic coordinates). Solving the phase problem requires the initial phases to be estimated. There are several methods available that allow the indirect determination of the phase angles but, in this chapter, only the molecular replacement (MR) method is used Long et al. (2008). This is the method of choice when a structure of a similar protein to the one of interest is available. Similarity is evaluated by the primary sequence similarity between the two proteins; the higher the sequence similarity, the higher the chance that the proteins will share a common fold. The known structure is referred to as the search model. It is used to locate the position of the protein of interest in the unit cell. Particular electron density maps (where the phases are attributed a value of 0, known as Patterson maps) are calculated for both proteins and superposed for comparison to find a match. This is usually done in two steps: (1) rotation of the search model to find the molecule's orientation and (2) translation of the model to find the position in the unit cell. The correct positioning of the search model in the unit cell provides the phase estimates that solve the phase problem and enable calculation of the first electron density map. After this, model adjustments will have to be made, in order to bring the model structure as close as possible to the structure of the protein of interest. In an iterative way, new electron density maps are calculated after model adjustments and these should improve with the addition of correct features to the model. Validation methods are used to monitor every step of model building and adjustment, comparing calculated structure factors $|F_{calc}|$, from the model, with observed structure factors $|F_{obs}|$, measured in the diffraction experiment, according to the expression:

$$\frac{\sum_{hkl} ||F_{obs,hkl}| - |F_{calc,hkl}||}{\sum_{hkl} F_{obs,hkl}} \quad (21.3)$$

This is known as the R factor. For cross-validation, 5–10% of the unique reflections are arbitrarily excluded from the refinement process and used to calculate the free R factor (R_{free}). This is used to prevent overfitting of the model, as R and R_{free} should not differ by more than 5%. Other validation criteria ensure that the final structure is in agreement with known chemical and geometrical parameters. In the following sections, a step-by-step procedure to solve the crystal structure of cohesin–dockerin complexes is presented.

5.1. Crystallization

The goal of crystallization is to obtain a well-ordered crystal, capable of producing a diffraction pattern, when subjected to X-rays. In the crystallization process, the purified protein gradually precipitates in an aqueous solution and, under the appropriate conditions, protein molecules adopt a consistent orientation, aligning themselves in repeating blocks of "unit cells". Protein crystallization is difficult because of the fragile nature of protein crystals, dominated by large channels of solvent, and where noncovalent interactions are responsible for sustaining the crystal lattice (Matthews, 1968). However, the fragility of protein crystals is not the only problem to overcome in crystallization. Since so much variation exists, each individual protein requires very specific and unique conditions to produce a well-ordered crystal. Many environmental factors have to be considered, like protein purity and concentration, pH, temperature, and precipitants.

Vapor diffusion is usually the method of choice for protein crystallization and the one we describe here. The two variations of the method are known as the hanging drop and sitting drop techniques. Basically, a drop containing purified protein, buffer, and precipitant is allowed to equilibrate against a similar reservoir solution, containing the precipitant in higher concentrations. In a sealed environment, water vapor diffuses from the drop to the reservoir, bringing the concentration of the precipitant in the drop closer to its concentration in the reservoir. The drop slowly loses volume, concentrating the protein slowly enough to permit the consistent orientation of protein molecules to form a crystal. The search for initial crystallization conditions is usually done using commercial screens, which provide trial formulations, often selected from known crystallization conditions for proteins.

Screening for suitable crystallization conditions requires the set up of many different conditions. Therefore, it is a common practice nowadays to use automated systems, known as crystallization robots. Still based on the vapor diffusion method, these systems have the major advantage of screening a large number of crystallization conditions while using a much smaller amount of protein solution, when compared with the traditional setups. Besides, setting up drops with a robotic system is a quick, reproducible, and less error-prone procedure. These equipments are computer-controlled, running software that not only gives instructions to operate the system, but can also be used to program different experiments, from screening to optimization. Drop volume can vary from 0.2 to 10 μl, using, for example, plates of 96 wells. An imaging system is useful for monitoring subsequent crystal growth.

The crystallization of a new cohesin–dockerin complex will require the search for initial conditions, which is most efficiently done using an automated nano-drop dispensing system, as described in the following protocol:

1. Prepare your protein complex for crystallization by buffer exchanging it into 10–20 mM Tris–HCl, pH 8, and 2 mM CaCl$_2$. Low concentration of buffer guaranties that the crystallization pH will be determined by the candidate precipitant solution and not the protein solution. Different pHs can cause different packings. It is very important that the complex is close to 97% pure. CaCl$_2$ is required for stabilization of the dockerin module.
2. Preload the plate's reservoirs with the solutions of your screens of choice.
3. If you are using an automated system, follow the equipment's instructions to load the protein. The volume of dispensed protein can vary from 0.1 to 10 μl.
4. Program the software to perform your screening. Depending on the system, it will dispense nano-drops of precipitant solutions, mixing them with the protein solution.
5. Seal the plate, visualize it under the microscope, take note of any early precipitation and store the plate at the chosen temperature (4 or 20 °C are commonly used, but different temperatures may be explored). Depending on the amount of purified complex, many screening trials can be set up.

6. If you are setting up your drops "by hand," you can use the procedure described below to prepare a screening of crystallization conditions, with one different condition per well.

Once the preliminary conditions are found, they may need optimization (which can be performed automatically in some systems) or scaling up to produce crystals with bigger dimensions (Fig. 21.3). Larger crystals may be more suitable for subsequent manipulation.

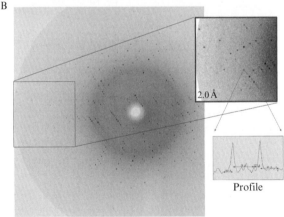

Figure 21.3 Crystals of cohesin–dockerin complexes and X-ray analysis. In panel A, cubic-shaped native Coh–Doc type I complex crystals, around $0.2 \times 0.2 \times 0.2$ mm, obtained in PEG/Ion Screen HR2-126 from Hampton Research (condition 18). In panel B, diffraction pattern obtained from a mutant cohesin–dockerin type I complex crystal, using graphite monochromated CuK_α radiation from an Enraf-Nonius FR591 rotating anode generator and a MAR-Research 300 mm imaging plate detector. Inset image depicts high resolution reflections and their profile along the dotted red line. (For the interpretation of the references to color in this figure legend, the reader is referred to the Web version of this chapter.)

1. Scaling up usually means preparing the drops using bigger volumes of protein and precipitant solutions, but it may also require adjustment of the preliminary crystallization conditions. At least, drop proportion should be assayed, but pH conditions may also change. Prepare a plate for crystallization (Linbro plates are commonly used). Grease the top of the reservoir well with high-vacuum grease.
2. Fill the reservoir with 500–700 µl of precipitant solution (found in the automated screening experiments).
3. In a lamella, dispense 2 µl of protein solution (different volumes can be used). Since lamellas are big enough, one suggestion would be to assay two different protein concentrations per lamella.
4. Pipette 2 µl of reservoir solution and mix it in the protein drop.
5. Flip the lamella, putting the fresh drop facing the reservoir solution. This describes the hanging drop method. In the sitting drop technique, the drop is dispensed on top of a support inside the reservoir. In the process of optimizing crystallization conditions, both techniques can be used.
6. Seal off the system using the high-vacuum grease to make it airtight.
7. Visualize the prepared drops under the microscope, take note of any early precipitation and store the plate at the temperature of interest.

5.2. X-ray data collection and reduction

After a period of 5–6 days for crystal growth at room temperature (Fig. 21.3), crystals can be cryo-protected, mounted, and flash-cooled in a dry nitrogen stream or directly in liquid nitrogen. Exposure to intense X-rays can cause radiation damage to the crystals, which is minimized by lowering the temperature to approximately 100 K. The cryo-protectant prevents the formation of disordered ice in the crystal's solvent channels, which can damage the crystal or interfere in diffraction by forming ice rings (visible in the diffraction pattern).

1. Stabilize the crystal by transferring it to a harvesting solution, similar to the crystallization solution, but where the precipitant is in higher concentration; this prevents dissolution of crystals.
2. Transfer the crystal to a second drop (~10 µl) containing 20–40% (v/v) of glycerol added to the harvesting solution, used previously.
3. Crystals can now be flash-cooled in a dry nitrogen stream or directly in liquid nitrogen, bringing the crystals to very low temperatures (around 100 K) and facilitating crystal handling and storage. These procedures require the use of specific cryo-tools (Pflugrath, 2004).
4. Crystal testing and data collection can be performed using graphite monochromated CuK_α radiation from a rotating anode generator and an imaging plate system as detector. Under this radiation, cohesin–dockerin crystals typically diffract beyond 2.5 Å resolution.

5. For higher resolution, X-ray diffraction data can be collected using a synchrotron radiation source in a beamline dedicated to macromolecular crystallography (e.g., ID14 at the European Synchrotron Radiation Facility).
6. The availability of 3D structures of the isolated cohesin (e.g., PDB ID code 1ANU) means that structure solution can be attempted by the well-established method of MR. Therefore, a single-wavelength data collection with high completeness is, in theory, sufficient to solve the 3D structure. Start by collecting a few frames, each with 1° oscillation angle (this is just a suggested starting value).
7. Using program MOSFLM (Leslie, 2006), autoindex the frames, determining the symmetry and unit cell parameters of the crystal, and the orientation of the crystal relative to the beam. Depending on the complex, space groups may differ (e.g., cubic $P2_13$ in the case of native *C. thermocellum* type I cohesin–dockerin complex, and monoclinic $P2_1$ in the case of S45A, T46A mutant) and the amount of unique data will depend on the inner symmetry of the crystals.
8. Redundancy of data may also be crucial for solving a structure by MR methods. However, in high-intensity beams, such as synchrotron sources, crystal decay can occur due to radiation damage, even for cryo-protected crystals, so the optimal strategy of data collection will have to take into account several of these factors. The strategy option in the program MOSFLM is helpful to calculate the best data collection strategy, determining the best phi angles and ranges to obtain a complete data set. This is followed by refinement of crystal parameters (unit cell and mosaic spread), detector, and beam parameters.
9. Proceed to obtain a data set as complete as possible, by collecting the necessary number of frames. Depending on the cell constants and the visible resolution limits, the oscillation angle and the detector-to-crystal distance will have to be adjusted, such that good spot separation is obtained and a significant number of full reflections are measured in each diffraction image.
10. After a complete data collection, use MOSFLM to integrate the several diffraction patterns (obtained over a wide range of rotations) into a list of indices (*hkl*), each with a measured intensity and an associated uncertainty.
11. Data integration is followed by scaling with program SCALA (Evans, 2006) from the CCP4 suite of programs (Winn *et al.*, 2011), to put all observations on a common scale. Check the overall quality of the data by analyzing the agreement between equivalent reflections. Check the value of R_{merge} against batch number to detect outliers. At this stage, you can already select the arbitrary fraction of data (5–10% of the reflections) to be used to calculate R_{free} during structure refinement.
12. Use program TRUNCATE from the CCP4 suite of programs (Winn *et al.*, 2011), to convert intensities (I_{hkl}), to structure factor amplitudes

(F_{hkl}), and to generate useful intensity statistics. Data quality indicators are factors R_{merge} and R_{pim}, mean$[(I)/\sigma(I)]$, completeness and multiplicity of data (Weiss, 2001). Possible twinning problems or the presence of translational NCS (noncrystallographic symmetry) can also be diagnosed at this stage.

5.3. Structure determination

3D structure determination requires solving the phase problem and, in the case of cohesin-dockerin complexes, this is done by the method of molecular replacement (MR). This method is commonly used when a structure of a homologous protein is available. Presently, several crystal structures of different cohesin-dockerin complexes and of the isolated cohesin are available, and, due to high sequence identity, any of these structures (isolated or in combination) can be used. A good strategy is to search only for the cohesin module and, subsequently, use manual or automated model building to trace the dockerin's polypeptide chain. This helps to prevent the effects of model bias, due to the inner symmetry of the dockerin module, which can cause connectivity errors when building the polypeptide chain. Initial estimated phases are provided by the correct positioning of the cohesin model, which generally provides enough phases to calculate an initial electron density map for the whole asymmetric unit, including the part covering the dockerin module.

Considering the technical procedures, in this method, the homologous model (e.g. the 3D structure of the isolated cohesin: PDB ID code 1ANU) is used to try to locate the cohesin module in two steps: first, a rotational search of the molecule is performed in order to find the correct orientation of the model in the asymmetric unit; second, a translational search is performed to find the position, in the unit cell, of the previously rotated model. This can be done with several programs. For example, PHASER (McCoy et al., 2007), using maximum likelihood, or BALBES (Long et al., 2008), which uses Bayesian methods. A suggested procedure is:

1. Perform an analysis of the cell content to estimate the number of cohesin-dockerin complexes in the asymmetric unit and the solvent content. This is done by calculating the Matthews coefficient, V_M (Matthews, 1968), using the knowledge about unit cell parameters, space group and the molecular weight of the cohesin-dockerin complex. V_M is easily calculated as: (volume of unit cell)/[(molecular weight of complex) × Z], where Z is the number of asymmetric units in the unit cell (i.e. the number of symmetry operations). V_M takes values within an empirically observed range (although there are exceptions), related to the fractional volume occupied by the solvent. The typical observed range for V_M will help to estimate the number of cohesin-dockerin complexes in the asymmetric unit.

2. If two or more complexes are expected in the asymmetric unit, a self-rotation function should be calculated to detect the rotational symmetry axis that relates the complexes in the asymmetric unit, and how this axis is oriented in the unit cell. Program POLARRFN of the CCP4 package can be used to compute the self-rotation function.
3. To perform maximum-likelihood Molecular Replacement (MR), use program Phaser-MR (McCoy et al., 2007), in the PHENIX software suite (Adams et al.). Choose the option Automated Molecular Replacement.
4. Input the data file containing the observed structure factors (F_{obs}) and associated errors for each hkl reflection. The program will automatically read in information like unit cell parameters, space group and data labels.
5. Input the search model for the cohesin module in the form of a PDB file and specify the sequence identity.
6. Define the composition of the asymmetric unit: input the sequence for each chain (cohesin and dockerin) and specify the number of copies of each chain that you expect in the asymmetric unit (estimated from the value of V_M).
7. Define the Search Procedure by specifying the cohesin model as the search model for MR. PHASER will proceed with rotational, followed by translational search, providing Z-scores for each solution, the number of clashes in the packing and the refined log-likelihood gain (LLG). Typically, a correct solution will exhibit a translation function Z-score (TF Z-score) higher than 5 and be well separated from the other solutions, but a careful examination of the output may be necessary to identify the correct solution.
8. Experimental phases can be improved by density modification methods, which include solvent flattening, histogram matching and non-crystallographic symmetry (NCS) averaging. Estimation of the solvent content (given by the value of V_M) and the number of copies of the complex in the asymmetric unit are important at this stage. Programs like DM of the CCP4 suite (Winn et al.) are used to improve phases by density modification. This is done in a cyclic procedure where density modification is followed by phase combination steps. Model building is now possible in the improved electron density.

5.4. Model building and structure refinement

Molecular replacement provides the solution to the phase problem and it is now possible to calculate an electron density map, which should be inspected for secondary structure features. The quality of this electron density map depends on the quality of the initial phases, as well as the resolution and quality of the measured data. The correctly placed cohesin model will have to be adjusted to the features observed in the electron density, which usually includes adjusting side chains or any parts that are not conserved between the search model and the cohesin of interest (some amino acid residues may need to be altered, according to the primary sequence). Since MR was not performed using a search model for the dockerin, this smaller module can be built from the observation of the electron density map. This can be done by the experimenter, using a graphics program (e.g., COOT; Emsley and Cowtan, 2004) and the direct observation of the electron density, or automatically, using dedicated software (e.g., ARP/wARP; Langer et al., 2008) or AutoBuild in PHENIX (Terwilliger et al., 2008). Model improvement and addition of correct features will bring the model closer to the real structure and new improved phases can be calculated. These are used, in an iterative way, to calculate a new electron density map, which may reveal new features that will have to be added to the model. The iterative process is validated by comparing calculated structure factors $|F_{calc}|$, from the model, with observed structure factors $|F_{obs}|$, measured in the diffraction experiment. After adjusting and building missing parts of the complex, water molecules and/or other ligands will have to be identified and added. Overall, model building and improvement is done in real space, using graphics programs, while refinement of atomic positions, temperature factors (a measure that indicates how much an atom oscillates around a specific position), and occupancies (the fraction of molecules where the atom occupies that specific position in the crystal) is performed in the reciprocal space (e.g., program phenix. refine). A suggested procedure for model building and refinement of cohesin–dockerin complexes is presented. These steps are a simple guidance; crystallographic software includes many other options that can be used in different ways, depending on the model building and refinement context.

1. After structure solution with PHASER McCoy et al. (2007), the coordinates of the positioned cohesin can be loaded in COOT for visualization. Use the unit cell and space group information to visualize also the cell content and symmetry. Confirm that there are no clashes and there is enough space to fit the dockerin module.
2. Load the file output by PHASER containing the information to calculate electron density maps. PHASER outputs a file containing column labels FWT/PHWT (amplitude and phase for $2m|F_{obs}| - D|F_{calc}| \exp(i\alpha_{calc})$ map) and DELFWT/PHDELWT (amplitude and phase for $m|F_{obs}| - D|F_{calc}| \exp(i\alpha_{calc})$ map) to calculate sigmaA-weighted electron density maps. The difference map $mF_{obs} - DF_{calc}$ will show positive density where features must be added to the model and negative density where atoms have to be removed from the model. The double difference map $2mF_{obs} - DF_{calc}$ is used to minimize bias from the model.
3. Use the modeling tools from COOT to draw a skeleton in the $2mF_{obs} - DF_{calc}$ map. Try to identify the part of the electron density

corresponding to the dockerin module and the characteristic α-helices. Identify the N- and C-terminus and confirm the connectivity of the polypeptide chain, making sure that the correct identification of helices is not impaired by the inner symmetry of the dockerin modules (i.e., make sure you can unambiguously identify some amino acid residues in each α-helix). If resolution permits and phases are sufficient, most of the polypeptide chain can be identified and no connectivity errors will be introduced. Good enough resolution may allow identification of the side chains of several amino acid residues.

4. Use COOT to place Cα carbons at 3.8 Å intervals, followed by a main-chain trace of the polypeptide chain. The presence of some characteristic residues may help to identify local stretches of sequence and side chains may be introduced.

5. Several solvent molecules may be indicated by positive peaks in the $mF_{obs} - DF_{calc}$ map and should be added to the model. This can also be done automatically using dedicated software; nevertheless, solvent molecules should be visually inspected and kept if matching several criteria, like temperature factors, map sigma level, and contacts.

6. After this, the model can be refined with program phenix-refine from the PHENIX software suite (Adams *et al.*, 2010), by uploading the coordinate file of the improved model and the file containing observed structure factors $|F_{obs}|$ (and associated uncertainties), measured in the diffraction experiment.

7. In refinement settings, choose to refine individual sites and individuals atomic displacement parameters (ADPs), as refinement strategy. The model x, y, z coordinates and temperature factors (also known as ADPs) are refined and used to calculate new phase angles, hopefully more accurate than the experimental phases.

8. Considering that cohesin–dockerin complexes are composed of separate domains, choose option TLS (Translation/Libration/Screw) parameters to automatically find suitable TLS groups, using program phenix. find_tls_groups (Winn *et al.*, 2001). The program will analyze the crystal structure of the complex, searching for evidence of flexibility, for example, local or interdomain motions. Individual chains are partitioned into multiple segments that are modeled as rigid bodies undergoing TLS vibrational motion. Each group, having a different number of segments, is scored according to its ability to explain the observed ADPs ("B values") obtained in the crystallographic refinement.

9. If two or more copies of the complex are present, NCS restraints should also be imposed. NCS groups can be found automatically by phenix.refine or be defined by the user.

10. Phenix.refine will output several files, including the refined model, various maps, structure factors, and complete statistics. Load the model and maps in COOT and proceed with new adjustments to the model, according to the new features in the improved electron density (use the several validation tools provided by COOT).

6. Summary

Anaerobic microbes produce a remarkably efficient nanomachine to deconstruct plant cell wall polysaccharides, which was termed, when discovered more than 20 years ago, as the cellulosome. Cellulases and hemicellulases are assembled into multienzyme complexes through a high affinity interaction established between type I dockerin domains of the modular enzymes and type I cohesin modules of a noncatalytic scaffoldin (Fig. 21.4). It is believed that integration of the microbial

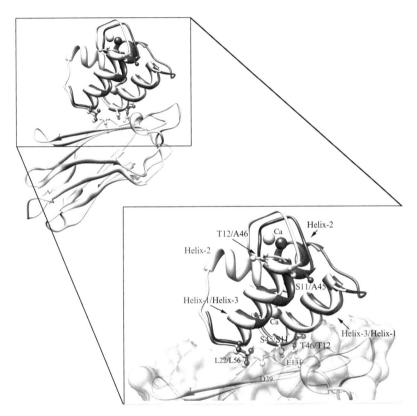

Figure 21.4 The dual binding mode of cohesin–dockerin complexes. Ribbon representation of the superposition of the dockerin modules of type I cohesin–dockerin native complex (light gray) with the S45A–T46A mutant complex (dark gray) in *C. thermocellum*. For simplification, only one cohesin module is represented. The inset shows a more detailed view of the cohesin–dockerin contacts and of the almost perfect superposition of helices 1 and 3 of both complexes. In the mutant complex, helix-1 (containing Ser-11 and Thr-12) dominates binding whereas, in the native complex, helix-3 (containing Ser-45 and Thr-46) plays a key role in ligand recognition. Ser-11, Thr-12, Ser-45, and Thr-46, which interact with the cohesin module, are depicted as ball-and-stick models.

biocatalysts into cellulosomes potentiates catalysis through the maximization of enzyme synergism afforded by enzyme proximity and efficient substrate targeting. Substantial structural and functional evidence exists, suggesting that cellulosomal dockerins display a dual cohesin binding interface. The dual binding mode expressed by cohesin–dockerin complexes may introduce enhanced flexibility in the quaternary organization of the multienzyme complex thus potentiating the hydrolysis of a predominantly insoluble substrate. Recently, it has become apparent that the cohesin–dockerin interaction is quite widespread in nature and may fulfill a large range of, mostly currently unknown, functions which remain to be described.

ACKNOWLEDGMENTS

This work has been supported by Fundação para a Ciência e a Tecnologia through grants PTDC/BIA-PRO/69732/2006, PTDC/QUI-BIQ/100359/2008, and PEst-C/EQB/LA0006/2011. J.A.B. and A.V. were supported by the individual fellowships SFRH/BD/38667/2007 and SFRH/BD/35992/2007, respectively.

REFERENCES

Adams, J. J., Pal, G., Jia, Z., and Smith, S. P. (2006). Mechanism of bacterial cell-surface attachment revealed by the structure of cellulosomal type II cohesin-dockerin complex. *Proc. Natl. Acad. Sci. USA* **103,** 305–310.

Adams, P. D., Afonine, P. V., Bunkoczi, G., Chen, V. B., Davis, I. W., Echols, N., Headd, J. J., Hung, L. W., Kapral, G. J., Grosse-Kunstleve, R. W., McCoy, A. J., Moriarty, N. W., et al. (2010). PHENIX: A comprehensive Python-based system for macromolecular structure solution. *Acta Crystallogr. D Biol. Crystallogr.* **66,** 213–221.

Bayer, E. A., Belaich, J. P., Shoham, Y., and Lamed, R. (2004). The cellulosomes: Multienzyme machines for degradation of plant cell wall polysaccharides. *Annu. Rev. Microbiol.* **58,** 521–554.

Carvalho, A. L., Dias, F. M. V., Prates, J. A. M., Nagy, T., Gilbert, H. J., Davies, G. J., Ferreira, L. M. A., Romao, M. J. R., and Fontes, C. M. G. A. (2003). Cellulosome assembly revealed by the crystal structure of the cohesin-dockerin complex. *Proc. Natl. Acad. Sci. USA* **100,** 13809–13814.

Carvalho, A. L., Dias, F. M. V., Nagy, T., Prates, J. A. M., Proctor, M. R., Smith, N., Bayer, E. A., Davies, G. J., Ferreira, L. M. A., Romao, M. J. R., Fontes, C. M. G. A., and Gilbert, H. J. (2007). Evidence for a dual binding mode of dockerin modules to cohesins. *Proc. Natl. Acad. Sci. USA* **104,** 3089–3094.

Choi, S. K., and Ljungdahl, L. G. (1996). Structural role of calcium for the organization of the cellulosome of *Clostridium thermocellum*. *Biochemistry* **35,** 4906–4910.

Emsley, P., and Cowtan, K. (2004). Coot: Model-building tools for molecular graphics. *Acta Crystallogr. D Biol. Crystallogr.* **60,** 2126–2132.

Evans, P. (2006). Scaling and assessment of data quality. *Acta Crystallogr. D Biol. Crystallogr.* **62,** 72–82.

Fontes, C. M. G. A., and Gilbert, H. J. (2010). Cellulosomes: Highly efficient nanomachines designed to deconstruct plant cell wall complex carbohydrates. *Annu. Rev. Biochem.* **79,** 655–681.

Langer, G., Cohen, S. X., Lamzin, V. S., and Perrakis, A. (2008). Automated macromolecular model building for X-ray crystallography using ARP/wARP version 7. *Nat. Protoc.* **3,** 1171–1179.

Leibovitz, E., and Beguin, P. (1996). A new type of cohesin domain that specifically binds the dockerin domain of the *Clostridium thermocellum* cellulosome-integrating protein CipA. *J. Bacteriol.* **178,** 3077–3084.

Leslie, A. G. (2006). The integration of macromolecular diffraction data. *Acta Crystallogr. D Biol. Crystallogr.* **62,** 48–57.

Long, F., Vagin, A. A., Young, P., and Murshudov, G. N. (2008). BALBES: A molecular-replacement pipeline. *Acta Crystallogr. D Biol. Crystallogr.* **64,** 125–132.

Matthews, B. W. (1968). Solvent content of protein crystals. *J. Mol. Biol.* **33,** 491–497.

McCoy, A. J., Grosse-Kunstleve, R. W., Adams, P. D., Winn, M. D., Storoni, L. C., and Read, R. J. (2007). Phaser crystallographic software. *J. Appl. Crystallogr.* **40,** 658–674.

Pages, S., Belaich, A., Belaich, J. P., Morag, E., Lamed, R., Shoham, Y., and Bayer, E. A. (1997). Species-specificity of the cohesin-dockerin interaction between *Clostridium thermocellum* and *Clostridium cellulolyticum*: Prediction of specificity determinants of the dockerin domain. *Proteins* **29,** 517–527.

Pflugrath, J. W. (2004). Macromolecular cryocrystallography methods for cooling and mounting protein crystals at cryogenic temperatures. *Methods* **34,** 415–423.

Pinheiro, B. A., Proctor, M. R., Fleites, C. M., Prates, J. A. M., Money, V. A., Davies, G. J., Bayer, E. A., Fontes, C. M. G. A., Fierobe, H. P., and Gilbert, H. J. (2008). The *Clostridium cellulolyticum* dockerin displays a dual binding mode for its cohesin partner. *J. Biol. Chem.* **283,** 18422–18430.

Rouillard, J. M., Lee, W., Truan, G., Gao, X., Zhou, X., and Gulari, E. (2004). Gene2Oligo: Oligonucleotide design for *in vitro* gene synthesis. *Nucleic Acids Res.* **32,** W176–W180.

Salamitou, S., Raynaud, O., Lemaire, M., Coughlan, M., Béguin, P., and Aubert, J. P. (1994). Recognition specificity of the duplicated segments present in *Clostridium thermocellum* endoglucanase CelD and in the cellulosome-integrating protein CipA. *J. Bacteriol.* **176,** 2822–2827.

Terwilliger, T. C., Grosse-Kunstleve, R. W., Afonine, P. V., Moriarty, N. W., Zwart, P. H., Hung, L. W., Read, R. J., and Adams, P. D. (2008). Iterative model building, structure refinement and density modification with the PHENIX AutoBuild wizard. *Acta Crystallogr. D Biol. Crystallogr.* **64,** 61–69.

Villalobos, A., Ness, J. E., Gustafsson, C., Minshull, J., and Govindarajan, S. (2006). Gene Designer: A synthetic biology tool for constructing artificial DNA segments. *BMC Bioinform.* **6,** 285.

Weiss, M. (2001). Global indicators of X-ray data quality. *J. Appl. Crystallogr.* **34,** 130–135.

Winn, M. D., Isupov, M. N., and Murshudov, G. N. (2001). Use of TLS parameters to model anisotropic displacements in macromolecular refinement. *Acta Crystallogr. D Biol. Crystallogr.* **57,** 122–133.

Winn, M. D., Ballard, C. C., Cowtan, K. D., Dodson, E. J., Emsley, P., Evans, P. R., Keegan, R. M., Krissinel, E. B., Leslie, A. G., McCoy, A., McNicholas, S. J., Murshudov, G. N., *et al.* (2011). Overview of the CCP4 suite and current developments. *Acta Crystallogr. D Biol. Crystallogr.* **67,** 235–242.

CHAPTER TWENTY-TWO

Measurements of Relative Binding of Cohesin and Dockerin Mutants Using an Advanced ELISA Technique for High-Affinity Interactions

Michal Slutzki,* Yoav Barak,[†] Dan Reshef,[‡]
Ora Schueler-Furman,[‡] Raphael Lamed,[§] *and* Edward A. Bayer*

Contents

1. Introduction	418
2. Preparation of Constructs for ELISA Procedure	420
2.1. CBM-fused cohesins	421
2.2. Expression and purification of xylanase-fused dockerins	422
3. Detection of Free Dockerins Using iELISA	422
3.1. Coating	424
3.2. Cohesin–dockerin equilibrated interaction	424
3.3. Detection of unbound dockerin concentration by iELISA	424
4. Data Analysis	425
5. Notes	426
6. Summary	427
References	427

Abstract

The cellulosome is a large bacterial extracellular multienzyme complex able to degrade crystalline cellulosic substrates. The complex contains catalytic and noncatalytic subunits, interconnected by high-affinity cohesin–dockerin interactions. In this chapter, we introduce an optimized method for comparative binding among different cohesins or cohesin mutants to the dockerin partner. This assay offers advantages over other methods (such as ELISA, cELIA, SPR, and ITC) for particularly high-affinity binding interactions. In this approach, the high-affinity interaction of interest occurs in the liquid phase during the

---

* Department of Biological Chemistry, The Weizmann Institute of Science, Rehovot, Israel
[†] Chemical Research Support, The Weizmann Institute of Science, Rehovot, Israel
[‡] Department of Microbiology and Molecular Genetics, Institute for Medical Research Israel-Canada, Hadassah Medical School, The Hebrew University, Jerusalem, Israel
[§] Department of Molecular Microbiology and Biotechnology, Tel Aviv University, Ramat Aviv, Israel

equilibrated binding step, whereas the interaction with the immobilized phase is used only for detection of the unbound dockerins that remain in the solution phase. Once equilibrium conditions are reached, the change in free energy of binding ($\Delta\Delta G_{binding}$), as well as the affinity constant of mutants, can be estimated against the known affinity constant of the wild-type interaction. In light of the above, we propose this method as a preferred alternative for the relative quantification of high-affinity protein interactions.

Abbreviations

BSA	bovine serum albumin
CBM	carbohydrate-binding module family 3a from *C. thermocellum*
cELIA	competitive enzyme-linked interaction assay
Coh	cohesin
Doc	dockerin
DSC	differential scanning calorimetry
ELISA	enzyme-linked immunosorbent assay
HRP	horseradish peroxidase
IPTG	isopropyl-1-thio-β-D-galactoside
ITC	isothermal titration calorimetry
MES	2-(*N*-morpholino) ethanesulfonic acid
SDS-PAGE	polyacrylamide gel electrophoresis in sodium dodecyl sulfate
SPR	surface-plasmon resonance
TBS	Tris-buffered saline
WT	wild type
Xyn	xylanase

1. Introduction

The ultra-tight interactions between cohesins and dockerins determine the organization of cellulosome (Bayer *et al.*, 1983), the multienzyme complex responsible for efficient degradation of cellulose in selected anaerobic bacteria. Cohesins and dockerins are known to interact with one another in a highly specific manner, depending on their function (i.e., binding to enzymatic, structural, or cell-surface anchoring subunits) and their microbial origin (e.g., dockerins from one bacterial species usually fail to bind cohesins from another). The nature of the specificity and promiscuity, which simultaneously exist in this system, is a very intriguing subject for scientific research. To investigate the possible role of different residues in recognition patterns, target amino acids can either be mutated to alanine or be modified to the corresponding residue from another cohesin/dockerin pair (usually from another species; Fierobe et al., 2001; Handelsman et al., 2004; Mechaly et al., 2000, 2001; Nakar et al., 2004; Pinheiro et al., 2008).

Several methods, described in the above publications, have been developed to measure the strength of the various cohesin–dockerin interactions, such as enzyme-linked immunosorbent assay (ELISA), competitive enzyme-linked interaction assay (cELIA), surface-plasmon resonance (SPR; e.g., BIAcore), isothermal titration calorimetry (ITC), and differential scanning calorimetry (DSC). Each of these methods, however, has limitations, either in general or in specific terms, for measuring high-affinity interactions. Moreover, some of these methods are not amenable for comparative analysis of multiple samples. Here, we specify the disadvantages of each method as compared to the indirect ELISA (iELISA)-based method that we describe in detail in this chapter.

(1) *Standard affinity-based ELISA* (Barak et al., 2005): In this method, the interaction occurs between two different phases: solid (immobilized) and liquid (protein in solution). Thus, we can control concentrations of only one binding partner (the one that is in solution), since during coating the immobilized protein undergoes partial denaturation. Clearly, binding constants cannot be determined using this technique, since concentrations of immobilized proteins are unknown, and the level of immobilization of different mutants on the plate is hard to compare.

(2) *cELIA* (Handelsman et al., 2004): At first glance, it seems as if the interaction between components occurs in the soluble phase in this assay. However, since the components (mutant cohesin and dockerin) are added simultaneously to the well coated with wild-type (WT) cohesin, the binding between the mutant cohesin and dockerin is accompanied by competition with the binding to the immobilized protein. Thus, not only is the immobilized protein involved in estimating the status of the interaction, but there is an uncertainty with respect to the equilibrium conditions, which interferes with calculations of the change in free energy of binding ($\Delta\Delta G_{binding}$).

(3) *SPR*: This methodology is limited by very slow dissociation rates for the cohesin–dockerin complex (undetectable, i.e., $<10^{-4}\,s^{-1}$), coupled with very high affinity. Therefore, low protein concentrations (below the sensitivity limits) had to be used to determine the association constants in a precise manner. In addition, since one of the interacting proteins is immobilized, the binding constants are apparently

overestimated relative to solution-based approaches, for example, ITC (Fierobe et al., 1999; Mechaly et al., 2001) vs. SPR (Carvalho et al., 2003, 2007; Miras et al., 2002), respectively.

(4) *ITC*: The binding affinity that can be accurately measured by this method is limited to a maximum of $K_a \sim 10^9 \, M^{-1}$ (Velazquez-Campoy et al., 2004). To estimate higher affinities, very low levels of protein concentrations are required, which are out of the sensitivity limit of the instrument. Alternatively, competitive binding can be measured in "displacement titration" (Velazquez-Campoy and Freire, 2006), but this would require an available ligand of lower binding affinity.

(5) *DSC*: This method can also measure very high-affinity interactions, but it cannot be used if the proteins are not soluble enough under the high concentrations required for this approach, or if thermal denaturation is not reversible. Another protein is frequently required as a carrier to improve solubility and expression of the target protein, which may interfere in distinguishing the peak of energy input contributed by the protein of interest.

Owing to the above-mentioned drawbacks, we therefore developed an alternative approach, based on iELISA, to overcome the above limitations. Our approach incorporates the previously described cELIA and ELISA methods, but modified by addition of a necessary equilibration step for interaction in solution. In this modified method, the estimated interaction takes place in the soluble phase (opposite to that of ELISA), which also renders this method more sensitive than cELIA, since the procedure is performed under conditions of much lower dockerin concentrations. The approach allows estimates of the change in free energy of binding ($\Delta\Delta G_{binding}$) between the test protein (e.g., mutant cohesin or dockerin module) and the reference protein (e.g., wild type). The present approach is similar to a method reported earlier (Friguet et al., 1985), with more accurate data analysis.

2. Preparation of Constructs for ELISA Procedure

When dockerins and cohesins are produced in their free form, they usually exhibit low expression levels, low solubilities, and a tendency to aggregate (Adams et al., 2005; Fierobe et al., 1999; Lytle et al., 2001). Therefore, we chose to fuse them to carrier proteins that are known for their good expression levels and solubility, when expressed in *Escherichia coli* (Barak et al., 2005). These carrier proteins were also selected from the original molecular context that is relevant to cellulose degradation, that is, the dockerin is fused to a particularly high-expressing xylan-hydrolyzing enzyme (xylanase)—a thermostable enzyme originating from the bacterium *Geobacillus stearothemophilus*, and the cohesin is expressed as a fusion protein together with a carbohydrate-binding module (CBM)—derived from the thermostable cellulosomal scaffoldin protein of *Clostridium thermocellum* (Morag et al., 1995). Using such protein fusions, we achieve standardization in purification procedures and in the assay itself. By adding a xylanase module on all dockerins, we were able to use the same antibodies to detect all the dockerins.

2.1. CBM-fused cohesins

Reagents

LB (Luria–Bertani) broth supplemented with 50 µg/ml kanamycin
IPTG (isopropyl-1-thio-β-D-galactoside) (Fermentas UAB, Vilnuis, Lithuania)
Amorphous cellulose: 7.5 g phosphoric acid-treated Avicel/liter DDW
Washing buffer #1: Tris-buffered saline (TBS), pH 7.4 with 1 M NaCl
Washing buffer #2: TBS, pH 7.4
Triethylamine: 1% (v/v)
2-(*N*-Morpholino) ethanesulfonic acid (MES): 1 M, pH 5.5

Protocol

The CBM-Coh gene cassette (Barak et al., 2005) consists of a family 3a CBM from the *C. thermocellum* CipA scaffoldin (Morag et al., 1995) cloned into plasmid pET28a (Novagen Inc., Madison, WI, USA), into which any cohesin gene can be introduced between *Bam*HI and *Xho*I restriction sites of the plasmid. For demonstration purposes, the Coh2, the second cohesin from the *C. thermocellum* CipA scaffoldin, was used in this chapter.

CBM-Cohs were thus expressed and purified, in a manner similar to the description of Barak et al. (2005). A cell culture was grown in LB medium with kanamycin. Upon induction of recombinant protein expression with IPTG, the culture was grown for 3 h at 37 °C. After sonication, 15 ml of amorphous cellulose was added to the supernatant and used for affinity purification. Binding between cellulose and CBM occurred during overnight incubation on a rotator at 4 °C. The amorphous cellulose was pelleted by centrifugation at 4000 rpm at 4 °C for 5 min and washed three times with 45 ml washing buffer #1, and three times with washing buffer #2. After each step, the suspension was pelleted again and the supernatant fluids were discarded. Finally, the CBM-Cohs were eluted with triethylamine and neutralized using MES until pH 7 was reached. Purity was verified using SDS-PAGE. The proteins were stored in 50% glycerol at −20 °C.

2.2. Expression and purification of xylanase-fused dockerins

Reagents

LB broth supplemented with 50 μg/ml kanamycin
IPTG (Fermentas UAB)
TBS, pH 7.4
Binding buffer: TBS, supplemented with 5 mM imidazole (Merck KGaA, Darmstadt, Germany)
Washing buffer: TBS, supplemented with 20 mM imidazole
Elution buffer #1: TBS, supplemented with 100 mM imidazole
Elution buffer #2: TBS, supplemented with 250 mM imidazole
Ni-NTA agarose (Qiagen GmbH, Hiden, Germany)

Protocol

The Xyn-Doc gene cassette (Barak *et al.*, 2005) consists of xylanase T6 from *G. stearothemophilus* with an N-terminal His-tag (Handelsman *et al.*, 2004; Lapidot *et al.*, 1996) cloned into plasmid pET9d (Novagen Inc.), into which any dockerin encoding sequence can be introduced between the *Kpn*I and *Bam*HI restriction sites of the plasmid. For demonstration purposes, the dockerin from cellulase S (DocS) of *C. thermocellum* was used in this chapter.

The dockerin-bearing xylanases (Xyn-Docs) were thus expressed and purified, in a manner similar to the description of Barak *et al.* (2005). A cell culture was grown in LB medium with kanamycin. Upon induction with IPTG, the culture was grown for 3 h at 37 °C. After sonication, the supernatant was applied to a Ni-NTA column, equilibrated with binding buffer. The column was washed with washing buffer on a rotator at 4 °C, and the Xyn-Docs were then released with elution buffers #1 and #2, sequentially. Purity was verified using SDS-PAGE. The proteins were stored in 50% glycerol at −20 °C.

3. Detection of Free Dockerins Using iELISA

Reagents

WT CBM-Coh (10 nM, the precise working concentration is determined by calibration as described in Barak *et al.*, 2005)
Coating buffer: 0.1 M Na$_2$CO$_3$, pH 9
Blocking buffer: TBS, 10 mM CaCl$_2$, 0.05% Tween 20, 2% BSA
Washing buffer: TBS, 10 mM CaCl$_2$, 0.05% Tween 20
Xyn-Doc: 600 pM (should be roughly in the range of K_D)
Polyclonal rabbit antibodies against Xyn-T6 (anti-Xyn antibody; Lapidot *et al.*, 1996): 1:10,000 (or according to a predetermined calibration)

Secondary antibody–enzyme conjugate (HRP-labeled goat anti-rabbit IgG) (Jackson ImmunoResearch Laboratories Inc., West Grove, PA, USA): 1:10,000 (or according to your predetermined calibration)
TMB (Dako 3,3′,5,5′-tetramethylbenzidine) + Substrate-Chromogen (Dako Corp., Carpinteria, CA, USA)

Protocol

See Fig. 22.1.

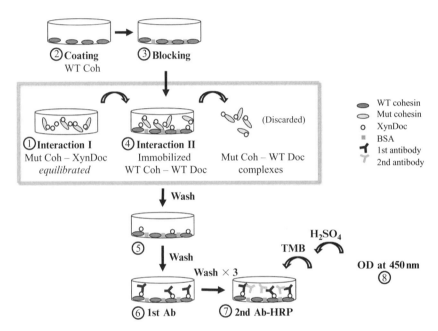

Figure 22.1 Flowchart of the indirect ELISA (iELISA)-based technique for semiquantitative measurement of high-affinity binding. *Note*: A pre-equilibration step (Step 3) is added to the conventional ELISA procedure, which allows complex formation between the mutant cohesin and dockerin modules prior to exposure to the coated plate (Step 4). Next, ELISA steps are used only for monitoring of the unbound dockerins (Steps 5–8). Detailed description: Steps 1 and 2: a 96-well microtiter plate is initially coated with wild-type *Ct* CBM-Coh overnight, then blocked with blocking buffer. Step 3: interaction between *Ct* Xyn-Doc and incremental concentrations of CBM-Coh mutants is performed for at least 1 h on a separate (nonadsorbing) plate. Then, the samples from the interaction plate are transferred to the CBM-Coh-coated plate for ∼10 min incubation (Step 4). Detection of Xyn-Doc bound to the coated CBM-Coh (Step 5) is achieved through sequential binding of rabbit anti-Xyn antibody (Step 6) and then horseradish peroxidase (HRP)-linked goat anti-rabbit antibody (Step 7). Washing is performed after each step. Tetramethylbenzidine is then added, the reaction is terminated with H_2SO_4 and product formation is determined at 450 nm using a microplate reader (Step 8).

3.1. Coating

In the first stage of this method, we expose the surface of the plates to our coating protein (Steps 1 and 2), in alkaline solution to promote its immobilization onto the wells of microtiter plates.

1. Dispense WT CBM-Coh in coating buffer (100 µl/well) into MaxiSorp 96-well plates.
2. Incubate overnight at 4 °C.
3. Discard the liquid contents of the well.
4. Block the wells with 100 µl/well of blocking buffer for 1 h at 37 °C with shaking.

3.2. Cohesin–dockerin equilibrated interaction

The equilibration step (Fig. 22.1, Step 3) is a critically important component of the iELISA approach.

1. Prepare 8–12 different concentrations (ranging from pM to μM) of the desired cohesins (e.g., both mutants and the wild-type reference) in blocking buffer (after addition of the dockerin (Xyn-Doc) the final concentration will be two-thirds of the original). Final volume in the well should be 100 µl.
2. Add 50 µl of 600 pM Xyn-Doc to each well (upon addition, will be diluted threefold to a final concentration of 200 pM), mix and incubate to reach equilibrium (different times should be checked in preliminary experiments until no change is observed. In this case approximately 1 h incubation is sufficient).

3.3. Detection of unbound dockerin concentration by iELISA

1. After the critical equilibrium step is achieved (Fig. 22.1, Step 3), the mixture is transferred to the coated plate (Step 4). The interaction at this stage should be as short as possible (e.g., 10 min). Only unbound dockerins are expected to react here, while those in the complex are kept intact due to the extremely low dissociation rates and very short exposure time.
2. Wash wells by adding washing buffer (Step 5). Previously formed cohesin–dockerin complexes will be washed out. No more intensive washing is needed until the last step since each of the subsequent steps is essentially a wash for the previous one.
3. Add 100 µl anti-Xyn antibody to each well (Step 6).
4. Incubate 40 min at 37 °C with shaking.
5. Wash wells by filling the wells with washing buffer and discarding it.

6. Add 100 μl goat anti-rabbit antibody–enzyme conjugate to each well (Step 7).
7. Incubate 40 min at 37 °C shaking.
8. Wash three times with washing buffer.
9. Add 100 μl TMB. Wait until the color develops (about 1–3 min).
10. Stop the reaction by addition of 50 μl 1 N H_2SO_4.
11. Read the optical density at 450 nm using a microplate reader.

4. Data Analysis

For each mutant, at least three iELISA experiments are carried out in duplicate. Absorbance data are plotted as a semilogarithmic graph as a function of the Xyn-Doc concentration and analyzed using GraphPad Prism (version 5.00 for Windows, GraphPad Software, San Diego, CA, USA). Curves are normalized to a scale of 0–1 indicating relative binding. Results from different experiments are standardized by dividing them by the wild-type values in each particular experiment and then multiplying the product by the average of all wild-type values. The results are fitted to the sigmoidal (dose–response) curve (Motulsky and Christopoulos, 2003), assuming 1:1 binding model:

$$Y = \text{Bottom} + \frac{\text{Top} - \text{Bottom}}{1 + 10^{X - \log IC_{50}}},$$

where Y is the relative binding of dockerin (Xyn-Doc) to the coated WT cohesin (CBM-Coh), "Bottom" is minimal and "Top" is maximal relative binding, X is the concentration of test cohesin in solution, and IC_{50} is the concentration of test cohesin that causes 50% inhibition of binding to the coated WT cohesin.

Changes in free energy of binding ($\Delta\Delta G_{binding}$) were calculated relative to wild type according to the equation:

$$\Delta\Delta G_{binding} = -RT \ln\left(\frac{IC_{50}WT}{IC_{50}mut}\right) = -RT \ln\left(\frac{K_D WT}{K_D mut}\right),$$

where R is the gas constant, T is the temperature, IC_{50} is the concentration of the test cohesin that causes 50% inhibition of binding, and K_D is the dissociation constant of either the wild type or the mutant. If the dissociation constant is known for the wild-type module (i.e., has been determined using an appropriate technique, such as SPR or ITC), then the value for the mutant is attained using the above equation.

Examples of calculation for change in free energy of binding and K_Ds for mutant from Fig. 22.2 are shown in Table 22.1. When $\Delta\Delta G_{binding}$ between

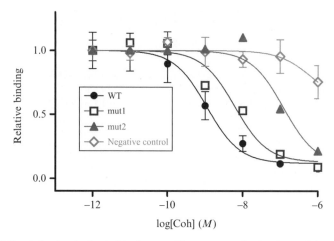

Figure 22.2 Cohesin–dockerin binding profiles, measured using iELISA by determining the amount of dockerin molecules which failed to bind to the cohesin mutant in the equilibration step. Cohesin mutants 1 and 2 show different binding affinities relative to the wild type.

Table 22.1 Change in free energy of binding ($\Delta\Delta G$) and dissociation constants (K_D) calculated from the half-maximal inhibitory concentration (IC_{50})

	IC_{50} (nM)	$K_D$mut/K_DWT	$\Delta\Delta G$ binding (kcal/mol)
WT	1	–	–
mut 1	5	5	1.0
mut 2	93	93	2.8

the wild-type and the mutant exceeds a threshold value (usually fixed at 1.5–2.0 kcal/mol), the mutated residue is considered to participate in the designated binding interaction (in this case, between the cohesin and the dockerin).

5. Notes

The description provided in this chapter focuses on mutant versus wild-type cohesin modules. If there is a need to compare many dockerin mutants, the procedure can be reversed, taking into account that immobilized Xyn-Docs provide a reduced response compared to the CBM-Cohs (Barak et al., 2005), presumably due to steric interference of the xylanase which may restrict access to the dockerin. These two modules have significant differences in their sizes: the molecular weight of the xylanase is approximately 44 versus 10 kDa of the dockerin. Alternatively, the difference may be due to the nature of the dockerin, such as its hydrophobicity and tendency to aggregate.

6. SUMMARY

The ELISA-based method, iELISA, described in this chapter is proposed to overcome several limitations of other methods used to determine binding affinity. It was optimized for the high-affinity cohesin–dockerin system to measure the effect of different mutations on cohesins or for comparison between affinities of different cohesins (e.g., from the same scaffoldin) toward the same dockerin, but can easily be adopted for measuring other high-affinity interactions.

REFERENCES

Adams, J. J., Webb, B. A., Spencer, H. L., and Smith, S. P. (2005). Structural characterization of type II dockerin module from the cellulosome of *Clostridium thermocellum*: Calcium-induced effects on conformation and target recognition. *Biochemistry* **44,** 2173–2182.

Barak, Y., Handelsman, T., Nakar, D., Mechaly, A., Lamed, R., Shoham, Y., and Bayer, E. A. (2005). Matching fusion protein systems for affinity analysis of two interacting families of proteins: The cohesin-dockerin interaction. *J. Mol. Recognit.* **18,** 491–501.

Bayer, E. A., Kenig, R., and Lamed, R. (1983). Adherence of *Clostridium thermocellum* to cellulose. *J. Bacteriol.* **156,** 818–827.

Carvalho, A. L., Dias, F. M., Prates, J. A., Nagy, T., Gilbert, H. J., Davies, G. J., Ferreira, L. M., Romão, M. J., and Fontes, C. M. (2003). Cellulosome assembly revealed by the crystal structure of the cohesin-dockerin complex. *Proc. Natl. Acad. Sci. USA* **100,** 13809–13814.

Carvalho, A. L., Dias, F. M. V., Nagy, T., Prates, J. A. M., Proctor, M. R., Smith, N., Bayer, E. A., Davies, G. J., Ferreira, L. M. A., Romão, M. J., Fontes, C. M. G. A., and Gilbert, M. J. (2007). Evidence for a dual binding mode of dockerin modules to cohesins. *Proc. Natl. Acad. Sci. USA* **104,** 3089–3094.

Fierobe, H. P., Pages, S., Belaich, A., Champ, S., Lexa, D., and Belaich, J. P. (1999). Cellulosome from *Clostridium cellulolyticum*: Molecular study of the dockerin/cohesin interaction. *Biochemistry* **38,** 12822–12832.

Fierobe, H. P., Mechaly, A., Tardif, C., Belaich, A., Lamed, R., Shoham, Y., Belaich, J. P., and Bayer, E. A. (2001). Design and production of active cellulosome chimeras. Selective incorporation of dockerin-containing enzymes into defined functional complexes. *J. Biol. Chem.* **276,** 21257–21261.

Friguet, B., Chaffotte, A. F., Djavadi-Ohaniance, L., and Goldberg, M. E. (1985). Measurements of the true affinity constant in solution of antigen-antibody complexes by enzyme-linked immunosorbent assay. *J. Immunol. Methods* **77,** 305–319.

Handelsman, T., Barak, Y., Nakar, D., Mechaly, A., Lamed, R., Shoham, Y., and Bayer, E. A. (2004). Cohesin-dockerin interaction in cellulosome assembly: A single Asp-to-Asn mutation disrupts high-affinity cohesin-dockerin binding. *FEBS Lett.* **572,** 195–200.

Lapidot, A., Mechaly, A., and Shoham, Y. (1996). Overexpression and single-step purification of a thermostable xylanase from *Bacillus stearothermophilus* T-6. *J. Biotechnol.* **51,** 259–264.

Lytle, B. L., Volkman, B. F., Westler, W. M., Heckman, M. P., and Wu, J. H. D. (2001). Solution structure of a type I dockerin domain, a novel prokaryotic, extracellular calcium-binding domain. *J. Mol. Biol.* **307,** 745–753.

Mechaly, A., Yaron, S., Lamed, R., Fierobe, H.-P., Belaich, A., Belaich, J.-P., Shoham, Y., and Bayer, E. A. (2000). Cohesin-dockerin recognition in cellulosome assembly: Experiment versus hypothesis. *Proteins* **39,** 170–177.

Mechaly, A., Fierobe, H. P., Belaich, A., Belaich, J. P., Lamed, R., Shoham, Y., and Bayer, E. A. (2001). Cohesin-dockerin interaction in cellulosome assembly: A single hydroxyl group of a dockerin domain distinguishes between nonrecognition and high affinity recognition. *J. Biol. Chem.* **276,** 9883–9888.

Miras, I., Schaeffer, F., Beguin, P., and Alzari, P. M. (2002). Mapping by site-directed mutagenesis of the region responsible for cohesin-dockerin interaction on the surface of the seventh cohesin domain of *Clostridium thermocellum* CipA. *Biochemistry* **41,** 2115–2119.

Morag, E., Lapidot, A., Govorko, D., Lamed, R., Wilchek, M., Bayer, E. A., and Shoham, Y. (1995). Expression, purification and characterization of the cellulose-binding domain of the scaffoldin subunit from the cellulosome of *Clostridium thermocellum*. *Appl. Environ. Microbiol.* **61,** 1980–1986.

Motulsky, H. J., and Christopoulos, A. (2003). Fitting Models to Biological Data Using Linear and Nonlinear Regression. A Practical Guide to Curve Fitting. GraphPad Software Inc., San Diego, CA, www.graphpad.com.

Nakar, D., Handelsman, T., Shoham, Y., Fierobe, H.-P., Belaich, J. P., Morag, E., Lamed, R., and Bayer, E. A. (2004). Pinpoint mapping of recognition residues on the cohesin surface by progressive homologue swapping. *J. Biol. Chem.* **279,** 42881–42888.

Pinheiro, B. A., Protctor, M. R., Martinez-Fleites, C., Prates, J. A., Money, V. A., Davies, G. J., Bayer, E. A., Fontesm, C. M., Fierobe, H. P., and Gilbert, H. J. (2008). The *Clostridium cellulolyticum* dockerin displays a dual binding mode for its cohesin partner. *J. Biol. Chem.* **283,** 18422–18430.

Velazquez-Campoy, A., and Freire, E. (2006). Isothermal titration calorimetry to determine association constants for high-affinity ligands. *Nat. Protoc.* **1,** 186–191.

Velazquez-Campoy, A., Ohtaka, H., Nezami, A., Muzammil, S., and Freire, E. (2004). Isothermal titration calorimetry. *Curr. Protoc. Cell Biol.* **17.8,** 1–24.

CHAPTER TWENTY-THREE

Designer Cellulosomes for Enhanced Hydrolysis of Cellulosic Substrates

Yael Vazana,* Sarah Moraïs,* Yoav Barak,[†] Raphael Lamed,[‡] *and* Edward A. Bayer*

Contents

1. Construction of Designer Cellulosomes	430
1.1. Chimeric scaffoldins	430
1.2. Chimeric enzymes	431
2. Scaffoldin Testing and Analysis of Cohesin–Dockerin Interaction	433
2.1. Reagents	433
2.2. Procedures	434
3. Chimeric Enzyme Testing	441
3.1. CBM-binding ability	442
3.2. Enzymatic activity	443
4. Sugar Analysis	449
4.1. Reagents	449
4.2. Procedures	450
5. Summary	450
References	451

Abstract

During the past several years, major progress has been accomplished in the production of "designer cellulosomes," artificial enzymatic complexes that were demonstrated to efficiently degrade crystalline cellulose. This progress is part of a global attempt to promote biomass waste solutions and biofuel production. In designer cellulosomes, each enzyme is equipped with a dockerin module that interacts specifically with one of the cohesin modules of the chimeric scaffoldin. Artificial scaffoldins serve as docking backbones and contain a cellulose-specific carbohydrate-binding module that directs the enzymatic complex to the cellulosic substrate, and one or more cohesin modules from different natural cellulosomal species, each exhibiting a different specificity, that allows the specific incorporation of the desired matching dockerin-bearing

---

* Department of Biological Chemistry, The Weizmann Institute of Science, Rehovot, Israel
[†] Chemical Research Support, The Weizmann Institute of Science, Rehovot, Israel
[‡] Department of Molecular Microbiology and Biotechnology, Tel Aviv University, Ramat Aviv, Israel

enzymes. With natural cellulosomal components, the insertion of the enzymes in the scaffold would presumably be random, and we would not be able to control the contents of the resulting artificial cellulosome.

There are an increasing number of papers describing the production of designer cellulosomes either *in vitro*, *ex vivo*, or *in vivo*. These types of studies are particularly intricate, and a number of such publications are less meaningful in the final analysis, as important controls are frequently excluded. In this chapter, we hope to give a complete overview of the methodologies essential for designing and examining cellulosome complexes.

1. Construction of Designer Cellulosomes

The employment of designer cellulosomes for management of cellulosic materials was first suggested in 1994 (Bayer *et al.*, 1994), and the first demonstration of their construction, characteristics, and use was accomplished in 2001 (Fierobe *et al.*, 2001).

1.1. Chimeric scaffoldins

In the design of the chimeric scaffoldin, the following aspects are taken into consideration. The arrangement of the cellulases in a cellulosome and the distance between the enzymes is controlled by the order of the specific divergent cohesin modules on the chimeric scaffoldin and by the length of the linker segments that separate them. In some bacteria, such as *Clostridium thermocellum*, *Bacteroides cellulosolvens*, *Acetivibrio cellulolyticus*, and *Ruminococcus flavefaciens*, the linkers are commonly 20–40 residues in length, whereas other cellulosome-producing clostridia, such as *C. cellulovorans*, *C. cellulolyticum*, *C. josui*, and *C. acetobutylicum*, possess markedly shorter linkers (between five and eight amino acids; Bayer *et al.*, 2009).

Scaffoldin Scaf•AT was assembled from two portions of cloned genomic DNA containing selected modules (cohesins and carbohydrate-binding module (CBM)). Cohesin A, the third cohesin from *A. cellulolyticus* scaffoldin C, was cloned from *A. cellulolyticus* genomic DNA using forward and reverse primers, respectively: 5′-TTAGA**CCATGG**ATTTACAGGTTG ACATTGGAAGT-3′ and 5′- GTACG**GGTACC**GATGCAATTACCT CAATTTTCC-3′ (*Nco*I and *Kpn*I restrictions sites in boldface), and CBM-T (the segment of the *C. thermocellum* scaffoldin coding for the CBM3a and third cohesin and intervening linker) was cloned from *C. thermocellum* YS genomic DNA using forward and reverse primers, respectively: 5′-CTAAG**GGTACC**GACAAACACACCGACAAACACA-3′ and 5′-CCTTA**CTCGAG**CTATATCTCCAACATTTACTCCAC-3′ (*Kpn*I and *Xho*I restrictions sites in boldface). The different modules were assembled in the linearized plasmid pET28a to form the chimeric scaffoldin. The amino acid sequence of the chimeric scaffoldin is described in Table 23.1. Expression and purification were previously described by Haimovitz *et al.* (2008). Determination of the scaffoldin concentration according to the absorbance at 280 nm should be taken with care, as the scaffoldin contains very few tryptophans, and hence has a particularly low extinction coefficient.

1.2. Chimeric enzymes

While designing a chimeric enzyme, an amino acid alignment between the characterized enzymes from the same glycoside hydrolase family is required to determine the estimated limits of the catalytic module, CBM and linker. The strategy is different for conversion of noncellulosomal enzymes to the cellulosomal mode or for switching the (cohesin) specificity of a cellulosomal enzyme.

1.2.1. Noncellulosomal enzymes

Depending on the modular architecture of the enzyme, several possible strategies are available. If the enzyme has a CBM, the dockerin module can be inserted at the CBM position (replacement strategy) or added to the whole enzyme (additive strategy) depending on the importance of the CBM to the catalytic activity of the enzyme (e.g., if the CBM modulates the character of

Table 23.1 Sequence of Scaf·AT modules and linkers

Module/linker	Sequence
CohA	MDLQVDIGSTSGKAGSVVSVPITFTNVPKSGIYALSFRTNFD PQKVTVASIDAGSLIENASDFTTYYNNENGFASMTFEAP VDRARIIDSDGVFATINFKVSDSAKVGELYNITTNSAYT SFYYSGTDEIKNVVYNDGKIEVIA
Linker	SVPTNTPTNTPANTPVS
CBM3a	GNLKVEFYNSNPSDTTNSINPQFKVTNTGSSAIDLSKLTLR YYYTVDGQKDQTFWCDHAAIIGSNGSYNGITSNVKGT FVKMSSSTNNADTYLEISFTGGTLEPGAHVQIQGRFAKN DWSNYTQSNDYSFKSASQFVEWDQVTAYLNGVLVW GKEP
Linker	GGSVVPSTQPVTTPPATTKPPATTIPPSDDP
CohT	NAIKIKVDTVNAKPGDTVNIPVRFSGIPSKGIANCDFVYSYD PNVLEIIEIKPGELIVDPNPDKSFDTAVYPDRKIIVFLFAED SGTGAYAITKDGVFATIVAKVKSGAPNGLSVIKFVEVGGF ANNDLVEQRTQFFDGGVNVGDI

the parent glycoside hydrolase or if it exhibits carbohydrate-binding ability to sugars other than cellulose, i.e., xylan, it is preferable to retain the CBM; Moraïs et al., 2010b). In the additive strategy, dockerin modules are generally placed at the C-terminus to mimic the modular organization of most natural cellulosomal enzymes, but in some cases, the N-terminal option has to be considered. In enzymes that are composed of a solitary catalytic module, dockerin modules are added at either the N- or C-terminus, and their natural linker would be included (6–10 amino acids).

In the following example, we describe the cloning of *Cel48Y-*t*, a chimeric enzyme, by replacing the native CBM of the noncellulosomal *C. thermocellum* Cel48Y cellulase with a dockerin from the same species (Vazana et al., 2010), thereby converting it into a cellulosomal enzyme. For this purpose, the DNA coding for the catalytic module of *C. thermocellum* Cel48Y was amplified using the following forward and reverse primers, respectively:

5′-TAAATA**GCTAGC**CACCATCACCATCACCATAATT-
CAAATGTTACATATTC-3′
5′-TATAAT**CTCGAG**TTATAT**GGTACC**CTGCTCAAA-
GAAAATGTGATAC-3′

(*Nhe*I, *Xho*I, and *Kpn*I restriction sites in boldface) and that for the Cel48S dockerin was amplified using forward and reverse primers, respectively:

5′-ATAATT**GGTACC**ACATATAAAGTACCTGGTA-3′
5′-ATATTA**CTCGAG**TTAGTTCTTGTACGGCAATGTATC-3′

(*Kpn*I and *Xho*I restriction sites in boldface) by PCR from the genomic DNA of *C. thermocellum* ATCC. The resultant DNA segments were cloned in two steps into a linearized pET21a plasmid. In the first step, the catalytic module was ligated with the linearized plasmid between the *Nhe*I and *Xho*I sites. In the second step, the dockerin was ligated with the linearized plasmid between the *Kpn*I site and the *Xho*I site. The amino acid sequence of the chimeric enzyme is shown in Table 23.2.

1.2.2. Cellulosomal enzymes

In the case of a native multi-modular cellulosomal enzyme that contains an "internal" dockerin module, which is destined for replacement, the strategy would be to retain its internal position and, hence, the natural overall modular arrangement of the resultant chimeric enzyme will match that of the wild-type protein.

Unless obstacles are encountered, standard purification techniques should be used for purification of the chimeric enzyme (i.e., affinity purification using an N- or C-terminus His-tag; Caspi et al., 2006).

Table 23.2 Sequence of *Cel48Y-t modules and linkers

Module/linker	Sequence
Cel48Y catalytic module	MASHHHHHHNSNVTYSKENVQSVYSDRFIALFEDIQK QGYLSEEGIPYHSIETLLVEAPDYGHLTTSEAMSYMV WLGATYGKLTGDWTYFKDAWDKTEQYIIPDPERDQ PGVNSYIPTQPAQYAPEADSPEKYPTPGDINAPTGIDP IADELASTYGTKAIYQMHWLLDVDNWYGYGNHGD GTSRCSYINTYQRGSGESVWETIPHPSWEDFRWGQ VNNGGFLKLFGNFGEPVRQWRYTSASDADARQIQ ATYWAYLWSKEQGKEKELQPYFEKAAKMGDYLRYT FFDKYFRPIGVQDSGRAGTGYDSCHYLLSWYASW GGDINGTWSWRIGSSHCHQGYQNPMAAYALAKES IFTPKSKNAKKDWEQSLDRQIELFLYLQSAEGAIAGG VTNSWSGAYGKYPEGTSTFYDMAYDPHPVYNDPPS NRWFGFQAWSMERIMEYYYLTGDSRVKELCKKWV SWAIENTRLKSDGTYEIPSTLEWSGQPDPWTGKPSE NKNLHCTVTEWTVDVGVTASYAKALIYYAAATE KHEKKIDDKARETAKQLLDRMWHNYRDKKGVAA KEPRADYKRFFDEVYIPHDFSGINAQGAEIKNGIT FIDLRPKYKEDKDYKMVEEAIKSGKDPVMTYHRYW AQAEVAMANAMYHIFFEQ
Cel48S linker	GTTYKVPGTPSTKLY
Cel48S dockerin	GDVNDDGKVNSTDAVALKRYVLRSGIS INTDNADLNEDGRVNSTDLGILKRYILKEIDTLPYKN

2. Scaffoldin Testing and Analysis of Cohesin–Dockerin Interaction

Once a chimeric scaffoldin is designed, cloned, expressed, and purified, it is imperative to test for its specific binding to the chimeric dockerin-containing cellulases. The binding of the chimeric scaffoldin to cellulose or hemicellulose via its CBM must also be examined; this task is carried out as described for the chimeric enzymes in Section 3.1.

2.1. Reagents

Coating solution: 0.1 M sodium carbonate, pH 9
Wash buffer: Tris–HCl buffered saline, pH 7.4 (TBS) buffer, supplemented with 10 mM $CaCl_2$ and 0.05% Tween 20

Blocking buffer: wash buffer, supplemented with 2% bovine serum albumin (BSA)

Interaction buffer: 50 mM citrate or acetate buffer, pH 5–6.0, supplemented with 12 mM CaCl$_2$ and 2 mM EDTA

Loading buffer: TBS buffer, supplemented with 2 mM CaCl$_2$

Primary antibody preparation: rabbit anti-xylanase T-6 antibody or rabbit anti-CBM antibody, diluted 1:10,000 in blocking buffer

Secondary antibody preparation: HRP-labeled anti-rabbit antibody diluted 1:10,000 in blocking buffer

TMB substrate-chromogen (Dako A/S, Glostrup, DK): 1 M H$_2$SO$_4$

Sample buffer: 0.25 M Tris–HCl, pH 6.8, 20% glycerol, 0.002% Bromophenol Blue

Xyn-Doc proteins were prepared as described in Barak et al. (2005). These proteins are composed of the xylanase T-6 from *Geobacillus stearothermophilus* fused to a dockerin module with the desired specificity.

2.2. Procedures

2.2.1. Enzyme-linked immunosorbent assay (ELISA)

The matching fusion-protein procedure of Barak et al. (2005) is followed to determine cohesin–dockerin specificity.

2.2.1.1. Analysis of divergent dockerins using immobilized cohesins (chimeric scaffoldins)

1. Coat MaxiSorp ELISA plates (Nunc A/S, Roskilde, Denmark) overnight at 4 °C with 1 μg/ml of the desired chimeric scaffoldin (100 μl/well) in coating solution.
2. The following steps are performed at room temperature with all reagents at a volume of 100 μl/well. Discard the coating solution.
3. Add blocking buffer and incubate for 1 h.
4. Discard the blocking buffer
5. Add incremental concentrations (0.01–100 ng/ml) of the desired Xyn-Doc in blocking buffer and incubate for 1 h. Wash the plates three times with wash buffer
6. Add the primary antibody preparation and incubate for 1 h. Wash the plates three times with wash buffer.
7. Add the secondary antibody preparation and incubate for 1 h. Wash the plates again (four times) with wash buffer.
8. To detect the interaction, add 100 μl/well TMB substrate-chromogen and terminate color formation by addition of 1 M H$_2$SO$_4$ (50 μl/well). Measure the absorbance at 450 nm using a tunable microplate reader. Plot the absorbance as a function of Xyn-Doc concentration, usually resulting in a sigmoidal (dose–response) curve.

Dockerin domains from two representative cellulosomal enzymes of *C. thermocellum* and *C. cellulolyticum* (Cel48S and Cel5A, respectively) were subjected to the affinity-based ELISA procedure using the corresponding immobilized cohesins (Fig. 23.1). The dockerins from these two enzymes are known to interact with their respective cohesins in a species-specific manner. Under the conditions of the assay, the *C. thermocellum* cohesin interacted exclusively with the *C. thermocellum* dockerin (Fig. 23.1A) and *vice versa* (Fig. 23.1B). The half-maximal response for the *C. thermocellum* interaction was about 6 pM as opposed to about 20 pM for the *C. cellulolyticum* interaction, reflecting the differences in affinities for these two species of cohesin–dockerin interactions (Barak *et al.*, 2005).

2.2.1.2. Analysis of divergent cohesins using immobilized dockerins (chimeric enzymes)
The assay for divergent cohesins is essentially the converse of that described above for the divergent dockerins: instead of immobilizing a cohesin construct, a suitable dockerin construct is substituted. The following modifications are then introduced:

1. Coating is performed with 20 nM of the desired chimeric enzyme.
2. The desired CBM-containing scaffoldin is diluted to concentrations of 10 pM to 10 nM.

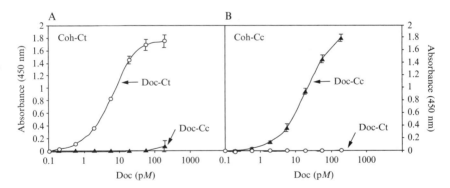

Figure 23.1 Affinity-based ELISA of divergent species of dockerins and cohesins: consequence of the immobilized component. (A and B) Assay of divergent dockerins on immobilized cohesins. (A) Immobilized Coh-Ct: ELISA plates were coated with the CBM-borne cohesin-2 from *C. thermocellum* at a concentration of 3 nM. Dockerin-containing xylanase-T6 fusion proteins, derived from either *C. thermocellum* Cel48S (Doc-Ct: ○) or *C. cellulolyticum* Cel5A (Doc-Cc: ▲), were added at incremented concentrations, and the resultant cohesin–dockerin interaction was detected using anti-xylanase primary antibody and HRP-labeled secondary antibody preparations. (B) Immobilized Coh-Cc: the ELISA assay was performed as in (A) except that the plates were coated with the CBM-borne cohesin-1 from *C. cellulolyticum* instead of Coh-Ct. (Barak *et al.*, 2005).

3. Rabbit anti-CBM antibody is used as the primary antibody preparation.
4. Subsequent steps are performed as described in the previous section.

2.2.2. Nondenaturating PAGE

The stoichiometric molar ratio of each enzyme with respect to the scaffoldin subunit is calibrated by titration using nondenaturing PAGE. The optimal interaction could form at a ratio somewhat different from the estimated 1:1 ratio, mainly due to impurities of the proteins that interfere with the true estimate of the concentration. Then, the full interaction between the chimeric scaffoldin and the various dockerin-containing enzymes can be examined according to the predetermined ratio obtained for each enzyme.

1. In a 30 μl reaction (containing 15 μl of wash buffer), equimolar amounts (4–8 μg) of each protein is added.
2. The 1.5 ml tubes are incubated 1–2 h at 37 °C to allow complex formation.
3. Sample buffer (7.5 μl, in the absence of SDS) is added to 15 μl of the reaction mixture, and the samples are loaded onto nondenaturating gels (4.3%-stacking/9%-separating gels).
4. A parallel SDS-PAGE gel (10%) is performed on the remaining 15 μl sample.

In Fig. 23.2, the trivalent scaffoldin was mixed with increasing amounts of the chimeric enzyme in molar ratios ranging from 1:0.3 to 1:2, and the

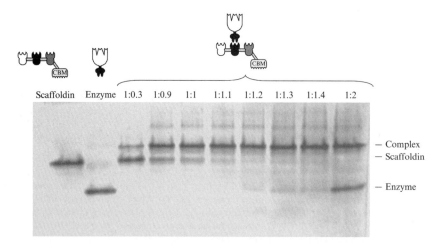

Figure 23.2 Nondenaturating PAGE analysis of mixtures of chimeric scaffoldin and chimeric enzyme (at the indicated molar ratio relative to scaffoldin). The equimolar ratio of each enzyme versus that of the scaffoldin is first estimated by spectrophotometric means, and the enzyme-scaffoldin concentration at and around this value is titrated in order to determine the effective equimolar concentration. The mobility pattern of each mixture was compared with that of the individual components alone.

mobility pattern of each mixture was compared with that of each component alone. As can be seen from the figure, between the ratios of 1:0.3 and 1:1.1, the scaffoldin band is in excess of that of the enzyme. The full interaction forms at a ratio of 1:1.2 (scaffoldin enzyme) as can be observed by the single band, shifted from the bands of each of the single proteins. By increasing the enzyme concentration to ratios of 1:1.3 up to 1:2, the enzyme band is in excess of that of the scaffoldin.

Figure 23.3 displays an example of complex formation between two chimeric dockerin-containing enzymes and a matching chimeric scaffoldin evaluated by nondenaturing PAGE. Denaturing SDS-PAGE was used as a control to verify sample content. Stoichiometric mixtures of the enzymes

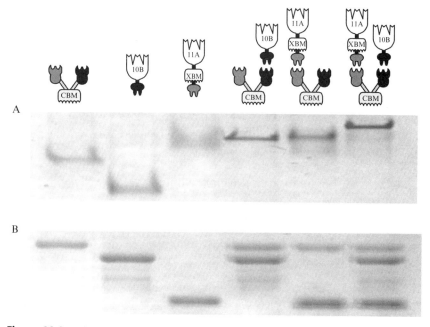

Figure 23.3 Electrophoretic mobility of components and assembled complexes on nondenaturing and denaturing gels. Equimolar concentrations of the chimeric enzymes (11A-XBM-*a* and 10B-*t*) and their matching divalent scaffoldin (Scaf·AT) were combined. Chimera 11A-XBM-*a* represents a chimeric form of the *Thermobifida fusca* xylanase 11A (Xyn11A), which contains a family 11 catalytic module and a xylan/cellulase-binding module (XBM) to which an *A. cellulolyticus* dockerin was attached. Chimera 10B-*t* is derived from Xyn10B of the same bacterium that lacks a native CBM, to which a *C. thermocellum* dockerin was grafted. The chimeric Scaf·AT contains cohesins which match the dockerin specificities of the chimeric enzymes. The single fusion proteins and the mixtures were subjected to nondenaturing PAGE (A) and denaturing SDS-PAGE (B). Analysis of the matching components by nondenaturing PAGE indicates their near-complete interaction as a single major band formed.

and the scaffoldin resulted in a single major band with altered mobility (compared to the progenitor proteins), indicating that complete, or near-complete, complexation is achieved in all cases.

2.2.3. Cellulose-binding assay

1. For incorporation into designer cellulosome complexes, equimolar mixtures of purified proteins (70 pmol each in interaction buffer) are incubated for 1–2 h at 37 °C and mixed with 20 mg of microcrystalline cellulose in a final volume of 200 μl.
2. The tubes are incubated at 4 °C for 1 h with gentle mixing before being centrifuged at 14,000 rpm for 2 min, and the supernatants (containing unbound protein) are carefully removed.
3. The cellulose pellet is washed twice by resuspension in 100 μl of the interaction buffer (supplemented with 0.05% of Tween 20 to eliminate nonspecific binding), centrifuged at 14,000 rpm for 2 min, resuspended in 60 μl of SDS-containing buffer, and heated at 100 °C for 10 min to dissociate any bound protein.

Equimolar mixtures of enzymes without the scaffoldin can be used as a negative control to ensure specificity of binding. The chimeric enzymes have to be tested for control of direct binding ability to cellulose. Bound, unbound fractions and individual proteins are analyzed by SDS-PAGE using a 10% polyacrylamide gel.

2.2.4. Size exclusion high-performance liquid chromatography (HPLC)

To check the full interaction between scaffoldin and enzymes, size exclusion HPLC can be performed.

1. Each enzyme or scaffoldin is prepared in 300 μl reaction mixture in loading buffer.
2. For formation of the designer cellulosome complex, equimolar concentrations of each protein are mixed and the reaction is supplemented with a similar volume of wash buffer. Loading buffer is added to a final volume of 300 μl.
3. The reaction tubes are incubated for 1–2 h at 37 °C to allow complex formation.
4. The reactions are injected onto an analytical Superdex 200 HR 10/30 column using an AKTA fast-performance liquid chromatography system (GE Healthcare, Uppsula, Sweden) using loading buffer at a flow rate of 0.5 ml/min.
5. Proteins are detected using a UV detector at a wavelength of 280 nm. The eluted fractions can be concentrated and analyzed using an SDS-PAGE gel.

The concentration of protein to be used for each injection should be calibrated according to the absorbance of the protein with the least number of tryptophan residues. In the following example (Fig. 23.4), 350 pmol of the scaffoldin and three chimeric cellulases were injected onto the column each in a separate run. The order in which the cellulases eluted from the column corresponds to their molecular weight: Cel9K (101.4 kDa), Cel48S (81.6 kDa), and Cel8A (51.6 kDa). The scaffoldin (68 kDa), however, eluted before Cel9K (101.4 kDa) probably due to its extended fold in comparison to the globular packed folding of the cellulases. Alternatively, the scaffoldin may run as a dimer (Fierobe *et al.*, 2001; Shimon *et al.*, 1997; Spinelli *et al.*, 2000; Tavares *et al.*, 1997). The trivalent designer cellulosome composed of the scaffoldin and three cellulases eluted in a major peak

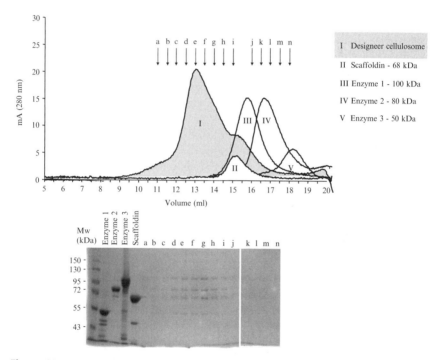

Figure 23.4 Superdex 200 gel filtration HPLC elution profile of a designer cellulosome complex. I—designer cellulosome composed of a trivalent chimeric scaffoldin and three chimeric *C. thermocellum* enzymes Cel9K, Cel48S, and Cel8A. The original *C. thermocellum* dockerin of Cel9K and Cel8A were replaced by dockerins from alternative species, in order to match the specificities of the cohesins of the chimeric scaffoldin, thereby ensuring precision incorporation of the enzyme into the complex. The elution profile of each of the single components was used as a marker, II—chimeric scaffoldin (68 kDa), III—Enzyme 1: Cel9K (101.4 kDa), IV—Enzyme 2: Cel48S (81.6 kDa), V—Enzyme 3: Cel8A (51.6 kDa). The gel at the bottom represents the SDS-PAGE analysis of the designated eluted fractions.

significantly faster than the single components of the complex. The shoulder in the designer cellulosome peak may indicate a population of designer-cellulosomes in which an enzyme is missing in the complex.

2.2.5. Comments

2.2.5.1. Stability assays Nondenaturating PAGE, cellulose-binding assays, and HPLC can also be performed to examine possible detrimental effects of various treatments (e.g., high temperature, solvents, etc.) following appropriate incubation periods. These procedures can be used to examine the stability of the different species of cohesin–dockerin interactions—notably those derived from mesophilic bacteria—under the conditions of the substrate-degrading activity assay, some of which are carried out at elevated temperatures for thermophilic systems.

2.2.5.2. Cellulose-based pull-down assay for examination of complex formation In many publications, designer cellulosome complexes were examined by nondenaturing PAGE (Caspi *et al.*, 2008; Fierobe *et al.*, 2005; Moraïs *et al.*, 2010a,b; Vazana *et al.*, 2010). Using this approach, complex formation is characterized by a single major band with minor altered mobility (band strengthened and usually shifted), indicating that complete or near-complete complexation is achieved. The position of a given band depends on the size, shape, and charge characteristics of the protein. In some cases, stoichiometric mixtures of the enzymes and the scaffoldin result in a complex characterized by a single major band, with only minor (or undetectable) altered mobility relative to the apo scaffoldin. The observed minor change in mobility of large protein complexes limits the efficacy of the nondenaturing PAGE technique; particularly as the number of protein components increase in higher-order complexes. Moreover, in a number of designer cellulosome studies, some of the chimeric scaffoldin and enzyme preparations are characterized by distinct, albeit very minor, contaminating bands, viewed either by denaturing and/or by nondenaturing PAGE, which tend to obscure the analysis. Therefore, as a complementary approach for complex formation, the ability of the complex to bind microcrystalline cellulose by virtue of its resident CBM can be employed (cellulose-binding assay). For this purpose, the designer cellulosome preparations are subjected to interaction with cellulose, and the bound and unbound fractions are examined by SDS-PAGE. Dockerin-bearing enzymes that remained in the unbound fractions indicated that they failed to interact properly with the matching cohesins of the scaffoldin protein. In Fig. 23.5, the complex formation of four enzymes and a chimeric scaffoldin was tested. It appears that complex formation was near-complete, as all of the designer cellulosome components were present in the bound fractions, and only negligible amounts of the proteins were found in the unbound fraction. In the absence of scaffoldin, the dockerin-bearing enzymes failed to bind to the cellulose matrix and remained in the unbound fraction (data not shown).

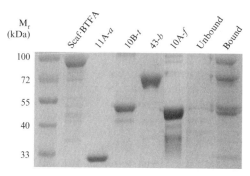

Figure 23.5 Cellulose-based pull-down assay: Cellulose-binding ability of the designer cellulosome complex. Samples included: lane 1, molecular weight marker; lane 2, Scaf·BTFA; lane 3, 11A-*a*; lane 4, 10B-*t*; lane 5, 43-*b*; lane 6, 10A-*f*; lanes 7 and 8, unbound and bound fractions immediately after complex formation.

3. Chimeric Enzyme Testing

Whenever a native enzyme is modified, all functional components must be examined and compared with that of the wild-type enzyme. Proper controls must always be included. When a dockerin is grafted onto a catalytic module, binding to the matching cohesin should be assayed (Section 2.2.1.1), while also including a nonmatching cohesin (e.g., from a different species) as a negative control.

If the native enzyme contains a CBM, it is important to determine its specificity (i.e., to which polysaccharides it binds). If the CBM serves as a simple targeting agent in delivering the catalytic module to cellulose, then the dockerin can replace the CBM, as the targeting function will be provided by the CBM on the scaffoldin component of the designer cellulosome. If there is no CBM on the scaffoldin, then it can be retained on one of the enzymes to provide the cellulose-targeting function in the designer cellulosome. It has been demonstrated earlier that more than one cellulose-binding CBM in a designer cellulosome may inhibit degradation of cellulosic substrates, and it is therefore advisable to limit the number of such modules to one (Mingardon *et al.*, 2007b).

If a CBM of a native enzyme (e.g., a xylanase) binds to a polysaccharide(s) other than cellulose (e.g., xylan), then its function may be crucial to the enzymatic activity, and the CBM should thus be retained in the final chimeric dockerin-containing enzyme. Again, it is critical to examine both the binding properties of the CBM and the enzymatic properties of the native enzyme along with those of a CBM-deficient derivative before preparing a dockerin-containing chimera. These steps will help determine the final design of the dockerin-containing enzyme. Once the latter has been prepared, the use of a monovalent CBM-cohesin scaffoldin can then serve as a control to determine whether the cellulose-targeting function restores enzymatic activity to the chimera, whether simple targeting is insufficient, and whether other considerations, such as the position of the CBM relative to the catalytic module (catalytic site), are important.

3.1. CBM-binding ability

In general, upon replacing a CBM by a dockerin module, we would expect the cellulose-binding properties to be abolished. In some cases, the catalytic module may exhibit inherent cellulose-binding capacity, and this function should thus be examined in the dockerin-containing chimera and compared to that of the wild-type enzyme. One should beware of eliminating carbohydrate-binding abilities to polysaccharides other than cellulose (e.g., xylan). Indeed, carbohydrate-binding abilities could be essential for some enzyme activities (Moraïs et al., 2010b), whereas the cellulose-binding function may be unnecessary to the action of the chimeric enzyme in designer cellulosomes, owing to the presence of a CBM in the chimeric scaffoldin. Interestingly, some native enzymes lack a CBM, and incorporation of a chimeric dockerin-containing form of these enzymes into a designer cellulosome would impart by default a cellulose-binding component (through the CBM-containing scaffoldin) to this type of enzyme.

3.1.1. Reagents

Nondenaturating PAGE buffer: 1.5 M Tris–HCl buffer, pH 8.2
BSA (MP Biomedicals, LLC, Solon, OH)
Interaction buffer: 50 mM citrate or acetate buffer pH 5–6.0, supplemented with 12 mM CaCl$_2$ and 2 mM EDTA

Insoluble xylan is prepared by the following protocol:

1. Oat-spelt xylan (Sigma Chemical Co., St. Louis, MO) is boiled for 30 min in distilled water and the insoluble residue is pelted by centrifugation.
2. Wash the pellet three times with distilled water.
3. Dry weight is determined and 0.5 mg of insoluble xylan is used for the binding assay.

Microcrystalline cellulose (Avicel) is purchased from FMC Biopolymer (Philadelphia, PA) and 10 mg of this polysaccharide is used to measure binding.

3.1.2. Procedures

3.1.2.1. Affinity electrophoresis-based carbohydrate-binding assay The procedure of Tomme *et al.* (2000) is followed with minor modifications.

1. Nondenaturing continuous PAGE (12.5% acrylamide gels) is performed in nondenaturating PAGE buffer.
2. For ligand-embedded gels, polysaccharide preparations containing soluble sugars or mixtures of soluble and insoluble sugars (i.e., oat spelt xylan, PASC) are added to the gel mixtures at a final concentration of 0.1% (w/v) prior to polymerization. Native gels without ligand are run simultaneously under the same conditions.
3. Electrophoresis is carried out at room temperature and 100 V.
4. BSA is used as a negative nonbinding control.
5. Separated proteins are revealed by staining with Coomassie Blue.

3.1.2.2. Binding to insoluble polysaccharides The binding of wild-type and chimeric enzyme to insoluble polysaccharides is determined qualitatively using SDS-PAGE.

1. Pure protein (5–10 μg in interaction buffer) is mixed with the insoluble polysaccharide in a final volume of 100 μl.
2. Tubes are incubated on ice for 1 h with gentle mixing before being centrifuged at 14,000 rpm for 1 min, and the supernatants (containing unbound protein) are carefully removed.
3. The polysaccharide pellet is washed once by resuspending in 100 μl of the interaction buffer (containing 0.05% Tween 20 to avoid nonspecific binding) and centrifuged before resuspending in 60 μl of SDS-containing interaction buffer and boiled for 10 min to dissociate any bound protein.

BSA can be used as negative control to ensure specificity of binding. Bound and unbound fractions are analyzed by SDS-PAGE.

3.2. Enzymatic activity

Each chimera is tested for its enzymatic activity on the relevant substrate and compared to the activity of the wild-type enzyme. In order to avoid erroneous conclusions, significant differences in substrate degradation have to be taken into consideration while studying the activity of the chimera as a part of designer cellulosome complexes.

Optimizing the conditions for designer cellulosome activity assay can greatly influence the results. Conditions such as the substrate, temperature, time course of the reaction, and pH are determined empirically. In addition, preliminary calibration of the linear range of individual chimeric and native enzyme activities are performed prior to testing the activity of the designer cellulosome complex.

Several substrates can be used in order to evaluate the activity of the designer cellulosomes, with varying degrees of crystallinity and complexity. Usually the simpler substrates, such as xylans and amorphous cellulose, are used to characterize the individual chimeric enzymes and compare their activities with those of the wild-type enzymes. The activity of the designer cellulosomes can be tested on these substrates as well; however, the proximity and targeting effects are usually greater on crystalline and more complex substrates, such as cell walls (e.g., wheat straw) or crystalline cellulose (Avicel, bacterial microcrystalline cellulose, filter paper; Fierobe et al., 2005).

During the course of cellulose hydrolysis, the rate of the reaction decreases as the hydrolysis progresses. It has been shown that the nonlinearity of the kinetics is caused by substrate heterogeneity of the insoluble cellulose (Zhang et al., 1999). During the fast initial phase of the reaction the exposed amorphous component is degraded. However, as the exposed portions diminish, a decrease in the rate of the reaction is observed, reflecting the degradation of crystalline and more complex components of the cellulose. The experiment should be performed until the plot approaches the plateau or after reaching about 4–5% hydrolysis of the total (crystalline) substrate, in order to enable degradation of a portion of the crystalline part of the substrate.

In the cellulosome field, the "targeting effect" refers to the increase in activity of the dockerin-bearing enzymes due to the binding of the CBM, contained in the chimeric scaffoldin, to its ligand. The use of a scaffoldin that lacks a CBM would be a proper control in this case. The "proximity effect" refers to the increase in activity that is caused by bringing the enzymes into close contact with each other on the scaffoldin subunit, thus allowing the products of one enzyme to become the substrate for the proximal enzyme ("substrate channeling"; Zhang, 2011). The increase in activity that is caused by the "proximity effect" can be distinguished from that caused by the "targeting effect" by using proper experimental controls, such as the single dockerin-containing enzymes preattached to a matching monovalent CBM-cohesin. If the designer cellulosome exhibits enhanced activity relative to that of the mixture of single enzymes preattached to individual CBM-cohesin, this would indicate that the enhancement in activity is due to the proximity effect (Mingardon et al., 2007a; Moraïs et al., 2010a). If their activities are the same, then the increase in activity would be due to the CBM targeting effect alone (Mingardon et al., 2007b; Vazana et al., 2010). It was also demonstrated that in some cases both proximity and CBM targeting effects can occur simultaneously in designer cellulosomes and the effects were demonstrated to be cumulative (Caspi et al., 2008, 2009, 2010; Fierobe et al., 2001, 2002, 2005; Moraïs et al., 2010b). In others, the free mode of action is the preferred choice, whereby enzymes preattached to monovalent CBM-cohesins show higher activities than the designer cellulosome (Caspi et al., 2006).

In summary, any given designer cellulosome complex may exhibit either of these effects, either singly or in combination, depending on the characteristics (specific enzymes, composition and organization of scaffoldin, linker regions, etc.) of the individual system. The phenomena that cause the synergistic effect seems to depend on the characteristics of the specific enzyme combination used to fabricate the designer cellulosome, and the properties of the component parts should be examined carefully in every study. See the summary in Table 23.3.

The controlled incorporation of cellulases into artificial designer cellulosomes was shown to induce enhanced synergism among the cellulases via targeting to the substrate and/or by the proximity of the cellulases in the complex (Fierobe et al., 2005).

An example of proximity effect in designer cellulosomes, without targeting, is given in Fig. 23.6 (Moraïs et al., 2010a). A tetravalent designer cellulosome containing two xylanases and two cellulases was analyzed for wheat straw degradation and compared to the mixtures of free wild-type enzymes, free chimeric enzymes, and the four monovalent designer cellulosomes (each chimaric enzyme attached to the corresponding monovalent scaffoldin—as targeting effect control). The tetravalent cellulose/xylanase-containing designer cellulosome complex exhibited an ∼2.4-fold enhancement after 20 h, compared to the combined action of the various mixtures of free enzyme. No significant difference was observed in the activity on the wheat straw substrate of the wild-type enzymes, the free dockerin-containing chimeric enzymes, and the monovalent scaffoldin-bearing enzymes, after 7 h of degradation. Enzyme proximity (monitored by designer cellulosome degradation) thus appeared to be mainly responsible for the improvement of substrate degradation, and little or no contribution could be attributed to the substrate-targeting effect.

Table 23.3 Relative degradative activities of designer cellulosomes versus those of other formats and the consequent effects

Comparison of activities	Effect
Designer-cellulosome > CBM-cohesin > free chimeric enzymes	Targeting + proximity
Designer-cellulosome > CBM-cohesin ≈ free chimeric enzymes	Proximity without targeting
Designer-cellulosome ≈ CBM-cohesin > free chimeric enzymes	Targeting without proximity
Designer-cellulosome < CBM-cohesin	Anti-proximity

CBM-cohesin designates the combination of monovalent scaffoldins (single cohesin) with the matching dockerin-containing enzymes. Free chimeric enzymes indicate the dockerin-containing enzymes (without CBM).

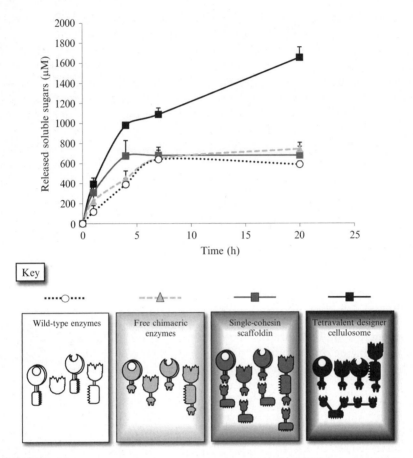

Figure 23.6 Kinetics studies of wheat straw hydrolysis by the free enzyme systems versus the tetravalent cellulosome. Curves are labeled as follows: tetravalent scaffoldin-bound enzyme complex (designer cellulosomes) (■), free chimaeric enzyme system (lacking CBMs) (▲), chimeric enzymes attached to monovalent scaffoldins (CBM-bearing) (■), and wild-type enzymes (containing native CBMs) (○). Triplicates of each reaction were carried out. Standard deviations are indicated (Moraïs et al., 2010a).

An example of substrate targeting without proximity effect is given in Fig. 23.7, where the designer cellulosome is composed of family-48 and a family-9 glycoside hydrolases from *C. thermocellum* (Vazana et al., 2010). The free dockerin-bearing enzymes were less active than the monovalent scaffoldin-bearing enzymes due to the targeting effect. However, the incorporation of the enzymes into a bivalent designer scaffoldin did not further improve their activity.

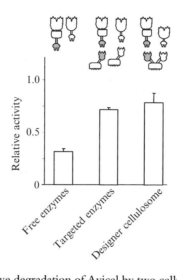

Figure 23.7 Comparative degradation of Avicel by two cellulases as free enzymes, as targeted enzymes attached to monovalent scaffoldins (CBM-bearing) or as a designer cellulosome complex attached to a divalent scaffoldin. Triplicates of each reaction were carried out. Standard deviations are indicated (Vazana et al., 2010).

3.2.1. Reagents

Wash buffer: TBS buffer, supplemented with 10 mM $CaCl_2$ and 0.05% Tween 20

Reaction Buffer 1: 50 mM acetate buffer, pH 5.0, 24 mM $CaCl_2$, 4 mM EDTA

Avicel solution: Avicel 10–20% (w/v) (Sigma-Aldrich) suspended in 100 mM acetate buffer, pH 5.0

PASC solution: phosphoric acid swollen cellulose (Lamed et al., 1985) to a final concentration of 0.75% (w/v)

Filter paper disks: 5-mm diameter disks of filter paper (Whatman no. 1, Whatman Limited, Maidstone-Kent, UK)

Reaction Buffer 2: 50 mM citrate buffer, pH 6.0, 12 mM $CaCl_2$, 2 mM EDTA

Xylan solution: 2% xylan (birchwood, beechwood, or oat spelt from Sigma Chemical Co), suspended in 50 mM citrate buffer, pH 6.0

Hatched wheat straw (0.2–0.8 mm) provided by Valagro Carbone Renouvelable Poitou-Charentes (Poitiers, France) is treated as follows (Fierobe et al., 2005; Tabka et al., 2006):

1. The crude substrate is incubated in distilled water under mild stirring for 3 h at room temperature, vacuum filtered on 2.7-μm glass filter, resuspended in water, and incubated for 16 h under mild stirring at 4 °C.

2. The suspension is filtered and washed three times with water.
3. A sample is dried at 100 °C overnight for estimation of dry weight.

3.2.2. Procedures
All assays are performed in triplicate.

1. Before substrate hydrolysis, dockerin-containing enzymes are subjected to 1–2 h incubation (37 °C, in the absence of substrate) in the presence of equimolar concentrations of scaffoldin, prior to assay in wash buffer.
2. The temperature of the assays depends on the source of the enzymes (i.e., from mesophilic and thermophilic bacteria).
3. After termination of the reaction, samples are centrifuged at maximum speed to remove the substrate and the amount of reducing sugars can be determined as described in Section 4.2.1.

3.2.3. Avicel hydrolysis

1. A typical assay mixture consists of 160 µl Reaction Buffer 1 with enzyme (0–1 µM).
2. The reaction is initiated by the addition of 40 µl of Avicel solution.
3. The reaction is carried out for 18–72 h at the desired temperature in a shaking incubator.

3.2.4. Phosphoric acid swollen cellulose (PASC) hydrolysis

1. A typical assay mixture consists of 50 µl Reaction Buffer 1 with enzyme (0–1 µM).
2. The reaction is initiated by the addition of 150 µl of PASC solution.
3. The reaction is carried out for 30–90 min at the desired temperature in a shaking incubator.

3.2.5. Filter paper hydrolysis

1. The enzymes in Reaction Buffer 1 (0–1 µM) are tested in 200 µl final volume and filter paper disks.
2. Reactions are incubated for a minimum of 16 h at the desired temperature.

3.2.6. Xylan hydrolysis

1. A typical assay mixture consists of 100 µl Reaction Buffer 2 with enzyme (0–10 nM).
2. The reaction is commenced by adding 100 µl of xylan solution and the reaction is continued for 20–60 min at the desired temperature.
3. The reaction is stopped by transferring the tubes to an ice-water bath.

3.2.7. Wheat-straw hydrolysis

1. The composition of the assay mixture consists of 100 µl Reaction Buffer 1 or 2 (depending on the enzymes used), hatched wheat straw is used at the concentration of 3.5 g/l and the concentration of enzymes set at 0–1 µM (in a 200 µl final volume reaction).
2. Reactions are incubated for a minimum of 16 h at the desired temperature.

3.2.8. Comments

Attachment of a dockerin module to a protein sometimes results in a slightly improved enzymatic activity at the maximum rate of reaction (excess substrate), suggesting that dockerin addition may reduce enzyme affinity to its substrate (increase K_m) and thus improve the rate of product release which is rate limiting in many glycoside hydrolases (Asenjo, 1983).

4. Sugar Analysis

Cellulose has a simple chemical structure and while undergoing degradation by endoglucanase and cellobiohydrolase, the major product is cellobiose, together with other soluble cellodextrins and glucose. However, xylan has a more intricate chemical structure, as it comprises a main chain of xylose units largely branched with a variety of relatively low molecular weight heteropolysaccharides. Thus, sugar analysis of the products of xylan degradation is complex, as many possible products can be generated in variable concentrations.

Naturally, the number of possible products of degradation increase even more when natural composite substrates are used, which are composed of both cellulose and hemicelluloses, as well as other polymers.

When considering the analysis of sugar products, it is rather important to appreciate the potential difference in enzymatic mechanisms of complexed and free enzymes.

4.1. Reagents

DNS reagent: the procedure of Wood and Bhat was followed (Wood and Bhat, 1988).
Dinitrosalicylic acid: 40 g
Phenol: 8 g
Sodium sulfite (Na_2SO_3): 2 g
Na-K tartarate (Rochelle salts): 800 g

All are dissolved in 2 l of 2% (w/v) NaOH solution and then diluted to 4 l with distilled water. The container is covered with aluminum foil.
Elution buffer: 200 mM NaOH

4.2. Procedures

4.2.1. DNS method
Enzymatic activity is determined quantitatively by measuring the reducing sugars released from the substrate by the DNS method (Ghose, 1987; Miller, 1959).

1. Enzymatic reaction is terminated by transferring the tubes to an ice-water bath.
2. 100 µl of the supernatant are then added to 150 µl DNS reagent.
3. The tubes are boiled for 10 min.
4. The absorbance is measured at 540 nm.
5. The generation of reducing sugars is determined using glucose (or another simple reducing sugar) as a standard.

4.2.2. High-performance anion exchange chromatography (HPAEC)
Analysis of sugar content can be performed using a HPAEC system equipped with a PA1 column (Dionex, Sunnyvale, CA, USA).

1. Reaction mixtures are loaded onto the column and eluted with elution buffer.
2. Sugar concentrations are determined by integration of the chromatographic peaks, based on sugar reference standards (i.e., arabinose, xylose, xylobiose xylotriose, glucose, cellobiose, etc.).
3. Levels of sugars observed in blanks are deducted in the samples.

4.2.3. Commercial kits
Sugar concentration kits are commonly used, according to the manufacturer's instructions (i.e., xylose concentrations can be measured by a D-xylose assay kit (Megazyme, Wicklow, Ireland); glucose can be determined using a Glucose Assay Kit GAGO20 (Sigma)).

5. SUMMARY

Designer cellulosomes are a versatile research tool for analyzing and studying wild-type and chimeric enzymes in basic science. As such it provides a powerful tool when characterizing a new enzyme or studying synergism among enzymes in a complex. Once a synergistic combination is found, it is possible to further assemble the designer cellulosome using an incremental approach, in which a new enzyme is introduced to enhance the existing synergy. While the currently described designer cellulosomes are not as active as wild-type cellulosomes, they are the necessary step toward next-generation complexes and for engineering microbial cell-surface display of designer cellulosomes for cost-efficient biofuel production.

REFERENCES

Asenjo, J. A. (1983). Maximizing the formation of glucose in the enzymatic hydrolysis of insoluble cellulose. *Biotechnol. Bioeng.* **25,** 3185–3190.

Barak, Y., Handelsman, T., Nakar, D., Mechaly, A., Lamed, R., Shoham, Y., and Bayer, E. A. (2005). Matching fusion-protein systems for affinity analysis of two interacting families of proteins: The cohesin-dockerin interaction. *J. Mol. Recognit.* **18,** 491–501.

Bayer, E. A., Morag, E., and Lamed, R. (1994). The cellulosome—A treasure-trove for biotechnology. *Trends Biotechnol.* **12,** 378–386.

Bayer, E. A., Smith, S. P., Noach, I., Alber, O., Adams, J. J., Lamed, R., Shimon, L. J. W., and Frolow, F. (2009). Can we crystallize a cellulosome? *In* "Biotechnology of Lignocellulose Degradation and Biomass Utilization," (K. Sakka, S. Karita, T. Kimura, M. Sakka, H. Matsui, H. Miyake, and A. Tanaka, eds.), pp. 183–205. Ito Print Publishing Division, Tokyo, Japan.

Caspi, J., Irwin, D., Lamed, R., Shoham, Y., Fierobe, H.-P., Wilson, D. B., and Bayer, E. A. (2006). *Thermobifida fusca* family-6 cellulases as potential designer cellulosome components. *Biocatal. Biotransform.* **24,** 3–12.

Caspi, J., Irwin, D., Lamed, R., Fierobe, H.-P., Wilson, D. B., and Bayer, E. A. (2008). Conversion of noncellulosomal *Thermobifida fusca* free exoglucanases into cellulosomal components: Comparative impact on cellulose-degrading activity. *J. Biotechnol.* **135,** 351–357.

Caspi, J., Barak, Y., Haimovitz, R., Irwin, D., Lamed, R., Wilson, D. B., and Bayer, E. A. (2009). Effect of linker length and dockerin position on conversion of a *Thermobifida fusca* endoglucanase to the cellulosomal mode. *Appl. Environ. Microbiol.* **75,** 7335–7342.

Caspi, J., Barak, Y., Haimovitz, R., Gilary, H., Irwin, D., Lamed, R., Wilson, D. B., and Bayer, E. A. (2010). *Thermobifida fusca* exoglucanase Cel6B is incompatible with the cellulosomal mode in contrast to endoglucanase Cel6A. *Syst. Synth. Biol.* **4,** 193–201.

Fierobe, H.-P., Mechaly, A., Tardif, C., Belaich, A., Lamed, R., Shoham, Y., Belaich, J.-P., and Bayer, E. A. (2001). Design and production of active cellulosome chimeras: Selective incorporation of dockerin-containing enzymes into defined functional complexes. *J. Biol. Chem.* **276,** 21257–21261.

Fierobe, H.-P., Mechaly, A., Tardif, C., Belaich, A., Lamed, R., Shoham, Y., Belaich, J.-P., and Bayer, E. A. (2002). Designer nanosomes: Selective engineering of dockerin-containing enzymes into chimeric scaffoldins to form defined nanoreactors. *In* "Carbohydrate Bioengineering: Interdisciplinary Approaches," (T. T. Teeri, B. Svensson, H. J. Gilbert, and T. Feizi, eds.), pp. 113–123. The Royal Society of Chemistry, Cambridge.

Fierobe, H.-P., Mingardon, F., Mechaly, A., Belaich, A., Rincon, M. T., Lamed, R., Tardif, C., Belaich, J.-P., and Bayer, E. A. (2005). Action of designer cellulosomes on homogeneous versus complex substrates: Controlled incorporation of three distinct enzymes into a defined tri-functional scaffoldin. *J. Biol. Chem.* **280,** 16325–16334.

Ghose, T. K. (1987). Measurements of cellulase activity. *Pure Appl. Chem.* **59,** 257–268.

Haimovitz, R., Barak, Y., Morag, E., Voronov-Goldman, M., Lamed, R., and Bayer, E. A. (2008). Cohesin-dockerin microarray: Diverse specificities between two complementary families of interacting protein modules. *Proteomics* **8,** 968–979.

Lamed, R., Kenig, R., Setter, E., and Bayer, E. A. (1985). Major characteristics of the cellulolytic system of *Clostridium thermocellum* coincide with those of the purified cellulosome. *Enzyme Microb. Technol.* **7,** 37–41.

Miller, G. L. (1959). Use of dinitrosalicylic acid reagent for determination of reducing sugar. *Anal. Biochem.* **31,** 426–428.

Mingardon, F., Chanal, A., López-Contreras, A. M., Dray, C., Bayer, E. A., and Fierobe, H.-P. (2007a). Incorporation of fungal cellulases in bacterial minicellulosomes yields viable, synergistically acting cellulolytic complexes. *Appl. Environ. Microbiol.* **73,** 3822–3832.

Mingardon, F., Chanal, A., Tardif, C., Bayer, E. A., and Fierobe, H.-P. (2007b). Exploration of new geometries in cellulosome-like chimeras. *Appl. Environ. Microbiol.* **73,** 7138–7149.

Moraïs, S., Barak, Y., Caspi, J., Hadar, Y., Lamed, R., Shoham, Y., Wilson, D. B., and Bayer, E. A. (2010a). Cellulase-xylanase synergy in designer cellulosomes for enhanced degradation of a complex cellulosic substrate. *MBio* **1,** e00285–e00310.

Moraïs, S., Barak, Y., Caspi, J., Hadar, Y., Lamed, R., Shoham, Y., Wilson, D. B., and Bayer, E. A. (2010b). Contribution of a xylan-binding module to the degradation of a complex cellulosic substrate by designer cellulosomes. *Appl. Environ. Microbiol.* **76,** 3787–3796.

Shimon, L. J. W., Bayer, E. A., Morag, E., Lamed, R., Yaron, S., Shoham, Y., and Frolow, F. (1997). A cohesin domain from *Clostridium thermocellum:* The crystal structure provides new insights into cellulosome assembly. *Structure* **5,** 381–390.

Spinelli, S., Fierobe, H. P., Belaich, A., Belaich, J. P., Henrissat, B., and Cambillau, C. (2000). Crystal structure of a cohesin module from *Clostridium cellulolyticum*: Implications for dockerin recognition. *J. Mol. Biol.* **304,** 189–200.

Tabka, M. G., Herpoël-Gimbert, I., Monod, F., Asther, M., and Sigoillot, J. C. (2006). Enzymatic saccharification of wheat straw for bioethanol production by a combined cellulase xylanase and feruloyl esterase treatment. *Enzyme Microb. Technol.* **39,** 897–902.

Tavares, G. A., Béguin, P., and Alzari, P. M. (1997). The crystal structure of a type I cohesin domain at 1.7 Å resolution. *J. Mol. Biol.* **273,** 701–713.

Tomme, P., Boraston, A., Kormos, J. M., Warren, R. A., and Kilburn, D. G. (2000). Affinity electrophoresis for the identification and characterization of soluble sugar binding by carbohydrate-binding modules. *Enzyme Microb. Technol.* **27,** 453–458.

Vazana, Y., Moraïs, S., Barak, Y., Lamed, R., and Bayer, E. A. (2010). Interplay between *Clostridium thermocellum* family-48 and family-9 cellulases in the cellulosomal versus noncellulosomal states. *Appl. Environ. Microbiol.* **76,** 3236–3243.

Wood, T. M., and Bhat, K. M. (1988). Methods for measuring cellulases activities. *Methods Enzymol.* **160,** 87–112.

Zhang, Y. H. (2011). Substrate channeling and enzyme complexes for biotechnological applications. *Biotechnol. Adv.* **29,** 715–725.

Zhang, S., Wolfgang, D. E., and Wilson, D. B. (1999). Substrate heterogeneity causes the nonlinear kinetics of insoluble cellulose hydrolysis. *Biotechnol. Bioeng.* **66,** 35–41.

CHAPTER TWENTY-FOUR

HIGH-THROUGHPUT SCREENING OF COHESIN MUTANT LIBRARIES ON CELLULOSE MICROARRAYS

Michal Slutzki,* Vered Ruimy,* Ely Morag,[†] Yoav Barak,[‡] Rachel Haimovitz,* Raphael Lamed,[§] and Edward A. Bayer*

Contents

1. Introduction	454
2. Cohesin Mutants: Library Preparation	456
3. Screening Small Cohesin Libraries	456
4. Screening Large Cohesin Libraries	458
5. Microarray Probing	459
6. DNA Extraction	460
7. Notes	461
8. Summary	461
References	462

Abstract

The specificity of cohesin–dockerin interactions is critically important for the assembly of cellulosomal enzymes into the multienzyme cellulolytic complex (cellulosome). In order to investigate the origins of the observed specificity, a variety of selected amino acid positions at the cohesin–dockerin interface can be subjected to mutagenesis, and a library of mutants can be constructed. In this chapter, we describe a protein–protein microarray technique based on the high affinity of a carbohydrate-binding module (CBM), attached to mutant cohesins. Using cellulose-coated glass slides, libraries of mutants can be screened for binding to complementary partners. The advantages of this tool are that crude cell lysate can be used without additional purification, and the microarray can be used for screening both large libraries as initial scanning for "positive" plates, and for small libraries, wherein individual colonies are printed

---

* Department of Biological Chemistry, The Weizmann Institute of Science, Rehovot, Israel
† Designer Energy, Rehovot Science Park, Rehovot, Israel
‡ Chemical Research Support, The Weizmann Institute of Science, Rehovot, Israel
§ Department of Molecular Microbiology and Biotechnology, Tel Aviv University, Ramat Aviv, Israel

on the slide. Since the time-consuming step of purifying proteins can be circumvented, the approach is also appropriate for providing molecular insight into the multicomponent organization of complex cellulosomes.

ABBREVIATIONS

BSA	bovine serum albumin
CBM	family 3a carbohydrate-binding module from *Clostridium thermocellum*
Coh	cohesin
Doc	dockerin
ELISA	enzyme-linked immunosorbent assay
IPTG	isopropyl-1-thio-β-D-galactoside
ITC	isothermal titration calorimetry
LB	Luria-Bertani Broth
TBS	Tris-buffered saline
WT	wild type
Xyn	xylanase

1. INTRODUCTION

The discovery of cellulosomes, the multienzyme cellulose-degrading systems produced by a variety of anaerobic bacteria, has revealed the enormous repertoires of cohesins and dockerins found in nature (Bayer *et al.*, 2004, 2008). This tenacious type of protein–protein interaction is not limited to cellulosome-producing bacteria but appears to be prevalent in non-cellulosomal proteins produced in the bacterial and archaeal forms of life, as well as in primitive Eukarya (Bayer *et al.*, 1999; Chitayat *et al.*, 2008a,b; Peer *et al.*, 2009). The high-affinity interaction between these two protein modules is responsible for assembling all of the cellulosomal components into a single entity. By deciphering the interaction network between the cohesin and the dockerin, we will gain insight into the origins of the observed specificity. This will contribute to our knowledge of cellulosome organizations and will also enrich significantly our reservoir of specific cohesin–dockerin pairs for biotechnology uses (such as designer cellulosomes; Bayer and Lamed, 1992; Bayer *et al.*, 1994; Ohmiya *et al.*, 2003). Unfortunately, specificity in most cases cannot be reliably predicted from sequences, and even the limited structures of cohesin and dockerin complexes have not led to the design of novel cohesin–dockerin interactions. The best way to address the question regarding specificity determinants in the two proteins is by mutating them. Decisions as to which residue(s) should be mutated can be predicted from existing knowledge (e.g., residues found in the same positions in orthologous proteins) or through the use of prediction software. The mutations can be performed individually, simultaneously into groups, or even an entire spectrum of amino acids ("saturation mutagenesis") at defined positions. Saturation mutagenesis at a single position in a protein sequence requires screening of about 100–200 colonies (Georgescu et al., 2003); for two positions, above 10,000 colonies should be screened. Starting with two mutated positions, automated equipment is necessary, which is expensive, laborious, wasteful, and requires access to specialized instrumentation and appropriate skills.

In this chapter, we present protocols for cellulose microarray technology (Haimovitz et al., 2008; Ofir et al., 2005) with a new approach for high-throughput screening of cohesins to determine their binding specificities to different dockerins. For experimental purposes, cohesins are fused to a microcrystalline cellulose-specific carbohydrate-binding module (CBM; Morag et al., 1995). Consequently, this method can also be implemented without extended purification procedures for cohesins, as affinity purification is implicit in a printing procedure, whereby the CBM binds specifically to the cellulose-coated microarray (Ofir et al., 2005). An additional advantage of this approach is that during the immobilization step the fusion of the interacting protein (cohesin) to the CBM results in its favorable orientation for dockerin binding (Fig. 24.1). This setup differs from the standard ELISA approach, whereby the cohesin is attached to the surface of microtiter plate wells, and its capacity for binding the dockerin is hindered (Friguet et al., 1984).

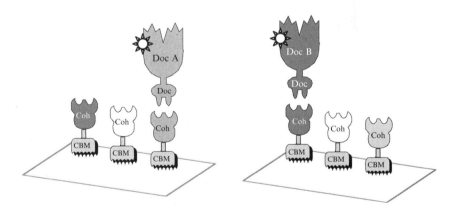

Figure 24.1 Schematic representation of a cellulose microarray experiment. Cohesins interact with the cellulose surface of the slide through their carrier CBM (different cohesins or mutants of the same cohesin are represented in different shades of gray). Different fluorescently labeled dockerins (A and B in this scheme) interact with their complementary cohesin in a specific manner.

The approach exploits the inherent affinity characteristics of the cohesin–dockerin interaction and the targeted interaction of the CBM to cellulose. The protocols involved are simple, economical, and avoid the necessity of robotic instrumentation.

2. Cohesin Mutants: Library Preparation

CBM-fused cohesin (CBM-Coh) and xylanase-bearing dockerin (Xyn-Doc) clones were prepared as described earlier (Barak et al., 2005; Slutzki et al., in press). We performed saturation mutagenesis using sequence overlap extension (SOE; Georgescu et al., 2003). This technique included two PCR cycles: in the first, two DNA fragments, overlapping in the region of mutation, were amplified using mutagenic primers (containing a randomized NNS nucleotide triad, where, by convention, N is any of the four nucleotides and S is either C or G). The two fragments were then used as "mega-primers" to amplify the entire plasmid, which will contain randomized amino acids in the selected position.

3. Screening Small Cohesin Libraries

A small library is defined as one that is possible to screen by printing colonies individually. Each colony represents a cohesin mutant from the library. The number of colonies to be screened can differ accordingly (whether they are printed manually or by using an automated robotic system).

Materials

LB agar plates, supplemented with 50 µg/ml kanamycin
LB medium, supplemented with 50 µg/ml kanamycin and 4 µM isopropyl-1-thio-β-D-galactoside (IPTG)
96-well microtiter plates (Nunc, Roskilde, Denmark)
PopCulture Reagent (Novagen Inc., Darmstadt, Germany), supplemented with 20 µg/ml deoxyribonuclease I (DNase I) from bovine pancreas (Sigma-Aldrich, St. Louis, MO) and 20 µg/ml lysozyme from chicken egg white (Sigma-Aldrich)
Cellulose-coated glass slides (Zephyr ProteomiX, Rehovot, Israel)
Manual MicroCASTer spotter (Schleicher & Schuell, Dassel, Germany)

Protocol

1. *Step 1* (Fig. 24.2A): Transform the plasmids containing the cohesin library into *Escherichia coli* BL21. Plate the cells on LB agar to achieve separate colonies and grow overnight at 37 °C.

(A) Small Coh library (~10^2 colonies)

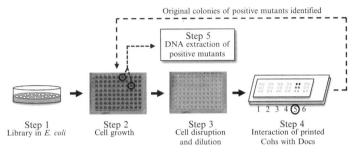

(B) Large Coh library (~10^4 colonies)

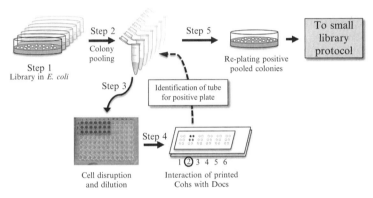

Figure 24.2 Schematic presentation of protocols for screening mutant libraries. (A) Small cohesin library. *Step 1*: Transformed *E. coli* cells harboring a library of cohesin mutants are plated. *Step 2*: Individual colonies (100–200) of cohesin mutants are selected from the LB plate and transferred to LB medium in wells of a 96-well microtiter plate. *Step 3*: Following growth at 37 °C for 2 h, the cells are transferred into a new plate containing PopCulture reagent for cell lysis. *Step 4*: Dilutions of the cell lysates are printed on a cellulose slide (only CBM-containing cohesin will bind to the cellulose) and interacted with the fluorescently labeled dockerin. The appearance of a fluorescent signal provides evidence of cohesin binding to dockerin. *Step 5*: The relevant positive colonies are identified on the microtiter plate in Step 2, and they are propagated further for DNA extraction. (B) Large cohesin mutant library. *Step 1*: Transformed *E. coli* cells harboring a large library (~10,000 colonies) of cohesin mutants are plated into multiple LB plates. *Step 2*: The colonies of each plate are transferred to a tube. *Step 3*: A portion of the cell suspension in each tube is transferred into wells of a microtiter plate containing PopCulture reagent for cell disruption. Each well represents the cell population of the original LB plate. *Step 4*: Dilutions of the cell lysates are printed on a cellulose slide and interacted with the fluorescently labeled dockerin. The appearance of a fluorescent signal indicates that one or more of the colonies in the original plate exhibits cohesin binding to dockerin. *Step 5*: The tubes representing the plates associated with a positive signal are identified and its contents plated. The resultant colonies are then subjected to the protocol in panel A for small cohesin libraries.

2. *Step 2*: Select 100–200 colonies, and transfer individually into wells of a microtiter plate containing LB medium supplemented with kanamycin and IPTG. Save this plate for subsequent DNA extraction (Section 6).
3. Grow for ~2 h at 37 °C.
4. Measure OD of each well at 600 nm.
5. *Step 3*: Transfer 50–100 μl of the culture to new wells, and add 1:10 (v/v) PopCulture mix to the *E. coli* culture.
6. Incubate for 15 min to allow the PopCulture to lyse the bacterial cells.
7. Prepare several 10-fold dilutions.
8. *Step 4*: Print the lysate and its dilutions in duplicate on the cellulose-coated slide with a manual MicroCASTer spotter.
9. It is important to include positive controls (e.g., wild-type cohesins) and negative controls (e.g., divergent cohesins) from other species or nonbinding mutant cohesins. Positive controls are especially important, as they allow orientation of the printed slide. Purified protein can also be used as controls.
10. Proceed to protocol in Section 5 for continuation of Step 4 and interaction of the microarray with a dockerin of choice (Xyn-Doc in the present description).

4. Screening Large Cohesin Libraries

When the number of colonies is too large to be screened individually and when the number of expected positive colonies is significantly smaller than negative ones, it is possible to perform preliminary screening to sort the plates that contain positive colonies from those that do not (see scheme in Fig. 24.2B). In the experimental setup, all of the plates containing a library are first screened to filter out all totally negative plates, and we then proceed only with the positive plates. In the following screening procedure, the population of each plate is pooled and its contents are represented by a single dot (and its dilutions) on the cellulose slide. The binding of the fluorescent dockerin probe indicates the presence of one or more positive mutants in the original plate. The protocol for small library screening (Section 3) is then implemented for positive plates.

Materials

LB agar plates, containing 50 μg/ml kanamycin and 4 μM IPTG
LB medium, supplemented with 50 μg/ml kanamycin
TBS (Tris-buffered saline), pH 7.4
96-well microtiter plates
PopCulture, supplemented with 20 μg/ml DNase I and 20 μg/ml lysozyme
Cellulose-coated glass slides
Manual MicroCASTer spotter

Protocol

1. *Step 1* (Fig. 24.2B): Transform the plasmids containing the cohesin library into *E. coli* BL21. Plate the cells on LB agar to achieve separate colonies and grow overnight at 37 °C. For about 10^4 colonies (e.g., for saturation mutagenesis of two positions in a protein sequence), about 20 plates are required.
2. *Step 2*: Scrape the plates containing the bacterial colonies, and suspend cells from each plate into ∼4 ml of LB medium. Distribute the pooled cell suspension from each plate into two tubes. At least one tube is preserved at 4 °C.
3. *Step 3*: Transfer 50–100 µl of the suspended cells to wells of a microtiter plate, and add 1:10 (v/v) PopCulture mix to the suspension.
4. Incubate for 15 min to lyse cells.
5. Make six to eight serial 10-fold dilutions of the resuspended cells in TBS from each plate.
6. *Step 4*: Starting with the 100-fold dilution well, print in duplicate on the cellulose-coated slide using a manual spotter.
7. As stated above (Section 3), it is important to add positive and negative controls.
8. Proceed to protocol in Section 5 for continuation of Step 4 and interaction of the microarray with a dockerin of choice (Xyn-Doc in the present description).

5. Microarray Probing

In the following protocol, the printed microarray, either from the small or the large cohesin library, is subjected to probing using a fluorescently labeled dockerin derivative (Xyn-Doc).

Materials

Blocking buffer: TBS, supplemented with 10 mM $CaCl_2$, 0.05% Tween 20, and 1% (w/v) BSA

Washing buffer: TBS, 10 mM $CaCl_2$, 0.05% Tween 20

Dockerin (or construct containing dockerin), labeled with Cy3 Mono-Reactive Dye (GE Healthcare Bio-Sciences AB, Uppsala, Sweden) according to the manufacturer's instructions: A 1-ml solution of Xyn-Doc (0.34 mg) in 0.1 M sodium carbonate buffer (pH 9.3) is added to a vial of Cy3 reagent, and the reaction mixture is incubated for 0.5 h at room temperature. The solution is then dialyzed for 8 h against 1 l TBS buffer with two buffer changes.

Protocol

1. Wash printed slide in a suitable container with blocking buffer for 30 min with gentle shaking. The dimensions of the container determine the amount of solution required for this and subsequent steps. In our setup, we use 6 ml of each solution which covers the top of the printed slide by about 0.5–1.0 cm.
2. Incubate the slide with Cy3-fluorescently labeled dockerin for 30 min with shaking. Discard the dockerin solution.
3. Wash three times, 10 min each, with washing buffer, shaking.
4. Dry the slide with a hair dryer.
5. Scan the slide using a fluorescent image analyzer (e.g., FLA-5000, FujiFilm, LPG laser, 532 nm wavelength).
6. *In the case of small libraries*, when a positive signal is obtained for a given colony, proceed to Section 6 (DNA extraction).
7. *In the case of large libraries*, when a positive signal is obtained for a given sample plate, the pooled cell suspension (Section 4) is again plated using several 10-fold dilutions to obtain 200–400 separate colonies per plate (Fig. 24.2B, Step 5). This plate is then screened again according to the protocol for small libraries (Section 3).

6. DNA Extraction

When a specific colony from a small library is identified as positive, the culture (Section 3, Step 2) is inoculated into new medium and the DNA is extracted for sequencing in order to determine the specific mutation(s) that are associated with the desired effect.

Materials

LB medium, supplemented with 50 µg/ml kanamycin
QIAprep Spin Miniprep Kit (QIAgen GmbH, Hilden, Germany)

Protocol

1. *Step 5* (Fig. 24.2A): For each positive mutant, identify its well on the growth plate (Step 2) and transfer its contents into 10 ml of fresh LB media in a 50-ml flask.
2. Grow the culture overnight at 37 °C.
3. Centrifuge the tubes to pellet the cells (at $2500 \times g$ for 15 min).
4. Discard the supernatant.
5. Extract DNA by Miniprep procedure according to manufacturer's instructions.

6. Send the DNA to be sequenced in order to determine the amino acid composition of the mutants that demonstrated the desired binding.

7. NOTES

(1) Long incubation in IPTG should be avoided, since it may be toxic to the cells and inhibit subsequent bacteria growth for DNA extraction.
(2) All the results should be standardized, if there are significant differences in optical densities between the wells.
(3) The spots on the microarray can also be quantified by densitometry, but such data cannot be used for calculations of dissociation constant, as the component with varied concentration is printed on the slide and not in solution.

8. SUMMARY

Screening of protein libraries for binding can be easily performed on cellulose-coated plates via their fusion to a CBM. Specific binding to cellulose provides both an inherent purification step and efficient immobilization of the molecule in a favorable orientation for dockerin recognition. In addition, large libraries (in which the number of positive binders is about two orders of magnitude lower than that of the negatives) can be analyzed using this technique by adding a preliminary screening step. This technology can also be used for other interacting pairs as long as one of the partners is fused to a CBM.

For an example of a small library, the pattern of binding by the mutant cohesins from the screened library is demonstrated in Fig. 24.3. Similarly,

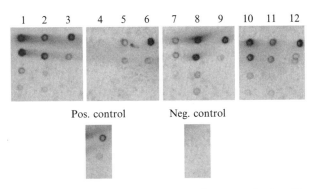

Figure 24.3 Typical microarray, printed from discrete colonies. Dots represent sequential 10-fold dilutions from top to bottom. The strength of binding can be estimated visually from the intensity of the dots.

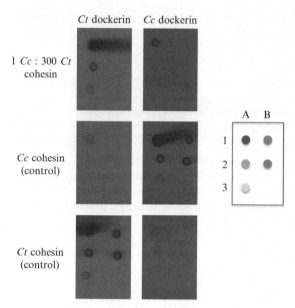

Figure 24.4 Prototype for screening of a large library. A mixture containing a 1:300 ratio of *Clostridium cellulolyticum* (*Cc*) and *Clostridium thermocellum* (*Ct*) CBM-fused cohesins are printed in serial 10-fold dilutions (A1–A3) on cellulose-coated slides. The slide is treated with fluorescently labeled dockerins (Xyn-Docs) from these two species. At this ratio, the presence of *Cc* can be clearly detected. Controls: A1–A3—10-fold dilutions of uncontaminated *Cc* and *Ct* (crude extracts); B1–B2—duplicates, printed from purified proteins.

the library can be screened for binding to several types of dockerins, and their putative preferences for a particular dockerin can be assessed. For large libraries, the ability to distinguish a positive colony from 300 negatives is demonstrated in Fig. 24.4. As can be judged from the intensity of the dots, even smaller ratios of positives to negatives can be revealed. In the subsequent step, positive plates are then subjected to the protocol for small libraries (Section 3).

REFERENCES

Barak, Y., Handelsman, T., Nakar, D., Mechaly, A., Lamed, R., Shoham, Y., and Bayer, E. A. (2005). Matching fusion protein systems for affinity analysis of two interacting families of proteins: The cohesin-dockerin interaction. *J. Mol. Recogn.* **18,** 491–501.

Bayer, E. A., and Lamed, R. (1992). The cellulose paradox: Pollutant *par excellence* and/or a reclaimable natural resource? *Biodegradation* **3,** 171–188.

Bayer, E. A., Morag, E., and Lamed, R. (1994). The cellulosome—A treasure-trove for biotechnology. *Trends Biotechnol.* **12,** 378–386.

Bayer, E. A., Coutinho, P. M., and Henrissat, B. (1999). Cellulosome-like sequences in *Archaeoglobus fulgidus*: An enigmatic vestige of cohesin and dockerin domains. *FEBS Lett.* **463,** 277–280.

Bayer, E. A., Belaich, J.-P., Shoham, Y., and Lamed, R. (2004). The cellulosomes: Multi-enzyme machines for degradation of plant cell wall polysaccharides. *Annu. Rev. Microbiol.* **58,** 521–554.

Bayer, E. A., Lamed, R., White, B. A., and Flint, H. J. (2008). From cellulosomes to cellulosomics. *Chem. Rec.* **8,** 364–377.

Chitayat, S., Adams, J. J., Furness, H. S., Bayer, E. A., and Smith, S. P. (2008a). The solution structure of the C-terminal modular pair from *Clostridium perfringens* reveals a non-cellulosomal dockerin module. *J. Mol. Biol.* **381,** 1202–1212.

Chitayat, S., Gregg, K., Adams, J. J., Ficko-Blean, E., Bayer, E. A., Boraston, A. B., and Smith, S. P. (2008b). Three-dimensional structure of a putative non-cellulosomal cohesin module from a *Clostridium perfringens* family 84 glycoside hydrolase. *J. Mol. Biol.* **375,** 20–28.

Friguet, B., Djavadi-Ohaniance, L., and Goldberg, M. E. (1984). Some monoclonal antibodies raised with a native protein bind preferentially to the denatured antigen. *Mol. Immunol.* **21,** 673–677.

Georgescu, R., Bandara, G., and Sun, L. (2003). Saturation mutagenesis. *In* "Directed Evolution Library Creation: Methods and Protocols," (F. H. Arnold and G. Georgiou, eds.), .Humana Press Inc., Totowa, NJ.

Haimovitz, R., Barak, Y., Morag, E., Voronov-Goldman, M., Shoham, Y., Lamed, R., and Bayer, E. A. (2008). Cohesin-dockerin microarray: Diverse specificities between two complementary families of interacting protein modules. *Proteomics* **8,** 968–979.

Morag, E., Lapidot, A., Govorko, D., Lamed, R., Wilchek, M., Bayer, E. A., and Shoham, Y. (1995). Expression, purification and characterization of the cellulose-binding domain of the scaffoldin subunit from the cellulosome of *Clostridium thermocellum*. *Appl. Environ. Microbiol.* **61,** 1980–1986.

Ofir, K., Berdichevsky, Y., Benhar, I., Azriel-Rosenfeld, R., Lamed, R., Barak, Y., Bayer, E. A., and Morag, E. (2005). Versatile protein microarray based on carbohydrate-binding modules. *Proteomics* **5,** 1806–1814.

Ohmiya, K., Sakka, K., Kimura, T., and Morimoto, K. (2003). Application of microbial genes to recalcitrant biomass utilization and environmental conservation. *J. Biosci. Bioeng.* **95,** 549–561.

Peer, A., Smith, S. P., Bayer, E. A., Lamed, R., and Borovok, I. (2009). Non-cellulosomal cohesin and dockerin-like modules in the three domains of life. *FEMS Microbiol. Lett.* **291,** 1–16.

Slutzki, M., Barak, Y., Reshef, D., Schueler-Furman, O., Lamed, R., and Bayer, E. A. (in press). Measurements of relative binding of cohesin and dockerin mutants using an advanced ELISA technique for high-affinity interactions. *Methods Enzymol.* Chapter 22.

Author Index

Note: Page numbers followed by "*f*" indicate figures, and "*t*" indicate tables.

A

Aam, B. B., 78–79, 79*f*
Abad-Zapatero, C., 150
Abbate, E., 122–123
Abbott, D. W., 211–231
Abou Hachem, M., 217–218
Aboushadi, N., 364–365
Abuja, P. M., 170–171, 184–186, 192
Achouak, W., 367
Acinas, S. G., 386
Adams, J. J., 143–144, 184–186, 396, 420–421, 430, 454–455
Adams, P. D., 409–412
Adney, W. S., 116, 184–186, 204–205
Aebersold, R., 155–156
Aehle, W., 143–144
Aerts, G., 114–115
Afonine, P. V., 410–412
Aghajari, N., 70, 192, 200, 201
Aguilar, C., 276
Akhmetova, L., 234–235
Alam, M. M., 224, 226*f*, 228
Alapuranen, M., 83–85
Alber, O., 195–196, 430
Albersheim, P., 98, 99*f*, 103, 105–106, 107, 111–112, 116, 123, 124, 126, 129, 131, 133
Ali, S., 184–186
Allison, H. E., 351, 352, 362–365, 367, 376, 382, 385
Allouch, J., 196–197
Alm, E. J., 382
Altschul, S. F., 383
Alves, V. D., 395–415
Alzari, P. M., 160–161, 184–186, 196–197, 419, 439–440
Amaya, M. F., 160–161
Amin, N., 263
Ammiraju, J. S., 379–380
Ammons, J. T., 385–386
Anbar, M., 261–271
Andersen, G. L., 382–383
Andersen, L. N., 123, 131
Andersen, N., 59
Anderson, C. T., 234–235
Ando, S., 350–351
Ando, T., 70, 80, 85, 89–90, 169–182
Andreu, J. M., 190
Anthon, G. E., 43–44, 109

Anthonsen, A., 38–39, 52–53, 54, 59–60, 70, 122–123
Anthonsen, M. W., 76
Antonio Gavira, J., 149–150
Antranikian, G., 350–351
Appel, R. D., 215–216
Arai, M., 87–88
Araki, J., 171
Archer, D. B., 363–364
Argyros, D. A., 318, 322–327, 328
Ariza, A., 99–100, 106–107, 108, 111–112, 114–115, 123, 157–159
Armand, S., 54
Arvas, M., 376
Asai, T., 333
Asel, E., 146
Asenjo, J. A., 449
Ashford, D., 46
Asther, M., 38, 447–448
Aten, J., 157
Attwell, R. W., 366–367
Aubert, J. P., 396
Auch, A. F., 382–383
Ausili, A., 276
Ausio, J., 217–218
Aux, G., 39
Awano, T., 98
Ayame, S., 363–364
Aylward, F. O., 350–351
Azin, M., 361–362
Azriel-Rosenfeld, R., 455–456

B

Baasov, T., 278–279
Babé, L., 263
Bacwaden, J., 39
Bahl, H., 302–303
Bai, X. L., 376–377, 384
Bai, Z. H., 88
Baik, K. S., 384–385
Bailey, M. J., 377–378
Bairoch, A., 215–216
Baker, D., 146
Baker, S. E., 376
Balan, V., 39
Balesdent, J., 367
Ball, A. S., 354, 362
Ball, C. L., 362

Ballard, C. C., 408
Banasiak, A., 98, 115
Bandara, G., 454–455, 456
Bandela, A. M., 382–383
Bansal, P., 87
Baraberato, C., 190–191, 197–200
Barak, Y., 417–428, 429–452, 453–463
Barakate, A., 38, 39–40, 43–44
Baranauskiene, L., 224
Bardavid, R. E., 350–351
Barlaz, M. A., 363–364
Barllori, X., 278–279
Barr, B. K., 75, 249
Barrett, D. M., 43–44, 109
Barrett, T. F., 318, 322–327, 328
Barron, A. E., 382, 385
Bartelmes, G., 376
Barton, C. J., 53–54, 58
Bartonek-Roxa, E., 217–218
Bashford, D., 197–200
Basle, A., 228–229
Basler, J., 263
Bathgate, G. N., 5–6
Batt, S. B., 100
Bauchop, T., 360–361
Bauer, S., 123, 131
Baumann, M. J., 87, 100, 101–102, 103–105, 106–107, 108, 111–112, 113, 114–115
Bauzan, M., 312, 313–314, 313f
Bayer, E. A., 52–53, 70, 90, 184–186, 194–196, 200, 201–203, 204–205, 204f, 247–259, 261–271, 314, 318, 350–351, 396, 397, 401, 402, 417–428, 429–452, 453–463
Beadle, I. R., 364–365, 376
Beane, J., 361
Becker, D., 70, 74, 89, 170–171
Beckham, G. T., 70, 90, 116, 203–205
Bedini, E., 274–276, 280t, 285, 288, 289–292, 293, 294, 295
Beever, D. E., 360–361
Begley, T., 382–383
Béguin, P., 184–186, 196–197, 396, 419, 439–440
Behe, M., 20
Belaich, A., 70, 184–186, 194–196, 249–250, 302–303, 304, 304f, 306, 307–310, 309f, 311–312, 397, 418–419, 420–421, 430, 439–440, 444, 445, 447–448
Belaich, J.-P., 70, 155, 184–186, 194–196, 248, 249–250, 303, 304, 304f, 306, 307–310, 309f, 311–312, 396, 397, 418–419, 420–421, 430, 439–440, 444, 445, 447–448, 454–455
Belakhov, V., 278–279
Beldman, G., 98, 99–100, 99f, 111–112
Belli, M. L., 365
Bellott, M., 197–200
Beloqui, A., 379–380

Ben Jelloul, M., 147–149
Benelli, D., 285
Benhar, I., 455–456
Benians, T. A. S., 236, 237f
Benjaminson, M. A., 234–235
Benkö, Z., 99–100
Bennet, A. J., 212
Benoit, L., 363–364
Benvenuti, M., 147
Beppu, T., 363–364
Berchtold, M., 365
Bercovici, J., 376–377
Berdichevsky, Y., 455–456
Berg, J., 112–113
Berg, R. H., 234–235
Berge, O., 367
Bergenstrahle, M., 203–204
Bergfors, T. M., 70, 71f, 144–145, 147, 150–152
Berglund, H., 144, 146
Bergmann, A., 190
Bergquist, P. L., 217–218
Bergstrom, G. C., 38, 39
Berka, R. M., 376
Berkelman, T., 345
Bernadac, A., 311–312
Bernalier-Donadille, A., 365–366
Bernard, T., 52–53, 70, 122–123, 142–144, 213t, 274–276, 383
Bernardo, P., 190, 200
Bernfeld, P., 109
Berrin, J.-G., 87
Berry, I., 147–149
Bertani, G., 334–335
Bertilsson, S., 386
Bertoli, E., 276
Bertozzi, C. R., 157
Besenmatter, W., 99–100, 106–107, 108, 111–112, 114–115, 123, 157–159
Bessueille, L., 107
Betian, H. G., 365–366
Beylot, M. H., 333–335, 337–338, 339, 341
Bhat, K., 265
Bhat, K. M., 449
Bhikhabhai, R., 143–144, 172–173, 184–186
Biely, P., 10–11, 112–113
Bignon, C., 38, 285, 292
Birbir, M., 350–351
Bird, P., 155–156
Bjornvad, M. E., 123
Bjørnvad, M. E., 100, 102f, 104f, 106–107, 111–112, 116f, 158–159
Black, G. W., 196–197, 333–334
Blackall, L. L., 363–364
Blackburn, A. R., 99, 99f, 105–107, 123, 124, 133
Blackburn, M. C., 382
Blackburn, R. S., 236, 237f
Blackshields, G., 254
Blake, A. W., 234–235, 237f, 241

Author Index

Blakeney, A. B., 127–129
Blanch, H. W., 72–73, 74, 376–377
Bleijlevens, B., 157
Bloch, P. L., 334–335
Blouzard, J. C., 194
Blow, D. M., 149–150
Blum, D. L., 196–197
Bobrov, K. S., 292–293
Boden, R., 378–379
Boer, H., 70, 74, 89, 170–171
Boerjan, W., 332
Bohn, A., 52–53
Bohringer, M., 305
Boisset, C., 63, 70, 75, 172–173
Bojarová, P., 274–276
Bolam, D. N., 87, 143–144, 196–197, 212, 217–218, 223–224, 228–229, 234–235, 237f, 241
Bolewski, J. C., 88
Bolivar, J. L., 38, 39
Bolognesi, M., 285
Bomble, Y. J., 70, 90, 203–205
Bommarius, A. S., 87, 263
Bonfante-Fasolo, P., 234–235
Boone, D. R., 363–364
Boraston, A. B., 87, 143–144, 145f, 151f, 211–231, 234–235, 237f, 241, 443, 454–455
Boraston, C. M., 145f
Borch, K., 87, 88, 90
Borchert, T. V., 53, 100, 102f, 104f, 106–107, 111–112, 116f, 123, 158–159
Borjesson, J., 59–60, 62, 63
Bornet, O., 184–186, 196–197
Borovok, I., 454–455
Bose, R. G., 352, 363
Bothwell, M. K., 88
Bott, R., 143–144
Bourne, Y., 285, 292
Bourquin, V., 98
Bowles, D. J., 160
Boyer, V., 70, 71f, 155
Boyum, J., 350–351
Brady, J. W., 116, 204–205
Brandl, M., 220–221
Brando, T., 53
Brandts, J. F., 224, 225–227
Brás, J. L. A., 395–415
Brauman, A., 365
Bray, M. R., 215–216
Brennan, Y., 54
Breslauer, K. J., 223–224
Breton, C. I. G., 87–88
Brewer, C. F., 223–224
Broda, P., 354, 362
Brodie, E. L., 382–383
Bromley, J. R., 51–67
Brooks, B. R., 197–200
Brooks, C. L. I., 204–205

Brown, D. M., 53–54, 55–56, 58
Brown, E. L., 385–386
Brown, K. A., 190–191, 201
Brown, N. P., 254
Brown, R. D. Jr., 234–235
Bruccoleri, R. E., 197–200
Brumer, H. III., 97–120, 123, 131–135, 144, 157–159, 292–293
Brumm, P., 39
Brunecky, R., 39–40
Bryant, M. P., 355–356, 363–364, 365–366
Bryder, M. J., 361
Brzozowski, A. M., 147–149, 154, 155–156
Bu, L., 203–204
Buckeridge, M. S., 98
Bulone, V., 107, 122–123, 274–276
Bunæs, A. C., 70
Bunkoczi, G., 412
Burke, E., 39
Burkey, L. A., 355
Burrell, P. C., 363–364
Buschiazzo, A., 160–161
Busse-Wicher, M., 51–67
Butler, R. M., 382–383

C

Caiazza, N. C., 318, 322–327, 328
Cailliez, C., 363–364
Callebaut, I., 152
Callen, N. C., 54
Callen, W., 292
Calli, B., 350–351
Cambillau, C., 197, 439–440
Campbell, N. L., 361
Canaleparola, E., 361
Canale-Parola, E., 194, 361
Canals, F., 278–279
Canevascini, G., 2
Cantaert, T., 100, 107, 112
Cantarel, B. L., 52–53, 70, 122–123, 142–144, 213t, 274–276, 376–377, 383
Cantu, S. L., 38, 39
Cantwell, B. J., 383–384
Cao, H., 376–377
Capalbo, L., 276
Cardenas, E., 382–383
Carillo, S., 274–276, 280t, 292, 293, 294, 295
Carnicero, A., 354
Carpita, N. C., 122, 127–130
Carrard, G., 88–89
Carroll, A., 234–235
Carter, C. W., 147–149
Carter, G. P., 302–303
Cartman, S. T., 302–303, 314–315
Cartmell, A., 155–156
Carvalho, A. L., 184–186, 194, 196–197, 376, 395–415, 419

Carver, T., 146
Casalot, L., 303, 304, 304f, 306, 307, 308, 312
Cash, M., 99, 99f, 105–107, 123, 124, 133
Caspi, J., 431–432, 440, 442, 444, 445, 446f
Cavalier, D. M., 101–102
Cayouette, M., 364–365
Cederkvist, J. B., 78–79, 80f
Chacon, P., 190
Chaffotte, A. F., 420
Chai, B., 382–383
Champ, S., 311–312, 419, 420–421
Chan, V. J., 100
Chanal, A., 301–316, 441, 444
Chang, D. C., 318
Chanliaud, E., 116
Chanzy, H., 52–53, 63, 70, 75, 170
Chaparro-Riggers, J. F., 263
Chaplin, J. A., 379–380
Chapman, J., 376
Charnock, S. J., 143–144
Chassard, C., 365–366
Chassy, B. M., 318
Chauvaux, S., 184–186, 196–197
Chayen, N. E., 149–152
Chebrou, M. C., 196–197
Chen, E. M. W., 53
Chen, G., 216f, 384–385
Chen, H., 376
Chen, J., 379–380
Chen, L., 4
Chen, V. B., 412
Chen, X., 383
Chen, Y.-C., 360–361, 367
Cheng, H.-L., 360–361
Cheng, M., 99–100, 116, 123
Cheng, T.-Y., 360–361
Chenna, R., 254
Chernikova, T. N., 376–377
Cherry, J., 70, 99–100
Chertkov, O., 350–351, 376
Chervenak, M. C., 225
Chi, E., 379–380
Chi, Y. I., 277–278
Chiba, S., 296
Chidthaisong, A., 361
Chien, C.-Y., 360–361
Chin, R., 263
Chirino, A. J., 184–186, 248, 249–250
Chisholm, S. W., 382, 385–386, 388
Chitayat, S., 454–455
Chiu, M. L., 150
Chiu, P., 217–218, 221–222, 221f, 223
Choi, S. K., 376, 397
Chokhawala, H. A., 376–377
Choudhuri, J. V., 382–383
Chow, Y., 86–87
Chowdhry, B. Z., 258
Christen, R., 367

Christopoulos, A., 425
Chuang, H. Y., 382–383
Chundawat, S. P. S., 39, 116
Ciaramella, M., 274, 276, 277, 278, 279, 280–283, 280t, 283t, 284t
Cid, M., 226f, 227–228
Ciucanu, I., 130
Claassen, P. A., 302–303
Claeyssens, M., 70, 71f, 75, 112–113, 114–115, 143–144, 155, 170–171, 172–173, 184–186, 192, 212
Clark, D. S., 72–73, 74, 376–377
Clark, M. E., 376–377
Clarke, J. D., 39
Clarke, W. P., 363–364
Cleary, J. M., 204–205
Clinch, K., 159–160
Cobucci-Ponzano, B., 273–300
Cohen, F. E., 190
Cohen, S. X., 410–412
Cohoon, M., 382–383
Colbert, T., 39
Cole, J. R., 382–383
Coleman, M. L., 385–386, 388
Collen, A., 212
Colman, P. M., 4
Colwell, R. R., 361–362
Comai, L., 39
Comfort, D. A., 274–276, 280t, 292–293, 294, 295
Condouret, P., 53
Connolly, J., 352, 376–377, 382, 383
Conrad, R., 361, 367
Conte, F., 274–276, 280t, 285, 288, 289–292
Cooksley, C. M., 302–303, 314–315
Cooper, A., 225–227
Correia, M. A., 212, 228–229
Corsaro, M. M., 274–276, 280t, 285, 288, 289–292, 293, 294, 295
Cosgrove, D. J., 98
Costello, C. E., 130, 135–137
Cotta, M. A., 302–303
Cottaz, S., 70, 71f, 100, 106–107, 108, 115
Cotton, D., 39
Couger, M. B., 360–361
Coughlan, M., 396
Coutinho, P. M., 52–53, 70, 122–123, 142–144, 194, 213t, 226f, 227–229, 274–276, 350–351, 376–377, 383, 454–455
Couturier, M., 38
Cowtan, K. D., 408, 410–412
Cox, M. J., 351, 352, 362–363, 364–365, 367, 376–377, 378–379, 382, 383, 385
Crawford, D. L., 362
Crawford, R. L., 362
Creagh, A. L., 217, 224, 225–227, 226f, 228
Cremona, M. L., 160–161
Cresswell, M. A., 366–367

Author Index

Croux, C., 302–303, 311
Crowley, M. F., 70, 90, 204–205
Cruys-Bagger, N., 87
Cudney, B., 150
Cudney, R., 150
Cuevas, W., 143–144
Cui, J. B., 55–56, 59, 82–83
Cui, Z. L., 376–377
Cullen, D., 376
Cummings, S. P., 333–334
Currie, C. R., 350–351
Currie, M. A., 184–186
Czjzek, M., 100, 101–102, 106–107, 108, 113, 183–210, 249

D

da Silva, G. G. D., 38
Dai, Z., 305
Dal Piaz, F., 274–276, 280t, 285, 288, 289–292
Dale, B. E., 116, 332–333
Dale, D. E., 39
Dalevi, D., 382–383
Dam, T. K., 223–224
Damgaard, H. D., 87
Danchin, E. G., 376
Daniel, G., 235
Daniel, R., 378–380
Danielsen, S., 100, 102f, 104f, 106–107, 111–112, 116f, 123, 158–159
Darbon, H., 184–186, 196–197
D'Arcy, A., 144–145, 149–152
Darvill, A. G., 98, 99, 99f, 103, 105–107, 111–112, 116, 123, 124, 126, 129, 131, 133
Dauter, M., 154, 155–156
Dauter, Z., 59–60, 161–162, 184–186, 196–197
David, G., 187
Davies, D. R., 350–351, 354–355, 360–361
Davies, G. J., 59–60, 70, 71f, 75, 87, 99–100, 103–105, 106–107, 108, 111–112, 114–115, 123, 141–168, 184–186, 194, 196–197, 212, 223–224, 234–235, 396, 397, 401, 402, 418–419
Davis, B. G., 144, 160
Davis, I. W., 412
Day, J. S., 150
de Crecy-Lagard, V., 382–383
De Gori, R., 154
de Jong, L. E., 99, 99f, 103, 105–106
de Lorenzo, V., 338–339
de Menezes, A. B., 351, 352, 362–363, 364–365, 367, 376
de Philip, P., 194
de Ramon, E., 278–279
De Silva, J., 116
de Vos, W. M., 277–278, 302–303
de Vries, E. J., 379–380
Deblaere, R., 342

DeBolt, S., 122–123
Decker, S. R., 39–40, 203–204
Deery, M. J., 53–54, 58
Deevong, P., 364–365
Dekker, N., 146
DelaFuente, G., 354
Del'homme, C., 365–366
Delmas, E., 365–366
DeLong, E. F., 385–388, 387f
Demain, A. L., 194, 248, 318–319
Demedts, B., 332
Dempsey, M. J., 366–367
Deneke, J., 350–351
Denman, S. E., 106–107
der Oost, J., 277–278
Derba-Maceluch, M., 98
Derewenda, Z. S., 160–161
Desai, S. G., 318, 361
DeSantis, G., 379–380
DeSantis, T. Z., 382–383
Deshpande, M. V., 2
Desmet, T., 100, 107, 112
Desnos, T., 53
Desprez, T., 53
DeTitta, G. T., 146
Di Lauro, B., 276
Dias, F. M. V., 184–186, 194, 196–197, 376, 396, 397, 402, 419
Diaz, J. F., 190
Diaz, N., 382–383
Dickerman, J. M., 352–353
Dideberg, O., 75
Dieterich, C., 376
Dietrich, B., 39
Dijkstra, B. W., 76
Dill, K. A., 225–227
Ding, S.-Y., 116, 122–123, 249
Ding, Y. F., 361
Disz, T., 382–383
Divne, C., 70, 74, 89, 170–171, 184–186
Djavadi-Ohaniance, L., 420, 455–456
Djordjevic, G., 364–365
Dobson, R. C., 187
Dodson, E. J., 144, 408
Dodson, G. G., 59–60, 184–186
Doi, R. H., 194, 248
Dolphin, D. H., 155–156
Dominguez, R., 184–186, 196–197
Domon, B., 135–137
Domozych, D. S., 98
Donati, C., 385
Doner, L. W., 75
Dong, A., 147–149
Dong, H., 305
Doniach, S., 190
Donker-Koopman, W. E., 157
Dore, J., 376–377
dos Santos, H. P., 98

Dower, W. J., 318
Doyle, J. D., 361–362
Draga, A., 1–17
Drakenberg, T., 70, 73, 78–79, 88, 133–135
Dray, C., 444
Driguez, H., 70, 71f, 74, 75, 89, 100, 106–107, 108, 115, 154, 155, 170–171, 195, 274–276
Drinkwater, C., 350–351
Drummond, C. J., 150
D'Souza, M., 382–383
Duan, C. J., 376–377, 379–380, 384
Dubois, M., 367, 368
Dubrovsky, G., 318, 322–325
Ducros, V. M.-A., 151f, 154, 155–156, 184–186
Dumon, C, 212
Dunbrack, R. L. J., 197–200
Duncan, S. H., 365–366
Dunman, P. M., 385–386
Dunstan, P. K., 382
Dupont, C., 155–156
Dupree, P., 51–67
Dupuy, N. F., 302–303
Durand, D., 190–191
Durany, O., 278–279
Dürre, P., 305
Duvaud, S., 215–216
Dykes, G. A., 352–353
Dyson, H. J., 201–203

E

Eady, N. A., 190–191, 201
Eaton, J. T., 192, 200, 201
Eberhard, S., 123
Echols, N., 412
Eck, J., 376
Edgcomb, S. P., 225–227
Edward, A. M., 147–149
Edwards, C., 352, 376–377, 382, 383
Edwards, J. L., 352, 376–377, 382, 383
Edwards, R. A., 382–383, 388, 389
Eftink, M. R., 223
Egan, D., 147–149
Eggins, H. O. W., 363
Ehsaan, M., 302–303, 314–315
Eijsink, V. G. H., 69–95
Eirin-Lopez, J. M., 212
Eisen, J. A., 382–383
Eklöf, J. M., 97–120, 123, 157–159
Elander, R., 332–333
Elberson, M. A., 361–362
Elborough, K., 376–377
Elmerdahl, J., 87, 88, 90
Elmore, C., 149–150
Elshahed, M. S., 360–361
Elzer, P. H., 336–337
Emami, K., 333–335, 337–339, 341–343
Emsley, P., 408, 410–412

Engel, K. C., 335
Engstrom, A., 170
Erdos, G. W., 234–235
Ericsson, U. B., 146
Eriksson, K.-E. L., 2, 172–173
Eriksson, L., 30
Eriksson, T., 212
Escovar-Kousen, J., 70, 74–75
Esparis, A., 362
Estabrook, M., 263
Esterbauer, H., 170–171, 184–186
Evans, G. B., 159–160
Evans, P. R., 408
Evenseck, J. D., 197–200

F

Fabrega, S., 152
Fabregas, J., 362
Faccio, A., 234–235
Fagard, M., 53
Fairweather, J. K., 274–276
Falbel, T. G., 38, 39
Falcon, M. A., 354
Falkow, S., 385
Farkaš, V., 106–107, 112–113, 241
Farmer, J., 32
Farris, R. J., 382–383
Faure, R., 100, 106–107, 108, 115
Fazio, V. J., 147–149
Febbraio, F., 276
Feinberg, L. F., 318, 325–327
Feller, G., 192, 200, 201
Feng, J. X., 376–377, 379–380, 384
Feng, Y., 376–377, 379–380, 384
Ferreira, L. M. A., 100, 103–105, 106–107, 111–112, 184–186, 194, 196–197, 337–339, 341–343, 395–415, 419
Ferrer, M., 376–377, 379–380
Ficher, S., 197–200
Ficko-Blean, E., 212, 215–216, 454–455
Field, D., 388, 389
Field, M. J., 197–200
Fierobe, H.-P., 183–210, 249, 301–316, 396, 401, 402, 418–419, 420–421, 430, 432, 439–440, 441, 444, 445, 447–448
Figurski, D. H., 337–338
Fincher, G. B., 4, 154, 274–276
Finet, S., 195–196, 200, 201–203, 204f, 249
Fink, H.-P., 52–53
Finn, R., 212, 217–218
Firbank, S. J., 212
Fischer, C. J., 86–87
Fischle, W., 258
Flagiello, A., 285
Flanagan, B. M., 352–353
Fleites, C. M., 396, 401, 402
Flint, H. J., 248, 365–366, 454–455

Author Index

Flint, J. E., 144, 212, 228–229, 235, 237f, 241
Florea, B. I., 157
Flors, A., 363–364
Foden, J. M., 318, 325–327
Fonstein, M., 382–383
Fontes, C. M. G. A., 100, 103–105, 106–107, 111–112, 155–156, 184–186, 194, 196–197, 228–229, 248, 337–339, 341–343, 376, 395–415, 419
Fontesm, C. M., 194, 418–419
Ford, C. M., 160
Forsberg, C. W., 302–303
Forsberg, Z., 70
Forster, F., 190–191
Forster, R., 389–390
Foster, S. D., 382
Foust, T. D., 116
Fouts, D. E., 382–383
Fox, J. M., 72–73, 74
Frandsen, T. B., 155, 192, 200, 201
Frandsen, T. P., 70, 71f, 155
Frank, E. D., 382–383
Frasch, A. C., 160–161
Fraschini, C., 63, 70, 75
Freelove, A. C., 217–218, 228–229
Freire, E., 223–224, 253–254, 420
French, D., 70
Frey, G., 54
Frias-Lopez, J., 385–386, 388
Friguet, B., 420, 455–456
Friis, E. P., 99–100, 106–107, 108, 111–112, 114–115, 123, 157–159
Fritsch, E. F., 322, 341
Fritz, G., 190
Frolow, F., 184–186, 195–196, 247–259, 430, 439–440
Fruchard, S., 154, 155
Fry, E. H., 150
Fry, S. C., 53, 99f, 101–102, 107, 112–113, 123
Fuglsang, C. C., 53
Fujii, K., 363
Fujimoto, E. K., 109
Fujimoto, Z., 143–144
Fukuma, T., 171
Funada, R., 98
Furness, H. S., 454–455

G

Gabor, E. M., 376, 379–380
Gal, L., 195, 274–276, 311–312
Galili, U., 295, 296
Gambacorta, A., 276
Gao, D., 39
Gao, J., 197–200
Gao, P., 362
Gao, W., 249
Gao, X., 398

Garcia, J. L., 365
Garcia-Alvarez, B., 376
Garcia-Ruiz, J. M., 149–150
Gardner, J. G., 331–347
Garman, E. F., 161
Garrett, T. P. J., 4
Garrity, G. M., 382–383
Gartner, F. H., 109
Gåseidnes, S., 76, 78, 79f
Gasteiger, E., 215–216
Gattiker, A., 215–216
Gattlen, C., 2
Gaudin, C., 184–186, 194, 195, 311–312
Gautier, C., 386
Gavira, J. A., 149–150
Gay, P., 345
Gelfand, D., 59–60
Gelhaye, E., 363–364
Gelman, R. A., 353
Georgescu, R., 454–455, 456
Gerday, C., 192, 200, 201
Gerngross, U. T., 194, 248
Ghassemian, M., 364–365
Ghazi, A., 379–380
Ghinet, M. G., 212
Ghose, T. K., 450
Ghose, T. M., 8–9
Giannone, R. J., 318, 322–325
Gibbs, M. D., 217–218
Gibeaut, D. M., 122
Gibson, D. M., 38, 39
Gibson, T. J., 254
Gidley, M. J., 116, 352–353
Gilary, H., 444
Gilbert, H. J., 38–39, 53–54, 87, 99–100, 107, 112–113, 122–123, 143–144, 155–156, 157–158, 184–186, 194, 196–197, 212, 217–218, 223–224, 228–229, 234–235, 237f, 241, 248, 332, 333–335, 337–339, 341–343, 350–351, 395–415, 418–419
Gilbert, J. A., 388, 389
Gilbert, M. J., 419
Gilkes, N. R., 184–186, 215–216, 228
Gill, M., 360–361
Giller, K. E., 342
Gilles, K. A., 367, 368
Gilna, P., 388, 389
Giordano, A., 276, 277–278, 279, 280–283, 280t, 282t, 283t, 284t, 285
Girbal, L., 302–303, 311
Gitton, J., 363–364
Glass, E. M., 382–383
Glatter, O., 190
Gloster, T. M., 100, 102f, 104f, 106–107, 111–112, 116f, 123, 143–144, 145f, 157–159, 212
Goeke, N. M., 109
Goethals, K., 342

Goicoechea, J. L., 379–380
Goldberg, M. E., 420, 455–456
Golyshina, O. V., 376–377, 379–380
Gomez, L. D., 37–50
Goodacre, R., 53–54, 58
Goodwin, L. A., 350–351
Gorman, M. A., 187
Gorrec, F., 150
Goubet, F., 53–54, 58, 236, 237f
Gouet, P., 192, 200, 201
Goumy, V., 365–366
Govindarajan, S., 398
Govorko, D., 420–421, 455–456
Gowda, K., 39
Graf, E., 146
Graham, J. E., 376–377
Grasdalen, H., 76
Graycar, T., 263
Greenwood, D. J., 385–386
Greenwood, J. M., 215
Greffe, L., 100, 107, 114–115
Gregg, K. J., 212, 454–455
Grepinet, O., 196–197
Griffith, G. W., 360–361
Griffiths, R. I., 377–378
Grigoriev, I. V., 376
Grishutin, S. G., 100, 104f
Gritzali, M., 234–235
Grondin, G. G., 212
Gross, C. A., 342–343, 344
Gross, L., 100, 107, 112
Grosse-Kunstleve, R. W., 409–412
Gruno, M., 85
Gualfetti, P., 100, 107, 112, 263
Guazzaroni, M. E., 379–380
Guerinot, M. L., 361
Guerreiro, C. I., 100, 103–105, 106–107, 111–112
Guerreiro, C. I. P. D., 155–156
Guillen, R., 105, 107, 111–112
Gulari, E., 398
Gull, K., 350–351, 354–355
Gullfot, F., 106–107
Gunn, N. J., 187
Guo, H., 197–200
Gusakov, A. V., 100, 104f
Guss, A. M., 318, 322–325
Gustafsson, C., 398

H

Ha, S., 197–200
Haas, D., 336–337
Haber, A., 149–150
Hada, N., 295
Hadar, Y., 431–432, 440, 442, 444, 445, 446f
Hagglund, P., 212
Haichar, F. E. Z., 367

Haimovitz, R., 430–431, 444, 453–463
Hall, J., 196–197
Hall, M., 87
Hallberg, B. M., 146
Halpern, A. L., 382–383
Halpin, C., 38, 39–40, 43–44
Ham, R. K., 363–364
Hamilton, J. K., 367, 368
Hammel, M., 195–196, 200, 201–203, 204–205, 204f, 249
Han, B. R., 379–380
Han, M. J., 87–88
Hancock, S. M., 274–276
Hand, G., 278–279
Handelsman, T., 418–419, 420–421, 422, 426–427, 434, 435, 435f, 456
Haney, S., 385–386
Hansen, P. I., 204–205
Hansson, H., 143–144
Harjunpää, V., 70, 73, 78–79, 88, 133–135
Harle, R., 70, 74, 89, 170–171
Harper, A., 99, 99f, 105–107, 123, 124, 133
Harrington, J., 2
Harris, D., 122–123
Harris, J. M., 292–293
Harris, M., 70, 74, 89, 170–171
Harris, P. J., 59–60, 62, 63, 99, 99f, 103, 105–107, 127–129
Harvey, D. J., 103, 105
Harvey, L. K., 105, 107, 111–112
Haser, R., 70, 155, 184–186, 192, 195, 200, 201
Hastrup, S., 70, 75
Hattori, S., 364–365
Hauser, L., 350–351
Havukainen, R., 157
Hawkins, R., 99–100
Hayashi, T., 98, 99f, 123
Haynes, C. A., 215–216, 217, 224, 225–227, 226f, 228
Hazlewood, G. P., 54, 184–186, 194, 196–197, 217–218, 228–229, 248
He, M., 389–390
He, S., 285
He, Y. L., 361
Headd, J. J., 412
Healey, M., 212, 215–216
Healey, S., 54
Heap, J. T., 302–303, 314–315
Heckman, M. P., 194, 420–421
Heidelberg, J. F., 382–383
Heightman, T. D., 154
Heimann, M., 332–333
Heinze, T., 135
Helbert, W., 75
Helinski, D. R., 337–338
Hendricks, C. W., 361–362
Henikoff, S., 39
Henriksson, H., 106–107

Author Index

Henrissat, B., 52–53, 63, 70, 75, 100, 114–115, 122–123, 142–144, 152, 154, 155, 184–186, 192–194, 196–197, 200, 201, 213*t*, 215, 226*f*, 227–229, 274–276, 285, 292, 350–351, 376–377, 383, 439–440, 454–455
Henry, R. J., 127–129
Herman, A., 263–264
Hermanson, G. T., 109
Hermanson, S., 39
Herpoël-Gimbert, I., 447–448
Herrero, M., 338–339
Hervé, C., 98, 212, 234–235, 241
Hess, M., 376
Hethener, P., 365
Hettich, R. L., 318, 322–325
Heublein, B., 52–53
Heulin, T., 367
Higgins, M. A., 221*f*
Higuchi, R., 277–278
Hildén, L., 235
Hilgenfeld, R., 220–221
Hilz, H., 99, 99*f*, 103, 105–106
Himmel, M. E., 39–40, 70, 90, 116, 184–186, 204–205, 262, 368, 384
Hinsa, S. M., 322
Hirst, B. H., 196–197
Hiyoshi, A., 100, 111–112
Hjort, C., 59–60, 184–186
Ho, P., 39
Hobson, P. N., 354–355, 356
Hodge, S. K., 112–113
Hoffman, H., 234–235
Hoffman, M., 99, 99*f*, 105–107, 123, 124, 133
Hofmann, L., 361
Höfte, H., 53
Hogg, D., 54
Hogsett, D. A., 318, 322–327, 328
Hoj, P. B., 4
Hojer-Pedersen, J., 59–60, 62, 63
Holdeman, L. V., 365–366
Hollak, C. E. M., 157
Holst, O., 217–218
Holsters, M., 342
Holtzapple, M., 332–333
Hon, D. N.-S., 170
Honda, M., 143–144
Honda, Y., 296
Hong, C., 147–149
Hongoh, Y., 364–365
Hoogland, C., 215–216
Hooibrink, B., 157
Hooman, S., 83–85
Hoos, R., 157–158
Hori, C., 174–175
Hori, R., 87, 171, 172–173
Horn, S. J., 69–95
Hosono, K., 363–364
Houghton, J. N. I., 363–365, 376, 385

Hrmova, M., 154, 274–276
Hrynuik, S., 217–218, 227–229
Hsieh, Y. S. Y., 99, 99*f*, 103, 105–107
Hu, C.-Y., 360–361
Hu, P., 382–383
Hu, Q., 88
Hu, T. C., 147–149
Hu, Y., 379–380
Hu, Y. L., 379–380
Huang, J. S., 10–11
Huang, Y., 388, 389
Hubbard, R. E., 59–60, 184–186
Huber, T., 382–383
Huffer, S., 376–377
Hugenholtz, P., 382–383
Hugley, B. M., 361–362
Hult, E. L., 75
Humphreys, P., 364–365, 376
Humphry, D. R., 333–334
Hung, L. W., 410–412
Hungate, R. E., 351–352, 354–356, 357, 361, 363–365, 366
Hunt, D. E., 386
Hura, G. L., 200, 204–205
Huskisson, N. S., 194, 248
Huson, D. H., 382–383, 389
Hutcheson, S. W., 74–75
Hyman, L., 385–386
Hyrnuik, S., 221*f*

I

Ibatullin, F. M., 98, 100, 102*f*, 104*f*, 106–107, 111–112, 114–115, 116*f*, 123, 158–159
Ichinose, H., 53–54, 64–65, 143–144
Igarashi, K., 70, 75, 80, 85, 87, 89–90, 169–182
Iino, R., 171
Imai, T., 70, 75, 172–173, 274–276
Improta, R., 276
Innis, M., 59–60
Inoko, Y., 187
Inoue, T., 364–365
Ireton, G. C., 150–152
Irwin, D. C., 70, 74–75, 81, 88, 99–100, 116, 123, 184–186, 249, 432, 440, 444
Irwin, P. L., 75
Ishida, H., 350–351
Ishida, T., 70, 174–175
Ishikawa, K., 350–351
Ishikawa, R., 171
Ishikawa, S., 333
Ismagilov, R. F., 149–150
Isupov, M. N., 412
Itami, T., 333
Itoh, Y., 336–337
Ivanen, D. R., 292–293
Ivanova, N. N., 350–351, 364–365
Iwakura, Y., 76

Iwasaki, N., 126
Izumi, A., 107

J

Jabs, A., 220–221
Jackson, P., 53–54, 58
Jackson, W. M., 225–227
Jacobs, S. A., 258
Jacques, D. A., 187–188, 204–205
Jaeger, K. E., 379–380
Jager, W., 339–340
Jakeman, D. L., 274–276
Jalak, J., 70, 83–85, 87
Jamal, S., 151f
James, D. C., 53
Jamshidian, M., 147
Jancarik, J., 147–149, 277–278
Janssen, D. B., 379–380
Jarvis, G., 376–377
Jarvis, M. C., 53
Jaskolski, M., 161–162
Jefferson, R. A., 342
Jenkins, N., 53
Jenkins, S. N., 367
Jensen, H. M., 20–21
Jeronimidis, G., 116
Jervis, E., 215, 217, 225–227, 228
Jia, Z. C., 123, 124, 133, 143–144, 184–186, 396
Jia, Z. H., 99, 99f, 105–107
Jindou, S., 249–250
Joblin, K. N., 360–361
Johansen, K. S., 38–39, 42, 52–53, 54, 59–60, 62, 63, 70, 100, 102f, 104f, 106–107, 111–112, 116f, 122–123, 158–159
Johansson, G., 70, 80, 81, 85, 87, 90, 170–171, 235
Johansson, P., 106–107
Johns, K., 155–156
Johnson, D. K., 116
Johnson, E. A., 318–319
Johnson, P. E., 184–186, 278–279
Johnston, J. E., 149–150
Joint, I., 352, 376–377, 382, 383, 388, 389
Jones, D. L., 364–365
Jones, L., 46
Jones, T. A., 70, 71f, 73, 75, 78–79, 106–107, 155, 170–171, 184–186
Jørgensen, C. I., 38–39, 52–53, 54, 59–60, 70, 122–123
Jørgensen, P. L., 100, 102f, 104f, 106–107, 111–112, 116f, 123, 158–159
Joseleau, J. P., 99f, 123
Joseph-McCarthy, D., 197–200
Joshi, M. D., 184–186, 278–279
Judge, R. A., 150
Juers, D. H., 154
Jun, H. K., 379–380
Juy, M., 70, 155, 195

K

Kabel, M. A., 99, 99f, 103, 105–106
Kado, C. I., 345
Kadouri, D. E., 322
Kaewthai, N., 114f
Kahn, R., 75
Kalinowski, J., 339–340
Kallas, Å .M., 100, 101–102, 106–107, 108, 113
Kallemeijn, W. W., 157
Kallio, J., 83–85
Kameranpour, N., 361–362
Kamerling, J. P., 112–113
Kampfer, P., 365
Kandori, H., 171
Kaneko, S., 53–54, 64–65, 143–144, 226f, 227–228
Kang, Y. E., 54
Kantardjieff, K. A., 147
Kaper, T., 276, 277–278, 279, 280–283, 280t, 282t, 283t, 284t
Kapral, G. J., 412
Karjala, G., 274–276
Karkehabadi, S., 143–144
Karlsson, J., 72–73
Karpinets, T. V., 385–386
Karplus, M., 197–200
Karplus, P. A., 75, 184–186, 249
Kasarskis, A., 385
Kataeva, I. A., 196–197, 376
Kataoka, N., 363–364
Katayama, T., 296
Katayose, Y., 379–380
Kato, R., 296
Kato, Y., 99f, 123
Katouno, F., 75, 78–79
Kauppinen, S., 123, 131
Keating, D. H., 331–347
Keegan, R. M., 408
Keller, K., 382–383
Kelly, P. M., 159–160
Kelly, R. M., 274–276, 280t, 292–293, 294, 295
Kemp, P., 354–355
Kenig, R., 249, 418–419, 447
Kenne, L., 131–135
Kennedy, D. F., 150
Kerby, M. B., 382, 385
Kerek, F., 130
Khachane, A. N., 376–377
Khorasanizadeh, S., 258
Kiat-Lim, E., 160
Kiel, J. T., 302–303
Kilburn, D. G., 184–186, 215–216, 216f, 217–218, 219f, 220, 221–222, 221f, 223, 224, 225–227, 226f, 228, 443
Kim, D., 384–385
Kim, H., 379–380
Kim, J. O., 384–385

Kim, K. M., 146
Kim, M. Y., 379–380
Kim, R., 147–149
Kim, S. H., 147–149, 277–278
Kim, S. J., 379–380
Kim, S. N., 384–385
Kimberley, E. S., 292
Kimura, A., 296
Kimura, S., 70, 80, 85, 89–90, 169–182
Kimura, T., 454–455
King, B. C., 38, 72–73
Kingspooner, C., 360–361
Kinnari, T., 70, 73, 78–79
Kipper, K., 70, 80, 81, 87
Kirk, O., 53
Kitaoka, M., 143–144, 296
Klassen, J., 215–216
Klemm, D., 52–53
Klenk, D. C., 109
Klepac-Ceraj, V., 386
Kleywegt, G. J., 70, 71f, 75, 155, 184–186
Kljun, A., 236, 237f
Kloareg, B., 75, 98
Klopp, C., 376–377
Knowles, J. K. C., 70, 71f, 143–144, 170–171, 184–186
Knox, J. P., 233–245
Ko, C., 39
Kobayashi, H., 143–144
Kobayashi, T., 194, 248
Koch, M. H., 190–191, 197–200
Kodera, N., 171
Kofod, L. V., 123, 131
Koivula, A., 70, 71f, 73, 74, 75, 78–79, 80, 83–85, 88, 89–90, 133–135, 155, 169–182
Kolli, B. K., 365
Komander, D., 76, 78, 79f
Kondo, H., 100, 111–112
Kong, T., 389–390
Kongruang, S., 87–88
Konig, H., 365
Koo, B. S., 379–380
Kopp, J., 277–278
Korczynska, J., 147–149
Kormos, J. M., 217–218, 219f, 220, 224, 226f, 228, 443
Korner, M., 278–279
Kosik, O., 51–67
Kostylev, M., 87
Kosugi, A., 248
Kosugi, Y., 350–351
Kotake, T., 53–54, 64–65
Kovach, M. E., 336–337
Kozin, M. B., 197
Kramer, G., 157
Kramer, R., 20–21
Kren, V., 274–276
Kretz, K. A., 54

Krissinel, E. B., 408
Krogh, K., 59–60, 62, 63, 70
Krogh, K. B. R. M., 38–39, 52–53, 54, 59–60, 122–123
Krohn, R. I., 109
Krummel, B., 277–278
Kuang, L. R., 55–56, 59, 82–83
Kubal, M., 382–383
Kuchnir, L., 197–200
Kudo, T., 364–365
Kudol, T., 364–365
Kudrna, D., 379–380
Kuehne, S. A., 302–303, 314–315
Kuga, S., 171
Kulam-Syed-Mohideen, A. S., 382–383
Kulminskaya, A. A., 292–293
Kumagai, H., 296
Kumar Kolli, V. S., 123, 126
Kuno, A., 143–144
Kurašin, M., 70, 74, 75, 80, 81, 82, 82f, 85, 86–90, 86f, 170–171
Kurita, K., 76
Kurkal, V., 195–196, 200, 201–203, 204f, 249
Kusakabe, I., 143–144
Kuwahara, A., 363
Kwasnoski, J. D., 146
Kwok, S., 59–60
Kyriacou, A., 88

L

La Cara, F., 276
Laatsch, H., 20
Ladbury, J., 258
Ladisch, M., 332–333
Ladner, M., 59–60
Laine, R., 155–156
Laitinen, T., 157
Lam, D., 292
Lamed, R., 52–53, 184–186, 194–196, 247–259, 263–264, 318, 396, 397, 417–428, 429–452, 453–463
Lammerts van Bueren, A., 215–216, 217–218, 221f, 225–227
Lamzin, V. S., 410–412
Land, M., 350–351
Lander, D. J., 354–355
Lane, P., 146
Langan, P., 170
Langer, G., 410–412
Langer, M., 376
Langston, J. A., 122–123
Lapidot, A., 420–421, 422, 455–456
Lappalainen, A., 83–85
Larkin, M. A., 254
Larsbrink, J., 107
Larsen, N., 382–383
Larson, S. B., 150

Lassen, S. F., 87, 263
Latham, M. J., 364–365
Laurie, J. I., 184–186
Law, V., 215–216
Lawrence, M. I. G., 350–351, 354–355, 360–361
Lawson, B., 147–149
Lawson, P. A., 365–366
Le Coq, D., 345
Leavitt, R. I., 337–338
Lebbink, J. H., 277–278
Leclerc, M., 365–366, 376–377
Lednicka, M., 106–107, 112–113
Lee, C. M., 379–380
Lee, D., 72–73
Lee, E., 54
Lee, J. H., 87
Lee, S. F., 302–303
Lee, W., 398
Lee, Y. C., 126
Lee, Y. Y., 332–333
Lehmann, M., 263
Lehn, P., 152
Leibovitz, E., 396
Leininger, S., 389
Lemaire, M., 396
Lepore, L., 274–276, 280t, 285, 288, 289–292
Leschine, S. B., 194, 332–333, 353–354, 361
Leslie, A. G., 408
Lever, M., 7, 8–9, 109
Levine, S. E., 72–73, 74
Levinson, H. S., 184–186
Levy, I., 215
Levy, S., 382–383
Levy-Assraf, M., 247–259
Lewis, R. J., 212, 228–229
Lexa, D., 419, 420–421
Leydier, S., 157–158
Li, A., 216f
Li, A. Y., 379–380
Li, F., 384–385
Li, K.-Y., 157
Li, L., 159–160
Li, W., 388, 389
Li, X. L., 362, 376
Li, Y. C., 70, 74–75, 305
Liang, H.-C., 53–54, 64–65
Liebert, T., 135
Liebeton, K., 376
Liebl, W., 292
Liepman, A. H., 53–54, 58
Liggenstoffer, A. S., 360–361
Lignon, S., 194
Lim, C. H., 384–385
Lin, L. N., 224
Lindeberg, G., 170
Linder, M., 88–89, 170
Linehan, B. A., 365–366
Lipman, D. J., 383

Lipski, A., 274–276, 280t, 285, 288, 289–292
Liu, A., 263
Liu, J., 376–377
Liu, W. D., 376–377, 384–385
Liu, W. F., 88
Liu, Z. L., 70
Ljungdahl, L. G., 196–197, 376, 397
Llorca, O., 376
Lo, J., 318, 322–325
Lo Leggio, L., 38–39, 52–53, 54, 59–60, 70, 122–123
Lobanov, V. S., 146
Loch, C., 263
Lockhart, R. J., 351, 352, 362–363, 364–365, 367, 376, 385
Lohman, T. M., 86–87
Lombard, V., 52–53, 70, 122–123, 142–144, 213t, 274–276, 383
Londei, P., 285
Long, F., 402–404, 409–410
Long, Y. Q., 361
Longenecker, K. L., 150
Longmire, A. G., 74–75, 350–351
Lopez, R., 254
López-Contreras, A. M., 302–303, 444
Lopez-Cortes, N., 379–380
Lorences, E. P., 99f, 123
Lorenz, P., 376
Louis, P., 365–366
Lovegrove, A., 53–54, 64–65
Lu, Y., 367
Lucius, A. L., 86–87
Luginbuhl, P., 364–365
Luo, M., 379–380
Lutke-Eversloh, T., 302–303
Ly, H. D., 278–279
Lynd, L. R., 55–56, 59, 74, 82–83, 87–88, 317–330, 332–333, 350–351, 361
Lytle, B. L., 194, 420–421

M

Ma, A. Z., 88
Ma, B., 296
Ma, L. X., 379–380
Ma, Q. S., 376–377, 384
Macauley, K., 100, 102f, 104f, 106–107, 111–112, 116f, 123, 158–159
Macdonald, J. M., 145f, 157–158
MacEachran, D. P., 322
Macher, B. A., 295, 296
MacKenzie, A., 70
Mackenzie, C. R., 88
Mackenzie, L. F., 154, 155–156, 274
MacKerell, J. A. D., 197–200
Maclachlan, G. A., 99f, 123
MacLeod, A. M., 114–115
MacSweeney, A., 149–150

Madden, M., 379–380
Madden, R. H., 361
Madden, T. L., 383
Madia, A., 318–319
Madsen, R., 157
Maeder, D. L., 149–150
Maehara, T., 143–144
Mah, M., 274–276
Mah, R. A., 363–364
Mahajan, S., 108
Mahdi, S., 70, 74, 89, 170–171
Majima, T., 126
Makarov, D. E., 204–205
Malekzadeh, F., 361–362
Malet, C., 274
Malhotra, A. V., 105–106
Mallia, A. K., 109
Malmstrom, R. R., 382
Malten, M., 19–36
Maluf, N. K., 86–87
Mandelman, D., 70
Mangani, S., 147
Maniatis, T., 322, 341
Mann, S. O., 351, 356, 367
Mansfield, H. R., 365–366
Manuel Garcia-Ruiz, J., 149–150
Marais, M.-F., 367
Marcus, S. E., 234–235, 241
Mark, P., 106–107
Markov, A. V., 100, 104f
Markovic, O., 10–11
Markwell, J., 110
Marol, C., 367
Marry, M., 101–102
Marsh, J., 53–54, 64–65
Marsh, M., 150–152
Marsh, T. L., 386
Martin, D. J., 112–113
Martinez, D., 376
Martinez-Cruz, L. A., 277–278
Martinez-Fleites, C., 100, 103–105, 106–107, 111–112, 144, 160, 194, 418–419
Martin-Orue, S. M., 54
Maruyama, D., 171
Master, E. R., 100, 108, 131–135
Masuko, T., 126
Materna, A. C., 382
Matheson, N. K., 4, 9–10
Mathur, E. J., 292
Matsumoto, T., 379–380
Matte, M. H., 361–362
Matthews, B. W., 154, 404, 409
Matthews, J. F., 203–205
Matthews, K. J., 112–113
Matuliene, J., 224
Matulis, D., 224
Mayer, C., 274–276
Mayer, W. E., 376

Mazumder, K., 135–137
Mazzone, M., 285
McAllister, T., 389–390
McBee, R. H., 360
McCabe, C., 90, 203–204
McCallum, C. M., 39
McCann, M. C., 53
McCarter, J. D., 154
McCarthy, A. J., 349–374, 375–394
McCartney, L., 235, 237f, 241
McCleary, B. V., 1–17
McClure, W. R., 86–87
McCool, J. D., 318, 322–325, 327, 328
McCormack, A. A., 100
McCoy, A. J., 408, 409–410, 411, 412
McCrae, S. I., 360–361
McDonald, I. R., 378–379
McDonald, J. E., 349–374, 375–394
McDougall, G. J., 101–102, 107
McElver, J., 39
McFarland, K. C., 19–36, 87, 88, 90
McFeeters, R. F., 109
McGarrell, D. M., 382–383
McGettigan, P. A., 254
McHardy, A. C., 364–365
McIntosh, L. P., 184–186, 203–204, 278–279
McKee, L. S., 228–229
McKie, V. A., 1–17, 333–335, 337–338, 339, 341
McLean, B. W., 215–216, 216f
McMillan, J., 32
McNeil, M., 99f, 123, 124, 129
McNicholas, S. J., 408
McPherson, A., 146–147, 150
McQuaid, J., 379–380
McQueen-Mason, S. J., 37–50
Mcsg, Sgc., 147–149
McWilliam, H., 254
Mead, D., 39
Mechaly, A., 194–196, 418–419, 420–421, 422, 426–427, 430, 434, 435, 435f, 439–440, 444, 445, 447–448, 456
Medini, D., 385
Medve, J., 72–73
Meged, R., 147–149
Melero, R., 376
Mellerowicz, E. J., 98, 115
Mello, L. V., 383
Meloncelli, P. J., 157–158
Merino, S., 70
Mermelstein, L. D., 304, 305
Mertoglu, B., 350–351
Methe, B. A., 361
Meulewaeter, F., 236, 237f
Meurer, G., 376
Meyer, A. S., 42
Meyer, F., 382–383, 389
Michel, G., 75, 98, 100, 101–102, 106–107, 108, 113

Michelsen, M., 59
Middendorf, A., 263
Mielenz, J. R., 262, 368, 384
Mikaelian, I., 276
Mikhailova, N., 350–351
Mikkelsen, D., 352–353
Mikkelsen, J. M., 59–60, 184–186
Milan, A., 379–380
Milanova, D., 382, 385
Miles, G. P., 53–54, 55–56, 58
Miller, B. B., 318, 322–327, 328
Miller, E. S. Jr., 292
Miller, G. L., 8–9, 267, 367, 450
Miller, J. F., 318
Miller, J. H., 334–335
Miller, M., 379–380
Miller, R. C. Jr., 184–186, 215, 228
Miller, T. L., 359
Miller, W., 383
Milne, J. L., 46
Minami, A., 126
Mingardon, F., 301–316, 440, 441, 444, 445, 447–448
Minor, W., 161–162
Minshull, J., 398
Minton, N. P., 302–303, 314–315
Miras, I., 419
Mishra, A., 105–106
Mislovicová, D., 10–11
Mitchell, E. P., 161
Mitchinson, C., 100, 107, 112
Mitra, S., 389
Mitsuishi, Y., 100, 111–112, 123
Miya, A., 363–364
Miyagi, A., 171
Miyazaki, K., 100, 111–112, 379–380, 384
Mizuno, H., 143–144
Mo, X. C., 379–380
Mohnen, D., 332
Molinier, A. L., 201–203
Money, V. A., 155–156, 157–158, 194, 396, 401, 402, 418–419
Monod, F., 447–448
Monsan, P., 376–377
Monsarrat, B., 53
Monserrate, E., 361
Montanier, C. Y., 212, 228–229
Montgomery, L., 365–366
Monti, M., 285
Mooibroek, H., 302–303
Moon, Y. H., 74–75
Moore, C., 385–386
Moracci, M., 273–300, 350–351
Moraccim, M., 285
Morag, E., 184–186, 196–197, 248, 249–250, 397, 418–419, 420–421, 430–431, 439–440, 453–463
Moraïs, S., 429–452

Moran, F., 190
Moran-Mirabal, J. M., 88
Mori, H., 296
Moriarty, N. W., 410–412
Morimoto, K., 454–455
Moriya, S., 364–365
Mornon, J. P., 152
Morreel, K., 332
Mort, A. J., 53, 99f, 123, 131
Mortier-Barriere, I., 302–303, 311
Mortimer, J. C., 53–54, 55–56, 58
Mosbah, A., 184–186, 196–197
Mosier, N., 332–333
Motulsky, H. J., 425
Mougel, C., 367
Mouille, G., 53
Moxon, R., 385
Mulhern, T. D., 187
Mullings, R., 354, 361–362
Murao, S., 87–88
Murgola, E. J., 153–154
Murphy, D. M., 318, 322–325, 327, 328
Murphy, E., 385–386
Murphy, K. P., 211–231
Murphy, L., 87, 88, 90
Murray, W. D., 361
Murrell, J. C., 367, 378–379
Murshudov, G. N., 155–156, 402–404, 408, 409–410, 412
Muzammil, S., 253–254, 420
Myambo, K., 59–60
Myslik, J., 146

N

Nadler, D. C., 376–377
Nagae, M., 296
Nagy, T., 184–186, 194, 196–197, 337–339, 341–343, 396, 397, 402, 419
Najmudin, S., 395–415
Nakamura, K., 363
Nakar, D., 418–419, 420–421, 422, 426–427, 434, 435, 435f, 456
Nakhai, A., 107
Nakotte, S., 305
Nall, B. T., 258
Navarro, D., 38
Nechitaylo, T. Y., 379–380
Neef, A., 376–377
Neidhardt, F. C., 334–335
Nelson, K. E., 382–383
Nelson, N., 6, 8–9, 109
Nelson, W., 382–383
Nemerow, G. R., 147–149
Nenni, N. V., 38, 39
Nerinckx, W., 70, 71f, 75, 100, 107, 112, 155, 212
Ness, J. E., 398

Netland, J., 101–102
Neufeld, J. D., 378–379
Neufeld, R. J., 88
Newcomb, M., 318
Newhouse, Y., 150
Newman, J., 147–149, 150
Newman, M., 32
Newstead, S. L., 212
Nezami, A., 253–254, 420
Ng, J. D., 149–150
Nguyen, C., 150
Nicol, G. W., 389
Niedringhaus, T. P., 382, 385
Niehaus, F., 376
Nielsen, B. B., 350–351, 354–355
Niesen, F. H., 146
Nigou, J., 53
Nikaidou, N., 78–79
Nimlos, M. R., 116, 204–205
Nishikori, S., 171
Nishikubo, N., 98
Nishimura, S.-I., 126
Nishimura, T., 285, 292
Nishiyama, A., 295
Nishiyama, Y., 170
Noach, I., 195–196, 430
Noda, S., 364–365
Noelting, G., 109
Noji, H., 171
Nonaka, T., 78–79
Noori-Daloii, M. R., 361–362
Noparatnaraporn, N., 364–365
Norberg, A. L., 78–79
Nordberg Karlsson, E., 217–218
Nordlund, P., 146
Noro, N., 100, 111–112
Notenboom, V., 157–158
Nouailler, M., 201–203
Nucci, R., 276
Nutt, A., 90
Nymand, S., 87, 88, 90

O

O'Brien, R., 258
O'Donnell, A. G., 367, 377–378
Offen, W. A., 99–100, 106–107, 108, 111–112, 114–115, 123, 157–159, 160
Ofir, K., 455–456
Ogan, A., 350–351
Ogmen, M. N., 350–351
Ogura, T., 171
Oh, M. J., 384–385
Ohkuma, M., 364–365
Ohmiya, K., 454–455
Ohnuki, T., 379–380
Ohtaka, H., 253–254, 420
Oka, J., 295

Okamoto, T., 70, 80, 85, 89–90, 171, 179
Okano, T., 171
Okuyama, M., 296
Olafson, B. D., 197–200
Oliver, R. P., 342
Olsen, C., 59–60, 62, 63
Olsen, S. N., 87
Olson, B. J. S. C., 109, 110
Olson, D. G., 317–330
Olson, R., 382–383
Olsson, L., 59–60, 62, 63
O'Neill, M. A., 121–139
Ong, E., 215, 217, 225–227, 228
Oppenheim, A. B., 79f
Ordaz-Ortiz, J. J., 241
Oren, A., 350–351
Orlando, R., 123, 126
Orpin, C. G., 184–186, 350–351, 354–355, 360–361
O'Sullivan, C., 363–364
Otalora, F., 149–150
O'Toole, G. A., 322
Otten, H., 38–39, 52–53, 54, 59–60, 70, 122–123
Ottesen, E. A., 385–388, 387f
Ottosson, J., 144
Overbeek, R., 382–383
Ozkan, M., 361
Ozkose, E., 360–361

P

Paarmann, D., 382–383
Paczian, T., 382–383
Pages, S., 194, 195, 249–250, 311–312, 397, 419, 420–421
Pakula, T. M., 350–351
Pal, G., 143–144, 396
Palacios, D., 385–386
Palackal, N., 54
Paley, M. S., 150
Palonen, H., 88
Palop, M. L., 363–364
Pang, H., 376–377, 379–380, 384
Panine, P., 192–194, 200, 201
Pantoliano, M. W., 146
Pantos, E., 190
Papanikolau, Y., 79f
Papoutsakis, E. T., 304, 305
Parish, J. H., 354, 361–362
Park, S. C., 384–385
Parker, M. W., 187
Parker, N., 292
Parkhill, J., 385
Parrilli, M., 274–276, 280t, 285, 288, 289–292
Parsiegla, G., 70, 155, 195
Pasamontes, L., 263
Pasti, M. B., 365
Patel, S., 150

Patterson, S. E., 38, 39
Patton, D., 39
Paul, J., 365
Paulsen, I., 382–383
Pauly, M., 38, 39, 98, 101–102, 116, 123, 131
Payne, C. M., 90
Pearl, L. H., 276
Peat, T. S., 147–149, 150
Pedersen, H. L., 226f, 227–228, 241
Pedersen, M., 42
Peer, A., 454–455
Pell, G., 54, 143–144, 157–158, 333–335, 337–338, 339, 341
Pelosi, L., 274–276
Peña, M. J., 99, 99f, 105–107, 121–139
Penner, M. H., 87–88
Pennington, O. J., 302–303
Penttilä, M., 70, 80, 85, 89–90, 169–182
Perestelo, F., 354
Perez, J., 187
Perez, R., 354
Perkins, S. J., 192, 200, 201
Perotto, S., 234–235
Perrakis, A., 76, 147–149, 410–412
Perret, S., 303, 304, 304f, 306, 307–310, 309f, 312
Persson, S., 131
Perugino, G., 273–300
Peter, M. G., 76, 78, 79f
Peterson, K. M., 336–337
Petitdemange, E., 363–364
Petitdemange, H., 363–364
Petkun, S., 249–250
Petoukhov, M. V., 190–191, 197, 201, 285
Petratos, K., 79f
Petrella, E. C., 146
Petrikaite, V., 224
Pettersson, G., 2, 85, 87, 90, 143–144, 170–171, 172–173, 184–186
Pezeshk, V., 101–102
Pflugrath, J. W., 407
Pham, T. M., 147–149
Phelan, M. B., 362
Phillips, R. W., 336–337
Piens, K., 70, 74, 89, 98, 100, 107, 112, 131–135, 170–171
Pierce, M. M., 258
Pilz, I., 170–171, 184–186, 192
Pinaga, F., 363–364
Pinheiro, B. A., 194, 396, 401, 402, 418–419
Pipelier, M., 157–158
Pizzut-Serin, S., 376–377
Planas, A., 100, 106–107, 108, 115, 154, 155, 274, 278–279
Plesniak, L. A., 278–279
Plückthun, A., 263
Pluvinage, B., 221f
Podar, M., 379–380, 383–384
Pohlschroder, M., 194

Polaina, J., 379–380
Polizzi, K. M., 263
Polizzi, S. J., 228–229
Polz, M. F., 386
Pometto, A. L., 362
Poole, D. M., 194, 248
Poole, N. J., 361
Poon, D. K., 203–204
Popper, Z. A., 98
Poulsen, J.-C. N., 38–39, 52–53, 54, 59–60, 70, 122–123
Pouwels, J., 277–278
Powlowski, J., 100, 108
Praestgaard, E., 87, 88, 90
Prag, G., 79f
Prates, J. A. M., 100, 103–105, 106–107, 111–112, 184–186, 194, 196–197, 212, 228–229, 376, 395–415, 418–419
Presting, G., 39
Preston, C. M., 386
Pretorius, I. S., 332–333, 350–351
Proctor, M. R., 194, 396, 397, 401, 402, 419
Projan, S. J., 385–386
Prosser, J. I., 389
Protctor, M. R., 418–419
Provenzano, M. D., 109
Pucci, P., 285
Pufan, R., 147–149
Pugh, G. J. F., 363
Puhler, A., 339–340
Punt, P. J., 342
Puranen, T., 83–85
Purdy, K. J., 378
Pusey, M. L., 150
Putnam, C. D., 200, 204–205
Puzo, G., 53

Q

Qi, J., 382–383, 389
Qu, Y. B., 88
Quinlan, R. J., 38–39, 52–53, 54, 59–60, 70, 122–123
Qureshi, N., 302–303

R

Raghothama, S., 217–218
Ragsdale, C. W., 318
Rajgarhia, V. B., 318, 322–325, 327, 328
Ralph, J., 332
Raman, C. S., 258
Rancurel, C., 52–53, 70, 122–123, 142–144, 213t, 274–276, 350–351, 383
Ranjard, L., 367
Ransom-Jones, E., 364–365
Rappuoli, R., 385
Rashid, M. H., 263
Rasmussen, G., 59–60, 184–186

Rattray, J. B., 302–303
Raval, G., 363–364
Raynaud, F., 302–303, 311
Raynaud, O., 396
Rayon, C., 53
Read, R. J., 409–412
Realff, M. J., 87
Rebers, P. A., 367, 368
Receveur, V., 192–194, 201
Receveur-Bréchot, V., 183–210, 249
Reczey, K., 99–100
Reese, E. T., 184–186
Refregier, G., 53
Regalado, V., 354
Reichardt, W., 361–362
Reichstein, M., 332–333
Reid, J. S. G., 99f, 123
Reinikainen, T., 170–171, 184–186
Relman, D. A., 385
Remington, K., 382–383
Rempel, B. P., 155–156
Reno, H., 363–364
Renwick, K. F., 112–113
Resch, M. G., 203–204
Reshef, D., 417–428, 456
Reva, O., 379–380
Reverbel, C., 70, 155
Reverbel-Leroy, C., 70, 155, 194, 195, 311–312
Reyes-Duarte, D., 376–377
Reynolds, S., 39
Rhodes, G., 161–162
Rhodius, V. A., 342–343, 344
Richardson, E. A., 126
Richardson, T. H., 364–365, 379–380
Richens, J., 53–54, 58
Richter, D. C., 389
Rickers, A., 350–351, 354–355
Rigden, D. J., 383
Rincon, M. T., 440, 444, 445, 447–448
Roach, L. S., 149–150
Robb, F. T., 376–377
Robe, P., 376–377
Roberge, M., 263
Robert, C., 365–366
Roberts, I. N., 342
Roberts, S. M., 99–100, 102f, 104f, 106–107, 108, 111–112, 114–115, 116f, 123, 141–168
Robertson, D. E., 277–278, 379–380
Robinson, R., 363–364
Robyt, J. F., 70
Rodrigue, S., 382
Rodriguez, A., 354, 382–383
Rogers, R. D., 150
Rogers, S. R., 318, 325–327
Rogowski, A., 234–235, 241
Rojas, M., 382–383
Romaniec, M. P., 194, 248
Romao, M. J., 184–186, 194, 196–197, 376

Romão, M. J. R., 395–415, 419
Rooks, D. J., 349–374, 375–394
Roop, R. M. II., 336–337
Roos, A. A., 98
Rosano, C., 285
Rose, D. R., 155–156, 157–158
Rose, J. K. C., 99–100, 116, 123
Rose, J. P., 196–197
Rosenheck, S., 249–250
Rossi, M., 273–300, 350–351
Rouanet, C., 302–303, 311
Rouau, X., 38
Rouillard, J. M., 398
Roussel, A., 197
Rouvinen, J., 70, 71f, 73, 78–79, 157
Rovira, C., 154, 155
Rowland, S. E., 376–377
Ruchel, R., 20
Ruda, M. C., 97–120
Rudsander, U. J., 131–135
Ruimy, V., 453–463
Ruiz, R., 32, 33
Ruohonen, L., 70, 73, 78–79, 88, 133–135, 170–171
Rupek, P., 389
Rupitz, K., 114–115
Rupp, B., 147, 161–162
Rusch, D., 382–383
Rutten, S. J., 274–276

S

Sabathe, F., 302–303, 304, 304f, 306, 307, 308, 312
Saha, B. C., 302–303
Saiki, R. K., 277–278
Saito, K., 171
Sakai, N., 171
Sakamoto, R., 87–88
Sakka, K., 454–455
Sakon, J., 75, 184–186, 249
Salamitou, S., 396
Saldajeno, S., 143–144
Salemme, F. R., 146
Saloheimo, M., 350–351, 376
Sambrook, J., 322, 341
Samejima, M., 70, 75, 80, 85, 87, 89–90, 169–182
Sams, C. E., 385–386
Sanderson, I., 276
Sandgren, M., 70, 143–144
Sandstrom, C., 131–135
Sannan, T., 76
Santoro, N., 38, 39
Santos, V. A., 376–377
Saridakis, E., 150–152
Sasaki, T., 379–380
Sathitsuksanoh, N., 74, 88
Saunders, E. H., 350–351

Saunders, S., 318
Saura-Valls, M., 100, 106–107, 108, 115
Saxena, S., 365
Scarlata, C., 32, 33
Schadt, E. E., 385
Schaefer, D. M., 363–364
Schaeffer, F., 419
Schafer, A., 339–340
Schafer, H., 378–379
Schaffer, A. A., 383
Schaffer, S., 305
Schapira, M., 108
Scheffers, M., 143–144
Schell, D., 32
Schellenberger, V., 263
Schiller, B., 263
Schimmel, P., 153–154
Schleper, C., 389
Schloter, M., 389
Schmidt, R. R., 276
Schmuck, M., 170–171, 184–186
Schnurr, J. K., 101–102
Schols, H. A., 99, 99f, 103, 105–106
Schonig, I., 365
Schramm, V. L., 159–160
Schubot, F. D., 196–197
Schueler-Furman, O., 417–428, 456
Schülein, M., 59–60, 63, 70, 71f, 75, 99–100, 154, 155–156, 157–158, 184–186, 192–194, 200, 201
Schulz, G. E., 277–278
Schulze, R., 376
Schuster, L. N., 376
Schuster, S. C., 382–383, 385–386, 388, 389
Schwark, L., 389
Schwarz, W. H., 249
Schweickart, V., 59–60
Scirè, A., 276
Scott, J. C., 302–303, 314–315
Seffen, K. A., 53–54, 55–56, 58
Segura, M. P., 53–54, 55–56, 58
Seibert, G. R., 234–235
Seifert, M., 135
Selig, M., 39–40
Semenova, M. V., 100, 104f
Seong, C. N., 384–385
Sergeant, A., 276
Serruto, D., 385
Servinsky, M. D., 302–303
Sessions, A., 39
Setter, E., 184–186, 249, 318, 447
Seviour, R., 389–390
Shabalin, K. A., 292–293
Shaghasi, T., 122–123
Shah, A. K., 196–197
Shahamat, M., 361–362
Shallom, D., 278–279
Shani, Z., 215

Shanks, R. M. Q., 322
Sharpe, M. E., 364–365
Sharrock, K. R., 2
Shea, E. M., 127–130
Shen, W. J., 376–377
Shewry, P. R., 53–54, 64–65
Shi, Y., 385–386, 388
Shibata, M., 171
Shimon, L. J. W., 184–186, 195–196, 249–250, 430, 439–440
Shin, D. H., 75, 249
Shin, T. S., 384–385
Shiratori, H., 363–364
Shlaes, D., 385–386
Shoemaker, S., 59–60
Shoham, G., 278–279
Shoham, Y., 52–53, 184–186, 194–196, 248, 249–250, 278–279, 396, 397, 418–419, 420–421, 422, 426–427, 430, 431–432, 434, 435, 435f, 439–440, 442, 444, 445, 446f, 454–456
Short, J. M., 292
Shoseyov, O., 194, 215
Shulman, M., 249
Sigoillot, J. C., 447–448
Siika-Aho, M., 83–85, 99–100
Sikorski, P., 74–75, 76–79, 77f, 80f, 90
Sild, V., 90
Silverman, L., 367, 368
Simala-Grant, J. L., 296
Simon, C., 378–379
Simpson, P. J., 217–218
Sinitsyn, A. P., 100, 104f
Sinning, I., 184–186
Siu, R. G., 184–186
Sjöström, M., 30
Skinner, F. A., 361
Skøjt, M., 99–100, 106–107, 108, 111–112, 114–115, 123, 157–159
Slaytor, M., 365
Sleat, R., 363–364
Sluiter, A., 32, 33
Sluiter, J., 32, 33
Slutzki, M., 417–428, 453–463
Smidsrød, O., 76
Smidt, H., 302–303
Smith, B. J., 154
Smith, D. F., 334–335
Smith, D. K., 147–149
Smith, D. L., 352, 376–377, 382, 383
Smith, F., 367, 368
Smith, J. C., 195–196, 200, 201–203, 204f, 249
Smith, N. L., 196–197, 396, 397, 402, 419
Smith, P. K., 109
Smith, R. C., 112–113
Smith, R. E., 366–367
Smith, S. P., 143–144, 184–186, 376, 396, 420–421, 430, 454–455

Snead, M. A., 292
Snyder, M. P., 382, 385
Solomon, D., 278–279
Somerville, C., 234–235
Somerville, C. R., 123, 131
Sommer, R. J., 376
Somogyi, M., 6, 8–9, 367
Sonan, G. K., 192, 200, 201
Song, H., 363–364
Sørbotten, A., 74–75, 76–78, 77f, 90
Sorek, R., 364–365
Sørlie, M., 69–95
Soucaille, P., 302–303, 311
Souchon, H., 184–186, 196–197
Sowers, K. R., 361–362
Spadiut, O., 99–100, 106–107, 108, 111–112, 114–115, 123, 157–159
Spencer, H. L., 420–421
Spezio, M., 74–75, 81
Spinelli, S., 184–186, 196–197, 439–440
Springer, B. A., 146
Srisodsuk, M., 184–186
Ståhlberg, J., 70, 71f, 74, 75, 87, 89, 155, 170–171, 184–186
Stalbrand, H., 98, 99–100, 112–113, 212
Starr, T. J., 352–353
States, D. J., 197–200
Steedman, H. F., 239
Steele, H. L., 379–380, 384
Steele-King, C. G., 38
Stege, J. T., 364–365
Steinbacher, S., 263
Steine, M. N., 39
Steinmetz, M., 345
Steipe, B., 263
Steitz, T. A., 184–186, 248, 249–250
Stenby, E. H., 59
Stengel, D. B., 98
Stenstrøm, Y., 70
Stenvall, M., 144
Stephens, E., 53–54, 55–56, 58, 64–65
Stevens, R., 382–383
Stevenson, D. M., 350–351, 361
Stevenson, Y. T., 124
Stewart, C. S., 356
Stewart, F. J., 385–388, 387f
Stewart, P. D. S., 149–150
Stick, R. V., 145f, 157–158
Stihle, M., 149–150
Stoddard, B. L., 150–152
Stokke, B. T., 90
Stoll, D., 155–156
Stone, B. A., 127–129, 234
Storms, R., 100
Storoni, L. C., 409–410, 411
Straume, M., 223–224
Strazzulli, A., 273–300
Street, I. P., 155–156

Streit, W. R., 379–380, 384
Strijland, A., 157
Strittmatter, A. W., 379–380
Strompl, C., 376–377
Strongitharm, B., 116
Stuart, D. I., 147–149
Studer, D., 263
Stura, E. A., 147–149, 150–152
Suen, G., 350–351
Suenaga, H., 379–380
Sugimoto, H., 100, 111–112
Sugiyama, J., 70, 75, 78–79, 170, 172–173
Suhnel, J., 220–221
Sukharnikov, L. O., 383–384
Sulová, Z., 106–107, 112–113
Sulzenbacher, G., 155–156, 157, 274–276, 280t, 285, 288, 289–292
Sun, L., 454–455, 456
Sund, C. J., 302–303
Sundberg, B., 98
Sundqvist, G., 144
Sunna, A., 217–218, 350–351
Sussman, J. L., 147–149
Sutton, J. D., 364–365
Svergun, D. I., 190–191, 197–200, 201, 285
Swaminathan, S., 197–200
Swanson, R. V., 277–278
Sweeney, M. D., 38–39, 52–53, 54, 59–60, 70, 87, 122–123
Synstad, B., 76, 77f, 78–79, 79f, 80f
Szabo, L., 143–144, 155–156

T

Tabka, M. G., 447–448
Tailford, L. E., 53–54, 144
Tainer, J. A., 200, 204–205
Takahashi, J., 98, 115
Takahashi, S., 150
Takai, E., 171
Takeda, T., 295
Takeo, K., 217–218
Takeyasu, K., 171
Tan, X., 54
Tanfani, F., 276
Tang, J. L., 10–11, 376–377, 379–380, 384
Taniguchi, H., 296
Taniguchi, M., 171
Tap, J., 376–377
Tarbouriech, N., 151f
Tardif, C., 70, 155, 194–196, 201–203, 303, 304, 304f, 306, 307–310, 309f, 311–312, 313–314, 313f, 418–419, 430, 439–440, 441, 444, 445, 447–448
Tarling, C. A., 157–158, 160, 285, 292
Tasse, L., 376–377
Tatsumi, H., 87
Tauch, A., 339–340

Tavares, G. A., 439–440
Tavlas, G., 79f
Tawfik, D. S., 263–264
Taylor, C. B., 90, 203–204
Taylor, D. E., 296
Taylor, E. A., 144, 159–160
Taylor, E. J., 100, 103–105, 106–107, 111–112
Taylor, G., 212
Taylor, L. E. II., 350–351
Taylor, L. T., 386
Teather, R. M., 111, 353, 384
Teeri, T. T., 70, 71f, 72–73, 75, 78–79, 88, 98, 100, 101–102, 106–107, 108, 113, 115, 131–135, 143–144, 155, 170–171, 184–186
Tekant, B., 215, 228
Teleman, A., 70, 73, 78–79, 88, 133–135
Teleman, O., 70, 88, 133–135
Templeton, D., 32, 33
Tenkanen, M., 88
Terwilliger, T. C., 410–412
Teugjas, H., 81, 82–83, 87–88
Tews, I., 76
Thayer, D. W., 365
Theodorou, M. K., 350–351, 354–355, 360–361
Thierbach, G., 339–340
Thomas, B. J., 360–361
Thomas, S. R., 184–186
Thompson, J. D., 254
Tiedje, J. M., 382–383
Tilbeurgh, H., 172–173
Tilbrook, M. G., 145f
Till, B. J., 39
Timberlake, S. C., 382
Timmis, K. N., 338–339, 376–377
Tine, M. A. S., 98
Ting, C. L., 204–205
Ting, T. Y., 225–227
Tjerneld, F., 72–73
Toda, A., 171
Toft, M., 19–36
Tokuda, G., 365
Tolley, S. P., 59–60, 184–186
Tomme, P., 143–144, 170–171, 184–186, 192, 217–218, 219f, 220, 224, 225–227, 226f, 228, 443
Tompa, R., 39
Toone, E. J., 225
Tormo, J., 184–186, 248, 249–250
Tornqvist, C.-E., 38, 39
Torp, B., 20
Torronen, A., 157
Toth, M. J., 153–154
Toutain, C. M., 322
Tovborg, M., 38–39, 52–53, 54, 59–60, 70, 122–123
Trakulnaleamsai, S., 364–365
Trewhella, J., 187–188, 204–205
Trinci, A. P. J., 350–351, 354–355, 360–361

Trincone, A., 274–276, 277–278, 279, 280–283, 280t, 282t, 283t, 284t, 285
Tringe, S. G., 364–365
Tripathi, S. A., 318, 322–327, 328
Truan, G., 398
Tryfona, T., 38–39, 52–54, 59–60, 64–65, 70, 122–123
Tsai, S.-D., 360–361
Tsang, A., 100
Tsuda, S., 100, 111–112
Tsumuraya, Y., 53–54, 64–65
Tucker, M., 39–40
Tucker-Kellogg, G., 385–386
Tull, D., 114–115, 155–156
Tuohy, M. G., 98
Tuomivaara, S. T., 121–139
Turkenburg, J. P., 100, 102f, 104f, 106–107, 111–112, 116f, 123, 151f, 158–159
Turner, M. B., 150
Turner, S. R., 53–54, 55–56, 58, 385
Tyler, P. C., 159–160
Tyson, G. W., 385–386, 388
Tyurin, M. V., 318

U

Uchihashi, T., 70, 80, 85, 89–90, 169–182
Uchiyama, T., 75, 78–79, 384
Ueda, K., 333, 363–364
Umbarger, H. E., 337–338
Upppugundla, N., 39
Urbanowicz, B. R., 121–139
Urich, T., 389
Ustinov, B. B., 100, 104f

V

Vaaje-Kolstad, G., 70, 72f, 78–79, 80f
Vaccaro, C., 276
Vagin, A. A., 402–404, 409–410
Valentin, F., 254
Valette, O., 201–203
Väljamäe, P., 69–95, 170–171
Valles, S., 363–364
van Aalten, D. M. F., 76, 78, 79f
van Beilen, J. B., 384
van Bueren, A. L., 212
Van Damme, J., 170–171, 184–186
Van den Eede, G., 342
van den Hondel, C. A., 342
van den Nieuwendijk, A. M. C. H., 157
van der Oost, J., 276, 277–278, 279, 280–283, 280t, 282t, 283t, 284t, 302–303
Van Dyke, M. I., 363–365, 376
van Halbeek, H., 123, 133
van Heusden, H. H., 277–278
van Loo, B., 277–278
van Loon, A. P., 263
van Montagu, M., 342

van Scheltinga, A. C. T., 76
van Sint Fiet, S., 384
van Tilbeurgh, H., 112–113, 143–144, 170–171, 184–186
van Zyl, W. H., 332–333, 350–351
Vandamme, E. J., 296
Vandamme, J., 143–144
Vandekerckhove, J., 143–144, 170–171, 184–186
Vangylsw, N. O., 355
Vanhalbeek, H., 103, 105–106, 107
Vanholme, R., 332
Vanhooren, P., 296
Varghese, J. N., 4, 154
Varma, A., 365
Varrot, A., 70, 71f, 75, 154, 155–156, 157–158
Vårum, K. M., 69–95
Vasella, A., 154, 157–158, 277–278
Vasu, P., 123, 131
Vaughan, M. D., 274–276
Vazana, Y., 429–452
Vazquez-Figueroa, E., 263
Vedadi, M., 146
Vehmaanpera, J., 83–85
Veiga, M., 362
Velazquez-Campoy, A., 253–254, 420
Velleste, R., 81, 82–83, 87–88
Venter, J. C., 382–383
Verhertbruggen, Y., 241
Verhoef, R., 99, 99f, 103, 105–106
Vernhettes, S., 53
Vespa, N., 276
Vian, B., 234–235
Vicky, W. W. A., 53–54, 58
Viegas, A., 395–415
Viikari, L., 99–100
Viikri, L., 83–85
Viladot, J. L., 278–279
Viles, F. J., 367, 368
Villa, E., 190–191
Villalobos, A., 398
Villard, F., 150–152
Vincken, J. P., 98, 99–100, 99f, 111–112
Violot, S., 192, 200, 201
Virden, R., 194
Vlasenko, E., 99–100, 122–123
Vocadlo, D. J., 153–154, 155–156, 157
Vogel, J., 98
Voget, S., 379–380, 384
Volkman, B. F., 194, 420–421
von Ossowski, I., 70, 71f, 74, 89, 155, 170–171, 192, 200, 201
Vongkaluang, C., 364–365
Voragen, A. G. J., 98, 99–100, 99f, 103, 105–106, 111–112
Vorgias, C. E., 79f
Voronov-Goldman, M., 430–431, 455–456
Voutilainen, S. P., 83–85
Vriend, G., 78–79, 80f

Vrsanska, M., 112–113
Vuong, T. V., 70, 74–75

W

Wada, C., 171
Wada, J., 296
Wada, M., 70, 80, 85, 87, 89–90, 169–182
Waeghe, T., 129
Wakarchuk, W. W., 278–279
Wakatsuki, S., 296
Waliczek, A., 379–380
Walker, L. P., 38, 39, 74–75, 81, 88
Walker, R. C., 204–205
Wallace, I. M., 254
Wallace, R. J., 354–355
Walter, C. P., 38–39, 52–53, 54, 59–60, 70, 122–123
Walter, T. S., 147–149
Walther, D., 190
Walton, J. D., 38, 39
Wang, B. C., 196–197
Wang, F., 384–385
Wang, J. W., 147
Wang, Q., 274, 382–383
Wang, Z. G., 204–205
Warcup, J. H., 363
Ward, D., 143–144
Warnecke, F., 364–365, 376
Warner, A. K., 318, 322–325, 327, 328
Warnick, T. A., 361
Warren, A. J., 215
Warren, R. A. J., 114–115, 184–186, 215, 216f, 217–218, 219f, 220, 221–222, 221f, 223, 224, 226f, 228, 274–276, 443
Watanabe, H., 365
Watanabe, K., 296
Watanabe, T., 75, 78–79
Watanabe, Y., 187
Watson, B. J., 74–75
Watson, J. N., 212
Waxman, K. D., 38, 39
Webb, B. A., 420–421
Wedekind, K. J., 365–366
Wei, J., 379–380
Weimar, T., 53–54, 55–56, 58
Weimer, P. J., 332–333, 350–351
Weiner, D. P., 379–380
Weiner, R. M., 350–351
Weisgraber, K., 150
Weiss, M. S., 220–221, 408
Welchman, H., 53–54
Wenzel, M., 365
Westereng, B., 70
Westh, P., 87, 88, 90
Westlake, K., 363–364
Westler, W. M., 194, 420–421
White, A. R., 101–102, 155–156

White, B. A., 248, 454–455
White, E. A., 361
White, P., 217–218, 228–229
Whitehead, C., 37–50
Whiteley, A. S., 367, 377–378
Wigstrom, W. J., 335
Wilchek, M., 420–421, 455–456
Wilke, A., 382–383, 389
Wilkening, J., 382–383
Wilkins, M. R., 215–216
Willats, W. G. T., 98, 226f, 227–228, 241
Williams, S. J., 157–158
Williams, S. T., 362
Williamson, M. P., 217–218
Williston, S., 224
Wilm, A., 254
Wilson, C. A., 360–361
Wilson, D. B., 38–39, 53, 70, 74–75, 81, 87, 88, 99–100, 116, 123, 131–135, 184–186, 249, 332–333, 350–351, 431–432, 440, 442, 444, 445, 446f
Wilson, I. A., 147–149, 150–152
Wilson, K. J., 342
Wilson, K. S., 59–60, 76, 99–100, 102f, 104f, 106–107, 108, 111–112, 114–115, 116f, 123, 154, 155–156, 157–159, 184–186, 196–197
Wing, R. A., 379–380
Winn, M. D., 408, 409–410, 411, 412
Wischmann, B., 19–36
Wiseman, T., 224
Withers, S. G., 114–115, 153–154, 155–156, 157–158, 203–204, 274–276, 278–279, 285, 292
Witholt, B., 384
Witte, M. D., 157
Wlodawer, A., 161–162
Wohlert, J., 203–204
Wohlfahrt, G., 170
Wold, S., 30
Wolfenden, R., 170
Wolfgang, D. E., 444
Wolin, M. J., 359
Wong, D., 100
Wong, K., 379–380
Wong, S. B., 263
Wood, P. J., 111, 353, 384
Wood, T. M., 265, 352, 360–361, 449
Woodward, M. J., 382
Wright, P. E., 201–203
Wu, C. F., 379–380
Wu, D., 382–383
Wu, J. H. D., 194, 318, 379–380, 420–421
Wu, M., 70
Wu, S., 385–386
Wyman, C., 332–333
Wyss, M., 263

X

Xia, Y., 389–390
Xian, L., 376–377, 384
Xiang, B. S., 99–100, 116, 123
Xie, H., 143–144
Xu, F., 99–100, 122–123
Xu, Q., 249
Xu, X., 147–149
Xue, G. P., 184–186

Y

Yagi, A., 171
Yakimov, M. M., 376–377, 379–380
Yamamoto, D., 171
Yamamoto, K., 296
Yamashita, H., 171
Yamashita, Y., 363
Yang, M., 144, 160
Yang, Y. F., 74
Yang, Z. Y., 101–102
Yaniv, O., 247–259
Yaoi, K., 100, 111–112, 123
Yarbrough, J. M., 203–204
Yaron, S., 184–186, 249, 418–419, 439–440
Yazdi, M. T., 361–362
Ye, Z.-H., 126
Yeo, Y. S., 379–380
Yeung, A. T., 39
Yi, E. C., 146
Yokokawa, M., 171
Yoon, S. H., 379–380
York, W. S., 98, 99–100, 99f, 101–102, 103, 105–107, 111–112, 116, 121–139
Yoshida, S., 143–144
Yoshimura, S. H., 171
Young, P., 402–404, 409–410
Youssef, N. H., 360–361
Yu, X., 39, 53–54, 55–56, 58
Yu, Y., 379–380
Yuan, Y., 170

Z

Zagursky, R. J., 385–386
Zakariassen, H., 78–79, 79f
Zamboni, V., 196–197
Zechel, D. L., 145f, 153–154, 155–156, 157–158
Zeitler, L. A., 335
Zhai, H., 70
Zhang, G. M., 379–380
Zhang, H. T., 74–75, 350–351
Zhang, J., 383
Zhang, K. Y. J., 146
Zhang, P., 379–380
Zhang, S., 74–75, 249, 444
Zhang, X. Z., 74

Zhang, Y. H. P., 55–56, 59, 74, 82–83, 87–88, 262, 305, 361, 368, 384, 444
Zhang, Z., 51–67, 383
Zhang, Z. M., 74
Zhang, Z. N., 53–54
Zhao, G. C., 376–377, 384
Zhao, J., 385–386
Zhao, X. L., 203–204, 376–377
Zhao, Y. X., 70, 74, 89, 170–171
Zheng, B., 149–150
Zheng, Y., 100
Zhong, L., 204–205
Zhong, R., 126
Zhou, G.-K., 126
Zhou, W. L., 70, 74–75
Zhou, X., 398
Zhu, X., 276
Zhu, Y., 228–229
Zhu, Z. G., 74, 88
Zhuang, G. Q., 88
Zhulin, I. B., 383–384
Ziser, L., 278–279
Zorzetti, C., 274–276, 280t, 292, 293, 294, 295
Zou, J.-Y., 70, 71f, 75, 155, 184–186
Zuccotti, S., 285
Zulauf, M., 144–145
Zwart, P. H., 410–412

Subject Index

Note: Page numbers followed by "*f*" indicate figures, and "*t*" indicate tables.

A

Affinity gel electrophoresis (AGE)
  description, 217–218
  experimental setup, 218
  polysaccharide-infused gel, 217–218, 219*f*
  quantification, constants
    isotherm, 218–220
    normalized values, 218–220
    protein detection, 220
AGE. *See* Affinity gel electrophoresis (AGE)
Apparent processivity (P^{app}) measurement
  intrinsic processitivity, 86–89
  single-hit approach (*see* Single-hit approach, P^{app} measurement)
  single-turnover approach (*see* Single-turnover approach, P^{app} measurement)
Avicel, 353
Avicel hydrolysis, 448

B

Bacterial cohesin-dockerin complexes
  anaerobic ecosystems, 396
  dual binding mode and crystallization, 402
  in *E. coli*
    expression, 400
    purification, 399*f*, 401
  prokaryotic expression vectors
    coexpression, 397
    gene cloning, PCR, 397–398
    gene construction, protein coexpression, 398–400, 399*f*
    noncatalytic domains, 397
    synthetic production, 398
  ribbon representation, 413–414, 413*f*
  scaffoldin, 396
  structural and functional evidence, 413–414
  X-ray crystallography (*see* X-ray crystallography)
Bacterial microcrystalline cellulose (BMCC), 228
Biofuel
  and biorenewables, 49
  production, 38, 39
  saccharification, plant biomass, 39
Biomass
  enzyme systems, 20
  industry, 20
  substrates, 21

BMCC. *See* Bacterial microcrystalline cellulose (BMCC)

C

Carbohydrate-binding module (CBM) function. *See also* CBMs *vs.* cellulosomal linker peptides
  abundant molecules, 228–229
  AGE (*see* Affinity gel electrophoresis (AGE))
  cellulose-coated microarray, 455–456, 455*f*
  definition, 212
  derivation, constants, 215
  inherent affinity characteristics, 455–456
  ITC (*see* Isothermal titration calorimetry (ITC))
  polysaccharide interactions, 212
  protein libraries screening, 461
  quantitative approaches, 228–229
  solid state depletion assay (*see* Solid state depletion assay)
  structural and functional, 212, 213*t*
  UV difference
    experimental setup, 221–222
    fluorescence spectroscopy, 223
    interaction, 220–221
    quantification, constants, 222–223
    spectroscopic approaches, 220–221, 221*f*
  Xyn-Doc clones, 456
Carboxymethyl cellulose (CMC), 353
Carrier domains
  C. acetobutylicum scaffoldin CipA
    cargo module, 313–314
    recombinant strains, 313–314
  C. cellulolyticum scaffoldin CipC
    chaperone, 312
    cohesin and dockerin, 312
    deleterious cellulase, 312–313
    fusion enzyme, 312
    scaffoldin modules, 312, 313*f*
  cellulase activity, 314
  scaffoldins and enzymes, 311–312
  solventogenic bacterium, 311
  solventogenic strain, 311
  toxicity, 311
CBDs. *See* Cellulose-binding domains (CBDs)
CBH. *See* Cellobiohydrolase (CBH)
CBM. *See* Carbohydrate-binding module (CBM)
CBMs *vs.* cellulosomal linker peptides

489

CBMs vs. cellulosomal linker peptides (cont.)
  amino acid motifs, 249
  aromatic and polar residues, 248
  carbohydrate degradation, 249–250
  Clostridium thermocellum, 249
  ELISA, 250–253
  GH9s and endoglucanase, 249
  glycosylation, 249
  ITC, 253–258
  "scaffoldin" subunit, 248
  shallow-groove region, 249–250
cELIA. *See* Competitive enzyme-linked interaction assay (cELIA)
Cellobiohydrolase (CBH)
  natural cellulolytic systems, 70
  preparations, 81
  processitivity, 80
  property, 70
Cellobiose, 353
Cello-oligosaccharides
  and cellulose, 56
  DP, 2, 10
  EG, 10
  hydrolysis, 2, 63
Cellulase activity assay
  hydrolysis products, 367
  physical properties, substrates, 368
  reductions, 368
Cellulases
  paper recycling and cotton processing, 38–39
  XGs hydrolysis, 4
Cellulose. *See also* Fluorescent cellulose decay (FCD) assay
  cellulases, 87
  and chitin, 75
  ^{14}C-labeled, 82–83
  crystalline and amorphous, 2
  EG measurement, 10–11
  glycoside hydrolases, 70
  hydrolysis, 80
  native, 3–4
  processive degradation, 70
  reduced, 81
  separation, 87–88
Cellulose and xyloglucan-hydrolyzing enzymes
  feature, 122–123
  oligosaccharide products, 124–127
  polysaccharide-rich plant cell wall, 122
  reaction products chemical and structural analysis, 127–137
  substrates preparation, 124
  XGs, 123
Cellulose azure dye-release assay, 366–367
Cellulose-binding assay, 438
Cellulose-binding domains (CBDs), 234–235
Cellulose-degrading microorganisms
  anaerobic chytridiomycetes, 350–351
  biodegradable organic polymers, 363–364

in cultures
  cellulase activity assays, 367–368
  cellulose azure dye-release assay, 366–367
  clear and transparent zones, colonies, 366
  Congo red, 367
  filter paper strips, 367
  SIP (*see* Stable isotope probing (SIP))
  description, 350–351
  human feces, 365–366
  interdependency, microbial taxa, 368–369
  isolation
    aerobic bacteria, 361–363
    aerobic fungi, 363
    anaerobic bacteria, 361
  molecular biological and "omics" techniques, 368
  molecular evidence, *Fibrobacteres*, 364–365
  the rumen
    description, 354–355
    Hungate technique (*see* Hungate technique)
    isolation, 355
  substrates
    acid-treated, 351–352
    Avicel, 353
    bacterial, 352–353
    cellobiose, 353
    CMC (*see* Carboxymethyl cellulose (CMC))
    dewaxed cotton string, 352
    factors, 351
    filter paper and ball-milled, 351
    hemicelluloses, 353–354
    lignocelluloses, 354
  termite guts, 365
Cellulose enzymatic cleavage, PACE
  ANTS, 53–54
  applications, 64–65
  biofuels, 53
  cello-oligosaccharides separation, 54–55, 55f
  cellulolytic hydrolysis, PASC, 59–60
  characterization, hydrolytic enzymes, 54
  defined, 52–53
  GHs, 53
  hydrolysis mechanism, 63–64, 64f
  plant cell wall, 53
  poly, oligo and monosaccharide, ANTS, 56
  poly, oligo and monosaccharide sample preparation, 55–56
  preparation and interpretation
    polymerization, 56–57
    and quantification, 58
    resolving gel, 57
    samples analysis, 57–58
    separation, 57
  substrate disappearance assay
    derivation and quenching, conditions reduction, 61–62
    GH3 enzyme, 60–61, 62f

Subject Index

oligosaccharides digestion, 62–63
Cellulosomal, chimeric enzymes
  "internal" dockerin module, 432
  purification techniques, 432
Cellulosome. *See* Heterologous minicellulosome
Chimeric enzyme test
  activity, CBM, 441–442
  amorphous and xylans cellulose, 444
  Avicel hydrolysis, 448
  CBM-binding ability
    affinity electrophoresis-based carbohydrate-binding assay, 443
    description, 442
    insoluble polysaccharides, 443
    reagents, 442
  cellulose hydrolysis, 444
  comparative degradation, Avicel, 446, 447f
  designer cellulosome activity assay, 443
  dockerin attachment, 449
  filter paper hydrolysis, 448
  functional components, 441
  kinetic studies, 445, 446f
  PASC hydrolysis, 448
  procedures, 448
  reagents, 447–448
  relative degradative activity, 445, 445t
  substrate degradation, 443
  target and proximity effect, 444
  wheat-straw hydrolysis, 449
  xylan hydrolysis, 448
Chitinases
  and cellulose, 75
  defined, 76
  enzyme-substrate dissociation, 78
  homogenous deacetylation, 76
  processivity and directionality
    product profiles, 78–79, 80f
    substrate binding, chitobiohydrolases, 78–79, 79f
  processivity, chitosan measurement
    degradation experiments, 76–78
    diagnostic product patterns, 76
    mutants comparison, 78
    preparation, 76
Clostridium acetobutylicum
  "carrier" modules, 311–314
  "dockerin" module, 302–303
  electrotransformation
    C. acetobutylicum, 305
    plasmid preparation, 303–305
    recombinant strains storage, 305–306
  heterologous cellulosomal protein, 308
  heterologous minicellulosome, 308–310
  mono and disaccharides, 302–303
  separate hydrolysis and fermentation (SHF), 302–303
  xylan-degrading enzymes, 302–303
CMC. *See* Carboxymethyl cellulose (CMC)

Cohesin mutant libraries
  CBM, 461
  cellulose microarray experiment, 455–456, 455f
  DNA extraction
    materials, 460t
    protocol, 457f, 460t
  dots representation, 461–462, 461f
  high-affinity interaction, 454–455
  incubation, IPTG, 461
  inherent affinity characteristics, 455–456
  library preparation, 456
  microarray probing (*see* Microarray probing)
  optical density, 461
  prototype, screening, 461–462, 462f
  saturation mutagenesis, 454–455
  screening
    defined, small library, 456–458
    materials, 456t, 458t
    positive colonies, 457f, 458–459
    protocol, 456t, 457f, 459t
    spots, microarray, 461
Competitive enzyme-linked interaction assay (cELIA), 419
Congo red dye, 367
Consensus libraries
  DNA library construction, 263–264
  primers, 264
  selecting mutations, 263, 264f
  thermostability, 263
CopyControl^{TM} fosmid library production kit, 379–380
Corn, 20, 32
Cryo-electron microscopy (CryoEM), 190
Crystallization and structural analysis, cellulases
  buffers
    Alzari group, 160–161
    cryo-protectant problems, 159–160, 159f
    entropy reduction, 160–161
    positive transition state, 160
    tactics, 160
    TRIS, 159–160
  catalytic domain, 143–144
  commercial and bespoke
    ammonium sulfate, 147–149
    crystal growth, 147
    dyes, 146–147
    gel filtration (GF), 147–149
    index^{TM}, 147–149
    lysozyme, 147
    Phoenix, 147–149
    polarizer, 146–147
    protein characterization, 147–149
    SDS-PAGE, 147–149
  cryo-protectants, 161
  degrading enzymes, 162
  domain boundaries, 162
  3-D structure, enzyme, 142

Crystallization and structural analysis, cellulases (*cont.*)
  genetic construct, 143–144
  and glycosidase, 152–159
  goal, 142–143
  optimization
    micro-batch methods, 149–150
    microcrystals, 150–152
    Morpheus™, 150
    natural materials, 150–152
    PEGs, 149–150
    protein solution, 150–152
    seeding, 150–152, 151*f*
    soaking experiments, 149–150
    testing, 150
    "universal nucleant", 150–152
    vapor diffusion, 149–150
  problems, 161–162
  protein quality control
    description, 144
    native gels, 145–146, 145*f*
    polydispersity, 144–145
    qPCR, 146
    tags, 146
    thermal shift assay, 146

D

Designer cellulosomes
  chimeric enzymes
    cellulosomal, 432
    description, 431
    noncellulosomal, 431–432
  chimeric enzyme testing (*see* Chimeric enzyme test)
  chimeric scaffoldins
    description, 430
    forward and reverse primers, 430–431
    sequence, Scaf AT modules and linkers, 430–431, 431*t*
  scaffoldin testing and cohesin-dockerin interaction
    cellulose-based pull-down assay, 440, 441*f*
    cellulose-binding assay, 438
    ELISA (*see* Enzyme-linked immunosorbent assay (ELISA))
    nondenaturating PAGE, 436–438
    reagents, 433–434
    size exclusion HPLC, 438–440
    stability assays, 440
  sugar analysis
    analysis, 449
    cellulose structure, 449
    commercial kits, 450
    DNS method, 450
    HPAEC (*see* High-performance anion exchange chromatography (HPAEC))
    reagents, 449–450
Differential scanning calorimetry (DSC), 420

DNS method, 450
Dockerins (Doc), iELISA
  coating, 424
  equilibrated cohesion, 423*f*, 424
  protocol, 423*f*, 423*t*
  reagents, 422*t*
  unbound concentration, 423*f*, 424–425
DSC. *See* Differential scanning calorimetry (DSC)

E

Electroporation
  electrotransformation, 318
  materials
    autoclaved reverse, 319
    DSM122 and CTFUD, 318–319, 319*t*
    plasmids, 320
    stock solutions, selective agents, 320
    strains, 320
    vitamin solution, 318–319, 319*t*
  methods
    gene disruption, 322–325
    markerless deletion, 325–327
    plasmid construction, 322
    selective markers, 327–328
    transformation protocol, 320–322
  targeted gene deletion, 318
  troubleshooting, 328–329
Electrospray ionization (ESI), 103–105
Electrospray ionization mass spectrometry (ESI-MS)
  defined, 135–137
  per-O-methylated xyloglucan oligoglycosyl alditol, 135–137, 136*f*
ELISA. *See* Enzyme-linked immunosorbent assay (ELISA)
Embedding procedures
  description, 238
  LR white resin
    plant material, 240
    resin-embedded materials, 240
  protocol
    ligands and substrates, 239
    Steedman's wax, 239
    wax-embedded material, 239–240
Endo-1,4-β-glucanase (EG)
  enzymes purity, study, 2–3, 3*f*
  enzymic dissolution, crystalline and amorphous cellulose, 2
  insoluble chromogenic substrates
    enzyme activity, 14–15, 15*f*
    substrate preparation, 14
  measurement, 15–16
  polysaccharide and oligosaccharide, 3–4
  reducing-sugar assays, 6–10
  soluble chromogenic substrates
    enzyme activity, 12–13
    precipitant solutions, 12
    substrate preparation, 10–11

Subject Index 493

viscometric assay, 4–6
Endoglucanases screening
  CMC-plate assay, enzyme activity
    Cel8A enzymes, 265–266
    procedures, 266t, 267f
    reagents, 265t
  thermostability screen, endoglucanase
    endoglucanase activity, 266–267
    procedures, 266t
Enzyme characterization
  efficiency determination, 45–46
  enzyme filter paper units determination, 48, 48f
  monosaccharide analysis, 46–47
  time course analysis
    automated determination, 44–45, 45f
    enzyme mixtures, 44
    saccharification, *Arabidopsis* and *Miscanthus*, 44–45, 46f
Enzyme inhibitors
  covalent inactivators, 155–156, 156f
  nonhydrolyzable substrate mimics, 155
  structural interpretation, 154
  transition-state mimics and compounds, 157–158
Enzyme-linked immunosorbent assay (ELISA)
  cloning
    DNA fragments, 251
    genomic DNA, 250–251
    PCR products, 251
    pMAL-c2x expression, 251
  cohesins analysis, 435–436
  dockerins analysis, 434–435, 435f
  materials
    anti-MBP monoclonal antibody, 250
    disposal regulation, 250
    scaffoldin subunit, 250, 251f
    TMB reagent, 250
  protein expression
    centrifugation, 251–252
    recombinant expression vectors, 251–252
  protein purification, 252
  protocol, 252, 253f
  results, assay, 253, 254f
Error-prone PCR
  error rate resulting, 262
  Escherichia coli, 262–263
  stratagen's genemorph II, 262
  transformations and screening, 262–263
ESI. *See* Electrospray ionization (ESI)
ESI-MS. *See* Electrospray ionization mass spectrometry (ESI-MS)

F

FID. *See* Flame ionization detection (FID)
Filter paper hydrolysis, 448
Flame ionization detection (FID), 127–129
Fluorescent cellulose decay (FCD) assay
  efficiency, enzyme catalyzed reactions, 20
  enzyme activity calculation, 35
  FB28, 20
  and HPLC, 21–23
  hydrolysis, 20–21
  incubation and fluorescence reading, 34–35
  model validation, 32
  multicomponent biomass enzyme systems, 20
  optimization, 24–25
  performance correlation method, 20
  plate layout development
    CV, 27
    enzyme sample, 26–27
    individual contributions estimation, error, 25–26, 27t
    microtiter plate, 27–28, 28f
    prediction profiles, incubation times, 25–26, 26f
    96-well microtiter plates, 25, 25f
  protein determination, sample dilution and microtiter plates
    assay samples, 34
    concentration, enzyme samples, 33
    control sample, 33–34
    plate sealing and mixing, 34
  reduction, assay time
    data, 30
    model, 30, 31f
    normalization, 29
    substrate conversion, 28–29, 29f
  substrate preparation
    defined, 32
    loading, 32–33
Fluorophore
  PACE, 53–54
  uses, 54–55

G

Gas chromatography–mass spectrometry (GC-MS), 127–129
Gel filtration (GF), 147–149
Gel permeation chromatography (GPC)
  DMSO, 102–103
  representative chromatograms, 103, 104f
  water-based system, 103
Gene disruption, marker
  dilution factor, 325
  plasmid design, 322–323, 323f
  Tm and 5FOA selection, 323–325, 324f
Genetic manipulation, *C. japonicus*
  description, 336
  DNA
    conjugation, 337–338
    description, 337
    electroporation, 337
    plasmid construction, 336–337
GFP. *See* Green fluorescent protein (GFP)
GHs. *See* Glycoside hydrlases (GHs)

Global expression profiling
  cDNA synthesis, labeling, and microarray analysis, 344
  cell capture, soluble substrate, 344
  description, 342–343
  extraction, RNA, 344
  soluble substrate, cell capture
    cell removal, filtration, 344
    centrifugation, 344
    xyn11A and xyn11B, 342–343, 343f
β-Glucosidase screening
  DP cello-oligosaccharides, 2
  magenta Glc-plate
    procedures, 268t
    reagents, 268t
  thermostability screen, 96-plate format, 269
β-Glucuronidase (GUS) transcriptional fusions, 342
Glycosidase and cellulase
  "apo-enzyme", 152
  enzyme inhibitors (see Enzyme inhibitors)
  inhibitor soaking, 158
  oligosaccharides
    active-site spanning complex, 154
    enzyme inactive, 153–154
    2-fluoro glycosides, 154
    hydrolyze substrates, 153–154
    panning ligand mixtures, 158–159
    product/pseudo-product complexes, 152–153
Glycoside hydrlases (GHs)
  characterization, 63
  families, 56
  glycosynthase, 274–276
  linkage-specific, 122
  PACE, 53–54
  products generation, 53
  reaction mechanisms (see Reaction mechanisms)
Glycosyl-linkage composition analysis
  defined, 129–130
  methylation, oligosaccharides, 130–131
  reaction products, NMR spectroscopy, 131–135
  solid NaOH/DMSO reagent preparation, 130
GPC. See Gel permeation chromatography (GPC)
Green fluorescent protein (GFP), 235

H

Heterologous cellulosomal protein
  C. acetobutylicum, 308, 309f
  enzymes
    CMC plates, 307
    extracellular activity, 307–308
    quantification, secreted enzyme, 307–308
    SEC signal peptide, 307
    western blot analysis, 307
  scaffoldins
    culture supernatant, 306
    miniscaffoldins, 307
    SDS-PAGE, 306–307
Heterologous minicellulosome
  batch fermentations, 310
  dominant species, 310
  N-terminal modules, 308–310
  smaller elution volume, 310
  supernatants examination, 309f, 310
  western blot analyses, 310
High-affinity interactions, ELISA
  CBM-fused cohesins
    protocol, 421t
    reagents, 421t
  cellulose degradation, 420–421
  data analysis
    absorbance, 425
    cohesin–dockerin binding profile, 425–426, 426f
    free energy binding, 425
    half-maximal inhibitory concentration (IC_{50}), 425–426, 426t
  description, 418–419
  dockerins, iELISA (see Dockerins (Doc), iELISA)
  methods and disadvantages, 419–420
  mutant *vs.* wild-type cohesin modules, 426–427
  protein fusions, 420–421
  role, residues, 418–419
  test *vs.* reference protein, 420
  xylanase-fused dockerins
    protocol, 422t
    reagents, 422t
Highlyoriented pyrolytic graphite (HOPG), 173–174, 174f
High-performance anion exchange chromatography (HPAEC), 450
High-performance anion exchange chromatography with pulsed amperorometric detection (HPAEC-PAD)
  eluent gradients, 101–102
  and Glc_8-based XGOs
    GHs, 112–113
    hydrolysis and transglycosylation, 114
    initial rates *vs.* substrate concentration, 113, 114f
    tetradecasaccharide, kinetic studies extraction, 113–114
  tamarind XGOs separation, 101–102, 102f
High-performance liquid chromatography (HPLC)
  defined, 101
  GPC, SEC, 102–103
  HPAEC-PAD, 101–102
High-pressure liquid chromatography (HPLC)
  dose–response curves, 22, 24f
  enzyme hydrolysis, biomass substrates, 21

Subject Index

FCD assay, 22, 22f
normalization, 21, 22
High-speed atomic force microscopy (HS-AFM)
  cellobiohydrolases, 170
  cellulase molecules
    imaging, 175–176, 175f
    immobilization, 174–175
  cellulase purification
    chromatography, 172–173
    Phenyl-Toyopearl column, 173
    T. reesei, 172–173
  crystalline cellulose
    heterogeneity, 171
    purification, 171
    TEM, 171, 172f
  degradation, 179
  glycoside hydrolases, 170–171
  HOPG surface, 176–177
  hydrophobic surface, 179
  image analysis
    CBH molecules, 177f, 178, 179f
    procedures, 177–178
    ROI, 177, 177f
    second frame, 178
  optical beam deflection, 176–177
  pyrolytic graphite disk
    bare mica surface, 173
    HOPG, 173–174, 174f
    schematic representation, 174f
  *Tr*Cel7A molecules, 176–177, 176f
High-throughput system. *See* Saccharification, biomass
HOPG. *See* Highlyoriented pyrolytic graphite (HOPG)
HPAEC. *See* High-performance anion exchange chromatography (HPAEC)
HPAEC-PAD. *See* High-performance anion exchange chromatography with pulsed amperorometric detection (HPAEC-PAD)
HPLC. *See* High-performance liquid chromatography (HPLC). *See also* High-pressure liquid chromatography (HPLC)
HS-AFM. *See* High-speed atomic force microscopy (HS-AFM)
Hungate technique
  Agar roll tubes, 359
  anaerobic fungi, 360–361
  colonies, 359
  culture dilution, 358–359, 358f
  culture medium
    Agar, 357
    buffer, 356
    cellulose substrate, 357
    defined, 356
    inorganic mineral medium, 356
    procedure, 357
    redox indicator, 356–357
    reducing agent, 356
  description, 355–356
  fluid, 356
  metal gas hooks/Pasteur pipettes, 357–358
  purity, cultures, 360
Hydrolysis
  biomass substrates, 21
  cellulose, 20, 21, 34
  enzymatic, 20

I

Indirect ELISA (iELISA), 422–425, 426f, 427
Indirect fluorescence imaging methods, 242–243
In situ detection, carbohydrate-binding modules
  CBDs, 234–235
  cellulose, description, 234–235
  directly coupled CBMs, 241–242
  enzymatic pretreatments, 241
  fluorescence methods, 236
  fluorescent brightener Calcofluor White, 234–235
  genetic approach, 235
  GFP, 235
  indirect fluorescence imaging methods, 242–243
  indirect methodology, 236
  methods and protocols, 244
  molecular probes, 235
  plant materials preparation
    aldehyde fixatives, 237–238
    Arabidopsis roots, 236
    cellulose, 236, 237f
    embedding procedures (*see* Embedding procedures)
    glycans, 236
    incubation steps, 237–238
  post-CBM-labeling treatments
    counterstaining, 243
    TBO, 243–244
Isothermal titration calorimetry (ITC)
  BMCC, 228
  buffer solution, 258
  CBM–polysaccharide interactions, 227–228
  cloning procedure, 256
  entropy, 228
  experimental setup
    crystalline polysaccharide, 224
    dilution effects, 224
    ligand concentration, 224
    stable and soluble protein, 224
  expression and purification, CBM3, 256
  heat capacity, 223–224
  linker analysis
    CipA scaffoldin, 254, 255f
    homologous, 254
  materials, 256
  microcalorimeters, 223–224
  principle

Isothermal titration calorimetry (ITC) (*cont.*)
  adiabatic shield, 253–254, 255*f*
  binding energetics, 253
  linker segment, 254
  protocol, 257
  results, 257–258, 257*f*
  sensitivity, measurement, 258
  thermodynamic parameters
    burying apolar groups, 225–227
    CBM–polysaccharide interactions, 225, 226*f*
    enthalpy and entropy, 225–227
    exothermic and endothermic, 225
    Gibb's free energy, 225
    polar interactions, 225–227
ITC. *See* Isothermal titration calorimetry (ITC)

L

"Labeling" techniques
  defined, 80
  intrinsic processivity
    cellulases, 86–87
    dissociation probability, 86–87
    Kcat, 87–88
    Koff value, 88–89
  P^{app} measurement
    single-hit approach, 81–82
    single-turnover approach, 82–85
  processivity, cellobiohydrolase Cel7A, 80
Lignocellulose, 38
"Limit-digest" products, 111–112
Linker peptide. *see* CBMs *vs.* cellulosomal linker peptides
Liquid chromatography
  defined, 125–126
  RP-HPLC, 126–127
  SEC, 126

M

MALDI. *See* Matrix-assisted laser desorption/ionization (MALDI)
Markerless deletion, removable marker system
  gene deletion, TM/FUDR/8AZH, 326–327
  plasmid design, 325, 326*f*
Mass spectrometry (MS)
  defined, 103
  ESI, 103–105
  ESI-MS, 135–137
  MALDI, 105
  MALDI-TOF-MS, 131
Matrix-assisted laser desorption/ionization (MALDI), 105
Metagenomic approach, cellulase
  bioinformatics bottleneck, 389–390
  defined, genes, 376–377
  defined, metaproteomics, 389–390
  description, 376
  libraries (*see* Metagenomic libraries)
  metatranscriptomics (*see* Metatranscriptomics)
  NGS, 389
  nucleic acid extraction
    CTAB protocol, 377–378
    description, 377
    DNA, 378–379
  SEED and KEGG, 389
Metagenomic libraries
  description, 379–380
  fosmid cloning systems, 379–380
  kits, 379–380
  large-insert
    end repair, 380
    fosmids package, 381
    host cell transduction, 381
    ligation, 380–381
    size selection, 380
  screening, cellulases
    environmental DNA, 382
    expression-based, 384–385
    in silico, 382–383
    primers, qPCR, 382
    sequence-based, 383–384
Metatranscriptomics
  description, 385
  enrichment, mRNA
    cDNA synthesis, 388–389
    de novo transcriptome sequencing, 385–386
    methods, 388
    prokaryotic rRNA, 385–386
    subtractive hybridization, 386–388, 387*f*
Microarray probing
  materials, 459*t*
  protocol, 457*f*, 460*t*
Microscopy
  electron, 236, 239
  fluorescence, 243, 244
  light, 240
Microtiter
  plate-based method, 20
  polypropylene flat-bottomed, plates, 33
  protein determination, sample dilution and loading, 33–34
Miscanthus sinensis, 40
Molecular modeling, 197–200, 201
Molecular replacement (MR) method, 402–404, 409, 410–412
Morpheus^{TM}, 150
MS. *See* Mass spectrometry (MS)

N

Nelson-Somogyi redycing-sugar assays
  enzyme activity, 7
  reagents preparation, 6–7
Next-generation sequencing (NGS), 389
NGS. *See* Next-generation sequencing (NGS)

Noncellulosomal, chimeric enzymes
  CBM, 431–432
  forward and reverse primers, Cel48Y cellulase, 432
  N-and C-terminus, 431–432
  sequence, ★Cel 48 Y-t modules and linkers, 432, 433t
Nondenaturating PAGE
  analysis, 436–437, 436f
  electrophoretic mobility, 437–438, 437f
  stoichiometric molar ratio, 436
Nuclear magnetic resonance (NMR) spectroscopy
  and MS, 123
  reaction products analysis
    defined, 131–133
    hydrolysis stereochemistry, 133–135, 323f
    water-soluble polysaccharides/oligosaccharides, 133

O

Oligosaccharide products
  isolation, soluble products generation
    enzymatic hydrolysis, insoluble substrates, 124–125
    enzymatic hydrolysis, polymeric substrates, 125
  liquid chromatography, 125–127
Oligosaccharides synthesis
  conventional glycosynthases, 296
  glycoside hydrolase (GH) family, 274–276
  glycosynthases and glycoside hydrolases, 274, 275f
  protein denaturants, 276
  Sulfolobus solfataricus, 274
  thermophilic α-d-galactosynthase
    assays and chemical rescue, 293–294
    characterization, 295
    mutagenesis, expression and purification, 292–293
    synthesis and analysis, galactosylated oligosaccharides, 294–295
  thermophilic α-l-fucosyntases
    characterization, 288–292
    cloning, mutagenesis expression and purification, 285–286
    enzymatic assays and chemical rescue, 287
    fucosylated oligosaccharides, 287–288
  thermophilic β-glycosynthases
    characterization, 280–284
    cloning, mutagenesis and expression, 276–278
    enzymatic assays and chemical rescue, 278–279
    transglycosylation reactions, 279–280
  thermophilic glycosynthases, 295–296

P

PAHBAH reducing-sugar assay
  enzyme activity, 8
  reagents preparation, 7–8
PASC. *See* Phosphoric acid swollen cellulose (PASC)
Phosphoric acid swollen cellulose (PASC)
  cellulolytic hydrolysis
    and cello-oligosaccharides, 59–60, 60f
    digestion time course, 59–60, 61f
    preparation, 59
  hydrolysis, 448
Plant cell wall degradation, *C. japonicus*
  acrylamide-solidified medium, 335–336
  agar-solidified medium, 335
  construction, mutation
    gene inactivation, 339–340
    random mutagenesis, transposons, 338–339
    systems biology, 338
  culture and storage conditions, 333–334
  description, 332
  genetic manipulation. (*see* Genetic manipulation, *C. japonicus*)
  global expression profiling (*see* Global expression profiling)
  liquid medium, 335
  media, screening and propagation, 335–336
  medium, routine growth, 334–335
  recalcitrance, 332–333
  role, carbon cycle, 332–333
  strain acquisition, 333
  systems-biological analysis, 333
  transcriptomic analysis (*see* Transcriptomic analysis, *C. japonicus*)
Plasmid preparation
  C. acetobutylicum and *Bacillus substilis*, 304
  cloning region, pSOS95, 304, 304f
  DNA encoding, 303
  E. coli secretion, 303
  restriction endonucleas, 305
Polysaccharide analysis by carbohydrate gel electrophoresis (PACE). *See* Cellulose enzymatic cleavage, PACE
Processivity measurement
  Cel6A catalytic domain, *Trichodermareesei*, 70–71, 71f
  cellobiohydrolases, 70
  chitinases, 76–79
  endo *vs*. exo-action, 89
  glycoside, 70
  kinetics and thermodynamics, 89
  methods
    features, 71–72, 72f
    product ratios, 72–74
    soluble *vs*. insoluble-reducing ends, 74–75
  molecular dynamics, 90
  quantification, 89–90

Product ratios measurement
  degradation reaction, 73
  exo-acting non-processive enzyme, 73
  HPLC, 73
  intermediate products formation, 73
  mutational effects, 74
  trimeric and monomeric, 72–73
Protein crystallography, 186

Q

Quantitative polymerase chain reaction (qPCR), 146

R

Reaction mechanisms
  exclusive formation, 289
  glycoside hydrolases and glycosynthases, 274, 275f
  recombinant enzyme, 292
Reaction products chemical and structural analysis
  ESI-MS, 135–137
  glycosyl-linkage composition, 129–131
  glycosyl residue composition, 127–129
  NMR spectroscopy, 131–135
  oligosaccharides analysis, MALDI-TOF MS, 131, 132f
  oligosaccharides convertion, oligoglycosyl alditols, 129
Recombination
  procedure
    digested fragments, 270, 271f
    DNA, shuffling, 270
    reagents, 269t
Reducing-sugar assays
  bicinchoninic acid, 109–110
  EG activity
    absorbance values, 9, 9t
    borohydride reduced cello-oligosaccharides hydrolysis, 10, 11f
    cello-oligosaccharides, 10
    measurement, 8–9
    Nelson-Somogyi, 6–7
    PAHBAH, 7–8
    polysaccharides hydrolysis rates, 9–10, 10t
  transglycosylation rate measurement, 112–113
Region of interest (ROI), 177, 178
Reversed-phase high-performance liquid chromatography (RP-HPLC), 126–127
ROI. *See* Region of interest (ROI)
RP-HPLC. *See* Reversed-phase high-performance liquid chromatography (RP-HPLC)

S

Saccharification, biomass
  enzyme characterization, 44–48

enzyme discovery, 49
enzymes, biofuel production, 39
grinding and weighing robot
  formation, material, 39–40, 41f
  Miscanthus and *Arabidopsis*, 40
  vial, 40
hydrolysis, cellulose, 38–39
industrial, 38
lignocellulose, 38
liquid handling robot
  MBTH, 43–44
  station, 42, 42f
  volumes and incubation conditions, pretreatment, 42, 43f
plant genes identification, 39
Scaffoldin
  aromatic and polar residues, 248
  CBM3 and Coh3, 250
  Pro residues, 249–250
SDS-PAGE, 147–149
SEC. *See* Size-exclusion chromatography (SEC)
Single-hit approach, P^{app} measurement
  quantification, 81
  reduced cellulose, 82
  SRGs and IRGs, 81
Single-turnover approach, P^{app} measurement
  cellulases, 82–83
  determination, 83, 84f
  intrinsic processivity, 85, 86f
  time course, ^{14}CB formation, 83–85, 84f
SIP. *See* Stable isotope probing (SIP)
Size-exclusion chromatography (SEC)
  cello and XG oligosaccharides, 126
  protocol, 126
Size exclusion HPLC
  interactions, 438
  Superdex 200 gel filtration, 439–440, 439f
Small-angle X-ray scattering (SAXS) and crystallography
  analysis and structural values
    graphical representations, 189, 189f
    Guinier law, 190
    gyration, 189
    intensity, 190
    protein, 188
  biophysical and structural methods, 203
  bottlenecks, 204–205
  cellulases and enzymes
    glycosylation, 192
    molecular dimensions, 192–194
    properties, 192, 193t
  cellulose, 184–186
  computational modeling
    catalytic domains, 201
    dockerin module, 201–203
    enzymatic efficiency, 201–203, 204f
    full-length protein, 201
    molecular description, 200

Subject Index 499

nature, linker, 201
scattering profiles, 201, 202f
structural and dynamic propertie, 203
data and quality assessment, 187–188
3D constructing
  atomic models, 197–200
  CRYSOL, 197–200, 198f
  GNOM, 197
  hydrolysis, 197–200, 199f
  molecular dynamics, 197–200
  multimodular xylanase Z, 196–197
  specific activity, 197–200, 200t
  XynZfree and XynZcoh, 197, 198f
dissect and build approach
  C. cellulolyticum / C. thermocellum, 195
  cellulosomes, 194
  diversity implies, 195–196
  enzyme linker, 195, 196f
  intercohesins linker, 195–196
  "mini-cellulosomes", 194–195
  protein–protein interaction, 185f, 194
  scaffolding protein, 194
3D models
  CryoEM, 190
  Fourier transform, 190
  experimental requirements, 187
  Hypocrea jecorina, 184–186
  modules and domains, 191
  neutron scattering, 204–205
  polysaccharide-degrading enzymes, 184–186, 185f
  sample preparation, 186
  sophisticated experimental design, 186
Solid state depletion assay
  adsorption, 215, 216f
  description, 215
  experimental setup
    dialysis, 215–216
    sophisticated solid state, 215–216
  quantification, constants, 217
Soluble *vs.* insoluble-reducing ends measurement
  DNS assay, 75
  emzyme binding, 75
  "labeling" techniques, 75
  mutational effects, 74–75
  processivity determination, 74
SPR. *See* Surface-plasmon resonance (SPR)
Stable isotope probing (SIP), 367
Standard affinity-based ELISA, 419
Stover, 20, 32
Substrate preparation
  defined, 105–106
  Glc$_4$-based XGOs
    endo-(xylo)glucanases, 106–107
    isolation, 107
  XGOs, 108
  xyloglucan solubilization, 106
Surface-plasmon resonance (SPR), 419

T

TBO. *See* Toluidine Blue O (TBO)
TEM. *See* Transmission electron microscopy (TEM)
Thermophilic α-d-galactosynthase
  assays and chemical rescue, TmGalA, 293–294
  characterization
    catalytic center, 295
    site-directed mutagenesis removal, 295
    transgalactosylation, 295
  mutagenesis, expression and purification, 292–293
  reactions, transgalactosylation, 294
  transgalactosylation efficiency, 294–295
Thermophilic α-l-fucosyntases
  cloning and mutagenesis, Ssαfuc, 285
  enzymatic assays and chemical rescue, 287
  expression and purification, Ssαfuc wild type and mutants, 286
  kinetic characterization, 288, 289t
  synthesis and analysis, fucosylated oligosaccharides
    reaction products and determination, 288
    transfucosylation reactions, 287
  Tmα-fuc cloning and mutagenesis, 285
  Tmα-fuc expression and purification, 286
  transglycosylation efficiency
    homo-condensation, 289
    mutant synthesized disaccharides, 289–292
    synthetic products, 289–292, 290t
  TmD224G, 292
Thermophilic β-glycosynthases
  cloning, mutagenesis expression and purification
    expression and purification, *Ta*βgly and CelB wild type and mutants, 278
    Ssβgly, 276
    Ssβgly wild type and mutants, 276–277
    *Ta*βgly and CelB, 277–278
  enzymatic assays and chemical rescue
    mutagenesis, 278–279
    nucleophile mutants, GHs, 278–279
  kinetic characterization
    molar extinction coefficients, 280–281, 280t
    *Ss*E387G and *Ta*E387G, 280–281
  optimization, hydrolytic activity, 281, 282t
  transglycosylation efficiency
    catalyst converting, 281
    NMR analysis, 281–283
    synthetic products, 283, 284t
    TLC analysis, 281–283
  transglycosylation reactions
    mutants prepare, 279–280
    NMR analysis, 280
    oxidase-peroxidase enzymatic assay, 280
Thermostability
  cellulase libraries

Thermostability (cont.)
  alignment-guided consensus libraries, 263–264
  error-prone PCR, 262–263
  in vitro recombination, 269–270
  screening β-glucosidase, 267–269
  screening endoglucanases, 265–267
Thin layer chromatography, 101
Toluidine Blue O (TBO), 243–244
Transcriptomic analysis, C. japonicus
  description, 341
  β-glucuronidase (GUS) transcriptional fusions, 342
  northern blots, 341
  real-time PCR, 341–342
Transformation
  cell growth (anaerobic), 320
  cell harvesting and washing (aerobic), 321
  Melt CTFUD agar, 321f, 322
  pulse application (anaerobic), 321
  recovery (anaerobic), 322
  transformants (anaerobic), 322
Transmission electron microscopy (TEM), 171, 172f
Trichoderma reesei
  Cel7A, 172–173
  enzymes hydrolyzing, 170–171
Troubleshooting
  colony PCR, 329
  electric pulse, 328
  lawn, colonies, 329
  Tm and negative selection, 328
  transformation, 328

V

Viscometric assay, EG activity
  enzyme, 5
  hydrolysis, 5–6, 6f
  substrate preparation, 4–5

W

Wheat-straw hydrolysis, 449

X

XGs. See Xyloglucans (XGs)
X-ray crystallography
  crystallization
    crystals, 406–407, 406f
  goal, 404
  protocols, nano-drop dispensing system, 405–406
  screening, 405
  vapor diffusion, 404–405
  data collection and reduction, 406f, 407–409
  defined, R factor, 404
  description, 402–404
  diffraction data, 402–404
  electron density function, 402–404
  electron density maps, 402–404
  method, MR, 402–404
  model building and structure refinement, 410–412
  structure determination
    Bayesian method, 409–410
    3D, 409
  validation methods, 402–404
Xylan hydrolysis, 448
Xyloglucanase activity, endo-β(1r 4)glucanases
  blocking loops, 100
  "cellulose" families, 100
  classes, 99
  defined, 98
  hydrolytic activities, 99–100
  protein, 100
  structures, variable side chain substituents, 98
  substrate preparation, 105–108
  xyloglucan
    activity, glucanase, 109–115
    hydrolysis, 116
    oligo and polysaccharide analysis, 101–105
Xyloglucans (XGs)
  activity, glucanase
    β-glycosides, XGOs, 114–115, 116f
    bicinchoninic acid reducing-sugar assay, 109–110
    bond cleavage specificity, 111–112
    endo vs. exo-cleavage preference, 110–111
    specificity, GH components, 109
    transglycosylation vs. hydrolysis, 112–114
  defined, 123
  hydrolysis, 116
  oligo and polysaccharide analysis
    defined, 101
    HPLC, 101–103
    MS, 103–105
    NMR spectroscopy, 105
    thin layer chromatography, 101
  solubilization, 106

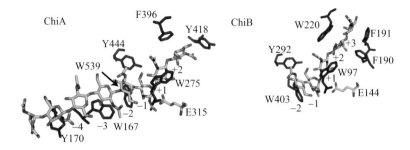

Svein J. Horn et al., Figure 5.4 Substrate binding in processive chitobiohydrolases ChiA and ChiB. The pictures show details of the crystal structures of ChiA in complex with an octameric substrate (Papanikolau *et al.*, 2001) and ChiB in complex with a pentameric substrate (van Aalten *et al.*, 2001). In addition to the substrate, the pictures show the side chains of the catalytic acids (E315 in ChiA, E144 in ChiB) and the side chains of aromatic residues interacting with the substrate. In ChiA, the +1 and +2 subsites are the "product" sites, where dimeric products are released. In ChiB, these product subsites are −1 and −2. Picture adapted from Zakariassen *et al.* (2009).

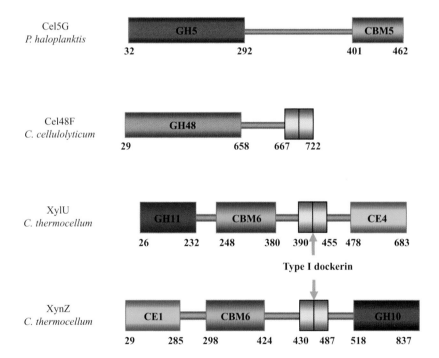

Mirjam Czjzek et al., Figure 10.1 Schematic representation of the modular composition of various carbohydrate-degrading enzymes. The names and the organisms of origin are given to the left of each polypeptide representation. The numbers underneath each protein correspond to the residue numbers delimiting the different modules, as determined by multiple sequence alignments and/or experimental evidence.

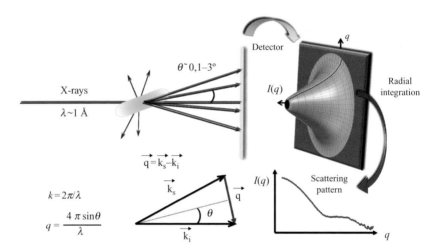

Mirjam Czjzek et al., Figure 10.2 Schematic representation of the setup and analyses of a small-angle scattering experiment. The scattering vector q is defined as the difference between the scattered wavevector k_s and the incident wavevector k_i.

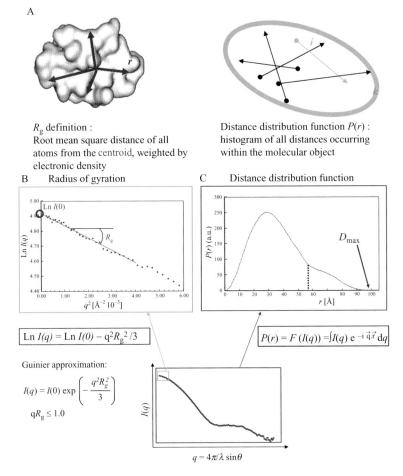

Mirjam Czjzek et al., Figure 10.3 (A) Graphical representations illustrating the definitions of the radius of gyration (R_g) and the distance distribution function ($P(r)$). (B, C) Graphical representations of how the values R_g, $P(r)$, and D_{max} are extracted from the experimental data, respectively.

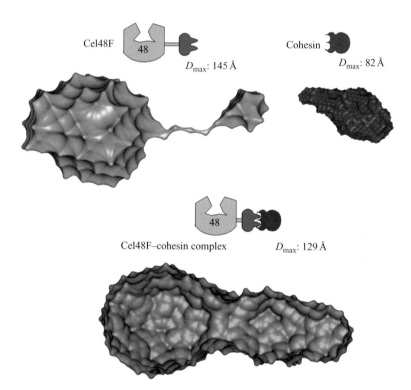

Mirjam Czjzek et al., Figure 10.4 Schematic icons and calculated envelopes using GASBOR corresponding to full-length Cel48F (top left), the cognate cohesin (top right), and the complex of both (bottom), as obtained by SAXS experiments. The experimental D_{max} values obtained for each molecule in solution are indicated.

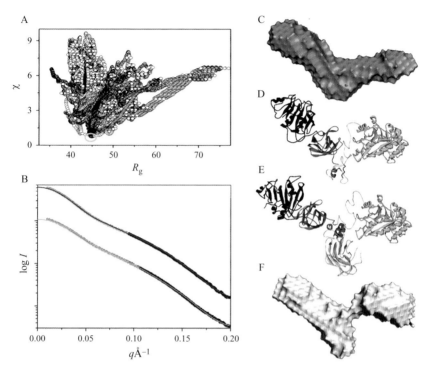

Mirjam Czjzek et al., Figure 10.5 SAXS analyses of XynZfree and XynZcoh. (A) The graph represents the discrepancy (ξ) values for 15,000 models as a function of their R_g values, calculated for XynZfree (XynZcoh, data not shown). The best-fitting model is indicated by a yellow circle. (B) Experimental SAXS curves and scattering profiles computed from the models of XynZfree (top) and XynZcoh (bottom). Black circles: experimental data; dark blue line (top): theoretical scattering curve from best-fit model obtained by rigid body modeling of XynZfree using the program CRYSOL; light blue line (bottom): theoretical scattering curve from best-fit model obtained by rigid body modeling of XynZcoh using the program CRYSOL. (C) The molecular envelope shape in surface representation of XynZfree calculated with GASBOR and (D) the corresponding best-fit model obtained from the molecular dynamics simulation in ribbon representation. The modules are colored as follows: CE1 in dark red, CBM6 in blue, dockerin in red, and the GH10 xylanase in green. (E) The best-fit model of XynZcoh obtained from the molecular dynamics simulation in ribbon representation. The modules are colored as above and the cohesin in beige and (F) the corresponding molecular envelope shape in surface representation calculated with GASBOR.

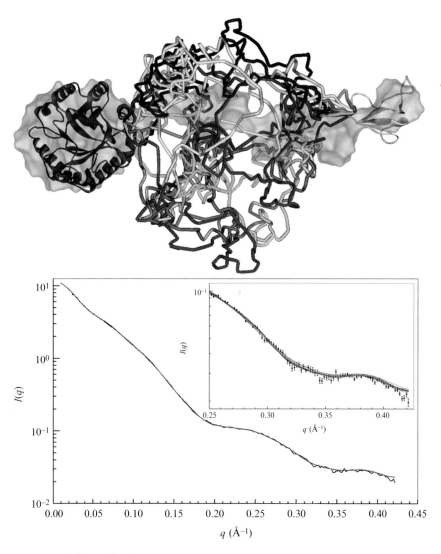

Mirjam Czjzek et al., Figure 10.7 (Top) Shape calculated with GASBOR (blue) superimposed with 10 different models of Cel5G provided by GLOOPY represented by secondary structure element type. (Bottom) Fit on the experimental scattering curve obtained with the average form factor of the different models provided by GLOOPY. (Inset) Fits on the experimental scattering curve obtained with individual form factors of the different models provided by GLOOPY.

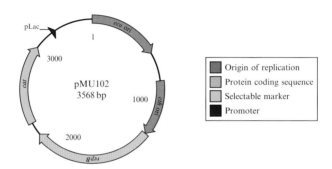

Name	Type	Region	Orientation	Description
eco ori	Replication origin	0001..0589	fwd	pUC19 origin of replication for plasmid propagation in *E. coli*
cth ori	Replication origin	0826..1280	fwd	pNW33N origin of replication for plasmid propagation in *C. thermocellum*
repB	CDS	1281..2285	fwd	Replication protein for pNW33N origin of replication
cat	CDS	2392..3042	fwd	Chloramphenicol acetyl-transferase, provides resistance to chloramphenicol and thiamphenicol
pLac	Promoter	3208..3244	fwd	Lactose inducible promoter from pUC19

Daniel G. Olson and Lee R. Lynd, Figure 17.1 Diagram of plasmid pMU102.

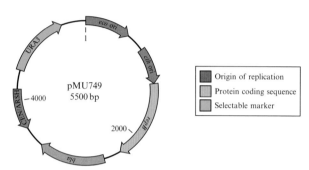

Name	Type	Region	Orientation	Description
eco ori	Replication origin	0001..0589	fwd	pUC19 origin of replication for plasmid propagation in *E. coli*
cth ori	Replication origin	0826..1280	fwd	pNW33N origin of replication for plasmid propagation in *C. thermocellum*
repB	CDS	1281..2285	fwd	Replication protein for pNW33N origin of replication
bla	CDS	2505..3362	fwd	Beta lactamase, provides resistance to ampicillin
CEN6/ARSH	Replication origin	3631..4144	rev	Yeast origin of replication, used for yeast mediated ligation cloning
URA3	CDS	4383..5186	fwd	Provides uracil prototrophy for selection of plasmids during yeast mediated ligation cloning

Daniel G. Olson and Lee R. Lynd, Figure 17.2 Diagram of plasmid pMU749.

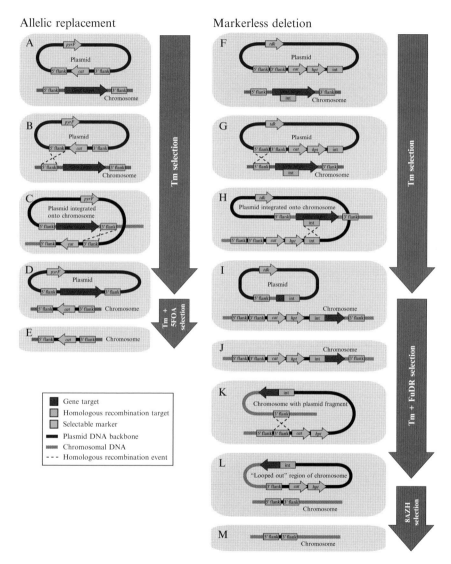

Daniel G. Olson and Lee R. Lynd, Figure 17.3 Selection scheme diagram. The steps involved in gene disruption by allelic replacement or markerless gene deletion are described.

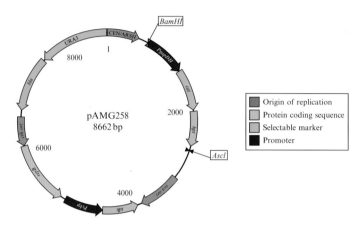

Name	Type	Region	Orientation	Description
CEN6/ARSH4	Replication origin	0001..0519	fwd	Yeast origin of replication, used for yeast mediated ligation cloning
PgapD	Promoter	0692..1339	fwd	Promoter from glyceraldehyde-3-phosphate dehydrogenase gene
cat	CDS	1340..1990	fwd	Chloramphenicol acetyl-transferase, provides resistance to chloramphenicol and thiamphenicol
hpt	CDS	2008..2553	fwd	Marker used for counterselection in *C. thermocellum* with 8AZH
eco ori	Replication origin	3099..3772	fwd	pUC19 origin of replication for plasmid propagation in *E. coli*
tdk	CDS	3833..4411	rev	Marker used for counterselection in *C. thermocellum* with FuDR
Pcbp	Promoter	4412..5032	rev	Promoter from cellobiose phosphorylase (cbp) gene of *C. thermocellum*
repB	CDS	5108..6112	rev	Replication protein for pNW33N origin of replication
cth ori	Replication origin	6113..6567	rev	pNW33N origin of replication for plasmid propagation in *C. thermocellum*
bla	CDS	6670..7530	rev	Beta lactamase, provides resistance to ampicillin
URA3	CDS	7548..8650	rev	Provides uracil prototrophy for selection of plasmids during yeast mediated ligation cloning

Daniel G. Olson and Lee R. Lynd, Figure 17.4 Diagram of plasmid pAMG258.